Springer Proceedings in Business and Economics

Springer Proceedings in Business and Economics brings the most current research presented at conferences and workshops to a global readership. The series features volumes (in electronic and print formats) of selected contributions from conferences in all areas of economics, business, management, and finance. In addition to an overall evaluation by the publisher of the topical interest, scientific quality, and timeliness of each volume, each contribution is refereed to standards comparable to those of leading journals, resulting in authoritative contributions to the respective fields. Springer's production and distribution infrastructure ensures rapid publication and wide circulation of the latest developments in the most compelling and promising areas of research today.

The editorial development of volumes may be managed using Springer's innovative Online Conference Service (OCS), a proven online manuscript management and review system. This system is designed to ensure an efficient timeline for your publication, making Springer Proceedings in Business and Economics the premier series to publish your workshop or conference volume.

More information about this series at http://www.springer.com/series/11960

Hui Yang • Robin Qiu • Weiwei Chen
Editors

AI and Analytics for Public Health

Proceedings of the 2020 INFORMS International Conference on Service Science

Springer

Editors
Hui Yang
Department of Industrial Engineering
Pennsylvania State University
University Park, PA, USA

Robin Qiu
Division of Engineering & Info Sci
Pennsylvania State University
Malvern, PA, USA

Weiwei Chen
Department of Supply Chain Management
Rutgers, The State University of New Jer
Piscataway, NJ, USA

ISSN 2198-7246 ISSN 2198-7254 (electronic)
Springer Proceedings in Business and Economics
ISBN 978-3-030-75165-4 ISBN 978-3-030-75166-1 (eBook)
https://doi.org/10.1007/978-3-030-75166-1

This Springer imprint is published by the registered company Springer Nature Switzerland AG
The registered company address is: Gewerbestrasse 11, 6330 Cham, Switzerland

Preface

This proceeding publishes the papers submitted, peer-reviewed, and presented at the 2020 INFORMS Conference Service Science (ICSS 2020), held in the all-live and virtual format on Dec. 19–21, 2020. This conference provided an excellent opportunity for scholars and practitioners to present their service science related research and practice work, to learn about the emerging technologies and applications, and to network with each other for further collaborative opportunities.

2020 was a difficult and challenging year for the world. The COVID-19 pandemic was unprecedented. Containing the pandemic was and still is challenging to humanity. Contributing to combating the unprecedented COVID-19 crisis, the ICSS 2020 conference theme was *AI and Analytics for Public Health*, aimed at promoting and facilitating the development of healthy and strong communities where we live, work, learn, and play, and uncovering solutions to protect the health of people and the communities, nationally and internationally. This conference attracted scholars and practitioners around the world to come together virtually to share what had been found, helping each other by timely sharing solutions and stimulating new ideas, which further helped enhance the needed solutions and extend them to uncharted territories. We are confident that in the fight against any virus, humanity will and must prevail.

This year we had over 120 submissions from around the world. All full/short paper submissions were carefully peer reviewed. After the rigorous review and revision process, 37 papers were finally accepted to be included in this proceeding. The major areas covered at the conference and included in this proceeding include:

- Public Health Service, Policy, Administration, Response, and Systems
- Service Management, Operations, Engineering, Design, Innovations, and Marketing
- Smart Cities, Sustainable Systems, IT and Service System Analytics, and Self-service Systems
- Smart and Intelligent Service, Healthcare Analytics, FinTech, Learning Analytics, and Others

- Big Data, Machine Learning, Artificial Intelligence, and Data-Driven Decision Making
- Systems Modeling, Management, and Simulation in Manufacturing, Supply Chain, Logistics, and Others
- AI, Data Analytics, and Data-driven Applications in Health, Energy, Finance, Transportation, Sport, and Governmental/Public Services

In addition to the accepted research papers, ICSS 2020 provided an opportunity for scholar and practitioners to share their ongoing studies. We invited six well-known service science experts to deliver plenary speeches:

- Dr. Jim Spohrer, director of IBM Cognitive Opentech Group, presented "Future of AI and Post-Pandemic Society: A Service Science Perspective."
- Prof. Paul Maglio, University of California at Merced, former EIC of INFORMS *Service Science,* discussed "What is Service Science?"
- Prof. Weiwei Chen, Rutgers University, delivered "Improving Service Designs and Operations Using Analytics."
- Prof. Saif Benjaafar, Distinguished McKnight University Professor, University of Minnesota, EIC of INFORMS *Service Science*, articulated "Dimensioning On-Demand Vehicle Sharing Systems."
- Prof. Dmitry Ivanov, Berlin School of Economics and Law, explained "Supply Chain Resilience Theory and COVID-19 Pandemic: What We Know, Where We Failed, and How to Progress."
- Prof. Victor Chan, Tsinghua-Berkley Shenzhen Institute, reviewed "Recent Mathematical and Computational Studies of COVID-19."

The conference had 16 parallel sessions, including 77 presentations. ICSS 2020 also had successfully organized the best conference paper competition and the best student paper competition. We would like to thank all authors, speakers, track chairs, session chairs, reviewers, and participants.

Finally, we would like to thank all authors for submitting their high-quality works in the field of service science, and the conference organizing and program committee members, listed on the following pages, for their tireless efforts and time spent on reviewing submissions. We are very grateful to Springer's editors, Neil Levine and Faith Su, and the production editor, Shobha Karuppiah, who have contributed tremendously to the success of the ICSS 2020 conference proceedings. We would also like to acknowledge the NSF I/UCRC Center for Healthcare Organization Transformation (CHOT), NSF I/UCRC award IIP-1624727, for sponsoring the conference.

Co-Editors – Proceedings of 2020 INFORMS Conference on Service Science

Malvern, PA, USA — Robin Qiu

University Park, PA, USA — Hui Yang

Piscataway, NJ, USA — Weiwei Chen

ICSS 2020 Committees

Program Committee

- Ralph Badinelli, Virginia Tech, USA
- Victor Chan, Tsinghua University, China
- Ozgur Araz, University of Nebraska-Lincoln, USA
- Jenny Chen, Dalhousie University, Canada
- Weiwei Chen, Rutgers University, USA
- Hongyan Dai, Central University of Finance and Economics, China
- David Ding, Rutgers University, USA
- Qiang Duan, Penn State, USA
- Yucong Duan, Hainan University, China
- Tijun Fan, East China University of Science and Technology, China
- Siyang Gao, City University of Hong Kong, China
- Yan Gao, University of Shanghai for Science and Technology, China
- Dmitry Ivanov, Berlin School of Economics and Law, Germany
- Hai Jiang, Tsinghua University, China
- Zhibin Jiang, Shanghai Jiaotong University, China
- Haitao Li, University of Missouri–St. Louis, USA
- Zhenyuan Liu, Huazhong University of Science and Technology, China
- Kelly Lyons, University of Toronto, Canada
- Rym M'Hallah, Kuwait University, Kuwait
- Juan Ma, iHeartMedia, USA
- Xin Ma, Texas A&M University, USA
- Paul Maglio, UC Merced, USA
- Aly Megahed, IBM, USA
- Paul Messinger, University of Alberta, Canada
- Chuanmin Mi, Nanjing University of Aeronautics & Astronautics, China
- Ran Mo, Central China Normal University, China
- Ashkan Negahban, Penn State, USA
- Kai Pan, Hong Kong Polytechnic University, China

- Patrick Qiang, Penn State, USA
- Robin Qiu, Penn State, USA
- Lun Ran, Beijing Institute of Technology, China
- Tina Wang, University of Oxford, UK
- Hui Xiao, Southwestern University of Finance and Economics, China
- Xiaolei Xie, Tsinghua University, China
- Hui Yang, Penn State, USA
- Ming Yu, Tsinghua University, China
- Canrong Zhang, Tsinghua University, China

Conference Organizing Committee

- Conference Co-chair(s): Prof. Robin Qiu and Prof. Hui Yang
- Program Co-chair(s): Prof. Weiwei Chen
- Invited Tracks:
  - Special Sessions (Prof. Qiang Duan and Prof. Xiaolei Xie)
  - Sharing Economy (Prof. Ashkan Negahban)
  - Healthcare Service and Analytics (Prof. David Ding and Prof. Xiaolei Xie)
  - Service Design, Operations, and Analytics (Prof. Victor Chan and Prof. Canrong Zhang)
  - Service Economy in the Emerging Market (Prof. Qiang Qiang)

Best Student Paper Award Committee

- Laura Anderson, IBM, USA
- Clara Bassano, University of Salerno, Italy
- Chiehyeon Lim, UNIST, South Korea
- Kelly Lyons, University of Toronto, Canada (Chair)
- Eleni Stroulia, University of Alberta, Canada

Best Conference Paper Award Committee

- Weiwei Chen, Rutgers University, USA
- Dmitry Ivanov, Berlin School of Economics and Law, Germany
- Yingdong Lu, IBM, USA (Chair)
- Ashkan Nagahban, Penn State, USA
- Jie Song, Peking University, China

Contents

Epidemic Informatics and Control: A Review from System Informatics to Epidemic Response and Risk Management in Public Health

Hui Yang, Siqi Zhang, Runsang Liu, Alexander Krall, Yidan Wang, Marta Ventura, and Chris Deflitch

1 Introduction

Epidemic outbreaks impact the health of our society and bring significant disruptions to the US and the world. For example, Coronavirus Disease 2019 (COVID-19) is currently ravaging multiple countries and was declared as a global pandemic by the World Health Organization (WHO) in March 2020. COVID-19 has caused a total of approximately 7.82 million infected cases and 432 K deaths worldwide, as well as 2.17 million infected cases and 118 K deaths in the US by June 16, 2020 (CDC, 2019). The abrupt increase of cases quickly exceeds the capacity of health systems and highlights the shortages of workers, beds, medical supplies and equipment. Many governments have taken a variety of actions (e.g., lockdown, large-scale testing, stay-at-home) to flatten the curve and avoid overwhelming health systems, but these reactionary policies have resulted in great economic losses. The US unemployment rate has skyrocketed from 3.5% in February 2020 to 14.7% in April 2020 (The Bureau of Labor Statistics, n.d.). The number of unemployed persons has increased to 23.1 million, which is even worse than the Great Depression in 1930s. The economic uncertainty has caused US stock markets to trigger the circuit breakers to halt trading for a historical 4 times in the week of March 9–16, 2020 (Zhang et al., 2020). The US GDP shrunk 4.8% in the first quarter of 2020.

H. Yang (✉)
Department of Industrial Engineering, Pennsylvania State University, University Park, PA, USA
e-mail: huy25@psu.edu

S. Zhang · R. Liu · A. Krall · Y. Wang · M. Ventura
Center for Health Organization Transformation, The Pennsylvania State University, University Park, PA, USA

C. Deflitch
Department of Emergency Medicine, Penn State Health Milton S. Hershey Medical Center, Hershey, PA, USA

H. Yang et al. (eds.), *AI and Analytics for Public Health*, Springer Proceedings in Business and Economics, https://doi.org/10.1007/978-3-030-75166-1_1

When the COVID-19 epidemic emerged, it was not uncommon to encounter a misperception or misinformation that coronavirus is like the seasonal influenza (flu). Although there are similarities (e.g., causing respiratory illness) between coronavirus and flu virus, they are significantly different. COVID-19 or severe acute respiratory syndrome (SARS) is caused by the family of coronavirus, which is not the same as the flu virus. There are three major types of flu viruses – Types A, B and C. Type A flu virus caused many epidemics in the past 100 years (e.g., 1918 Spanish Flu (Trilla et al., 2008), 1968 H3N2 epidemic (Alling et al., 1981), and 2009 H1N1 epidemic (Sullivan et al., 2010)). It is worth mentioning that Type A flu virus infects a wide variety of animals (e.g., poultry, swine, aquatic birds) and easily evolves and mutates genes. Once transported and adapted to humans, it can evolve into an epidemic. Types B and C flu viruses infect only humans as the typical seasonal flu and has rarely been the cause of past epidemics (Taubenberger et al., 2005). It is estimated by Center for Disease Control and Prevention (CDC) that seasonal flu causes approximately 140,000–810,000 hospitalizations and 12,000–61,000 deaths annually since 2010 (Disease Burden of Influenza, n.d.). However, the death toll of 1918 Spanish Flu is about 50 million worldwide and 675,000 in the US.

Historically, epidemics are inevitable and recur at more or less near-periodic cycles. It is difficult to predict when a new virus will emerge and cause an epidemic. The infection rate of a virus is commonly measured by the basic reproduction number R_0, which characterizes how many people on average can be infected by one infected individual in a susceptible population. For COVID-19, R_0 is estimated to range from 1.4 to 6.49, with a mean of 3.28 (Liu et al., 2020). The potential transmission pathway can be either through air droplets, which are generated when infected individuals talk, cough, or sneeze, or through contact with an infected person or surface that is contaminated with the virus. At the start of an outbreak, antivirals and vaccines are often not available. People can only resort to non-pharmaceutical interventions (NPIs) for the control and containment of virus spread (Davies et al., 2020). Traditional NPI methods include the practice of good personal hygiene, the use of disinfectants, the isolation and quarantine of infected individuals, and the limitation of public gatherings. From 1918 Spanish flu epidemic to COVID-19, this situation does not change much although health systems become more advanced and medical resources are richer than before.

However, one thing that does change is the faster and augmented capability of medical testing and diagnostics, thanks to rapid advances of gene/DNA, microbiology, and imaging technologies (Ravi et al., 2020). As such, large amounts of data are collected in the evolving process of epidemic outbreaks. The availability of data calls upon the development of analytical methods and tools to gain a better understanding of virus spreading dynamics, optimize the design of healthcare policies for epidemic control, and improve the resilience of health systems. Therefore, this paper presents a review of the system informatics approach of **D**efine, **M**easure, **A**nalyze, **I**mprove, and **C**ontrol (DMAIC) for epidemic management through the intensive use of data, statistics and optimization. Despite the sustained successes of DMAIC in a variety of established industries such as manufacturing, logistics, services and

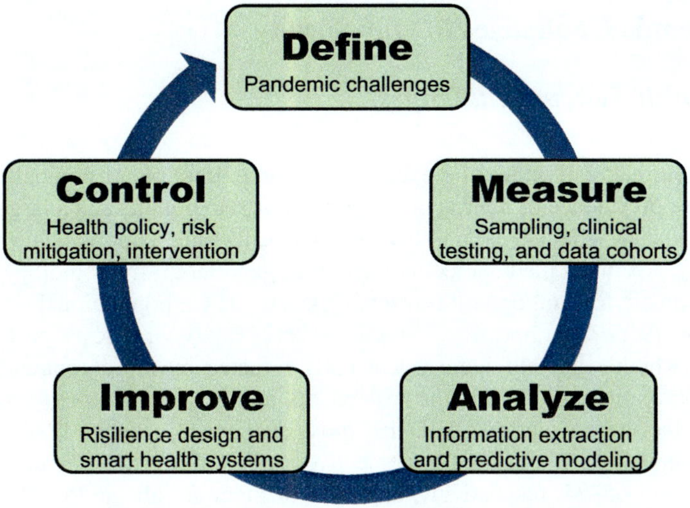

Fig. 1 The flowchart of system informatics for epidemic response and risk management

beyond (Yang et al., 2021; Knowles et al., 2005; Kumar et al., 2007), there is a dearth of concentrated review and application of the data-driven DMAIC approach in the context of epidemic outbreaks. As shown in Fig. 1, The DMAIC methodology consists of five phases: **(1) Define**: outline the societal challenges posed by the epidemic; **(2) Measure**: collect data about key variables in the epidemic process; **(3) Analyze**: extract useful information pertinent to the spread of epidemic; **(4) Improve**: design solutions and methods to improve the resilience of health systems; **(5) Control**: develop health policies, management plans, and intervention methods to control the spread of infectious diseases. The goal of this paper is to catalyze more in-depth investigations and multi-disciplinary research efforts to accelerate the application of system informatics methods and tools in epidemic response and risk management.

The rest of the paper is organized as follows: Section 2 discusses specific societal challenges arising from large-scale outbreaks of infectious diseases. Section 3 reviews the sampling and testing strategies to increase information visibility for epidemic management. Then, we present a review of analytical methods and tools for the extraction of useful information in Sect. 4. Continuous improvements and re-design to improve the resilience of health systems are discussed in Sects. 5 and 6 presents the health policies and intervention strategies for the control of virus spread. Section 7 discusses the system informatics approach for epidemic management and concludes this paper.

2 Epidemic Challenges to Our Society

2.1 Health System Challenges

Epidemic outreak calls upon the execution of large amounts of clinical testing to examine the prevalence of a virus in the population. No doubt, such a large demand poses significant challenges on the manufacturing and supply chain systems. Fortunately, advanced medical technology (e.g., gene/DNA, microbiology) enables the provision of viral and/or antibody testing kits to the US population. For example, as of June 19, 2020, there are a total of 26,781,666 viral tests performed to determine whether an individual is currently infected by the coronavirus (CDC, 2019). Approximately 10% of the test results are positive. Among a sample of 1,934,566 individuals with COVID-19, most of them are within 18–44 and 45–64 age groups (41.4% and 32.8%, respectively). For the rest, 5.1% and 9.5% are aged 0–17 and 65–74, respectively, and 11% of them are above 75 (CDC, 2019). In general, when the age of patients increases, the hospitalization rate also becomes higher. Hospitalization rate is the ratio between the number of individuals who are hospitalized within 14 days after a positive viral test and the total population in a spatial region. As shown in Table 1, the overall cumulative hospitalization rate is 94.5 per million (CDC, 2019). For people aged 50–64 and above 65, the rates increase to 143 and 286.9 per million, respectively. However, for people aged 0–4 and 5–17, the rates declined to 7.4 and 3.5, respectively.

The upsurge of positive cases poses significant challenges on the hospital capacity. As shown in Table 2, as of June 18, 2020, 70% of inpatient beds are occupied, in which 5% is used for COVID-19 patients. Also, nearly 63% of intensive care units (ICU) beds are occupied (CDC, 2019). In addition, the shortages of medical supplies (e.g., personal protection equipment (PPE)) become more and more prevalent in the health systems with a rising number of coronavirus cases and hospitalizations. In the era of globalization, US medical supplies are heavily dependent on importation, nearly 72% of active pharmaceutical ingredients (APIs) are imported from other countries. Specifically, approximately 13% of medical products are from China, and 18% of pharmaceutical imports are provided by India (COVID-19: Impact on Global Pharmaceutical and Medical Product Supply Chain Constraints U.S. Production, 2019). Also, generic drugs imported from these two countries account for about 90% of medicine supplies in the US. However,

Table 1 A summary of cumulative hospitalization rate for each age group

Age Group	Hospitalization rate per million
Overall	94.5
0–4 years	7.4
5–17 years	3.5
18–49 years	56.5
50–64 years	143.0
65+ years	286.9

Table 2 National estimates of hospital bed occupancy in the COVID-19 in the US

Estimates for June 18	Number (95% CI)	Percentage (95% CI)
Inpatient Beds Occupied (all Patients)	524,610 (500,844–548,376)	65% (64–66%)
Inpatient Beds Occupied (COVID-19)	40,112 (37,682–42,541)	5% (5–5%)
ICU Beds Occupied (all Patients)	77,029 (72,135–81,922)	63% (61–64%)

the COVID-19 outbreak in January shuts down almost all manufacturing facilities and non-essential businesses in China. Even though manufacturing activities were resumed in late February, the average capacity utilization at top 500 manufacturing enterprises in China was only 58.98% (Fernandes, 2020; ISM Report on Business, 2019). As such, a disrupted supply chain causes serious shortages of medical products in the US, which endangers the healthcare workers in the front line.

Indeed, healthcare workers are among the most vulnerable group of people who face a higher probability to get infected during the epidemic outbreak. The higher risk is due to their closer contact with patients, the shortage of PPEs, the delay of testing program in the early stage, and the high infection rate in the hospital. As the COVID-19 proliferates, healthcare workers suffer from occupational burnout and fatigue. The key factors include occupational hazards, emergence responses, process inefficiencies, and financial instability (Sasangohar et al., 2020; Shechter et al., 2020; Greenberg et al., 2020). During the period of February 12–April 9, 2020, approximately 19% of COVID-19 patients are healthcare workers. Therefore, this fact further exacerbates the shortage of staffing in the hospital. To avoid secondary infection in the hospital, screening and masks are required for all people upon entry into the hospital (Bartoszko et al., 2020). Patients with suspected or confirmed COVID-19 are placed in a single-occupancy room with a closed door and a separated bathroom. Also, all healthcare workers should wear PPE, isolation gowns and non-sterile gloves upon entering these patients' room. When transporting patients out of the room, both patients and healthcare workers should wear PPE. Moreover, hospitals conduct routine cleaning and disinfection procedures. Enhanced environmental cleaning and disinfection are preferred for rooms used by patients with suspected or confirmed COVID-19, and for areas used by healthcare workers who care for such patients (Chirico et al., 2020).

2.2 *Economic Challenges*

The COVID-19 epidemic made the nation shut down non-essential businesses, schools and instituted travel bans, which have greatly impacted the U.S. economy. The shocks to supply chain bring significant disruptions to manufacturing. Small and medium manufacturing enterprises faced unprecedented challenges, while some have to shut down entirely to mitigate the virus spread. With social distancing measures in place, many workers can only work from home. The production

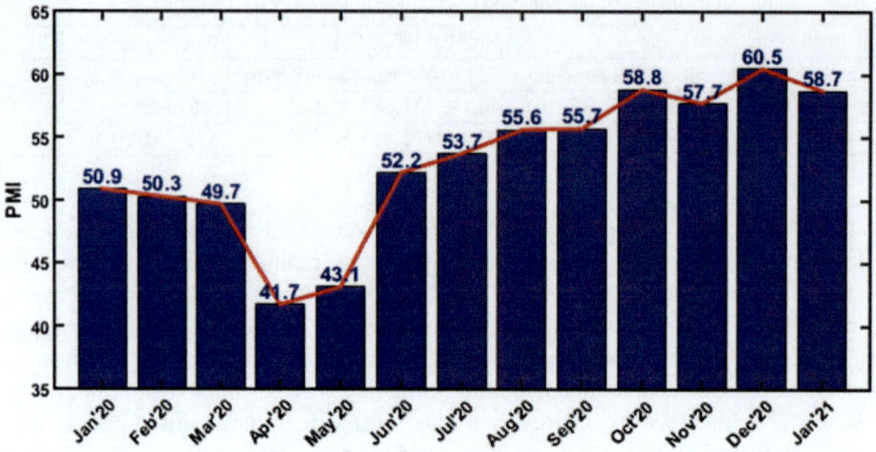

Fig. 2 The variations of Purchasing Manager's Index (PMI) from January to May 2020

lead time has doubled due to shortages of workers and materials (ISM Report on Business, 2019). Also, a limited number of products can be distributed worldwide by air or ocean because of trade wars, hiking tariffs, and importation restrictions. All these impacts of COVID-19 make companies question the just-in-time strategy and reconsider the design of supply chain. In March 2020, there was a 6.3% drop in manufacturing production, which was the largest 1-month drop since 1946 (ISM Report on Business, 2019; Bonaccorsi et al., 2020). The drop was even larger for April 2020. Note that the Purchasing Manager's Index (PMI) shows the impacts of COVID-19 on the economy. PMI is a composite index, ranging from 0 to 100, of economic activities including new orders, inventory levels, production, supplier deliveries, and employment. If the PMI is above 50, the manufacturing sector is generally expanding. If PMI is below 50, it is generally contracting. As shown in Fig. 2, US economic growth is strong in January 2020 with PMI 50.9, but decreases from January to April 2020 (ISM Report on Business, 2019; Bonaccorsi et al., 2020). When the COVID-19 outbreak occurred in March 2020, the PMI fell below 50, further dropped to 41.7 in April 2020, and then remained low through May 2020. From March to May 2020, COVID-19 poses significant challenges on the US economic activities due to unexpected outbreaks, lockdowns, and non-pharmaceutical interventions. After June 2020, the US economical activities recover with the rollout of stimulus plans, increasing manufacturing productions, and new modes for businesses such as teleconferencing, e-commerce and online learning.

A worse impact on the manufacturing industry during the epidemic would be caused by decreased spending because of job loss or reduced incomes. The disruption in the manufacturing industry and the tremendous drop in demand led to the layoff of workers. As of May 2020, the unemployment rate in the manufacturing industry increased to 11.6%. Table 3 summarizes the number of employees in the manufacturing sector as issued by the U.S. Bureau of Labor Statistics, for both the non-seasonally adjusted case and the seasonally adjusted case (Manufacturing: NAICS 31-33, n.d.). As shown in Table 3, when it is not seasonally adjusted, the number of employees in the manufacturing sector decreased by 1.32 million from March 2020 to April 2020, with about 0.90 million in durable goods manufacturing and 0.42 million in non-durable goods manufacturing. Meanwhile, there were about 1.13 million fewer jobs in May 2020, compared to May 2019. When it is seasonally adjusted, the U.S. manufacturing lost about 1.29 million jobs from March 2020 to April 2020. About 69% (0.91 million) of the job loss was in the durable good manufacturing, while the rest 31% (0.38 million) was in the non-durable good manufacturing. Compared to May 2019, there were 1.12 million fewer jobs in May 2020 (Manufacturing: NAICS 31-33, n.d.).

Schools and universities across the country have also been disrupted. In March 2020, most schools started to switch from in-person instruction to online-only instruction, which gave rise to the concerns about instruction quality (Crawford et al., 2020). Meanwhile, it is not uncommon that many universities faced financial challenges. As students moved out of on-campus housing, universities issued prorated refunds to them, which was a substantial amount of unexpected expenses. Also, universities needed to allocate additional funds for dorm cleaning and technology essentials for online classes. Moreover, due to the cancellation of college entrance exams worldwide and limitation on travel, the enrollment for the fall 2021 semester is likely to drop, which will also cause financial issues to universities.

These paramount challenges posed by epidemics call upon multiple scientific disciplines to design and develop new enabling methods and technological innovations for rapid response and management. For example, a complete picture of the new virus is urgently needed from the community of medical scientists. The manufacturing community should be agile to innovate the design and increase the production of personal protective equipment (PPE). In this paper, we propose a system informatics approach for data-driven epidemic response and operational management, thereby mitigating the risks and controlling the virus spread. In the following sections, "**Measure**" provides statistical methods for optimal sampling and testing of the population for the presence of virus, as well as a review of data management and data visualization methods. "**Analyze**" focuses on the handling and analysis of heterogeneous and interconnected datasets (e.g., from CDC, Census Bureau, Food and Drug Administration, state and federal health departments) that are collected during the epidemic lifecycle. "**Improve**" exploits data-driven knowledge to improve the resilience design of health systems, including healthcare capacity, resources, workflows, and operations. Further, "**Control**" focuses on the learning and optimization of health policies and action strategies for controlling the spread of virus. The system informatics methods and tools will complement medical, clinical and pharmaceutical research efforts, helping safeguard the population from infectious diseases and make health systems more resilient to overwhelming epidemic events.

Table 3 Employees on nonfarm payrolls in manufacturing (in thousands) (Manufacturing: NAICS 31-33, n.d.)

	Not seasonally adjusted				Seasonally adjusted				
	May 2019	Mar. 2020	Apr. 2020	May 2020	May 2019	Mar. 2020	Apr. 2020	May 2020	Change Apr.–May 2020
Manufacturing	12,810	12,747	11,427	11,677	12,829	12,806	11,482	11,707	225
Durable goods	8052	8013	7109	7234	8056	8031	7124	7243	119
Nondurable goods	4758	4734	4318	4443	4773	4775	4358	4464	106

3 Measure the Epidemic Dynamics

The "measure" step is directly aimed at testing the population for the prevalence of virus, which is critical to monitoring the temporal evolution of an epidemic in a spatial region. Rapid advances of gene, microbiology and imaging technologies have greatly improved the design and development of testing methods (e.g., speed and accuracy) of coronavirus and influenza. As discussed in Sect. 2, an epidemic poses paramount challenges on the health and economy of our society. The prevalence of a virus in a large population often incurs large amounts of testing, which leads to spatially-temporally big data. This provides an opportunity for the "analyze" step to develop an in-depth understanding of dynamically evolving statuses of an epidemic. Here, data could be collected in disparate efforts by private companies, research centers, universities, and government agencies, thereby leading to the formation of data cohorts to address issues of data management. Epidemic data can then be visualized in various ways to provide comprehensible information about the spatiotemporal variations of an epidemic. An effective visualization further helps the "analyze" step to estimate and extract salient features for the prediction of future trajectory or the monitoring of transmission risks.

3.1 Testing and Sampling

Clinical testing is a critical first step to stopping the spread, which consists of viral testing (i.e., examine whether an individual is currently infected or not) (Esbin et al., 2020) and antibody testing (i.e., check whether an individual was infected before and currently has the presence of antibodies in the blood) (Lipsitch et al., 2020). In the case of COVID-19, specimens are often collected through swabs in the nose or throat for the viral testing. If specimens show the existence of a virus's ribonucleic acid (RNA) or proteins, the test will be positive. The antibody testing is typically done by collecting a sample of blood serum and then examining the presence of antibodies. In order to monitor the prevalence of virus, testing can be performed in three different ways as follows:

- **100% testing**: Population is the entire collection of individuals of interests in a region of interest (e.g., university, city, county, or state). If the cost is not a concern, 100% testing makes sure everyone is tested and then all the infected individuals can be isolated and quarantined. This is an effective approach to stop the spread, but often encounters practical limitations such as inadequate supply of testing kits, prohibitive cost, and population instability due to mobility and immigration.
- **Acceptance sampling**: Sample is a representative subset of the population that can be tested for statistical inference. Acceptance sampling, also called Lot Quality Assurance Sampling (LQAS) (Hedt et al., 2012), is a middle ground between 0% and 100% testing and requires a small sample size for population

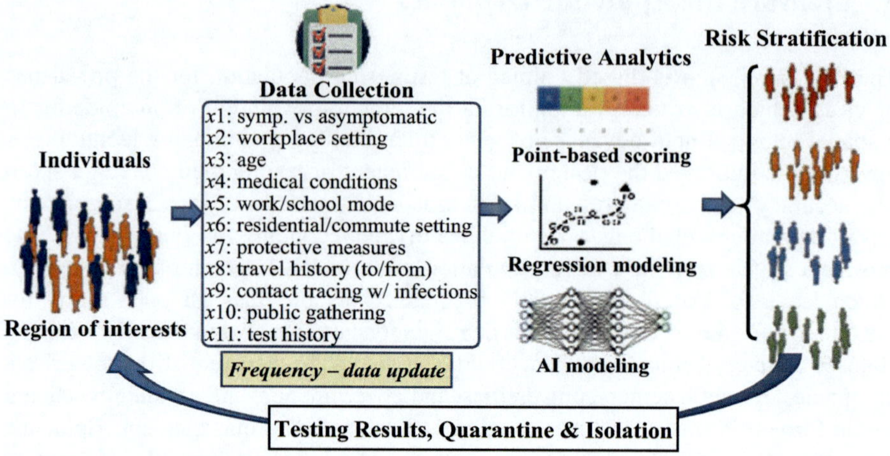

Fig. 3 Data-driven risk scoring systems for categorized sampling and testing

surveys. The population can be stratified into sub-groups (or lots), and each lot can be sampled for clinical testing so as to "accept" or "reject" the lot according to the risk tolerance levels. Also, these samples can be aggregated to establish the confidence interval of infected proportion for testing the hypothesis on the prevalence of an epidemic virus.

- **0% testing**: This means that no testing will be done for the individuals in a specific region. In the onset of an epidemic, few tests are performed because the new virus is just emerging and has not caught enough attention from the public. Once the epidemic virus is captured (e.g., genome sequenced and shared), testing kits can then be designed and developed.

Figure 3 shows that mobile or web-based applications can be used for data collection from individuals in a spatial region of interests, if the testing capacity is constrained and 100% testing cannot be implemented. Examples of the predictors may include x_1: symp. vs asymptomatic; x_2: workplace setting; $x3$: age; $x4$: medical/comorbidity conditions; $x5$: work/school mode; $x6$: residential/commute setting; $x7$: protective measures; $x8$: travel history (to/from); $x9$: contact tracing with infections; $x10$: public gathering; $x11$: test history; The response variable will be the risk probability of infection (range from 0 to 1). The data-driven decision support system helps stratify the individuals into groups (or lots) and then optimize the testing decisions. The risk scoring system categorizes the population into different groups with various levels of risk probability. For example, four groups can be stratified based on the risk probability, which helps further optimize the allocation of testing resources and identify the infected individuals for isolation and quarantine.

As shown in Fig. 3, risk scoring systems can be established in three different ways, namely point-based systems, regression modeling, or AI-based modeling. Such scoring systems help categorize the acuity levels of patients and then improve the quality of healthcare services (e.g., surgical procedures, medication usages, care guidelines, treatment plans, and resource allocations) (Chen & Yang, 2014; Imani et al., 2019). Point-based scoring systems use the simple points or weights, and can be easily implemented in questionnaire form. The points or weights can be adjusted for different predictors (or factors). For example, if the symptom is weighted more than other predictors, it may be assigned with a larger point (or weight). In clinical practice, point-based scoring systems are widely used to stratify the patients, e.g., Acute Physiology and Chronic Health Evaluation (APACHE) (Zimmerman et al., 2006), Sequential Organ Failure Assessment (SOFA) (Raith et al., 2017), Simplified Acute Physiology Score (SAPS) (Metnitz et al., 2005; Moreno et al., 2005), and Mini-mental state examination (MMSE) (Galasko et al., 1990). Figure 3 shows an example of risk factors for the design of point-based scoring systems, which also helps reduce the number of variables to compile into a short survey. An increasing score indicates a higher risk of infection. In addition, the infection risk can be derived using a multivariate logistic regression model as: $\log\left(\frac{risk}{1-risk}\right) = a + \sum_i b_i x_i$, where *Risk* is the risk of death, $\left(\frac{risk}{1-risk}\right)$ is the odds ratio, a is the intercept, b_i is the coefficients and $x_i^{'}s$ are independent predictors. Here, training data or medical domain knowledge can be used to adjust the regression coefficients for different predictors (or factors). Finally, it is not uncommon that AI modeling (e.g., neural networks) are utilized to learn from complex-structured data for risk stratification. AI models, however, need large amounts of data for training and learning the weights, and are difficult to implement for testing and sampling in an epidemic.

Statistical sampling is a cost-effective approach to survey the groups (or lots) of individuals when the testing capacity and supply chain are constrained. First, the confidence interval for the proportion of infections p can be estimated from testing data. If there are c infected individuals for a random sample of size n, then an approximate $100(1-\alpha)\%$ confidence interval for p is

$$\hat{p} - z_{\alpha/2}\sqrt{\frac{\hat{p}\left(1-\hat{p}\right)}{n}} < p < \hat{p} + z_{\alpha/2}\sqrt{\frac{\hat{p}\left(1-\hat{p}\right)}{n}} \tag{1}$$

where $\hat{p}$ is c/n, and $z_{\alpha/2}$ is the z value with an upper tail area of $\alpha/2$. This estimation tends to be more reliable when the number of confirmed individuals c is greater than 6 in the sample, and is also applicable in the case of hypergeometric distribution when the sample size n is small. Here, the choice of sample size is dependent on the significant level α and the margin of error (MOE), i.e., $z_{\alpha/2}\sqrt{\hat{p}\left(1-\hat{p}\right)/n}$. If a specific MOE value e is desired, then the sample size n is approximately $z_{\alpha/2}^2\hat{p}\left(1-\hat{p}\right)/e^2$. Note that the function $\hat{p}\left(1-\hat{p}\right)$ reaches the maximum 1/4 when $\hat{p} = 1/2$. Hence, the MOE is guaranteed not to exceed e if the sample size is chosen to be $z_{\alpha/2}^2/4e^2$. For example, it is 95% confident that the MOE will not exceed 0.02 when the sample size is $1.96^2/(4 \times 0.02^2) = 2401$.

Acceptance sampling is useful to help the decision-making process on whether or not to lockdown or reopen a region (or "lot") for regular businesses. As shown in Fig. 4a, the operating characteristic (OC) curve describes an acceptance sampling plan in terms of the probability of reopening versus the proportion infected. For example, the probability of reopening is $1 - \alpha$ if the region meets the acceptance risk level (ARL) p_{ARL}. The probability of reopening is β if the region is on the rejection risk level (RRL) p_{RRL}. Assuming a binomial distribution, the sample size n and acceptable number a can be obtained as:

$$1 - \alpha = \sum_{c=0}^{a} \frac{n!}{c!\,(n-c)!} p_{ARL}^{c} (1 - p_{ARL})^{n-c} \tag{2}$$

$$\beta = \sum_{c=0}^{a} \frac{n!}{c!\,(n-c)!} p_{RRL}^{c} (1 - p_{RRL})^{n-c} \tag{3}$$

Then, for this acceptance sampling plan, if there are more than a infections in the random sample of size n from the region, lockdown will be implemented. If there are less than or equal to a infections, the risk is below the ARL level and the region can be reopened. For example, Fig. 4b shows the acceptance sampling plans with $n = 2000$ and a is ranging from 15 to 95. When the acceptance number a increases, this does not significantly change the slope, but rather move the OC curves to the right. If the acceptance number a is small, the risk tolerance levels tend to be low. For larger values of a, both ARL and RRL levels are higher. If a region is above the RRL, NPIs such as lockdown and stay-at-home should be implemented. On the other hand, rectification testing programs can further screen individuals in the rejected region. Often, 100% testing can be performed to identify all the infected individuals, then isolate and quarantine them.

In the practice of clinical testing, acceptance sampling may have the following limitations. First, if the sample size is finite, then the distribution tends to be hypergeometric instead of binomial. However, binomial approximation of hypergeometric is valid if the ratio between sample size and lot size is less than 1/10. Second, acceptance sampling assumes the selection of samples at random from each region. Although clinical testing is prioritized for symptomatic cases or traced contacts of infected individuals, it can however assume that the infection of an individual is at random. Then, clinical testing can be assumed to be implemented on individuals who are infected at random, albeit with the introduction of bias to some extent. Third, individuals are assumed to be homogeneous in a region. In other words, homogeneity refers to the fact that the probability to get infected is approximately the same if in contact with pathogens. This is a reasonable assumption for a susceptible population, although there may be slight differences in the infection probabilities for uncontrollable factors such as age groups and blood types. These limitations and assumptions should be considered during the practice of acceptance sampling for clinical testing.

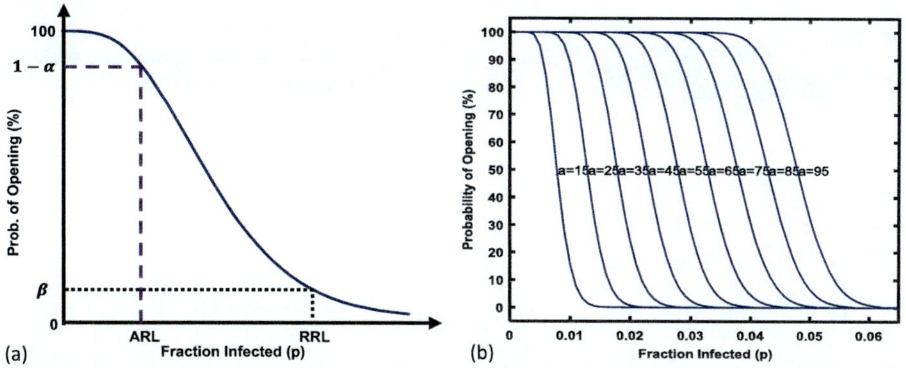

Fig. 4 (**a**) An illustration of operating characteristic (OC) curve, (**b**) OC curves of acceptance sampling plans with the sample size $n = 2000$ and the acceptance number a is ranging from 15 to 95

3.2 Spatiotemporal Surveillance of Epidemic Processes

Clinical testing brings significant amount of data pertinent to the evolution of an epidemic. The epidemic data may include total cumulative cases (or per capita), daily new cases, total deaths for multiple spatial regions (or lots) of interest and are dynamically changing over time. Therefore, the epidemic evolution is a spatiotemporal process, i.e., varying in both space and time. The availability of data provides a great opportunity to design monitoring charts and develop epidemiology surveillance programs. Statistical monitoring methods help health systems leverage sequentially observed data to trigger the alarms and identify the outbreak region. However, raw data are often not normalized and cannot be directly used to develop monitoring charts. For example, spatial regions often have different population sizes. Total cases should be adjusted for the population in a region. As such, features need to be extracted from the data to describe the epidemic characteristics in a region. Examples of features may include cases per million, the incidence rate, or transmission risk index that are characterized with data-driven models.

If the monitoring objective is to detect abnormal changes of incidence rates x_1, x_2, ..., x_k over k regions, then the feature vector will be $\boldsymbol{x} = [x_1, x_2, \ldots, x_k]^T$. The statistical test is aimed at setting up the null and alternative hypotheses, then seeking data-driven evidence to determine whether an anomaly is present in any dimension (i.e., a region) of the feature vector or not. Under the null hypothesis H_0, the incidence rates over k regions do not change over time. As such, the feature vector $\boldsymbol{x}$ is assumed to follow a multivariate normal distribution with population mean $\boldsymbol{\mu}$ and covariance matrix $\boldsymbol{\Sigma}$, i.e.,

$$f(\boldsymbol{x}) = \frac{1}{(2\pi)^{p/2}|\boldsymbol{\Sigma}|^{1/2}} \exp\left[-\frac{1}{2}(\boldsymbol{x}-\boldsymbol{\mu})'\boldsymbol{\Sigma}^{-1}(\boldsymbol{x}-\boldsymbol{\mu})\right] \tag{4}$$

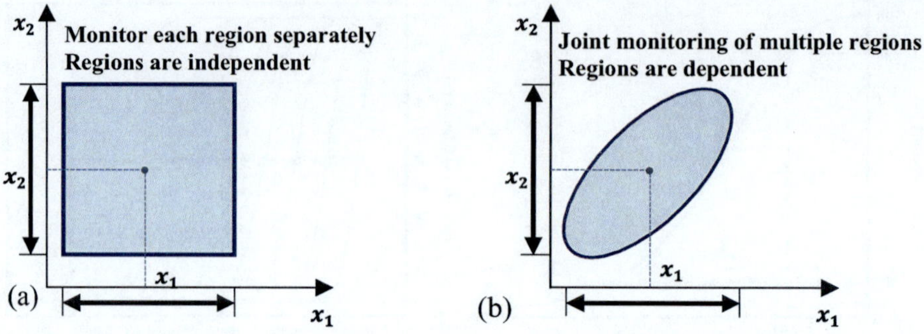

Fig. 5 Multivariate monitoring schemes for epidemic surveillance: (**a**) Monitor each region separately and regions are independent, (**b**) Joint monitoring of multiple regions and regions are dependent

If an outbreak occurs in one region or multiple adjacent regions, then the assumption of multivariate normal distribution is no longer valid. The alternative hypothesis H_1 that the joint distribution of multivariate features is non-normal will tend to hold. The hypothesis test accepts or rejects the null hypothesis H_0 at a significance level α. Although the assumption of multivariate normality is required to formally establish confidence limits in the statistical test, a slight deviation will not severely impact the results (Chen & Yang, 2016a). Here, multivariate normal probability plotting can be used to evaluate whether the extracted features of incidence rates are approximately normally distributed for multiple regions of interests.

As shown in Fig. 5a, most of traditional monitoring schemes assume that k regions are independent. Therefore, a common approach is to monitor each feature independently in the literature. In the bivariate case, control limits will form a rectangular region. If the pair of observations fall within this rectangular region, then the null hypothesis H_0 holds. If the pair of observations reside outside this region, then the null hypothesis H_0 is rejected. However, this monitoring scheme has limited applications due to the "curse of dimensionality". For example, if the probability of Type I error is α for each feature, then Type I error for monitoring k features independently is $1 - (1 - \alpha)^k$. The probability that all k observations fall within the confidence limits is $(1 - \alpha)^k$ if all the k regions are in control (Yang & Chen, 2014; Chen & Yang, 2015). Hence, the error is significant when the dimensionality of the feature vector increases. It may also be noted that k features are oftentimes not independent because adjacent regions tend to be correlated with each other in an epidemic situation.

Therefore, multivariate statistical methods that consider spatial correlations and jointly monitor these regions (or features) are urgently needed. As shown in Fig. 5b, due to the correlation among adjacent regions, the pair of observations now resides in the elliptical region for the bivariate case. Under the null hypothesis H_0, k regions will follow the multivariate normal distribution with the population covariance matrix $\boldsymbol{\Sigma}$. As such, the test statistic $\chi^2 = (\boldsymbol{x} - \boldsymbol{\mu})' \boldsymbol{\Sigma}^{-1} (\boldsymbol{x} - \boldsymbol{\mu})$ follows a chi-square distribution with k degrees of freedom. The joint distribution changes in the presence of regional anomalies. If there are shifts in at least one out of k regions, then χ^2 values will be above the upper control limit $UCL = \chi^2_{\alpha,p}$, where α is the significance level. If χ^2 values are below the upper control limit, then the null hypothesis H_0 holds and there will be no significant evidence of anomalies. The control ellipse of bivariate case in Fig. 5b is due to region-to-region correlations. Because off-diagonal elements are no longer zero in covariance matrix $\boldsymbol{\Sigma}$, the principal axes of the ellipse are not parallel to the $\overline{x}_1, \overline{x}_2$ axes any more.

In the real world, population mean $\boldsymbol{\mu}$ and covariance matrix $\boldsymbol{\Sigma}$ are often unknown and need to be estimated from the data. If the sample mean $\overline{\boldsymbol{x}}$ and covariance matrix $\boldsymbol{S}$ are used instead, then the test statistic becomes $T^2 = (\boldsymbol{x} - \overline{\boldsymbol{x}})'\boldsymbol{S}^{-1}(\boldsymbol{x} - \overline{\boldsymbol{x}})$, which is commonly called as the Hotelling T^2 statistic (Mason et al., 1997; Li et al., 2008). The new UCL for the Hotelling T^2 statistic is:

$$UCL = \frac{p\,(N+1)\,(N-1)}{N^2 - Nk} F_{\alpha,k,N-k} \tag{5}$$

where $\boldsymbol{x}_1, \boldsymbol{x}_2, \cdots, \boldsymbol{x}_N$ are N sequentially observed samples of epidemic data from k regions, $F_{\alpha,\,k,\,N-k}$ is the upper $100\alpha\%$ critical point of F distribution with k and $N - k$ degrees of freedom. Note that control limits are established in Phase I with in-control datasets (i.e., without the presence of anomalies). For Phase II monitoring, the control chart plots control limits and the test statistic $T^2(i)$, $i = 1, 2, \cdots N$ for each sample. When a new sample arrives, we will then compute the test statistic and check the conformance in the control chart. Note that it is not feasible to graphically construct the control ellipse for more than two regions as shown in Fig. 5b. The composite index (i.e., Hotelling T^2 statistic) helps characterize the multivariate distribution of k features (or regions), and further establish the control chart to effectively detect whether there are shifts in at least one out of k regions (i.e., multivariate epidemic monitoring and surveillance).

3.3 *Data Management and Visualization*

As the epidemic progresses, large amounts of data are organized in the form of data cohorts or lakes. Medical scientists collect pertinent data about the clinical picture of a new virus for the development of effective intervention methods, such as antivirals and vaccines. Epidemiologists and engineers leverage the public health data to develop analytical models for the prediction of virus spread dynamics. Real-time data of epidemic situations is critical to understand the spread, trace the contacts, and control the propagation. Data management is indispensable to integrate disparate data efforts from government agencies, universities, and private companies. Here, data cohort connects various organizations to manage the data using the defining characteristics, which help researchers save tremendous amount of time in finding, analyzing, evaluating and validating relevant data for useful information and insights to stop the epidemic. Nonetheless, data lake is a repository of unorganized data in the raw format. Data cohort may include necessary data from on-going and completed research, as well as contact tracing data. This type of data could contain the patient location, sociodemographic information, and the list of contacts during the elicitation window and where the patient has visited. When the number of infections become prevalent, data management gets increasingly complex. This is partly due to the large number of cases, as well as the long list of traced contacts of each positive case. Data management depends on the use of database systems to support such many-to-many relational tables and provide a higher level of flexibility of routine data storage, update, security, reporting, and On-Line Analytical Processing (OLAP).

Note that the epidemic data is varying in both space and time. Table 4 provides examples of data repositories and cohorts developed by government agencies, institutions, and private companies. These data cohorts are open access to the public or limited access by applications. The UN data lab, US CDC and European Centers for Diseases Control (ECDC) organize and publish the real-time position data of virus spread in either country level or county level. Such information can be used to study and track the spread of the disease. US National Science Foundation (NSF) supported a research project to develop the COVID Information Commons, which is an open website to promote data and knowledge sharing across different COVID research efforts. National Institute of Health (NIH) initiated an National COVID Cohort Collaborative (N3C) project for collaboration on data collection, sharing, and analytics, which also provides the open access to research literature about COVID-19 genomics, virus structures, and clinical studies.

Also, academic institutions such as John Hopkins University (JHU) and the University of Washington provides the organized COVID-19 data and popular dashboards for data visualization. This, in turn, greatly facilitates the general public in visualizing the spread and trend of epidemic, thereby promoting situational awareness. In addition, there are data cohorts from private companies and foundations that provide targeted information about the disease. For example, the COVID-19 tracking project assembles the testing data, hospitalization rates, treatment outcomes, race and ethnicity data for researchers to investigate the outbreak scale, the mortality rate, and regional effects of the disease. COVID-19 Open Research Dataset (CORD19) provides an application programming interface (API) to retrieve the infection data, research feed, and COVID related texts. This API can help researchers query data in a fast manner. Surgo Foundation provides the community vulnerability index, social distance tracking, and nurse sentiment data to help develop analytical methods and tools for epidemic response.

Large amounts of data are readily available from different sources. The next step is to visualize and represent the data so that useful information and salient features can be easily comprehensible by the audience. Data visualization focuses on compact representations of trends and patterns in the data with graphical methods and tools such as time series charts, density graphs, and heat maps. The human brain can perceive information in graphics and images better than pale texts or data tables. An effective visualization helps condense a thousand words in one picture.

Table 4 Examples of COVID-19 data repository/cohort and features

Data cohorts and repositories	Descriptions and features
Center for Disease Control and Prevention (CDC) https://www.cdc.gov/coronavirus/2019-ncov/cases-updates/	US infection data with cases, race, ethnicity, testing, hospital capacity and other data streams at local, state, and national levels
World Health Organization https://www.who.int/	Global case updates with total confirmed cases and deaths, new cases and deaths, and transmission classifications
European CDC https://www.ecdc.europa.eu/en/covid-19-Epidemic	COVID-19 situation updates, case counts and distributions for the EU/EEA, UK, and worldwide.
National Institutes of Health https://datascience.nih.gov/covid-19-open-access-resources	COVID-19 data and resources such as official data, related studies, and high-performance computing consortium
National COVID Cohort Collaborative (N3C) https://cd2h.org/	A very large patient-level COVID-19 clinical dataset shared by CTSA, CD2H and other distributed clinical data networks
Clinicaltrials.gov https://clinicaltrials.gov/ct2/results?cond=COVID-19	Detailed information about active and recruiting clinical trials such as intervention and phase
Johns Hopkins University https://github.com/CSSEGISandData/COVID-19	Global and US daily situation update at country and state level, along with time-series summary
NSF COVID Information Commons https://covid-info-commons.site.drupaldisttest.cc.columbia.edu/	Open website to facilitate knowledge sharing and collaboration focused on NSF funded COVID rapid response research projects
New York Times https://github.com/nytimes/covid-19-data	US state level and county level situation updates, with historical and live data
Twitter Dataset https://github.com/thepanacealab/COVID-19_twitter	Tweets and retweets data acquired from Twitter stream related to COVID-19 chatter with all languages
The COVID Tracking Project https://covidtracking.com/data	US infection data with cases, tests, hospitalized, severity (in ICU, on ventilator, etc.), and outcomes
CORD-19 https://www.kaggle.com/allen-institute-for-ai/CORD-19-research-challenge	A dataset of over 167,000 scholarly articles about COVID-19, SARS-CoV-2 and related coronavirus
Ding Xiang Yuan https://ncov.dxy.cn/	Global case updates with active, confirmed, recovered. China regional case updates with city level native/imported counts
OPENICPSR https://www.openicpsr.org/openicpsr/search/COVID-19/studies	Data cohort which contains links to US state policy database, government response dataset, and COVID-19 impact survey

There are a variety of visualization tools to represent the data in different ways, but it is important to choose the right tool to balance visual appearances and hidden information. The artisan spirit and craftsmanship help design better visualizations for the target users.

Table 5 provides examples of visualization dashboards available online for COVID-19, including the URL links and features. Most dashboards are developed with geographical maps and applications such as ArcGIS, as well as the COVID-19 data in the United States and worldwide. Figure 6 shows an illustration of the infection map in the county level of US from April 29 to September 23, 2020. The number of counties is close to 3141. Instead of pale numbers in the table, such a visualization quickly provides a sense of the current status and the virus spread across US counties. Spatial regions are often labeled with a color map or with markers whose sizes are proportional to the number of infected cases. This informs people quickly about the regions of interests and the current spread of virus in the world. The temporal variations are shown as trends about how the number of cases rises with respect to time. The dashboard can also include data-derived features such as incidence rate, case-fatality ratio, testing rate, and hospitalization rate. Examples of popular dashboards include the CDC, JHU, Google, Bing, and 1point3arc dashboards. Notably, ArcGIS Storymaps provide a visualization tool to depict how the disease is spread from a regional epidemic to pandemic in a time-lapsed manner. Pharmaintelligence visualizes the progress of drug discovery and clinical trials worldwide, which highlight endeavors that medical scientists made to control the epidemic.

4 Analyze the Data for Epidemic Insights

The "analyze" step focuses on the extraction of useful information from epidemic data collected in the "measure" step. There are a variety of factors (e.g., demographics, socioeconomic factors, education factors, economy factors, population health factors, and mobility index) that may be interrelated with epidemic characteristics (e.g., the growth of confirmed cases). Therefore, it is critical to delineate and determine salient factors that are sensitive to the response variable. Note that the evolution of an epidemic is highly nonlinear and nonstationary. Traditional linear methods tend to be limited in their ability to handle the nonlinearity. High level of spatial heterogeneity also leads to skewed datasets and non-normal distributions of factors. As such, data transformation is necessary to pre-process and transform the data into normal shape. It is also imperative to utilize statistical models to investigate the interrelationships between various factors and epidemic characteristics. Also, rich data from the "measure" step can be fed into the development of simulation models. This, in turn, will help the "improve" step (see Sect. 5) to forecast the real-time positions of virus spread and further run "what-if" analysis for the optimization of intervention strategies and healthcare policies. New experiments can then be

Table 5 Examples of COVID-19 visualization dashboards and features

Data visualization dashboard	Descriptions and features
Center for Disease Control and Prevention https://www.cdc.gov/covid-data-tracker	US infection maps with testing outcomes, forecasting, demographic trends of sex, race/ethnicity and age, and social impacts
World Health Organization https://COVID-19.who.int/	Country-level visualization of global trend of new cases, confirmed cases, and deaths
European CDC https://www.ecdc.europa.eu/en/covid-19-Epidemic	Interactive dashboard to show situation updates and case distributions for the EU/EEA, UK, worldwide
Johns Hopkins University https://coronavirus.jhu.edu/us-map	Interactive visualization of confirmed cases, deaths, and a status health report in the US and across the globe
Institute for Health Metrics and Evaluation https://COVID-19.healthdata.org/	Graphical visualization of deaths, infections and testing, and hospital resource utilization, predictions, and social distancing by country
Google https://news.google.com/COVID-19/map	A very high-level report of confirmed cases, recovered, deaths, and new cases (last 60 days) by country and worldwide
Facebook https://covid-survey.dataforgood.fb.com/	Interactive visualization of infection proportion, population density, and elderly population by country
Bing https://www.bing.com/covid	Visualization of confirmed cases, recovered, deaths, and relevant news by counties in the US
Worldometer https://www.worldometers.info/coronavirus/	Reported cases, deaths, and rankings by country or continent
ArcGIS COVID-19 hub https://coronavirus-disasterresponse.hub.arcgis.com/	Esri storymaps and visualization tools to create time-lapse animation of the spread and help guide decisions around health, racial, and economic equity
1 point 3 arc https://coronavirus.1point3acres.com/en	Interactive dashboard with a summary of the infected cases, deaths, recovered, and fatality rate
Pharmaintelligence https://pharmaintelligence.informa.com/resources/key-topics/coronavirus	Drug discovery and clinical trial visualization across the globe
The weather company https://accelerator.weather.com/bi/	High-level visualization of confirmed cases by day, by region (the last 14 days), deaths, rate of spread, rate of deaths, spread over time
Coronavirus3d https://coronavirus3d.org/	SARS-Cov-2 protein structure visualization
NextStrain https://nextstrain.org/ncov	Genomic epidemiology of novel coronavirus by region (Asia, Europe, North America, South America, etc), or by host, age, sex

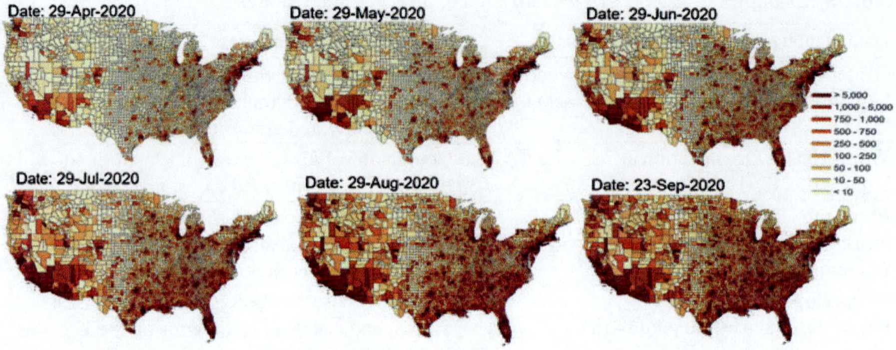

Fig. 6 The infection map of cumulative cases for 3141 counties in the United States

designed to test the effectiveness of these action strategies on either physical systems or computer simulation models.

4.1 Descriptive Analytics

This section of "descriptive analytics" aims to visualize the COVID-19 data and pertinent factors in an easily comprehensible form, and further investigate key predictors that are interrelated with the progress of infection situations in the US. In this study, the dependent variable (or responses) is set to be either cumulative $\boldsymbol{y_1}$ or weekly new cases $\boldsymbol{y_2}$ of COVID-19 infections at the county level,[1] which are retrieved from New York Times data repository (i.e., https://github.com/nytimes/covid-19-data as shown in Table 4). To avoid confounding effects by population sizes, we have also considered response variables that are averaged by the population, i.e., cumulative y_3 or weekly new cases y_4 per capita in each county. The data repository provides real-time updates coronavirus cases in the US since January 2020, and provides cumulative daily counts of cases at state and county levels, respectively. We leveraged and processed the data at county level from Mar. 29, 2020 (Week 1) to Aug. 22, 2020 (Week 21) for the cumulative and incremental new cases of coronavirus at each week. In total, this study includes pertinent data about 2781 counties from 50 US states with a time span of 21 weeks (Mar. 29–Aug. 22, 2020) for the descriptive analytics.

For the independent variables, we have extracted a total of 72 predictors[2] at the county level from Google COVID-19 community mobility reports, US Census

[1]Note that the distributions of response variables are highly skewed, and are therefore transformed to the log scale for descriptive analytics, i.e., $y' = \log(y+1)$.

[2]If predictors are approximately normally distributed, no transform is made. For positively skewed data, $\log(x+1)$ is used, while for negatively skewed data, $\log(\max(x+1) - x)$ is applied.

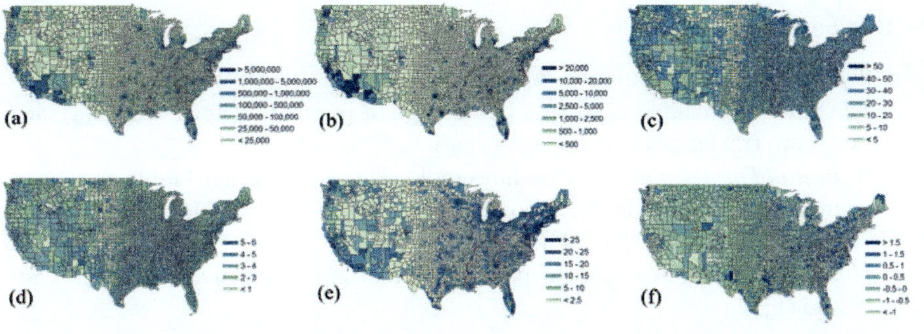

Fig. 7 The geographical distribution of (**a**) total population, (**b**) household with grandparents living with grandchildren, (**c**) percentage of people age >60, (**d**) average family size, (**e**) mean of mobility change in residential setting (Apr. 5) (**f**) skewness of mobility change in residential setting (Apr. 5)

database, and County Health Rankings reports. Figure 7 depicts the geographical distribution of some example predictors (e.g., total population, household with grandparents living with grandchildren, people aged >60) in 3141 US counties. These predictors are categorized into four groups, namely social-economy, health, demography, and mobility, as follows.

- **Social-economic predictors:** We have extracted pertinent data about 3 education variables, 4 economic variables, and 9 occupational variables from the US census database in 2018 at the county level. (1) ***Education*:** The percentage of population aged 25 and over who don't have a degree (x_1), have a bachelor's degree or higher (x_2), and have a graduate or professional degree (x_3), respectively. (2) ***Economy:*** The unemployment rate among population aged 16 years and over (x_4), median household income (x_5), median family income (x_6) and median earnings (x_7). (3) ***Occupation:*** Among the employed population aged 16 years and over, we consider the percentage of population who work in management, business, science, and arts (x_8); service (x_9); sales and office occupations (x_{10}); natural resources, construction, and maintenance (x_{11}); production, transportation, and material moving (x_{12}); manufacturing (x_{13}); wholesale trade (x_{14}); retail trade (x_{15}); educational services, and health care and social assistance (x_{16}).
- **Health predictors:** Moreover, we extracted the data about 10 health features from County Health Rankings reports as follows: percentage of population with disability (x_{17}), percentage of adults that report fair or poor health (x_{18}), average number of reported physically unhealthy days (x_{19}), percentage of adults that reported currently smoking (x_{20}), food environment index (x_{21}), percentage of adults that report no leisure-time physical activity (x_{22}), percentage of the population with access to places for physical activity (x_{23}), percentage of adults that report excessive drinking (x_{24}), percentage of people under age 65 without insurance (x_{25}), and primary care physician (PCP) rate (x_{26}).

- **Demography predictors:**

  Population: total population (x_{27})
  Age and sex: median age (x_{28}), percentage of population aged >60 (x_{29}), and sex ratio (males per 100 females) (x_{30}).
  Household and family: average household size (x_{31}), average family size (x_{32}), number of households with grandparents living with grandchildren (x_{33}), and percentage of single-parent households (x_{34}).
  Marital status: percentage of population married (x_{35}), divorced (x_{36}), widowed (x_{37}).

- **Mobility predictors:** (1) ***Commute mode:*** The US Census provides the percentage of population aged 16 and over who drive alone (x_{38}), or carpool (x_{39}) by car, truck, van, or use public transportation (excluding taxicab) (x_{40}), as well as the percentage of workers who commute in their car alone commute more than 30 min (x_{41}), work at home (x_{42}). (2) ***Community mobility***: Google provides the community mobility change from the baseline (in percentage) in 6 different types of places, namely retail and recreation, groceries and pharmacies, parks, transit stations, workplaces, and residential. The mobility data provide insights about how COVID-19 and related policies impact the population's mobility patterns in public places. The data were organized on a daily basis. For each week from Mar. 29 to Aug. 22, we calculated the average, median, variance, skewness, kurtosis of mobility variations in each type of place at the county level, thereby extracting a total of 30 mobility features (i.e., x_{43}~x_{72}: 5 features × 6 places).

4.1.1 Correlation Analysis

Figure 8a shows the Pearson correlations between 72 predictors (i.e., x_1~x_{72}, see details above) and cumulative confirmed cases y_1. In general, there are high correlations between COVID-19 situations and social-economic, demography and mobility predictors. The highest correlation (83.89%) is with the total population x_{27} in each county, also see the scatter plot in Fig. 9a. This shows the prevalence of COVID-19. The more population a county has, the more infections it will have. As of August 22, 2020, COVID-19 had spread over the whole US territory and few counties could be an exception.

The second highest (81.77%) is with the number of households with grandparents and grandchildren x_{33}; also see the scatter plot in Fig. 9b. Elderly people and children are both high-risk groups. When the number of households with grandparents living with grandchildren is high, these two groups of people are more vulnerable and more likely to transmit the virus to each other.

For some social-economic factors, the Pearson correlations are approximately in the range of 29–46%, also see Fig. 8. This is not as highly correlated as two demographic variables, but are sensitive to COVID-19 situations to some degree. Also, it may be noted that the Pearson correlations are approximately in the range of 42–63% for some mobility predictors. This is not surprising because the virus spread

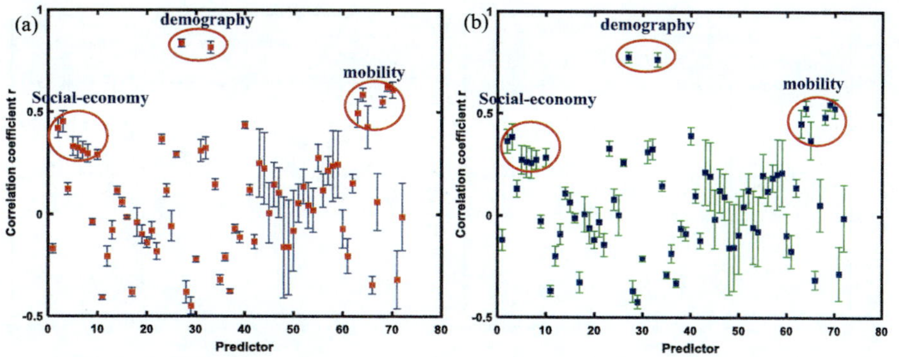

Fig. 8 The correlation between 72 predictors (i.e., $x_1 \sim x_{72}$) and cumulative confirmed cases y_1 (**a**) and weekly new cases y_2 of COVID-19, before period. The error bar represents the mean and standard deviation over 21 weeks

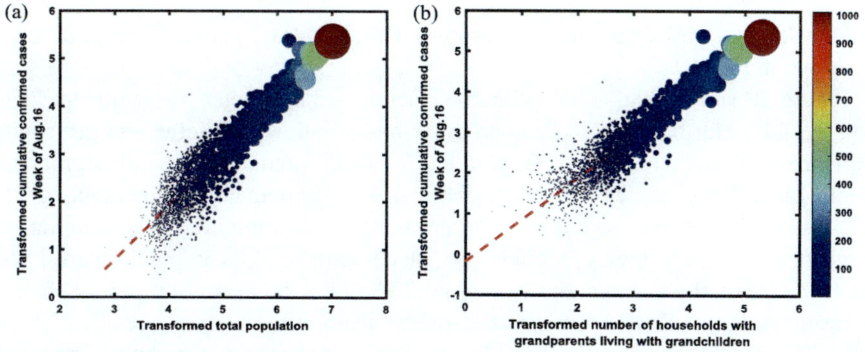

Fig. 9 Scatter plots of cumulative COVID-19 cases vs. total population (**a**) and the number of households with grandparents living with grandchildren (**b**). The circle size is proportional to the total population in each county

causes many businesses to shut down and people to stay at home. The variations of mobility patterns in community places are sensitive to the COVID-19 situations.

Further, we computed and compared with the Pearson correlations between 72 predictors (i.e., $x_1 \sim x_{72}$, see details above) and weekly new cases y_2 for 2781 counties from 50 US states, as shown in Fig. 8b. The results are similar to the cumulative confirmed cases in Fig. 8a, but with slight decreases in the magnitude of 3–10%. In other words, there are slightly higher correlations between 72 predictors and cumulative cases than weekly new cases.

The prevalence of coronavirus in the US leads to the highest correlation (i.e., 83.89%) with total population. This is conducive to building a regression model to forecast the growth of COVID-19 cases in each county. However, total population poses a confounding effect that dilutes the factorial effects from other predictors. Therefore, we have further examined each predictor's correlation with cumulative

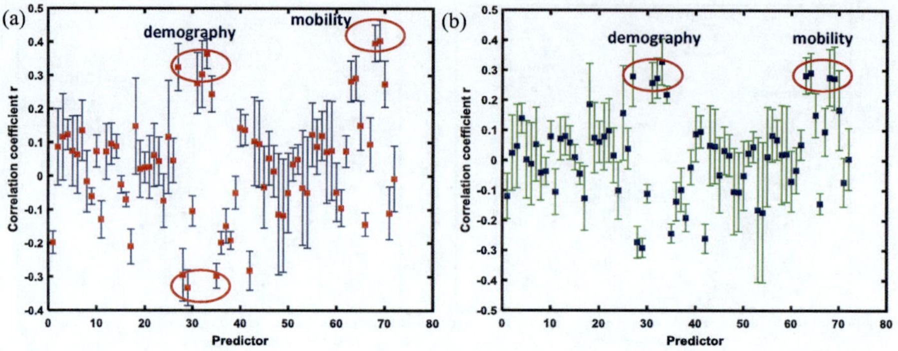

Fig. 10 The correlation between 72 predictors (i.e., $\boldsymbol{x_1}$~$\boldsymbol{x_{72}}$) and cumulative confirmed cases per capita $\boldsymbol{y_3}$ (**a**) and weekly new cases per capita $\boldsymbol{y_4}$ of COVID-19 (**b**). The error bar represents the mean and standard deviation over 21 weeks

and weekly new cases per capita as response variables (i.e., y_3 and y_4, respectively), as shown in Fig. 10.

Figure 10 shows that social-economic factors are no longer as significant as in Fig. 8, and yield the average Pearson correlations below 20% after the per capita adjustment. Nonetheless, demographic and mobility predictors are still significant among all, although their Pearson correlations are approximately in the range of 30–40%. As shown in Fig. 10a, the predictors with high correlations with cumulative confirmed cases per capita y_3 include: x_{29} the percentage of the population aged >60 x_{29} (−33.27%), the average family size x_{32} (30.32%), the mean of mobility change in residential x_{69} (40.4%), and the skewness of mobility change in residential x_{68} (39.67%). The scatter plots in Fig. 11 also show that there are correlations between these four predictors and the response variable (i.e., cumulative confirmed cases per capita y_3). However, neither positive nor negative correlations are as strong as the level of 83.89% in Fig. 9. Similarly, Fig. 10b shows the Pearson correlations between 72 predictors and weekly new cases per capita y_4 for 2781 counties from 50 US states. The results are similar to the cumulative cases per capita in Fig. 10a, but with slight decreases. In other words, weekly new cases per capita y_4 are essentially the week-by-week differences of y_3. Thus, there are slight decoupling of correlation effects.

4.1.2 Regression Modeling

Section 4.1.1 focuses on the relevancy between predictors and response variables. However, there is also redundancy (or multicollinearity) among the predictors that causes the regression model to be unstable and sensitive to external noises. A total of 72 predictors tend to bring the "curse of dimensionality" problem, and cause overfitting to the model. Therefore, we utilize the lasso regression model to shrink the number of predictors and further select a sparse set of significant variables. For

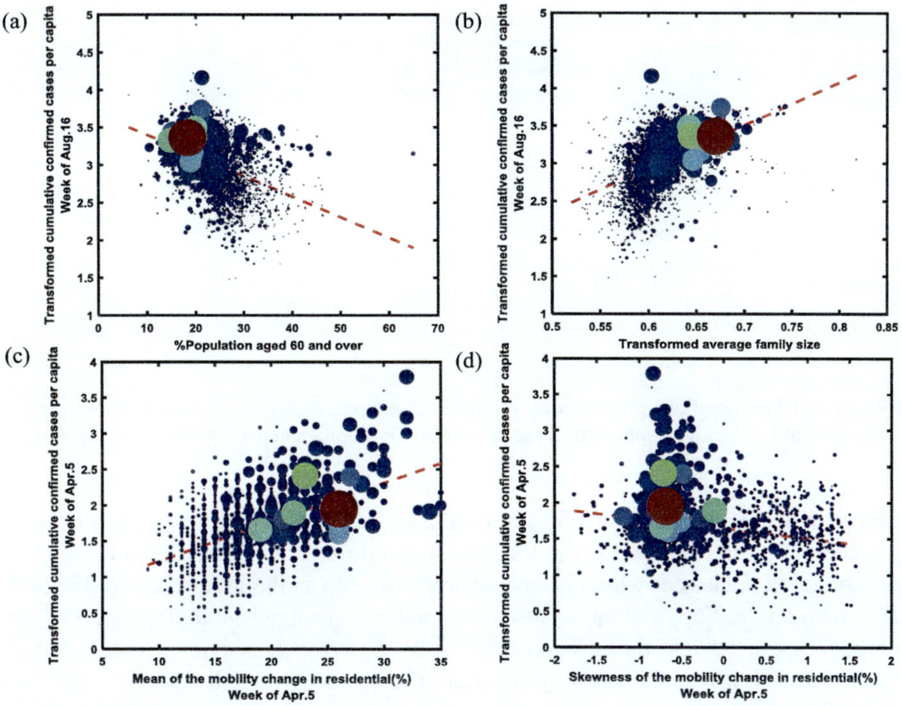

Fig. 11 Scatter plots of cumulative cases per capita vs. the percentage of the population aged >60 (**a**) the average family size in (**b**), the mean of mobility change in residential (%) (**c**), and the skewness of mobility change in residential (%) (**d**). The circle size is proportional to the total population in each county

a given value of λ, a nonnegative parameter, lasso regression penalizes the sum of L1 norm of regression parameters as:

$$\min_{\beta_0,\boldsymbol{\beta}} \left(\frac{1}{2N} \sum_{i=1}^{N} \left(y_i - \beta_0 - \boldsymbol{X}_i{}^T \boldsymbol{\beta} \right)^2 + \lambda \sum_{j=1}^{p} \left| \beta_j \right| \right) \tag{6}$$

where N is the number of observations, y_i is the response at observation i, $\boldsymbol{X}_i$ is a vector of predictor values at observation i, and p is the dimensionality of predictors. Lasso-penalized regression addresses the multicollinearity issue via regularized learning. A parsimonious set of predictors also helps increase the model interpretability, as opposed to a lower level of interpretability with the use of traditional dimensionality reduction methods (e.g., principal component analysis).

Figure 12a shows the variations of prediction errors with respect to the regularization parameter λ. The lasso experiment is performed with ten-fold cross validation for the response variable of cumulative cases per capita and 72 predictors. When λ decreases, the number of selected predictors increases. Note that the predic-

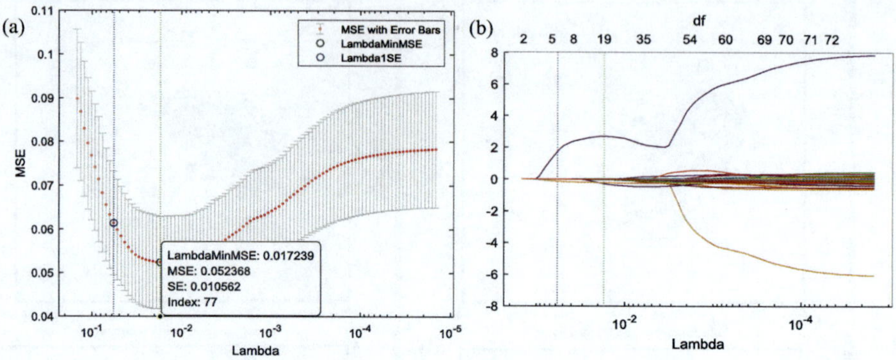

Fig. 12 (**a**) The variations of prediction errors vs. regularization parameter in Lasso regression with cumulative cases per capita and 72 predictors; (**b**) The coefficient path for the Lasso regression

tion error decreases to a local minimum and then increases. The optimal penalization parameter λ_{opt} is identified at the location with minimal cross-validation error plus one standard deviation, which is as shown in Fig. 12a as the green dashed line and the green circle. For cumulative cases per capita, λ_{opt} suggests the inclusion of 19 predictors which yield the lowest cross-validation error. It is evident that variable selection via Lasso penalization yields not only a sparser model, but also a smaller cross-validation error.

Figure 12b shows the coefficient paths of 72 predictors when the value of λ decreases. It may be noted that more and more predictors are included when λ decreases. The green dashed line locates an optimal regularization parameter for the selection of 19 predictors that is identified using the ten-fold cross validation. We have repeated the experiments for each of four response variables (y_1~y_4). The results are consistent with slight deviations because of the variations of correlations as in Figs. 8 and 10.

Furthermore, we use the selected set of 19 predictors to build the fixed-effect regression models and investigate the relationship between predictors and the temporal variations of four response variables (y_1~$y_4 \mid t$) over 21 weeks. The fixed-effect regression model is formulated as follows:

$$y_i \mid t = \beta_0 + \sum_{j=1}^{n} \beta_j x_{ij} + \sum_{k=1}^{m} \lambda_k I_{ik} + \varepsilon, \quad t = 1, 2, \cdots, 21 \tag{7}$$

where y_i is the number of cumulative (or weekly new) confirmed cases in county i, x_{ij} is county i's predictor j, n is the total number of predictors, and β_0 and β_j are parameter estimates. Also, λ_k is the fixed effect for state k, m is the number of states considered in the analysis, I_{ik} is an indicator function for county i and I_{ik}=1, if $i \in k$ (county i belongs to state k); otherwise, I_{ik}=0. We fitted the regression model on a weekly basis.

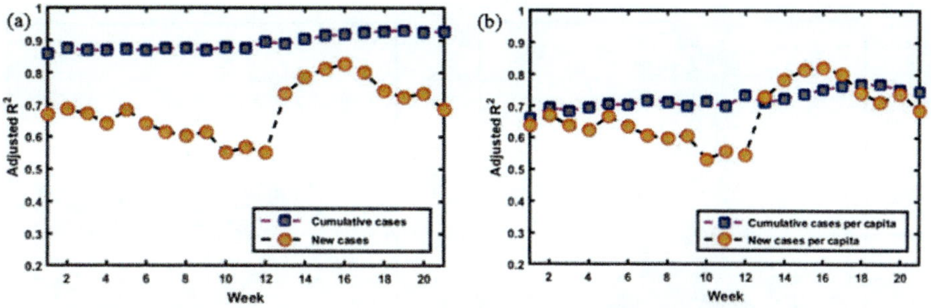

Fig. 13 The variations of adjusted R^2 for the fixed-effect models with response variables of (**a**) cumulative and weekly new cases, and (**b**) cumulative and weekly new cases per capita over the period of 21 weeks

Figure 13a shows the variations of adjusted R^2 for the fixed-effect models with response variables of cumulative and weekly new cases, respectively, over 21 weeks. Note that the presence of highly sensitive predictors (e.g., total population, 83.89% correlation) achieves the adjusted R^2 in the range from 88% to 94% for cumulative cases. However, for weekly new cases, there is a high level of fluctuation in the adjusted R^2 (i.e., approximately 55–82%) over 21 weeks. This is mainly due to policy adjustments from local and federal governments (e.g., reopen the economy), causing high variations of weekly new cases. Also, these policies are not consistent and sometimes heterogeneous in different US counties. Nonetheless, the high adjusted R^2 values show the predictability of fixed-effect models. Figure 13b shows the variations of adjusted R^2 for the fixed-effect models with response variables of cumulative and weekly new cases per capita, respectively, over 21 weeks. The results are consistent with correlation analysis in Sect. 4.1.1. Because of the decrease in variable correlation, the adjusted R^2 values are approximately in the range from 71% to 79% for cumulative cases per capita. Similarly, for weekly new cases per capita, the adjusted R^2 are still fluctuating due to policy adjustments.

Figure 14 shows an example of residual plots that provide diagnosis results of the fixed-effect regression model with the response variable of cumulative cases per capita. Note that no systematic patterns are discerned in the residual plots. The histogram plot in Fig. 14a shows that the normality assumption is valid. Figure 14b shows parallel bands centered around zero in the series of residuals.

4.2 *Spatiotemporal Analytics*

The outbreak of an epidemic is often spatially distributed and evolves over time, thereby generating spatially and temporally big data. For example, epidemic situation reports, in days, months and even years with multiple waves of infections, brings about large amounts of data. Infection dynamics can be visualized through

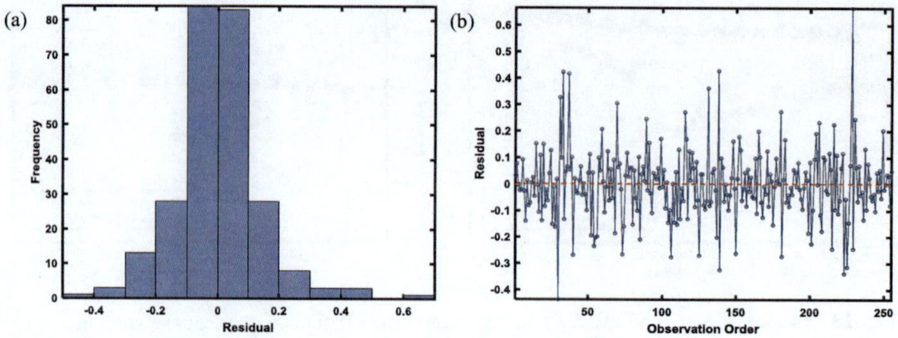

Fig. 14 The residual diagnosis of regression model with the response variable of cumulative cases per capita

data dashboards or time-lapsed visualization in geographically distributed regions, e.g., see Sect. 3.3. However, spatiotemporal data poses significant challenges for human experts to delineate key factor-to-factor interactions and predict the evolution of an epidemic. Fully utilizing the spatiotemporal data depends to a great extent on the development and implementation of information-processing methodologies. Only with effective analytical methods and tools, we can then enable and assist (i) the identification of key factors that are highly correlated with the epidemic growth, (ii) the development of spatiotemporal models for epidemic prediction and risk assessment, and (iii) the provision of decision-support tools for resource planning and intervention strategies towards smarter healthcare services.

Figure 15 illustrates spatiotemporal dynamics of epidemic data generated over geographical regions in the contiguous US. Each cross-section is a snapshot of the epidemic situation at a particular time point. As the infection dynamics evolve across both space and time, epidemiological surveillance systems produce spatiotemporal data: $\{\mathrm{Y}(\boldsymbol{s}, t) : \boldsymbol{s} \in \mathcal{S} \subset \mathbb{R}^d, t \in T\}$, where Y is dependent on both spatial domain $\mathcal{S}$ and time T symbolizes the spatiotemporal variations. Space and time dimensions are relevant but different in an epidemic. It may be noted that the time dimension includes the past, present, and future, which is not directly comparable to the space. Instead, the space dimension is indexed by spatial coordinates. Note that each spatial region can also be embodied with characteristic covariates, predictors or features, $\boldsymbol{x}_s$, such as demographics, socioeconomic factors, or mobility features. If two regions are close to each other, they tend to have a higher correlation. In general, the spatial "closeness" can be due to spatial distance, characteristic features, or high-level traffics (e.g., air transportation) between two regions. There is a need to investigate not only spatial correlation and temporal correlation, but also space-time interaction. Such spatiotemporal interactions bring substantial complexity in the scope of epidemic modeling and analytics.

In the past few decades, the proliferation of space-time data has fueled increased interests in spatiotemporal analytics. Examples of application areas include brain imaging (Bowman, 2007; Mark et al., 2004), public health (Waller et al., 1997;

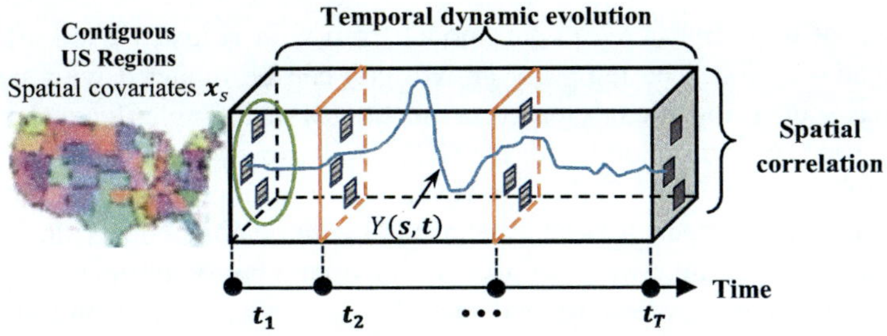

Fig. 15 An illustration of space-time dynamics in the evolution of an epidemic

Kelsall & Wakefield, 2002), service equity (Serban, 2011) and socio-economics (Mateu et al., 2004). The specific questions include the analysis of time-varying brain image and fMRI data, geographical diffusion of epidemic infectious diseases, and spatial equity of public services. Also, there are previous works that employ random fields in $\mathbb{R}^{d+1}$ to model space and time dependencies (Descombes et al., 1998). After a review of the literature on spatiotemporal modeling, we summarize four classical models for the analysis of space-time indexed data below. These models are not meant to be comprehensive or exclusive, but rather serve as initial ideas for spatiotemporal modeling of an epidemic.

- **Spatially-varying time series model**: $Y(s,t) = Y_s(t)$, which separates the temporal analysis for each location. This model $Y_s(t)$ shows specific interests in time-dependent patterns for each spatial location, and allows for location-to-location analysis between time series. For example, the time series of infection cases can be represented and characterized at each zip code, county, or state to investigate the variations of health policy and pertinent impacts on each location. In the literature, Yang et al. extracted patterns from ECG time series at each sensor location on the body surface and exploited the useful information for the identification of cardiovascular diseases (Liu & Yang, 2013; Yang et al., 2012, 2013).
- **Temporally-varying spatial model**: $Y(s,t) = Y_t(s)$, which separates spatial analysis for each time point. The model $Y_t(s)$ focuses more on space-dependent patterns at a particular time point. For example, spatial patterns of virus spread can be modeled at a specific time point; then how spatial patterns change over time. Yao et al. studied body-surface ECG images during the period of ventricular contraction for the detection of myocardial infarction sites (Yao et al., 2017; Yao & Yang, 2016). However, both $Y_s(t)$ and $Y_t(s)$ are conditional methods that investigate either the space given time or time given space, which tend to be limited in their ability to capture space-time correlations.
- **Space-time separation model**: This model separates the spatial and temporal components in the multiplicative form as $Y(s,t) = M(\boldsymbol{x}_s)g(t)$, where $\boldsymbol{x}_s$ are the characteristic covariates for each spatial region, $Y(s,t)$ can be the number of cumulative confirmed cases for a spatial region $\boldsymbol{s}$ at time t. Here, $M(\boldsymbol{x}_s)$ can take

the form of a nonlinear regression model form with adjusted fixed effects for each spatial region. The temporal growth $g(t)$ can be modeled with sigmoidal functions such as logistic or Gompertz functions. For example, Jia et al. presents a space-time separation model for the COVID-19 growth from Jan 24 to Feb 19, 2020 in China (Jia et al., 2020). However, this separation model only accounts for multiplicative effects between spatial and temporal components, and can only model the exponential growth with saturation after a period of time.

- **Parameter-driven spatiotemporal model**: To increase the flexibility to model spatiotemporal dynamics, at a particular time point t, a spatial model can be developed for the cross-section data to represent how epidemic patterns are correlated with characteristic covariates $\boldsymbol{x}_s$, i.e., $Y(s, t) = M(\boldsymbol{x}_s; \boldsymbol{\beta}_t) + \varepsilon$, where ε is the random noise and $M(\boldsymbol{x}_s; \boldsymbol{\beta}_t)$ is the parameterized model. As epidemic observations change over time, model parameters will also vary with respect to time, i.e., $M(\boldsymbol{x}_s; \boldsymbol{\beta}_{\mathbf{t}}), M(\boldsymbol{x}_s; \boldsymbol{\beta}_{\mathbf{t}+1}), \cdots$. Then, a state space model $\boldsymbol{\beta}_t = g(\boldsymbol{\beta}_{t-1}, \gamma)$ can be used to characterize temporal correlation and link the parameters over time, where $g(\cdot)$ is the nonlinear evolution model and γ is process noise. As such, spatial and temporal components interact with each other to sequentially update the model when new data are available at the next time point. For example, Yang et al. develop a sparse particle filtering approach for characterizing and modeling space-time dynamic data generated from stochastic sensor networks (Chen & Yang, 2016b).

4.3 *Privacy-Preserving Data Analytics*

As the epidemic data (e.g., contact tracing, quarantine) proliferate, people are increasingly concerned about privacy issues. When data resolution and dimensionality are high, each entry in a database is essentially unique. Hence, establishing a linkage with named individuals becomes a much simpler matter. In the traditional practice, data analytics tend to focus on the effectiveness and efficiency of models, but overlook privacy in the context of an epidemic. Privacy breaches can bring unexpected disruptions to health policies and mitigation efforts in the epidemic response. For example, data exfiltration of contact tracing endangers the privacy of pertinent individuals, thereby causing a trust crisis and potential failures to the execution of policies. It is estimated that healthcare systems suffer from the cost of approximately \$300 billion annually due to privacy and security threats (Walker-Roberts et al., 2018).

One immediate safeguard is data anonymization techniques, which unfortunately do not provide a substantial level of privacy protection to the patients while guaranteeing the performance of data analytics (Dwork & Roth, 2014). It is not uncommon to come across disturbing news about risks and vulnerabilities in anonymized data. For example, it is not a difficult task to "match known patients to anonymized health records in Washington state data" (Sweeney, 2013). Netflix is under fire because of the privacy concerns and lawsuits over the anonymized database of

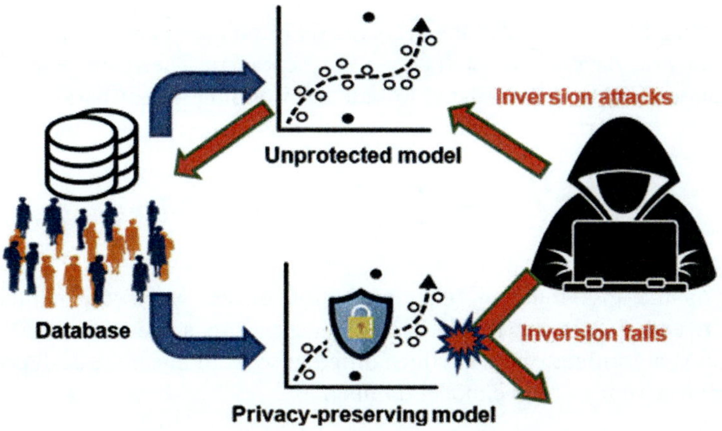

Fig. 16 An illustration of privacy-preserving predictive modeling

480,000 customers in the recommendation contest (Lohr, 2010). Achieving an optimal balance between model utility and data privacy is difficult when relying solely on data anonymization. Therefore, new privacy-preserving approaches are urgently needed to protect the privacy while capitalizing on the power of data analytics to build a smart and interconnected epidemic response system.

As shown in Fig. 16, differential privacy provides a viable solution to address the issue of data breaches, while realizing data analytics for smart health (Krall et al., 2020, 2021). A differential-privacy algorithm ensures that one's participation in a dataset, or lack thereof, will not be disclosed (Dwork & McSherry, 2010). Suppose that an epidemic database D contains n tuples, each with d input variables $\mathbf{x}_i = (x_{i1}, \ldots, x_{id})$ and the output space of a response variable y_i. The input space of $\boldsymbol{X}$ is assumed to be with symmetric neighboring relation $\mathbf{x} \cong \mathbf{x}'$. As shown in the Definition 1, privacy parameter $\epsilon \geq 0$ controls probability bounds about the level of privacy protection, which is the degree of difference allowed between output distributions by applying the function $\mathcal{F}(\cdot)$ onto databases D and D'.

Definition 1 *A randomized function $\mathcal{F} : X \to Y$ gives the ϵ-differential privacy if for all datasets D and D' differing by at most one row and for all $\xi \subseteq \mathrm{Range}(\mathcal{F})$, we have*

$$\Pr\{F(D) \in \xi\} \leq e^{\epsilon} \Pr\{F(D') \in \xi\} \tag{8}$$

Under differential privacy, one's inclusion within a dataset should make no statistical difference in an algorithm's output. Therefore, two databases that only differ by a single record of data should produce statistically similar results when running a differential-privacy algorithm (Dwork & Pottenger, 2013).

The "analyze" step develops and trains machine learning models (e.g., regression models, spatiotemporal statistical models, neural networks) by minimizing the objective function $J(\beta, D)$ with the available set of predictor and response variables in the epidemic database $D = \{(\boldsymbol{x}_1, y_1), \cdots, (\boldsymbol{x}_N, y_N)\}$. Therefore, empirical risk minimization (ERM) is formulated to search an optimal set of parameters β that minimize the regularized empirical loss function as

$$J(\beta, D) = \frac{1}{n} \sum_{(x_i, y_i) \in D} \ell(\beta, \boldsymbol{x}_i, y_i) + \Lambda R(\beta) \tag{9}$$

where ℓ is the loss function (e.g., prediction errors), Λ is the regularization parameter, and $R(\cdot)$ is the regularization function. To achieve the differentially private ERM algorithms, there are three different ways to inject the designed noises (e.g., Laplacian noises) in the model training.

- **Output perturbation**: add noises to the model's optimal coefficients $\beta^* = \text{argmin}_{\beta \in \mathcal{B}} J(\beta, D)$
- **Objective perturbation**: add noises to the objective function $J(\beta, D)$
- **Gradient perturbation**: optimize $J(\beta, D)$ with noisy gradients ∇J and stochastic gradient descent

In the literature, standard output and objective perturbation techniques for logistic regression models were developed by Chaudhuri and Monteleoni (2009). A variant of objective perturbation, known as the functional mechanism, was later introduced by Zhang et al. (2012). The functional mechanism works by injecting noise into the regressor coefficients. Furthermore, the sensitive mechanism, proposed by Wang et al. (2015), serves as an expansion to the functional mechanism. This new sensitive mechanism is capable of deterring against model inversion attacks by differentiating between sensitive and non-sensitive attributes when performing coefficient perturbation.

Traditionally, output and objective perturbation techniques are easy to implement and more suitable for centralized computing. Nonetheless, distributed processing has become a more dominant force in the era of big data. Gradient-based perturbation techniques provide a higher degree of flexibility in light of this distributed reality. Song et al. (2013) first proposed the gradient perturbation for differentially private updates, which however does not adaptively adjust the learning rate for fast convergence. As shown in Table 6, this paper presents a privacy-preserving algorithm with adaptive learning rate, which is also a newly revised implementation of gradient perturbation techniques.

At the beginning of this algorithm, several parameters (i.e., a privacy parameter ϵ, a regularization parameter Λ, the number of epochs K, and a batch size b) are firstly initialized. Note that the initial learning rate η_0 is calculated as $\sqrt{\frac{1}{\Lambda^{1/2}}}$. Next, an initial guess for regression coefficients $\beta^{(1)}$ is randomly generated. Before entering the main loop, iteration counter τ and epoch counter κ are both set to one. The parameter, τ_0, is an intermediate variable that is employed to determine $\eta^{(\tau)}$ at each iteration. The starting value of τ_0 is set as $1/\Lambda\eta_0$. Further, the Dataset D is divided into a set of batches B, each of size b.

Table 6 The gradient perturbation algorithm for privacy-preserving predictive modeling

Input: Data D, parameters ϵ, Λ, K, b, θ
Output: Approximate noisy minimizer $\overline{\beta}$
1: Initialize $\beta^{(1)}$, $\tau = 1$, $\kappa = 1$, $\eta_0 = \sqrt{\frac{1}{\Lambda^{1/2}}}$
2: Let $\tau_0 = \frac{1}{\Lambda\eta_0}$
3: Distribute D into a set of batches B, each of size b
4: **while**$\kappa \leq \mathsf{K}$
5: **for** each $j = 1, \ldots, \lvert\mathsf{B}\rvert$ **do**
6: Set $\eta^{(\tau)} = \frac{1}{\Lambda(\tau_0+\tau-1)}$
7: Set $\Delta^{(\tau)} = \frac{2\theta\eta^{(\tau)}}{b}$
8: Draw a vector $\mathbf{z}^{(\tau)} \sim Lap\left(\frac{\Delta^{(\tau)}}{\epsilon}\right)$
9: Set $\beta^{(\tau+1)} = \beta^{(\tau)} - \eta^{(\tau)}\left(\nabla J\left(\beta^{(\tau)}, \mathrm{B}_j\right) + \frac{1}{b}\mathbf{z}^{(\tau)}\right)$
10: Set $\tau = \tau + 1$
11: **end for**
12: Set $\kappa = \kappa + 1$
13: **If** $\lVert\beta^{(\tau+1)} - \beta^{(\tau)}\rVert < \delta$, break
14: **end while**
15: Let $\overline{\beta} = \beta^{(\tau)}$

For each epoch κ, the algorithm will process all batches B, i.e., $j = 1, \ldots, \lvert\mathsf{B}\rvert$. The processing of one batch constitutes a single iteration within the epoch. For each iteration τ, the learning rate $\eta^{(\tau)}$ is updated, whose value is then utilized to update the global sensitivity $\Delta^{(\tau)}$. Perturbation is carried forth by drawing a random vector $\mathbf{z}^{(\tau)}$~$Lap(\Delta^{(\tau)}/\epsilon)$, which is scaled by $1/b$. This scaled noise vector is injected into the gradient ∇J. Once the gradient has been perturbed, it is used to update β. The final step of each iteration entails updating the iteration counter τ by one. Once all batches are processed, the epoch counter κ is also incremented by one. This entire process continues until convergence or until $\kappa > \mathsf{K}$.

Figure 17 shows that decreasing ϵ causes both model and attack accuracies to degrade. Nonetheless, each will decay at different rates. Once epsilon falls beneath 10^{-2}, the attack accuracy experiences a substantial drop with minimal impact on the model accuracy. The attack accuracy approaches zero when ϵ draws closer to 10^{-4}. However, the model accuracy only decreases by ~5% from baseline when ϵ approaches 10^{-4}. Beyond this ϵ value, the degradation of model accuracy will accelerate. However, there is little utility in decreasing ϵ any further since the attack accuracy has already been reduced to zero. Privacy-preserving techniques provide an enabling tool to mitigate the risks and costs due to privacy breaches (e.g., model inversion attacks) while maintaining the performance of epidemic predictive models.

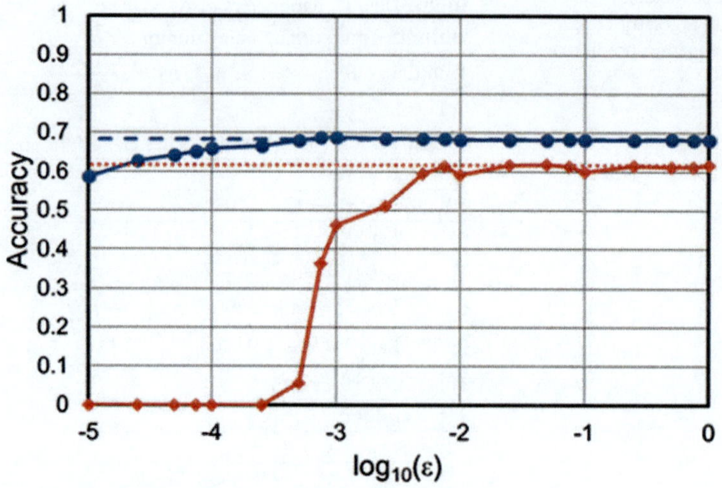

Fig. 17 Privacy model and attack accuracies with respect to varying ϵ

5 Improve the Resilience of Health Systems

Epidemic outbreaks demand medical resources in a short period of time. Such demands can outpace the supply for months, because the waves of an epidemic may recur in multiple years. The abrupt increase of infected cases quickly overwhelms health systems, causing the devastating shortages of staffs, beds, supplies and equipment. Also, if health systems are not prepared to handle highly contagious viruses, they will likely become "hotpots" and increase the spread of infectious diseases. The outbreak of COVID-19 urges changes and transformations of existing health systems to become a smarter and interconnected healthcare delivery system that is more resilient. The term "resilience" corresponds to the system's adaptiveness and robustness in the handling of unexpected events such as epidemic outbreaks or disasters (e.g., hurricanes, terrorist attacks, earthquakes), and can include multi-faceted definitions as follows:

- **Capacity resilience**: For increasing levels of demand, NPIs help flatten the curve to avoid the overload of health systems. On the other hand, if the curve is above capacity, a health system should be resilient to leverage the network for optimal capacity planning and allocation, as well as build up temporary capacity (e.g., field hospitals) to treat the patients and control the spread.
- **Resource resilience**: The supply chain should also be resilient to avoid shortages and provide sufficient medical resources (e.g., N95 masks, ventilators, antivirals) during epidemic events. Thus, a certain level of redundancy is needed in the design of supply chain. In addition, optimal resource allocation is urgently needed to ensure the equity and accessibility in the design of resource resilience.

- **Workflow resilience**: Traditional workflows in the hospital tend to cause secondary infections of healthcare workers and other susceptible patients. As a result, labor supply dwindles and more beds need to be allocated for the treatment of healthcare professionals. It is therefore imperative to re-design workflows and avoid secondary infections in case of an epidemic.
- **Operational resilience**: It is common that physicians see their patients in person for health care. However, with the increasing availability of wearable sensors, cloud computing, and information technology, such routine practices may be transformed to online delivery of health care or a hybrid online-onsite approach. Operational resilience calls upon the integration of telehealth systems with existing infrastructures and practices to advance the future of health care.

After all, rich data are provided in the "measure" step about the evolution of an epidemic. The "analyze" step extracts useful information from the data about epidemic characteristics. Now, the "improve" step exploits data-driven knowledge to improve the resilience design of health systems.

5.1 *Artificial Intelligence for Smart and Interconnected Health Systems*

Epidemic outbreaks call upon a resilient response from health systems. As shown in Fig. 18, artificial intelligence (AI) has a wide range of applications in various areas of health care, and can further promote the changes and transformation of existing healthcare practices. AI can help reduce the probability for a healthcare worker to get secondary infections via optimal allocation and use of PPEs, robot-assisted care, health informatics, and telehealth amongst many others. The provision of healthcare services includes a large number of healthcare professionals (e.g., physicians, nurses, radiologists) and medical technologies (e.g., wearable sensors, patient monitors, robotics, medical imaging). The future of work in health systems depends to a great extent on the seamless integration of human and technology. Figure 18 shows different application areas that AI has generated impacts or will bring transformations to the health systems:

- **Capacity planning**: The surge of hospitalizations is not uncommon during an epidemic outbreak. For a resilient preparation, AI tools can be developed to forecast the number of admissions, plan the medical resources, optimize the staffing level, and thereby improve availability of care. For example, the burn rate model can help health systems plan and optimize the use of PPE by predicting the trend and usage patterns during the COVID-19 pandemic (Raja et al., 2020). Also, data-driven models can be developed to facilitate hospital planning based on the estimated demand of ICU beds, non-ICU beds, COVID admissions, and ventilators (Klein et al., 2020).

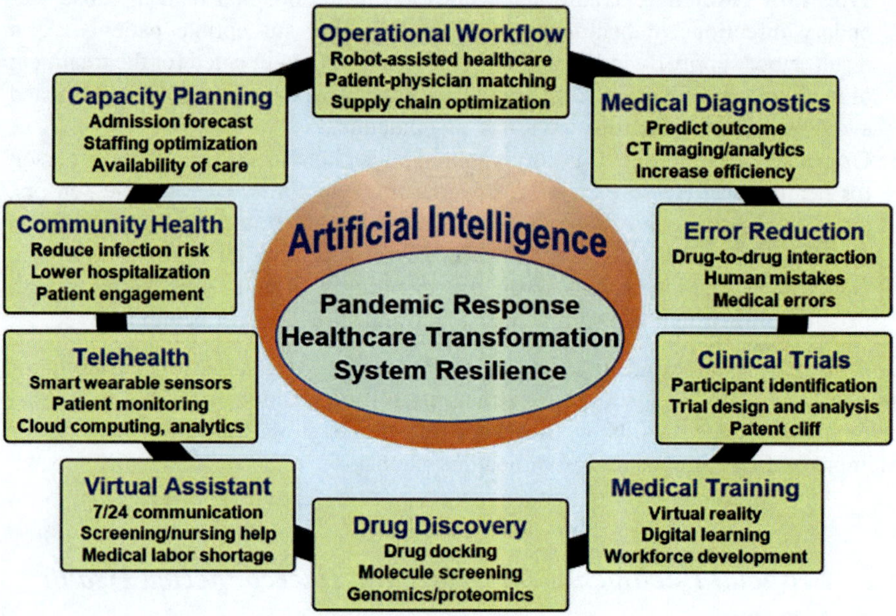

Fig. 18 AI-driven transformation of healthcare systems

- **Operational workflow**: Infectious diseases increase the transmission risks between human subjects. AI and robots can help improve the operational workflow by human-robot collaboration, thereby reducing the probability of secondary infections. For example, autonomous service robots can take over the labor-intensive logistic tasks to deliver medications, specimens, testing results throughout the hospital 7/24 (Ozkil et al., 2009; Rafflin & Fournier, 1996). AI algorithms can run the scheduling of jobs and tasks in an optimal manner. Also, operational workflow can be improved by adding an AI-supported telemedicine option, a redesign the triage process to separate infected patients from others, and smart sensing stations for symptom examination (e.g., wireless temperature sensors, thermal imaging cameras).
- **Medical diagnostics**: The increasing number of infections also brings significant amount of data in the care and treatment process. During an epidemic, a hospital is often expected to perform thousands of CT scans every day to check the status of lung infections. It is time consuming and labor intensive for human experts to visually inspect and interpret these CT images. However, AI screening is much faster than physicians, and significantly improves operational efficiency (Li et al., 2020). In addition, severely infected patients need advanced life support (e.g., ventilators), regular lab tests, and real-time monitoring in the ICUs. The on-hand availability of clinical data (e.g., blood pressure, heart rate, gas exchange, pulse oximetry and metabolic panel) provides an opportunity to establish AI models for the prediction of mortality and outcomes (Kim et al., 2020). This helps stratify

patients for better care and resource allocation, thereby reducing costs and the length of stay (LOS).

- **Error reduction**: To err is human, but medical errors could lead to tangible consequences. Human errors can be due to many aspects, e.g., physical condition, skill level, training, attitude, emotion and cognitive bias. AI tools can help reduce and minimize human errors, e.g., robust check of appropriate dosage levels, the reduction of diagnosis errors. For example, Liu et al. developed an AI model to offer warnings about serious side effects of drug-drug interactions (Liu et al., 2019). This model automatically labels data from thousands of drugs and screens millions of potential interactions among drugs, which helps reduce adverse drug events and improve patient safety.
- **Clinical trials**: AI is conducive to optimize the design of clinical trials to assess and evaluate the effectiveness of drugs or antivirals for the treatment of an infectious disease. For example, AI models can be used to facilitate the identification and grouping of participants. A poor design may not reveal the effectiveness of a drug or lead to wrong conclusions about an ineffective drug. Notably, Lancet retracts a study that reported an anti-malarial drug, named *hydroxychloroquine*, has little effects to curb COVID-19 (Funck-Brentano et al., 2020). AI-based design and analysis is critical to establish the statistical significance and validity of a clinical trial. Also, pharmaceutical companies can leverage AI methods and tools to circumvent the patent cliff for the new drug design and clinical trials (Topol, 2020; Kaitin, 2010).
- **Medical training**: Rapid advances of Virtual Reality (VR) technology have fueled increasing interests and steady growth in healthcare applications. For example, VR is conducive to improve medical training for decision making, and help patients to cope with pain, overcome anxiety and depression (Niederriter et al., 2020). VR has been used to evaluate different kinds of medical, surgical, psychiatric, and neurocognitive conditions, as well as to improve the effects of traditional therapies in current practices (Oyama et al., 1995; Bowman, 1997). VR provides an immersive 3D environment for active interactions and longer training sessions. AI models can integrate sensing data with user inputs to optimize the learning steps and improve the quality of medical training (Basdogan et al., 2007).
- **Drug discovery**: AI has also been extensively used for drug discovery, which is evident through the rising of spotlight companies such as Genesis Therapeutics, Atomwise, and Benevolent.ai. AI is used as a preliminary filtering step to screen potential molecules on how they interact and control the activity of a virus. As there are more than billions of molecules, it is impossible for biologists or chemists to test each one of them to characterize the effects in a short period of time. AI is integrated with molecular dynamics simulation and drug docking to identify the most sensitive and effective molecules (Smith et al., 2018; Smalley, 2017). Further, AI models can be developed for the analysis of genomics and proteomics, which will help gain a better epidemiological understanding of pathogen evolution and identify the origin host of a virus (Uddin et al., 2020; Libin et al., 2019).

- **Virtual assistant**: When infections and hospitalizations rise, health systems face increasing pressure due to the shortage of medical labor. There are growing concerns about the burnout and stress of physicians and nurses. Certainly, they cannot be readily available 24/7 to assist those people who are infected or are worried about getting infected by the virus. AI-embedded virtual assistants enable the patients to communicate with care providers at any time anywhere (Miner et al., 2020; Sezgin et al., 2020). The AI assistant can learn and understand the questions from a patient, screen the symptoms by steps, then guide the care.
- **Telehealth**: Telehealth provides an opportunity for healthcare professionals to deliver timely health care to patients through e-platforms (e.g., Teledoc, MeMD, iCliniq, Amwell) (Smith et al., 2020; Hollander & Carr, 2020). The remote interaction greatly helps preserve the PPEs and avoid secondary infections. There are three major types of telehealth modalities as follows: (i) *Synchronous*: This is conducted through real-time live audio-video interaction with smartphone, tablet, or computer. (ii) *Asynchronous*: This includes a "store and forward" technology where message, image, or data are collected first from patients, then interpreted or responded later. (iii) *Remote patient monitoring*: This includes smart sensors and internet-of-things technologies (Yang et al., 2020; Kan et al., 2015), where a patient's clinical data are measured and transmitted from a distance to healthcare providers. Telehealth helps screen symptoms of COVID-19, provide low-risk urgent care for non-COVID-19 patients, follow up with patients after discharge, and access to primary care or specialists for chronic disease management.
- **Community health**: AI models can be extended to the community. People live in the community and connect with social networks. Also, they commute through transportation networks. There are symptomatic and asymptomatic people who get infected, each can have a list of contacts with the risk to be infected. Contacts can be traced in the community with the use of smart sensors, mobile applications, and surveys (Kretzschmar et al., 2020; Budd et al., 2020), e.g., Sara Alert (https://saraalert.org/). AI models can be established to characterize the spread and propagation of infectious diseases, predict the future evolution for prevention and control. In the community, AI tools can help reduce the transmission risk, prevent unplanned hospitalization, and improve the patient engagement.

AI applications are not limited to those areas describe above, and can be extended to fitness, compliance, cybersecurity, data privacy and to name a few. Epidemic response and management depend on the realization of full potentials of AI and big data to build the next-generation health system.

5.2 Healthcare Resource Allocation for Coverage Control

Healthcare resources are critical to infection prevention and control during an epidemic. Examples of such resources include personal protective equipment (PPE), ventilators, medicines, antivirals, testing kits, and testing facilities (e.g., drive-thru testing sites). PPEs protect healthcare workers and patients from getting exposed to the virus. Ventilators are indispensable to saving the lives of patients with severe lung infections from coronaviruses (e.g., COVID-19) that cause excess fluid in the lungs and make patients experience difficulties to breathe on their own. Medicines and antivirals are also vital to stopping the spread of viruses and keep the mortality rate under control. Testing resources (e.g., test kits and sites) help identify infections in a timely manner and intervene as early as possible, e.g., quarantines or contact tracing to identify the patient's contacts and been-to. The availability of such resources directly determines the success or failure of virus containment before vaccines are available.

During the period of an epidemic or pandemic, the number of infected cases grows exponentially and yields heterogeneous distributions in multiple spatial regions. As such, the demand of healthcare resources is not uniformly distributed in spatial dimensions. Figure 20a shows the complex distribution for the number of confirmed cases in each zip code at the state of Pennsylvania. The demand varies with respect to the number of cases and population sizes across the regions, which is also called spatial demand heterogeneity. Such heterogeneity may be due to many factors, e.g., the demographic structure of the population, the infrastructure of the region, or the transportation in the area. The density of spatial demand (i.e., estimated from infected cases, demographics, or vulnerable population in each zip code) provide critical information to help optimize the allocation of healthcare resources, e.g., ventilators, testing kits, vaccines, drive-thru testing and/or vaccination sites.

The optimality, however, depends to a great extent on accessibility (e.g., shortest travel distances between demand and supply in each region) (Penchansky & Thomas, 1981) and equity (e.g., the distribution of demand density among coverage regions of testing sites) (Daskin, 1997). Specifically, resource accessibility refers to the ease of access to resources when the demand distribution is heterogeneous in a spatial region. Equity, on the other hand, is a coverage measure of heterogeneous demands over multiple regions. A high level of equity ensures equal coverage of healthcare resources. As shown in Fig. 19, let $\sigma(s)$ be the spatial demand function: $\Omega \rightarrow R^+$ that provides the demand density for each location s in the polygon space Ω. The objective function is to find optimal locations of resource facilities $\Theta = \{\theta_1, \theta_2, \ldots \theta_i\}$ in the space Ω that minimize the sum of weighted distance functions between supply and demand locations, which is formulated as a coverage control problem. The cost function is defined as the sum of the moment of inertia in the regions:

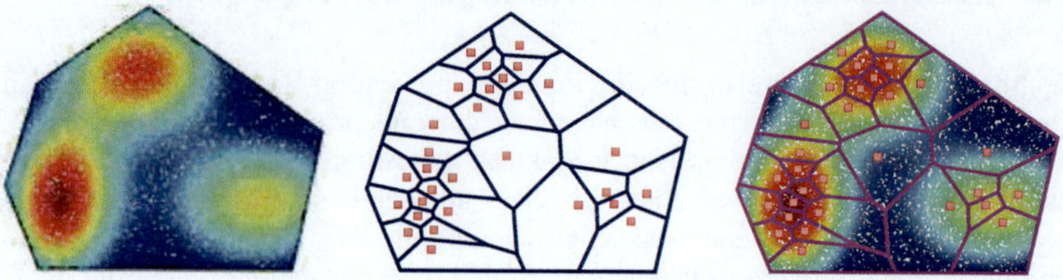

Fig. 19 Greedy-Voronoi tessellation with a heterogeneous demand function in the space

$$C(\boldsymbol{\Theta}, V) = \sum\nolimits_{i=1}^{I} \int_{V_i} d(\|s - \theta_i\|)\,\sigma(s)ds \tag{10}$$

where I is the number of resource facilities, V_i is the ith Voronoi region, θ_i is the ith Voronoi center, and $\sigma(s)$ can be the demand density of confirmed cases (or vulnerable population, virus spread situations) at location s. The location of each resource facility is determined by minimizing the cost function as:

$$c(V_i) = \text{argmin}_{\theta_i} C(\boldsymbol{\Theta}, \mathbf{V})$$

Voronoi tessellation guarantees that each resource facility is the closest to every location in its cell (Du et al., 1999). The moment of inertia loss function ensures the demand density within each region is taken into account. This, in turn, helps control the coverage of heterogeneous demand over all the Voronoi regions. Note that *this is different from traditional clustering problems, because of the need to consider both distance functions and the spatial demand function that can be highly heterogenous*. Table 7 shows the proposed algorithm of greedy-Voronoi tessellation, which includes two stages, namely sequential placement and global calibration. First, each facility is sequentially placed to minimize the cost function (i.e., defined as the sum of the moment of inertia in the regions). In each iteration, Voronoi tessellation is computed based on the locations of existing facilities. A new facility is then randomly placed in the Voronoi cell with the largest mass (i.e., the sum of the moment of inertia). The tessellation is then updated to re-evaluate the new cost function. The location of this newly placed facility is optimized step-by-step along the gradient direction. This process is repeated until convergence. Such a sequential formulation provides both monotonicity and submodularity properties, and yields a sub-optimal solution. After the sequential placement, the global calibration continues to search for the optimal solution by computing and updating the Voronoi tessellation to optimize locations of all I facilities. In each iteration, the locations of all facilities are adjusted and calibrated with the gradient descent algorithm. This process terminates until convergence. The algorithm then returns the optimized locations of all I facilities.

Table 7 The Greedy-Voronoi tessellation algorithm

Demand function $\sigma(s)$, Polygon space Ω, total number of facilities I	
1:	Place the first facility $\theta_1$at the center of mass of the Ω with density $\sigma(s)$
2:	**For**$i = 2$ to I
3:	Randomly place a new facility θ_j in the Voronoi cell with the largest mass
4:	**Compute** Voronoi tessellation V based on the location of current facilities
5:	**Compute** the cost function $C(\theta) = \sum_i \int_{V_i} dist(\theta_i, s)\,\sigma(s)ds$
6:	**Compute** the gradient $\frac{\partial C}{\partial \theta_i}$ for this newly added facility θ_i
7:	Update θ_j according to $\theta_i = \theta_i - \alpha \frac{\partial C}{\partial \theta_i}$
8:	**Repeat** 4-7 until convergence
9:	**Update** Voronoi tessellation V
10:	**End For**
11:	**Compute** the cost function $C(\theta) = \sum_{i=1}^{I} \int_{V_i} dist(\theta_i, s)\,\sigma(s)ds$
12:	**Compute** the gradient $\frac{\partial C}{\partial \theta_i}, i = 1, 2, \ldots, I$ for all facility locations
13:	Update all $\theta_i^{'}$s according to $\theta_i = \theta_i - \alpha \frac{\partial C}{\partial \theta_i}, i = 1, 2, \ldots, I$
14:	**Update** Voronoi tessellation V
15:	**Repeat** 11-14 until convergence
16:	**Return** facility locations $\theta_i, i = 1, 2, \ldots, I$

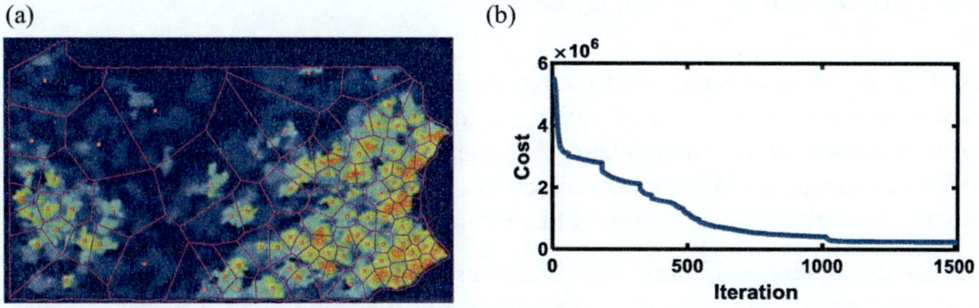

Fig. 20 (**a**) Greedy-Voronoi tessellation and (**b**) convergence curve of cost function for optimal allocation of 100 drive-thru testing sites in PA

We implemented and evaluated the proposed algorithm using a case study of the COVID-19 infection map in Pennsylvania, as shown in Fig. 20, where the red color indicates the high demand density and the blue is the low density. The proposed greedy-Voronoi tessellation helps balance between accessibility and equity for each region, allows the flexibility to dynamically adjust the tessellation based on real data (i.e., the number of available resource sites, or the variation of density in the PA map). The proposed algorithm is generally applicable to a variety of demand-driven allocation of resources in a spatial region.

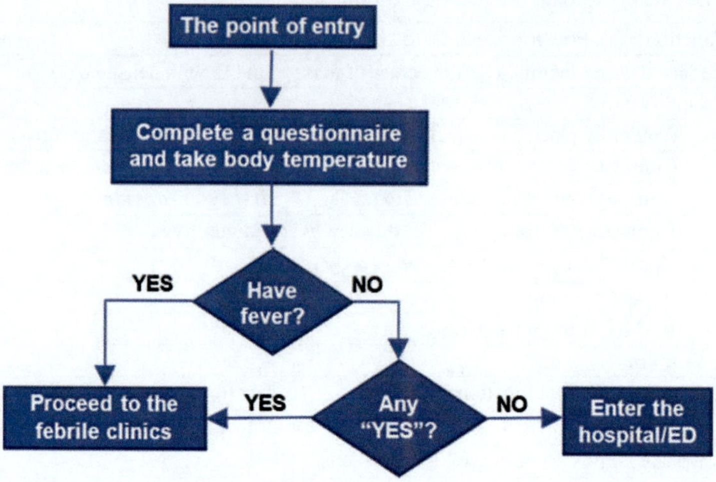

Fig. 21 The screening process and febrile clinics

5.3 *Re-design of Health Systems*

As COVID-19 cases continue to surge in the US, there is an urgent need to re-design health systems such as hospitals, medical clinics, and emergency rooms for better treatment and accommodation of patients. The main objective is to segregate infected patients, avoid secondary infections, and reduce transmission risks, thereby improving the safety and quality of healthcare services.

Temperature screening and febrile clinics An immediate change is to add a screening process for patients at the entrance of a hospital, which makes it easier and less costly to detect, monitor and control the infiltration of contagious diseases (Cameron et al., 2006; Li et al., 2005). As shown in Fig. 21, patients are required to complete a short questionnaire about their health conditions, which includes but not limited to their travel history, fever history, and symptoms. Meanwhile, their body temperature will be taken and recorded. If the answer to any of the questions on the questionnaire is "YES" or the body temperature is above the normal level, the patients will be guided to a febrile clinic, an isolated area with quarantine units at the hospital. Otherwise, they can enter the hospital/ Emergency Department (ED) and proceed to the triage area.

The identification and control of fever and high-risk patients during the screening process can separate them from other patients at an early stage, which reduces the risk of secondary infections in the hospital (Lateef, 2009; Improving Hospital Design for Better Infection Control, n.d.). Febrile clinics have a separate entrance and negative-pressure ventilation systems, which keep the air mix with other areas in the hospital at a minimum level. Also, febrile clinics have isolated exam rooms and its own pharmacy so that high-risk patients do not infect each other or travel to other areas of the hospital (Lateef, 2009). Depending on the diagnosis results, patients may either stay in the patient rooms in the febrile area to get further treatments or go home.

Environmental considerations to prevent infections Healthcare-associated infections have caused high morbidity and mortality during the epidemic outbreaks, which are mainly due to the contact between patients and healthcare workers, patients and staff, and patients and the environment. Here, we present a brief review of recommended designs and guidelines to minimize healthcare-associated infections:

Airflow system

- The difference of air pressure between isolation rooms and other areas should be about positive 15 Pa (Lateef, 2009).
- Room air should be changed 10–12 times every hour to sufficiently dilute the bacterial load around an infected patient (Eames et al., 2009).
- Equip ventilation, especially in communal areas (Eames et al., 2009).
- Install negative-airflow systems in areas where high-risk patients will be cared for (Noskin & Peterson, 2001).
- Isolation rooms should have negative airflow and frequent air exchanges. The air cannot be recirculated (Noskin & Peterson, 2001; Baker & Lamb Jr, 1992; Burmahl, 2000).

Hygiene and cleaning

- Install at least one sink in every patient room, examination room, procedure room and isolation room, which is close to the entrance of the room. Each sink should be with a hands-free control, soap dispenser, and paper towel holder (Noskin & Peterson, 2001; Stiller et al., 2016).
- Use information systems to monitor hand hygiene performance and provide feedback (Marques et al., 2017).
- Frequent disinfection of non-disposable material, equipment, work surfaces, wards, environment, facilities, horizontal surfaces, surfaces touched by patients and staff and toilet facilities using hypochlorite 1000 ppm (Stiller et al., 2016).

Room design

- Convert the patient rooms into single rooms with en suite toilets (Stiller et al., 2016; Bacon & Erickson, 1950).
- Recommend square footage for patient rooms in critical care units (ICU): 13.94 m^2 per bed for single-patient rooms and 11.15 m^2 for multiple-patient rooms (Facility Guidelines Institute, 2014).
- Add ante-rooms in negative pressure rooms to reduce the escape of droplet nuclei (Lateef, 2009).
- Equip the observation unit in the ED with isolation rooms, which have automatic doors (Lateef, 2009).

Drive-thru medical clinics As shown in Fig. 22, another idea is to transform the garage of a hospital into drive-thru medical clinics. This design is currently

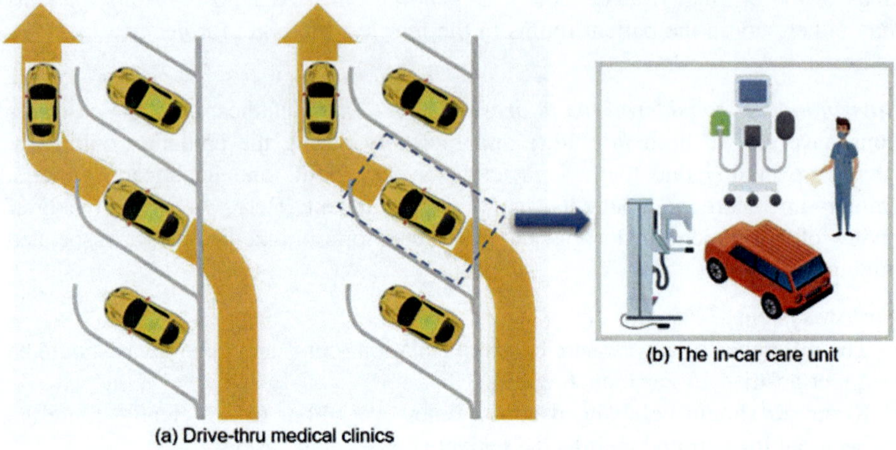

Fig. 22 An illustration of (**a**) drive-thru medical clinics and (**b**) in-car care units

proposed and evaluated by our University Medical Center in collaboration with NBBJ (i.e., an American global architecture, planning and design firm, http://www.nbbj.com/). Each in-car care unit will include necessary medical equipment and a paramedic. Drive-thru medical clinics bridge the gap between telehealth and in-room visits. It not only enables patients to get timely access to healthcare services, but also gain in-person interactions with healthcare providers for diagnosis and treatments. Furthermore, in-car care units provide medical monitoring that cannot be achieved otherwise via telehealth, which includes but not limited to vital signs, blood pressure, electrocardiogram, and auscultation.

In conventional medical visits, patients often go through a lengthy cycle that includes parking the car, screening in the main entrance, registration in the front desk, waiting to be called, entering the exam room, getting the services, waiting for the results, exiting the hospital, and then departing the garage. There is a higher risk for patients to get infected or infect others in each step of the process. Instead, drive-thru medical clinics follows a much shorter cycle, i.e., entering the in-car care units for examination or treatments and then departing the garage, as shown in Fig. 22. For a car with four seats, four people can get health care at the same time. In-car healthcare platforms bring benefits to both patients and healthcare providers. For patients, they can get the necessary exam or treatments more conveniently and faster, while avoiding secondary infection in the hospital. For healthcare providers, they can treat more patients while maintaining the hospital capacity and have a lower risk of secondary infection.

6 Prescriptive Analytics – Control the Spread

Predictive analytics extract useful information from the data to delineate key risk factors and predict real-time positions of virus spread. Further, this section will present the prescriptive analytics that exploit the knowledge from predictive analytics to identify the course of actions to control the spread of virus. Specifically, we will focus on the development of simulation models and computer experiments to benchmark the performances of health policies and action strategies.

6.1 Simulation Modeling and Computer Experiments

Simulation models provide a mathematical description about the physics of disease propagation and how infections are correlated with the dynamics of human movements in spatial regions. With rapid advances in epidemic surveillance systems, abundant infection data are collected. The availability of data offers an unprecedented opportunity to model human traffics and the progress of an epidemic from a dynamic, as opposed to a static sense. Fast and accurate simulation models are critically needed to: (1) analyze main effects and interaction effects of process parameters in an epidemic, (2) predict how these parameters of interests impact the resource allocation and epidemic outcomes, (3) aid the design of health policies and action plans, (4) compare and benchmark a variety of existing policies and strategies, and (5) augment real-world epidemic control by providing a model-based baseline for process adjustment.

Experiments, either physical or computer-based, are critical to the discovery of new insights and knowledge from epidemic processes. However, physical experiments, also called clinical surveys or trials, on the human population are often difficult, even with the approval of an internal review board (IRB). Note that there are many practical and ethical limitations pertinent to physical experiments of human subjects. Also, it is very expensive to design a comprehensive protocol to collect data from a large population. Computer experiments with simulation models (Du et al., 2016) are highly flexible and offers a great opportunity for the investigation of epidemic processes. As such, research communities have identified the urgent need to develop epidemic simulations and, more importantly, design computer experiments to accelerate prescriptive analytics and control the virus spread. This is essential for making the health system respond in a fast and proactive manner to disease variations and disruptive events.

From a broader vista, epidemic simulation can be categorized into two classes contingent on the level of modeling details of human behaviors, namely, continuous system dynamics modeling and discrete event simulation (DES). Continuous system dynamics models, e.g., "susceptible-infected-recovered" (or SIR) compartment models, are constructed with a set of differential equations (Prem et al., 2020; Chen et al., 2020). The population is assumed to be segregated into a variety of compartments (e.g., susceptible, exposed, infected, recovered), which represents different system states in an epidemic. The rates of change among these states are modeled with differential equations. Such continuous models operate at a much more aggregate level by concentrating on system states and the rates of change in sub-populations. As a result, they are more suitable to answering questions in the macro level instead of micro level. In other words, the large number of human subjects are represented as continuous states for a better description of aggregated behaviors, but individual activities cannot be tracked or modeled through the continuous system dynamics models. On the contrary, discrete-event simulation (DES) focuses more on detailed representations of individuals' activities and environments in the spread process of infectious diseases (Currie et al., 2020). DES models capture detailed behaviors on the individual level (e.g., movement behaviors, contact patterns, personal protective measures) and allow heterogeneity in the rates of change within sub-populations. Discrete models are generally more suitable when individual behaviors need to be modeled so that operational details are available to investigate health policies. Hence, DES models are conducive to answer specific questions in the operational or tactical level.

6.2 Epidemic Simulation in the Spatial Network

Figure 23 shows our proposed DES simulation of human movements and epidemic dynamics in a spatial network. This framework is embodied by five components, namely spatial data, network modeling, human traffic, infection modeling, and computer experiments, in a close loop to investigate detailed representations of individuals' activities in spatial environments. This modeling framework is designed to overcome the complexity to model human activities directly in a spatial environment, and leverage the extracted or derived network structure to model spatiotemporal dynamics of the virus spread.

(1) Spatial Data Daily activities are often inter-connected and happen in a spatial region with key activity locations such as schools, grocery stores, shopping malls, restaurants, and homes. Spatial data are readily available as geographical information system (GIS) mapping files from US Census Bureau and other geospatial service providers. Examples of mapping file formats include Environmental Systems Research Institute (ESRI) files (e.g., ArcGIS) (Kienberger & Tiede, 2008), Keyhole Markup Language (KML) files (Google) (Ballagh et al., 2011), and shapefiles that are in a geospatial vector data format with TIGER/Line and cartographic boundary (US Census Bureau, 2010). The shapefiles also include geographic entity codes (GEOIDs) that can be used to link with demographic data from the US Census Bureau. However, GIS map files contain geospatial details (e.g., forests, water wells, and rivers) that are not necessary for epidemic simulation. Although geospatial maps are static, the movement of human subjects is dynamic and tends to form traffic flows in a networked way.

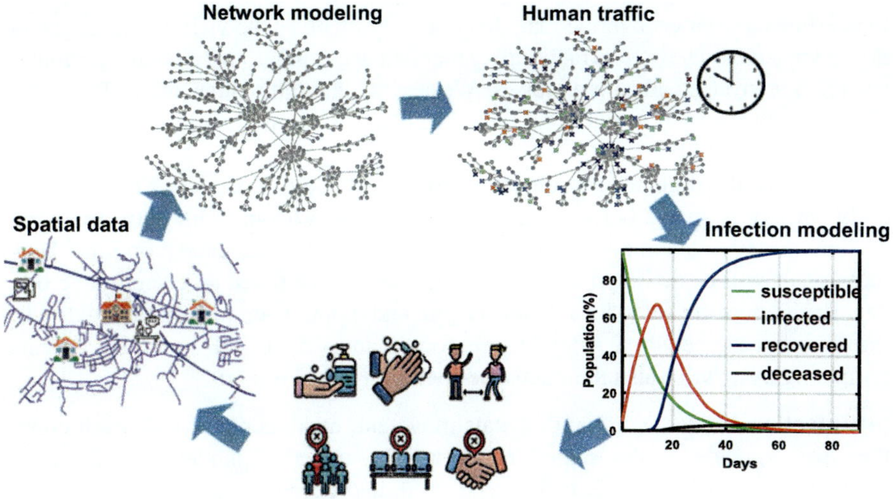

Fig. 23 The flow chart of epidemic simulation in a spatial network

(2) Network Model Indeed, many real-world systems can be represented by network models with a large number of nodes that are connected by edges or links. In a small-scale spatial environment, human subjects often visit a set of key locations (e.g., schools, stores, offices) daily that are connected by roads. In a large-scale spatial environment, people travel through a network of highways, or a network of airports. The spatial network is a graph representation with a set of nodes (i.e., key locations) that are linked by edges (i.e., spatial relationship via interconnected means of transportation). Based on real-world geospatial information, these nodes can be characterized with network features such as the degree, centrality, clustering coefficient (Yang & Liu, 2013; Albert & Barabási, 2002; Liu & Yang, 2017). In the state of the art, there are abundant literatures on network models (e.g., social network, citation network, neural network, sensor network). However, very little has been done to derive network models from GIS data in a spatial region of interest and then further investigate epidemic dynamics in the spatial network. It is imperative to model the movement dynamics of human subjects in a spatial network and further capture details of human contacts and interactions during the virus spread.

(3) Human Traffic Notably, geographic entity codes (GEOIDs) can be used to link with demographic data in a spatial region. This, in turn, helps simulate the number of human subjects with a diverse set of demographic information (e.g., age groups, population sizes). The population can also be divided into different activity levels, namely low, medium and high, which correspond to the number of nodes they are going to visit. For example, individuals with a high level of activity visit more places than low and medium sub-groups in a day. These individuals will then be assigned to nodes in the spatial network, and many can be placed in the same node due to the clustered nature of residences (e.g., family members in houses and roommates in apartments). For each individual, daily activity involves the visit to a sequence of nodes via edges. The path is randomly generated according to individuals' attributes such as age groups and activity levels. We assume that individuals often choose the shortest path for each activity and therefore plan the route by Dijkstra's algorithm (Zhan & Noon, 1998).

The schedule of human movements is simulated in a day of 24 h as follows: The daily activity is sparse before 8 am, but become busy from 8 am until midnight. The number of active individuals in the spatial network is dependent on time. New individuals will be activated and join the network traffic based on current time of a day. Rush hours are set to be at 8 am, 12 pm, and 6 pm, when more individuals will move within the network. After 11 pm, no new individuals will be added, and the remaining ones will finish their activities before a new day starts.

(4) Infection Model When individuals move and make contacts with each other, the virus spreads in the spatial network. The infection model provides real-time positions of healthy, infected, recovered, and deceased individuals based on human movement dynamics in the network. In this investigation, we assume that infections primarily occur in nodes, and rarely on the path. When individuals visit a sequence of nodes, they come across each other in the same node. When they share the same environment, infections occur with a certain probability by surrounded virus carriers. The infection probability is dependent on exposure time τ, virus transmissibility ρ, the infectivity level r of virus carriers, the number of surrounding carriers N_r, and the susceptibility level s_i of an individual i. The virus transmissibility ρ is a disease-specific property that defines how likely a susceptible person will be infected by the virus on average. For a virus carrier, the infectivity level r defines a person's capability to infect susceptible people. In other words, some virus carriers, also called super spreaders, may have a higher infectivity level than others (Gómez-Carballa et al., 2020). The susceptibility level s_i defines the degree of vulnerability of an individual getting infected that may vary due to risk factors such as age, gender, and comorbidity. As such, the model of infection probability is formulated as:

$$p_i = 1 - \exp\left(\tau \sum_{r \in \langle R \rangle} N_r \ln(1 - r s_i \rho)\right), \quad 0 < r, s_i, \rho < 1 \tag{11}$$

where R is the set of virus carriers, $\langle R \rangle$ is the set of infectivity levels from surrounding carriers, and N_r is the number of surrounding carriers at the infectivity level r. For super spreaders, $r \approx 1$.

The infected individuals can be either symptomatic or asymptomatic. For a specific infectious disease, this ratio between symptomatic and asymptomatic cases can be available when more data are collected from clinical studies (Nishiura et al., 2020). For symptomatic individuals, it takes a time lag (e.g., a random variable with the mean of one day) towards self-isolation or quarantine. They will stay isolated until recovery or deceased. Asymptomatic individuals are not aware that they are a carrier of a contagious virus and will continue their daily activities. There is little time lag for a susceptible individual to get infected. However, once someone gets infected, it will take a time lag (e.g., a random variable with the mean of 14 days) to either recover or become deceased (Pan et al., 2020; Baud et al., 2020). Once recovered, this individual will gain an increased level of immunity to the disease. Networked traffic of human movements is integrated with infection model to study spatiotemporal dynamics of the virus spread.

6.3 Computer Experiments of NPIs

The availability of simulation models enables "what-if" analysis that will help local authorities in a spatial region to dynamically adjust health policies, plan near-term health care capacity, and control virus spread with rapid and timely measures. The proposed DES simulation captures not only detailed behaviors in the individual level (e.g., movement behaviors, contact patterns, personal protective measures), but also the dynamics of population traffic for infection modeling in a spatial network. Further, we evaluate and benchmark alternative healthcare policies, akin to making informed decisions, such that the healthcare system is more resilient and can respond expeditiously and effectively to epidemic events. In the experimental setting, the spatial network contains 5000 nodes, 5000 edges, and a total number of 6000 individuals who interact with each other based on daily schedules. The total simulation time is 90 days.

How asymptomatic vs. symptomatic impact the virus spread? Figure 24 shows the time evolution of virus spread under three different ratios of asymptomatic vs. symptomatic cases. When the ratio decreases, there will be more symptomatic cases than asymptomatic ones. Because symptomatic cases can be quickly identified and then go into self-isolation or quarantine, the virus spreads at a slower rate. Specifically, when the ratio is reduced from 3 to 0.33, infection peaks decreases from 85.48% to 45.93%. After 90 days, the remaining susceptible populations are 0.4%, 2.18%, and 13.93% for three scenarios, respectively. At 30 days, the percentages of

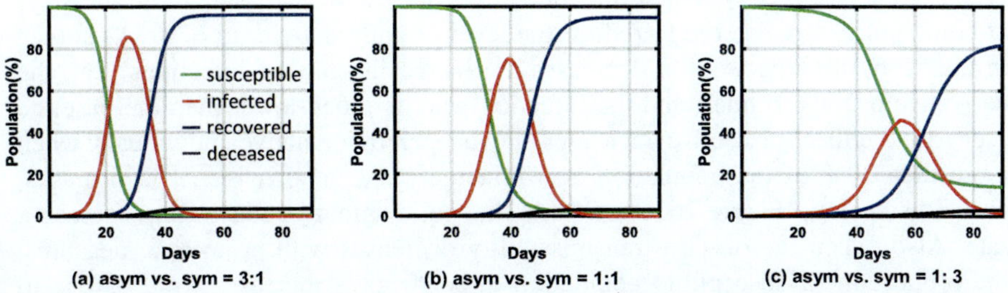

Fig. 24 Temporal characteristic curves of virus spread for different asym vs. sym ratios, (**a**) 3:1, (**b**) 1:1, (**c**) 1:3

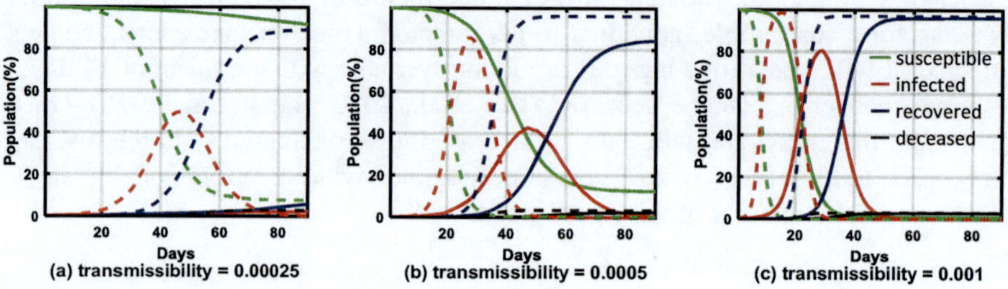

Fig. 25 Temporal characteristic curves of virus spread under stay at home policy when viral transmissibility is (**a**) 0.00025, (**b**) 0.0005, and (**c**) 0.001

infected population are 81.32%, 29.03% and 2.85%, respectively. As shown in Fig. 24a–c, the percentages of recovered population at 30 days are 14.07%, 0.53% and 1.10%, and the percentages of deceased population are 0.58%, 0% and 0% for three scenarios, respectively. Therefore, temporal infection characteristics are sensitive to the variations of asymptomatic vs. symptomatic ratio.

How stay-at-home impacts the virus spread? Figure 25 shows the impacts of the stay-at-home policy on the time evolution of virus spread with three different transmissibility values (i.e., $\rho = 0.00025$, 0.0005 and 0.001). The solid line represents the implementation of stay-at-home policy that reduce daily activities to 67%, while the dashed line represents the scenario with regular activities. After the stay-at-home policy is enforced, Fig. 25a–c shows that infection peaks drop dramatically from 49.83%, 85.48%% and 98.20% to 2.3%, 41.60% and 80.52%, respectively. When daily activities are reduced, the time to reach infection peak is also prolonged. This time delay decreases when the virus transmissibility increases. Hence, the stay-at-home policy is critical to stopping the virus spread and flattening the curve, which will provide tremendous help to avoid an overload on the healthcare systems.

How non-pharmaceutical interventions impact the virus spread? Figure 26 shows the impacts of NPIs on the virus spread that is benchmarked with the baseline scenario (i.e., regular daily activities without interventions, and the ratio of asym vs. sym is 3:1). When infections exceed 20% of the population, the stay-at-home policy is triggered to reduce the level of daily activity to 67%. As shown in Fig. 26b, this intervention decreases the increasing rate of infections (i.e., the derivative of the red line), and the infection peak is much lower than the baseline scenario. Further, protective measures are triggered for active individuals when more than 30% of the population gets infected (e.g., good hygiene, face masks, social distancing). Figure 26c shows that this policy greatly reduces the increasing rate. Also, when the disease transmissibility decreases with protective measures, the proportion of susceptible population after 90 days increases from 7.10% to 30.00%. Eventually, the proportion of deceased population is approximately 3%, 2.2% and 1.2%, respectively, as shown in Fig. 26a–c. NPIs reduce the transmission

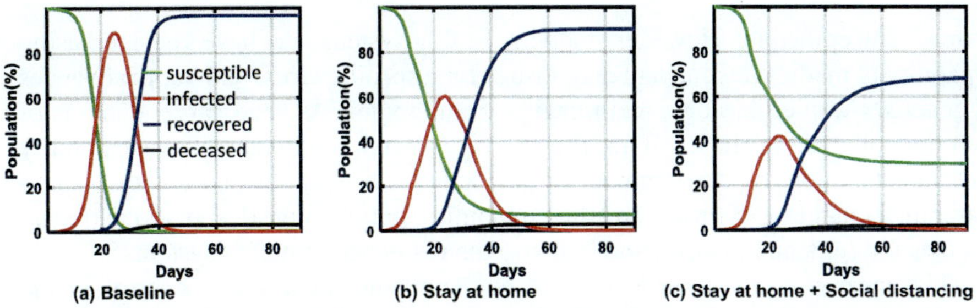

Fig. 26 Temporal characteristic curves of virus spread under NPIs, (**a**) baseline, (**b**) stay at home, (**c**) stay at home and social distancing

risks of infectious diseases. Therefore, a combination of NPI policies should be implemented to effectively lower the probability of infection and save more lives.

Simulation-based decision support provides an enabling tool to benchmark alternative healthcare policies and make health systems more resilient to coronavirus events, rather than relying solely on the experience and expertise of human experts. The proposed DES simulation provides detailed behaviors of individuals (e.g., movement behaviors, contact patterns, personal protective measures) in a spatial network. Such details are often not available in conventional DES, SEIR, or statistical models, and therefore can be used to help design and analyze clinical testing programs for the population in the future work. Furthermore, networked traffic of human movements offers a higher level of flexibility for future investigation of network interdiction models in the epidemic settings. In other words, public health experts will be able to investigate the traffic control through arc interdictions to stop the spread of infectious diseases. In summary, effective simulation analysis and prediction of virus positions in geographic regions will not only help optimize the design of healthcare policies to control the virus spread, but also help safeguard the population and make health systems more resilient to epidemic events. The proposed methodology can be applicable in general to a wide range of infectious diseases.

7 Conclusions

The broad spread of a highly infectious disease leads to an epidemic in a country and may also bring a global pandemic if ravaging over multiple countries. For example, COVID-19 changes everyone's daily life and poses significant challenges to health and economy of our society. Before vaccines or antivirals are available, non-pharmaceutical interventions (e.g., isolation, quarantine, hygiene, face masks and social distancing) are only effective means for the control and containment of virus spread. This does not change much in the twenty-first century, although health systems are equipped with more advanced technologies than the era of 1918

Spanish flu epidemic. However, modern health systems do have the increasing capability of medical testing and diagnostics for a specific virus, with rapid advances of gene/DNA, microbiology, and imaging technologies. As such, large amounts of data are collected in the evolving process of epidemic outbreaks. The availability of data calls upon the development of new analytical methods and tools to gain a better understanding of virus spreading dynamics, optimize the design of healthcare policies for epidemic control, and improve the resilience of health systems.

This paper presents a review about epidemic informatics and control in the framework of **D**efine**, M**easure, **A**nalyze, **I**mprove, and **C**ontrol (DMAIC), which focuses more on the intensive use of data, statistics and optimization. The proposed DMAIC framework integrates epidemic data with statistics, AI, privacy, system design, and simulation models to predict real-time positions of virus spread in the spatial network, simulate human traffic and virus spread dynamics, and provide decision support tools for the design of healthcare policies. As opposed to purely data-driven approaches, which cannot suggest action strategies, this DMAIC framework provides a higher level of flexibility to not only design computer experiments for the analysis of a variety of alternative health policies and strategies, but also augment real-world epidemic control by providing a model-based baseline for process adjustment. In addition, epidemic surges mandate the re-design of health systems such as hospitals, medical clinics, and emergency rooms for better treatment and accommodation of patients. Such re-designs help segregate infected patients, avoid secondary infections, and reduce transmission risks, thereby improving the safety and quality of healthcare services. System informatics show strong potentials to spur the growth of healthcare innovations in the US and the world, as well as complement the pharmaceutical and medical approaches to stop the spread. We hope this review can help catalyze more in-depth investigations and multi-disciplinary research efforts to advance the system informatics methods and tools for the future of healthcare.

Acknowledgements The authors would like to acknowledge the NSF I/UCRC Center for Healthcare Organization Transformation (CHOT), NSF I/UCRC award IIP-1624727, and NSF RAPID grant IIP-2026875 for funding this research.

References

Albert, R., & Barabási, A. (2002). Statistical mechanics of complex networks. *Reviews of Modern Physics, 74*(1), 47.

Alling, D., Blackwelder, W., & Stuart-Harris, C. (1981). A study of excess mortality during influenza epidemics in the United States, 1968–1976. *American Journal of Epidemiology, 113*(1), 30–43.

Bacon, A. S., & Erickson, C. A. (1950). Efficient hospitals. *Hospital Progress, 31*(6), 174–175.

Baker, J., & Lamb, C. W., Jr. (1992). Physical environment as a hospital marketing tool. *Journal of Hospital Marketing, 6*(2), 25–35.

Ballagh, L. M., Raup, B. H., Duerr, R. E., Khalsa, S. J. S., Helm, C., Fowler, D., & Gupte, A. (2011). Representing scientific data sets in KML: Methods and challenges. *Computational Geosciences, 37*(1), 57–64.

Bartoszko, J. J., Farooqi, M. A. M., Alhazzani, W., & Loeb, M. (2020). Medical masks vs N95 respirators for preventing COVID-19 in healthcare workers: A systematic review and meta-analysis of randomized trials. *Influenza and Other Respiratory Viruses, 14*(4), 365–373.

Basdogan, C., Sedef, M., Harders, M., & Wesarg, S. (2007). VR-based simulators for training in minimally invasive surgery. *IEEE Computer Graphics and Applications, 27*(2), 54–66.

Baud, D., Qi, X., Nielsen-Saines, K., Musso, D., Pomar, L., & Favre, G. (2020). Real estimates of mortality following COVID-19 infection. *The Lancet Infectious Diseases, 20*(7), 773-3099(20)30195-X. Epub 2020 Mar 12.

Bonaccorsi, G., Pierri, F., Cinelli, M., Flori, A., Galeazzi, A., Porcelli, F., Schmidt, A. L., Valensise, C. M., Scala, A., Quattrociocchi, W., & Pammolli, F. (2020). Economic and social consequences of human mobility restrictions under COVID-19. *Proceedings of the National Academy of Sciences of the United States of America, 117*(27), 15530–15535.

Bowman, T. (1997). VR meets physical therapy. *Communications of the ACM, 40*(8), 59–60.

Bowman, F. D. (2007). Spatiotemporal models for region of interest analyses of functional neuroimaging data. *Journal of the American Statistical Association, 102*(478), 442–453.

Budd, J., Miller, B. S., Manning, E. M., Lampos, V., Zhuang, M., Edelstein, M., Rees, G., Emery, V. C., Stevens, M. M., & Keegan, N. (2020). Digital technologies in the public-health response to COVID-19. *Nature Medicine, 26*, 1183–1192.

Burmahl, B. (2000). Facilities of the future: New designs put patients first. *Health Facilities Management, 13*(2), 30, 32, 34.

Cameron, P. A., Schull, M., & Cooke, M. (2006). The impending influenza pandemic: Lessons from SARS for hospital practice. *Medical Journal of Australia, 185*(4), 189–190.

Chaudhuri, K., & Monteleoni, C. (2009). Privacy-preserving logistic regression. *Advances in Neural Information Processing Systems*, 289–296.

Chen, Y., & Yang, H. (2014). Heterogeneous postsurgical data analytics for predictive modeling of mortality risks in intensive care unit. In *Engineering in Medicine and Biology Society (EMBC), Proceedings of 2014 Annual International Conference of the IEEE* (pp. 1–5).

Chen, Y., & Yang, H. (2015). Heterogeneous recurrence T^2 charts for monitoring and control of nonlinear dynamic processes. In *Proceedings of 2015 IEEE International Conference on Automation Science and Engineering (CASE)* (pp. 1066–1071), Gothenburg, Sweden.

Chen, Y., & Yang, H. (2016a). Heterogeneous recurrence representation and quantification of dynamic transitions in continuous nonlinear processes. *The European Physical Journal B, 89*(6), 155.

Chen, Y., & Yang, H. (2016b). Sparse modeling and recursive prediction of space–time dynamics in stochastic sensor networks. *IEEE Transactions on Automation Science and Engineering, 13*(1), 215–226.

Chen, Y.-C., Lu, P.-E., Chang, C.-S., & Liu, T.-H. (2020). A time-dependent SIR model for COVID-19 with undetectable infected persons. *IEEE Transactions on Network Science and Engineering, 7*, 3279–3294.

Chirico, F., Nucera, G., & Magnavita, N. (2020). COVID-19: Protecting healthcare workers is a priority. *Infection Control & Hospital Epidemiology, 41*, 1117.

Coronavirus Disease 2019 (COVID-19) in the U.S. (2019). https://www.cdc.gov/coronavirus/2019-ncov/

COVID-19: Impact on Global Pharmaceutical and Medical Product Supply Chain Constraints U.S. Production. (2019). https://www.fticonsulting.com/insights/articles/covid-19-impact-global-pharmaceutical-medical-product-supply-chain

Crawford, J., Butler-Henderson, K., Rudolph, J., Malkawi, B., Glowatz, M., Burton, R., Magni, P., & Lam, S. (2020). COVID-19: 20 countries' higher education intra-period digital pedagogy responses. *Journal of Applied Learning & Teaching, 3*(1), 1–20.

Currie, C. S. M., Fowler, J. W., Kotiadis, K., Monks, T., Onggo, B. S., Robertson, D. A., & Tako, A. A. (2020). How simulation modelling can help reduce the impact of COVID-19. *Journal of Simulation, 14*(2), 83–97.

Daskin, M. (1997). Network and discrete location: Models, algorithms and applications. *The Journal of the Operational Research Society, 48*(7), 763–764.

Davies, N. G., Kucharski, A. J., Eggo, R. M., et al. (2020). Effects of non-pharmaceutical interventions on COVID-19 cases, deaths, and demand for hospital services in the UK: A modelling study. *The Lancet Public Health, 5*(7), e375–e385.

Descombes, X., Kruggel, F., & Von Cramon, D. Y. (1998). Spatio-temporal fMRI analysis using Markov random fields. *IEEE Transactions on Medical Imaging, 17*(6), 1028–1039.

Disease Burden of Influenza. (n.d.). https://www.cdc.gov/flu/about/burden/index.html

Du, Q., Faber, V., & Gunzburger, M. (1999). Centroidal Voronoi tessellations: Applications and algorithms. *SIAM Review, 41*(4), 637–676.

Du, D., Yang, H., Ednie, A. R., & Bennett, E. (2016). Statistical metamodeling and sequential design of computer experiments to model glyco-altered gating of sodium channels in cardiac myocytes. *IEEE Journal of Biomedical and Health Informatics, 20*(5), 1439–1452.

Dwork, C., & McSherry, F. D. (2010). *Differential data privacy*. US Patent US7698250B2.

Dwork, C., & Pottenger, R. (2013). Toward practicing privacy. *Journal of the American Medical Informatics Association, 20*(1), 102–108.

Dwork, C., & Roth, A. (2014). The algorithmic foundations of differential privacy. *Foundations and Trends in Theoretical Computer Science, 9*(3–4), 211–407.

Eames, I., Tang, J., Li, Y., & Wilson, P. (2009). Airborne transmission of disease in hospitals. *The Journal of the Royal Society Interface, 6*(Suppl 6), S697–S702.

Esbin, M. N., Whitney, O. N., Chong, S., Maurer, A., Darzacq, X., & Tjian, R. (2020). Overcoming the bottleneck to widespread testing: A rapid review of nucleic acid testing approaches for COVID-19 detection. *RNA, 26*(7), 771–783.

Facility Guidelines Institute. (2014). *Guidelines for design and construction of hospitals and outpatient facilities*. American Hospital Association. American Society for Healthcare Engineering.

Fernandes, N. (2020). *Economic effects of coronavirus outbreak (COVID-19) on the world economy*. Available at SSRN 3557504.

Funck-Brentano, C., Nguyen, L. S., & Salem, J. E. (2020). Retraction and republication: Cardiac toxicity of hydroxychloroquine in COVID-19. *Lancet, 396*(10245), e2–e3.

Galasko, D., Klauber, M. R., Hofstetter, C. R., Salmon, D. P., Lasker, B., & Thal, L. J. (1990). The mini-mental state examination in the early diagnosis of Alzheimer's disease. *Archives of Neurology, 47*(1), 49–52.

Gómez-Carballa, A., Bello, X., Pardo-Seco, J., Martinón-Torres, F., & Salas, A. (2020). Mapping genome variation of SARS-CoV-2 worldwide highlights the impact of COVID-19 superspreaders. *Genome Research, 30*(10), 1434–1448.

Greenberg, N., Docherty, M., Gnanapragasam, S., & Wessely, S. (2020). Managing mental health challenges faced by healthcare workers during covid-19 pandemic. *BMJ, 368*, m1211.

Hedt, B. L., van Leth, F., Zignol, M., Cobelens, F., van Gemert, W., Nhung, N. V., Lyepshina, S., Egwaga, S., & Cohen, T. (2012). Multidrug resistance among new tuberculosis cases: Detecting local variation through lot quality-assurance sampling. *Epidemiology, 23*(2), 293–300.

Hollander, J. E., & Carr, B. G. (2020). Virtually perfect? Telemedicine for COVID-19. *The New England Journal of Medicine, 382*(18), 1679–1681.

Imani, F., Cheng, C., Chen, R., & Yang, H. (2019). Nested Gaussian process modeling and imputation of high-dimensional incomplete data under uncertainty. *IISE Transactions on Healthcare Systems Engineering, 9*(4), 315–326.

Improving Hospital Design for Better Infection Control. (n.d.). https://hmcarchitects.com/news/improving-hospital-design-for-better-infection-control-2020-04-15/

ISM Report on Business. (2019). https://www.ismworld.org/

Jia, J. S., Lu, X., Yuan, Y., Xu, G., Jia, J., & Christakis, N. A. (2020). Population flow drives spatio-temporal distribution of COVID-19 in China. *Nature, 582*, 1–5.

Kaitin, K. I. (2010). Deconstructing the drug development process: The new face of innovation. *Clinical Pharmacology & Therapeutics, 87*(3), 356–361.

Kan, C., Chen, Y., Leonelli, F. M., & Yang, H. (2015). Mobile sensing and network analytics for realizing smart automated systems towards health internet of things. In *Proceedings of 2015 IEEE International Conference on Automation Science and Engineering (CASE)* (pp. 1072–1077), Gothenburg, Sweden.

Kelsall, J., & Wakefield, J. (2002). Modeling spatial variation in disease risk: A Geostatistical approach. *Journal of the American Statistical Association, 97*(459), 692–701.
Kienberger, S., & Tiede, D. (2008). ArcGIS explorer review. *GEO Informatics, 11*(2), 42–47.
Kim, L., Garg, S., O'Halloran, A., Whitaker, M., Pham, H., Anderson, E. J., Armistead, I., Bennett, N. M., Billing, L., & Como-Sabetti, K. (2020). Risk factors for intensive care unit admission and in-hospital mortality among hospitalized adults identified through the US coronavirus disease 2019 (COVID-19)-associated hospitalization surveillance network (COVID-NET). *Clinical Infectious Diseases, 72*, 1–9.
Klein, M. G., Cheng, C. J., Lii, E., Mao, K., Mesbahi, H., Zhu, T., Muckstadt, J. A., & Hupert, N. (2020). COVID-19 models for hospital surge capacity planning: A systematic review. *Disaster Medicine and Public Health Preparedness*, 1–17.
Knowles, G., Whicker, L., Femat, J. H., & Canales, F. D. C. (2005). A conceptual model for the application of Six Sigma methodologies to supply chain improvement. *International Journal of Logistics: Research and Applications, 8*(1), 51–65.
Krall, A., Finke, D., & Yang, H. (2020). Gradient mechanism to preserve differential privacy and deter against model inversion attacks in healthcare analytics. In *2020 42nd Annual International Conference of the IEEE Engineering in Medicine & Biology Society (EMBC)* (pp. 5714–5717).
Krall, A., Finke, D., & Yang, H. (2021). Mosaic privacy-preserving mechanisms for healthcare analytics. *IEEE Journal of Biomedical and Health Informatics, 25*(6), 2184–2192.
Kretzschmar, M. E., Rozhnova, G., Bootsma, M. C., van Boven, M., van de Wijgert, J. H. H. M., & Bonten, M. J. (2020). Impact of delays on effectiveness of contact tracing strategies for COVID-19: A modelling study. *The Lancet Public Health, 5*(8), e452–e459.
Kumar, M., Antony, J., Antony, F. J., & Madu, C. N. (2007). Winning customer loyalty in an automotive company through Six Sigma: A case study. *Quality and Reliability Engineering International, 23*(7), 849–866.
Lateef, F. (2009). Hospital design for better infection control. *Journal of Emergencies, Trauma, and Shock, 2*(3), 175–179.
Li, Y., Huang, X., Yu, I., Wong, T., & Qian, H. (2005). Role of air distribution in SARS transmission during the largest nosocomial outbreak in Hong Kong. *Indoor Air, 15*(2), 83–95.
Li, J., Jin, J., & Shi, J. (2008). Causation-based T2 decomposition for multivariate process monitoring and diagnosis. *Journal of Quality Technology, 40*(1), 46–58.
Li, L., Qin, L., Xu, Z., Yin, Y., Wang, X., Kong, B., Bai, J., Lu, Y., Fang, Z., Song, Q., Cao, K., Liu, D., Wang, G., Xu, Q., Fang, X., Zhang, S., Xia, J., & Xia, J. (2020). Using artificial intelligence to detect COVID-19 and community-acquired pneumonia based on pulmonary CT: Evaluation of the diagnostic accuracy. *Radiology, 296*(2), E65–E71.
Libin, P. J., Deforche, K., Abecasis, A. B., & Theys, K. (2019). VIRULIGN: Fast codon-correct alignment and annotation of viral genomes. *Bioinformatics, 35*(10), 1763–1765.
Lipsitch, M., Kahn, R., & Mina, M. J. (2020). Antibody testing will enhance the power and accuracy of COVID-19-prevention trials. *Nature Medicine, 26*(6), 818–819.
Liu, G., & Yang, H. (2013). Multiscale adaptive basis function modeling of spatiotemporal cardiac electrical signals. *IEEE Journal of Biomedical and Health Informatics, 17*(2), 484–492.
Liu, G., & Yang, H. (2017). Self-organizing network for group variable selection and predictive modeling. *Annals of Operation Research, 263*, 119–140.
Liu, N., Chen, C., & Kumara, S. (2019). Semi-supervised learning algorithm for identifying high-priority drug–drug interactions through adverse event reports. *IEEE Journal of Biomedical and Health Informatics, 24*(1), 57–68.
Liu, Y., Gayle, A. A., Wilder-Smith, A., & Rocklöv, J. (2020). The reproductive number of COVID-19 is higher compared to SARS coronavirus. *Journal of Travel Medicine, 27*(2), taaa021.
Lohr, S. (2010, March 13). Netflix cancels contest after concerns are raised about privacy. *New York Times*, p. B3.
Manufacturing: NAICS 31-33. (n.d.). https://www.bls.gov/iag/tgs/iag31-33.htm#about
Mark, W. W., Mark, J., Michael, B. J., & Stephen, M. S. (2004). Fully Bayesian spatio-temporal modeling of FMRI data. *IEEE Transaction on Medical Imaging, 23*(2), 213–231.

Marques, R., Gregório, J., Pinheiro, F., Póvoa, P., Da Silva, M. M., & Lapão, L. V. (2017). How can information systems provide support to nurses' hand hygiene performance? Using gamification and indoor location to improve hand hygiene awareness and reduce hospital infections. *BMC Medical Informatics and Decision Making, 17*(1), 15.

Mason, R. L., Tracy, N. D., & Young, J. C. (1997). A practical approach for interpreting multivariate T2 control chart signals. *Journal of Quality Technology, 29*, 396–406.

Mateu, J., Montes, F., & Plaza, M. (2004). The 1970 US draft lottery revisited: A spatial analysis. *Journal of the Royal Statistical Society: Series C (Applied Statistics), 53*(1), 219–229.

Metnitz, P. G., Moreno, R. P., Almeida, E., Jordan, B., Bauer, P., Campos, R. A., Iapichino, G., Edbrooke, D., Capuzzo, M., & Le Gall, J. (2005). SAPS 3—From evaluation of the patient to evaluation of the intensive care unit. Part 1: Objectives, methods and cohort description. *Intensive Care Medicine, 31*(10), 1336–1344.

Miner, A. S., Laranjo, L., & Kocaballi, A. B. (2020). Chatbots in the fight against the COVID-19 pandemic. *NPJ Digital Medicine, 3*, 65-020-0280-0. eCollection.

Moreno, R. P., Metnitz, P. G., Almeida, E., Jordan, B., Bauer, P., Campos, R. A., Iapichino, G., Edbrooke, D., Capuzzo, M., & Le Gall, J. (2005). SAPS 3—From evaluation of the patient to evaluation of the intensive care unit. Part 2: Development of a prognostic model for hospital mortality at ICU admission. *Intensive Care Medicine, 31*(10), 1345–1355.

Niederriter, B., Rong, A., Aqlan, F., & Yang, H. (2020). Sensor-based virtual reality for clinical decision support in the assessment of mental disorders. In *2020 IEEE Conference on Games (CoG)* (pp. 666–669).

Nishiura, H., Kobayashi, T., Miyama, T., Suzuki, A., Jung, S., Hayashi, K., Kinoshita, R., Yang, Y., Yuan, B., Akhmetzhanov, A. R., & Linton, N. M. (2020). Estimation of the asymptomatic ratio of novel coronavirus infections (COVID-19). *International Journal of Infectious Diseases, 94*, 154–155.

Noskin, G. A., & Peterson, L. R. (2001). Engineering infection control through facility design. *Emerging Infectious Diseases, 7*(2), 354–357.

Oyama, H., Miyazawa, T., Aono, M., Ohbuchi, R., & Suda, S. (1995). VR medical support system for cancer patients. Cancer edutainment VR theater (CEVRT) and psychooncological VR therapy (POVRT). In *Interactive technology and healthcare* (pp. 433–438). IOS Press and Ohmsha.

Ozkil, A. G., Fan, Z., Dawids, S., Aanes, H., Kristensen, J. K., & Christensen, K. H. (2009). Service robots for hospitals: A case study of transportation tasks in a hospital. In *2009 IEEE International Conference on Automation and Logistics* (pp. 289–294).

Pan, F., Ye, T., Sun, P., Gui, S., Liang, B., Li, L., Zheng, D., Wang, J., Hesketh, R. L., Yang, L., & Zheng, C. (2020). Time course of lung changes at chest CT during recovery from coronavirus disease 2019 (COVID-19). *Radiology, 295*(3), 715–721.

Penchansky, R., & Thomas, J. W. (1981). The concept of access: Definition and relationship to consumer satisfaction. *Medical Care, 19*, 127–140.

Prem, K., Liu, Y., Russell, T. W., et al. (2020). The effect of control strategies to reduce social mixing on outcomes of the COVID-19 epidemic in Wuhan, China: A modelling study. *The Lancet Public Health, 5*(5), e261–e270.

Rafflin, C., & Fournier, A. (1996). Learning with a friendly interactive robot for service tasks in hospital environments. *Autonomous Robots, 3*(4), 399–414.

Raith, E. P., Udy, A. A., Bailey, M., McGloughlin, S., MacIsaac, C., Bellomo, R., & Pilcher, D. V. (2017). Prognostic accuracy of the SOFA score, SIRS criteria, and qSOFA score for in-hospital mortality among adults with suspected infection admitted to the intensive care unit. *Journal of the American Medical Association, 317*(3), 290–300.

Raja, S., Patolia, H. H., & Baffoe-Bonnie, A. W. (2020). Calculating an institutional personal protective equipment (PPE) burn rate to project future usage patterns during the 2020 COVID-19 pandemic. *Infection Control & Hospital Epidemiology, 41*(12), 1474–1475.

Ravi, N., Cortade, D. L., Ng, E., & Wang, S. X. (2020). Diagnostics for SARS-CoV-2 detection: A comprehensive review of the FDA-EUA COVID-19 testing landscape. *Biosensors and Bioelectronics, 165*, 112454.

Sasangohar, F., Jones, S. L., Masud, F. N., Vahidy, F. S., & Kash, B. A. (2020). Provider burnout and fatigue during the COVID-19 pandemic: Lessons learned from a high-volume intensive care unit. *Anesthesia and Analgesia, 131*(1), 106–111.
Serban, N. (2011). A space-time varying coefficient model: The equity of service accessibility. *The Annals of Applied Statistics, 5*, 2024–2051.
Sezgin, E., Huang, Y., Ramtekkar, U., & Lin, S. (2020). Readiness for voice assistants to support healthcare delivery during a health crisis and pandemic. *NPJ Digital Medicine, 3*(1), 1–4.
Shechter, A., Diaz, F., Moise, N., Anstey, D. E., Ye, S., Agarwal, S., Birk, J. L., Brodie, D., Cannone, D. E., & Chang, B. (2020). Psychological distress, coping behaviors, and preferences for support among New York healthcare workers during the COVID-19 pandemic. *General Hospital Psychiatry, 66*, 1–8.
Smalley, E. (2017). AI-powered drug discovery captures pharma interest. *Nature Biotechnology, 35*(7), 604–606.
Smith, J. S., Roitberg, A. E., & Isayev, O. (2018). Transforming computational drug discovery with machine learning and AI. *ACS Medicinal Chemistry Letters, 9*(11), 1065–1069.
Smith, A. C., Thomas, E., Snoswell, C. L., Haydon, H., Mehrotra, A., Clemensen, J., & Caffery, L. J. (2020). Telehealth for global emergencies: Implications for coronavirus disease 2019 (COVID-19). *Journal of Telemedicine and Telecare, 26*(5), 309–313.
Song, S., Chaudhuri, K., & Sarwate, A. D. (2013). Stochastic gradient descent with differentially private updates. In *2013 IEEE Global Conference on Signal and Information Processing* (pp. 245–248).
Stiller, A., Salm, F., Bischoff, P., & Gastmeier, P. (2016). Relationship between hospital ward design and healthcare-associated infection rates: A systematic review and meta-analysis. *Antimicrobial Resistance & Infection Control, 5*(1), 51.
Sullivan, S. J., Jacobson, R. M., Dowdle, W. R., & Poland, G. A. (2010). 2009 H1N1 Influenza. *Mayo Clinic Proceedings, 85*(1), 64–76.
Sweeney, L. (2013). *Matching known patients to health records in Washington state data*. Available at SSRN 2289850.
Taubenberger, J. K., Reid, A. H., & Fanning, T. G. (2005). Capturing a killer flu virus. *Scientific American, 292*(1), 62–71.
The Bureau of Labor Statistics: Supplemental data measuring the effects of the coronavirus (COVID-19) pandemic on the labor market. (n.d.). https://www.bls.gov/cps/effects-of-the-coronavirus-covid-19-pandemic.htm
Topol, E. J. (2020). Welcoming new guidelines for AI clinical research. *Nature Medicine, 26*(9), 1318–1320.
Trilla, A., Trilla, G., & Daer, C. (2008). The 1918 "Spanish flu" in Spain. *Clinical Infectious Diseases, 47*(5), 668–673.
Uddin, M., Mustafa, F., Rizvi, T. A., Loney, T., Suwaidi, H. A., Al-Marzouqi, A. H. H., Eldin, A. K., Alsabeeha, N., Adrian, T. E., & Stefanini, C. (2020). SARS-CoV-2/COVID-19: Viral genomics, epidemiology, vaccines, and therapeutic interventions. *Viruses, 12*(5), 526.
US Census Bureau. (2010). *TIGER/Line shapefiles*. US Census Bureau.
Walker-Roberts, S., Hammoudeh, M., & Dehghantanha, A. (2018). A systematic review of the availability and efficacy of countermeasures to internal threats in healthcare critical infrastructure. *IEEE Access, 6*, 25167–25177.
Waller, L. A., Carlin, B. P., Hong, X., & Gelfand, A. E. (1997). Hierarchical spatio-temporal mapping of disease rates. *Journal of the American Statistical Association, 92*(438), 607–615.
Wang, Y., Si, C., & Wu, X. (2015). Regression model fitting under differential privacy and model inversion attack. In *International Joint Conference on Artificial Intelligence* (pp. 1003–1009).
Yang, H., & Chen, Y. (2014). Heterogeneous recurrence monitoring and control of nonlinear stochastic processes. *Chaos, 24*(1), 013138.
Yang, H., & Liu, G. (2013). Self-organized topology of recurrence-based complex networks. *Chaos, 23*(4), 043116.
Yang, H., Bukkapatnam, S. T., & Komanduri, R. (2012). Spatiotemporal representation of cardiac vectorcardiogram (VCG) signals. *Biomedical Engineering Online, 11*(1), 1–15.

Yang, H., Kan, C., Chen, Y., & Liu, G. (2013). Spatiotemporal differentiation of myocardial infarctions. *IEEE Transactions on Automation Science and Engineering, 10*(4), 938–947.

Yang, H., Kan, C., Krall, A., & Finke, D. (2020). Network modeling and internet of things for smart and connected health systems—A case study for smart heart health monitoring and management. *IISE Transactions on Healthcare Systems Engineering, 10*(3), 159–171.

Yang, H., Rao, P., Simpson, T., Lu, Y., Witherell, P., Nassar, A. R., Reutzel, E., & Kumara, S. (2021). Six-Sigma quality management of additive manufacturing. *Proceedings of the IEEE, 109*, 347–376.

Yao, B., & Yang, H. (2016). Physics-driven spatiotemporal regularization for high-dimensional predictive modeling. *Scientific Reports, 6*, 39012.

Yao, B., Zhu, R., & Yang, H. (2017). Characterizing the location and extent of myocardial infarctions with inverse ECG modeling and spatiotemporal regularization. *IEEE Journal of Biomedical and Health Informatics, 22*(5), 1445–1455.

Zhan, F. B., & Noon, C. E. (1998). Shortest path algorithms: An evaluation using real road networks. *Transportation Science, 32*(1), 65–73.

Zhang, J., Zhang, Z., Xiao, X., Yang, Y., & Winslett, M. (2012). Functional mechanism: Regression analysis under differential privacy. *Proceedings of the VLDB Endowment, 5*(11), 1364–1375.

Zhang, D., Hu, M., & Ji, Q. (2020). Financial markets under the global pandemic of COVID-19. *Finance Research Letters, 36*, 101528.

Zimmerman, J. E., Kramer, A. A., McNair, D. S., & Malila, F. M. (2006). Acute Physiology and Chronic Health Evaluation (APACHE) IV: Hospital mortality assessment for today's critically ill patients. *Critical Care Medicine, 34*(5), 1297–1310.

Private vs. Pooled Transportation: Customer Preference and Congestion Management

Kashish Arora, Fanyin Zheng, and Karan Girotra

1 Introduction

Traffic congestion and its economic and social consequences plague most large urban areas. On-demand taxi services have done little to alleviate these concerns, and, on the contrary, have exacerbated congestion in some of the biggest cities. In recent years, *pooled transportation* has emerged as a cheaper and more environmentally and traffic friendly alternative to on-demand transportation services. Examples of such pooled transportation options range from Uber's UberPool service to the *shuttle services* offered by Via, Chariot, and others in major cities across countries. At the same time, most of these services face challenges as, many customers remain reluctant to switch to pooled services from private on-demand transportation options. The goal of this paper is, therefore, to understand customer preferences in choosing between private and pooled transportation services and to investigate the way that policies can be designed to incentivize customers to use pooled transportation and effectively manage congestion.

From the government's perspective, the most often used policies are congestion surcharges. Local governments in large cities such as New York, Singapore, and London have imposed congestion prices to incentivize increased usage of pooled transportation and less usage of the private rides. For instance, the city of London uses an *all-day* congestion surcharge policy, whereas Singapore uses a *peak hour* surcharge policy. A key challenge for many cities is to assess the potential impact and the relative effectiveness of these policies before they are implemented. Experi-

K. Arora (✉) · K. Girotra
Cornell University, Ithaca, NY, USA
e-mail: ka522@cornell.edu; girotra@cornell.edu

F. Zheng
Columbia University, New York, NY, USA
e-mail: fanyin.zheng@columbia.edu

H. Yang et al. (eds.), *AI and Analytics for Public Health*, Springer Proceedings in Business and Economics, https://doi.org/10.1007/978-3-030-75166-1_2

menting with different policies in practice is extremely expensive in this setting, and, hence, their implementation has been largely on an ad hoc basis. Our paper seeks to provide guidance to evaluate the performance of different policies in incentivizing customers to switch from private to pooled transportation. Specifically, our guidance to policy makers is in terms of two aspects. First, we evaluate the efficacy of three particular types of congestion strategies: (i) *price strategies* based on congestion surcharges applied in the city; (ii) *price strategies* based on providing discounts to pooled transportation services; and (iii) *operational strategies* based on improving the service features of pooled transportation services. Second, our findings highlight the importance of incorporating customer heterogeneity of preferences in designing effective policies to promote pooled transportation usage.

Our analysis proceeds in two steps. First, we estimate customer demand for pooled and private transportation with usage data from *Ola Cabs* in India. The form of pooled transportation we study is the shuttle service, and the private transportation we focus on is cabs. We adopt a structural modeling approach to estimate demand and recover the customer's preferences for price and other operational service features. Then, we evaluate various congestion management strategies using the estimated model via counterfactual analyses. We provide a brief description of each of the two steps and summarize our main findings next.The results from this analysis suggest a substantial degree of substitution between the two services. We then build a structural model to estimate customer preferences over prices and different service features for the cab and shuttle services. Our control function approach with Lasso selected instruments corrects for the bias in the customer price elasticity and wait time sensitivity estimates. The estimated price coefficient is lower than the estimate without the correction, suggesting the presence of route based discounting by the platform to grow the customer base. We estimate an average cost of ₹98.5 ($1.3) for walking an additional km to the shuttle stop ₹3.6 for traveling ten extra minutes on the shuttle. Using the estimates from the model, we provide prescriptive recommendations on reducing congestion through counterfactuals. In the first counterfactual, we apply differential percentage *congestion surcharges* to cab and shuttle services in a congestion zone of the city. We find that a 20% congestion surcharge on cabs achieves a 15.0% overall vehicle reduction on the road. The corresponding vehicle reduction due to customers substituting from cab to shuttle service is around 4.04%. We also apply congestion surcharges to the services in peak hours following similar policies implemented in Singapore. We find that, interestingly, surcharges applied to the evening rush hours achieve around 3 times the %age vehicle reduction as compared to the morning rush hours. Moreover, the evening rush hour surcharge policy achieves a higher %age vehicle reduction than an all-day surcharge policy. In the second counterfactual, we evaluate the impact of providing discounts to shuttle rides on customer choices. We find that a 20% discount on shuttle rides leads to around 1% reduction in vehicles due to customers substituting from cabs to the shuttle service. Moreover, the reduction disproportionately comes from new users with relatively low past shuttle usage and more room for usage growth in the future. Finally, we evaluate *operations based strategies* and estimate the change in congestion levels when the firm improves some of the key shuttle service features. We find that a city could reap a significant portion of the benefits obtained by applying congestion surcharge policies by utilizing the operational levers itself. More importantly, adopting these strategies avoids the deadweight losses associated with the price surcharges due to its tax nature. Specifically, we find that a 20% decrease in customers' walking distance to shuttle pick up stops can achieve 35% of the total effect achieved by the congestion surcharge in terms of the number of customers substituting from the cab to shuttle services. Similarly, a 20% decrease in the shuttle travel time can achieve 6.9% of the total substitution achieved by the congestion surcharge. This result shows that improving the service features of pooled ride services is an important alternative to price-based policies in managing congestion in big cities.

Our paper is closely related to the structural demand estimation of mobility services in operations management and economics. He et al. (2019) and Kabra et al. (2019) study customer demand in bike-share systems. Buchholz (2020) and Ata et al. (2019) study spatial demand for the taxi service. To the best of our knowledge, our work is the first empirical study to investigate customer choices between the on-demand and the pooled ride services, which allows us to quantify the impact of congestion policies on customer choices. Our paper also closely relates to the empirical literature on ride-sharing services in operations management. Cohen et al. (2020b) run field experiments to nudge commuters to carpool using in-app notifications. Also, Cohen et al. (2020a) document the frustrations caused by inconveniences such as longer travel time in pooled services. The main difference between these studies and our paper is that, instead of an experimental approach, we recover customer preferences for choosing the services while directly incorporating the inconveniences associated with the shuttle service in the model.

Our work also contributes to the literature on congestion management in transportation. Han et al. (2019) build a stochastic model to develop a road pricing scheme to curb congestion. Recent work by Ostrovsky and Schwarz (2018) studies the relationship of carpooling, road pricing, and autonomous transportation. The authors highlight the role of road pricing in the adoption of pooled transport. Almost all of these studies are either analytical or predictive, even though the topic is of high practical relevance to both policy makers and ride-sharing platforms. Our work complements the literature using data and empirical methods to estimate customer preferences and provide prescriptive recommendations for the design of congestion policies.

Our work also fits into the growing literature on structural estimation in operations management. We use a discrete choice model with a control function to estimate the customer preference parameters for different ride service features. Similar methods have been applied in Petrin and Train (2010) and Guajardo et al. (2012). To correct for endogeneity, we build on the network type of instrumental variable method used in prior work on demand estimation in operations management (He et al., 2019). To strengthen the relevance of our instruments, we employ selection methods commonly used in machine learning and the causal inference literature (Belloni et al. 2011, 2012).

2 Data

Our study uses data provided to us by the Indian ride-hailing company Ola Cabs.The firm competes directly with (i) other, similar platforms such as Uber; (ii) public transportation; and (iii) city taxis. Its market share in the Indian ride-hailing market was around 65% at the time we collected our data. financing. We use four different sources of data in our analysis: (i) cab rides data; (ii) shuttle rides and trips data; (iii) Google Places API data; and (iv) census data

The cab rides dataset contains over 25 million cab rides from Jan-Aug 2016 in Delhi. Each ride record contains information about (i) the customer's pickup and drop-off locations (latitudes and longitudes); (ii) anonymized ID; (ii) timestamps for the initial ride request, pickup, and drop-off; (iii) prices; and (iv) distance traveled. There are around 3.5 million unique users on the platform. The shuttle data contain information about customer rides and shuttle trips. About 76K unique users took 1.28 million rides in Delhi from Jan-Aug 2016. For each ride record, we observe (i) timestamps for the customer's initial booking request; (ii) anonymized ID; (iii) latitudes and longitudes for pickup and drop-off; (iv) prices; (v) distance traveled; and (vi) latitude and longitude of the customer's mobile device when she makes the initial booking request. We also obtained latitudes and longitudes of around 176K points of interest in the city, collected from Google Places API. The places are classified into 93 classes, including restaurants, museums, libraries, hospitals, and theaters. In addition, we obtain demographic information from the Indian census of 2011.

2.1 Descriptive Evidence

To motivate our main model, we document two sets of descriptive evidence to motivate our main analysis. First, we show that a large number of customers use both the on-demand cab service and the shuttle service, and, hence, the market is not segmented into cab-only and shuttle-only users. Moreover, there is a large heterogeneity in customers' preferences for choosing shuttles and cabs. This motivates the structure of our model in Sect. 3. We control for this heterogeneity by including variables that measure the customer's past usage metrics on the platform.

Second, using a difference-in-differences analysis, we estimate the degree of substitution between the two services, thereby establishing that the services are substitutes. We leverage the fact that the shuttle platform was adding routes as it was expanding over time. The addition of routes over time serves as a quasi-experiment for our difference-in-differences analysis. We run a two-way-fixed-effects (TWFE) model to identify the causal impact of opening up the shuttle route on cab ridership. There is a net reduction of 88 rides per route when the shuttle services are operating. The reduction in cab ridership shows directly that the two services act as substitutes. In summary, we show that customers choose between shuttle and cab services, and we provide evidence of substitution between the two services. In the next section, we outline a richer customer-level choice model that helps us understand customer preferences when they are making choices between the two services.

3 Choice Model

Customer's Utility Customer i, who wants to travel from origin location $j \in \{1, \ldots, L\}$ to destination location $k \in \{1, \ldots, L\}$, chooses to take one of the alternatives, $a \in \{\text{Shuttle}, \text{Cab}\}$ or an outside option. The customers who travel from j to k belong to one market. The utility that customer i traveling in market jk gets from choosing an alternative a at time t is:

$$U_{ijkta} = \alpha_a + \sum_r \beta_r p_{jka} d_{ir} + \gamma w_{jc} + \sum_r X^{1'}_{jka} \Theta_r d_{ir} + X^{2'}_{jk} \Delta$$
$$+ Q^{'}_{iw(t)} \Omega + \mathrm{T}_{as(t)} + \xi_{jk} + \xi_i + \epsilon_{ijkta}. \tag{1}$$

In Eq. (1), α_a represents the baseline preference for the shuttle and the cab service. p_{jka} denotes the average price for service a in market jk. w_{jc} denotes the average wait time for the cab service. We do not include the wait time for the shuttle in our model since the majority of rides arrive at the scheduled time. We allow for the price coefficient to vary across two groups of customers segmented by their total shuttle usage in the past. d_{ir} is an indicator that identifies whether customer i belongs to the low-usage or new users group ($r = 1$) or the high-usage or experienced users group ($r = 2$). We include four sets of service features and control variables in our model as follows:

Market-Alternative-Level Service Features X^1_{jka} is a vector of key service features other than prices and wait time that affect the customer's choice. First, it includes the time and distance traveled on the ride across the two services. For any origin-destination pair jk, the time and distance traveled in shuttles is larger than those in cabs. The extra time and distance traveled and walking are the inconveniences associated with the shuttle service. Like price, we also allow for the vector of sensitivities to the service features to vary across the two customer groups.

Market-Level Controls X^2_{jk} is the vector of market-level control variables. We include a rich set of market variables in the model. Specifically, we include the number of Google Places category counts at both j and k (20 in total), and demographic information such as population densities and working population at both the locations (eight in total).

Customer-Time-Level Controls $Q_{iw(t)}$ is the vector of the time-varying usage history of a customer on the two platforms. $w(t) \in \{1, \ldots, T\}$ is an operator that denotes the number of weeks starting from January 1, 2016. $Q_{iw(t)}$ includes two variables: (i) the cumulative number of shuttle rides taken by the customer up to last week, and (ii) the number of recent rides taken by the customer across the two services. The first usage variable controls for customers' familiarity with the experience of the shuttle service. Since the shuttle service was newly launched during the period of our data, new shuttle customers may not have been fully aware of the experience of using the shuttle service and, hence, may have been less likely to choose that service. Thus, it is important to control for this variable. The second variable captures the effect of the customer's recent activity on the platform. A customer may be more likely to choose the service if she was recently active on the platform. Both variables control for customer heterogeneity in terms of their awareness of the shuttle service.

Time-Level Controls $T_{as(t)}$ are the time fixed effects, and $s(t)$ is an operator that denotes the time slot of the day (morning, evening, night, etc).

Unobservables ξ_{jk} are the market-level unobservables that affect demand. Examples of some of these factors are the unobserved popularity of the route, the level of congestion on the route. ξ_i are the customer-level unobservables that affect demand. Examples of these unobservables include income, age, and any other factors that affect a customer's preference for choosing between the two services. ϵ_{ijkta} are independent and identically distributed idiosyncratic errors that follow extreme value type 1 distribution. Apart from the shuttles and cabs, the customer can also choose to take an outside option. The utility of the outside option o is defined as:

$$U_{ijkto} = u_o + \epsilon_{ijkto}. \tag{2}$$

where u_o is normalized to be zero. The customer chooses the alternative that maximizes her utility. The choice probability of customer i is given by :

$$P_{ijkta} = \frac{\begin{bmatrix} exp(\alpha_a + \sum_r \beta_r p_{jka} d_{ir} + \gamma w_{jc} + \sum_r X^{1'}_{jka} \Theta_r d_{ir} + X^{2'}_{jk} \Delta \\ + Q^{'}_{iw(t)} \Omega + \mathrm{T}_{as(t)} + \xi_{jk} + \xi_i) \end{bmatrix}}{\begin{bmatrix} 1 + \Sigma_{a \in \{c,s\}} exp(\alpha_a + \sum_r \beta_r p_{jka} d_{ir} + \gamma w_{jc} + \sum_r X^{1'}_{jka} \Theta_r d_{ir} \\ + X^{2'}_{jk} \Delta + Q^{'}_{iw(t)} \Omega + \mathrm{T}_{as(t)} + \xi_{jk} + \xi_i). \end{bmatrix}} \tag{3}$$

Our goal is to estimate the unknown scalars (β_r, γ) and vectors $(\Theta_r, \Delta, \Omega)$ in the model.

4 Estimation

4.1 Endogeneity

As in many discrete choice demand estimation settings, some of the observed product attributes, such as price, are often correlated with unobserved product characteristics such as quality and, hence, are endogenously determined. Specifically, a firm that tries to optimize profits or growth adjusts prices for products and services based on features that are observable to itself but not to the researcher. In our setting, routes may be priced by the platform managers based on their popularity. If the firm raises prices on the popular routes to optimize profits, without taking this into account in the estimation, the price coefficient obtained from the model will be underestimated. However, if the firm cuts prices on popular routes to grow the customer base, the price coefficient obtained from the model would be overestimated. In either case, using the biased estimate would lead to unreasonable prescriptive policy recommendations and managerial insights in the counterfactual analyses. Interestingly, in our setting, the shuttle service was in a phase of growth and expansion, whereas, by comparison, the cab service was in a more mature phase. This makes for an interesting setting in which to study the price endogeneity problem. Prices of the two services are not the only endogenous variables in our setting. Wait times for cabs are correlated with the unobserved popularity or congestion level of the route and are determined endogenously. When wait times increase with unobserved popularity or congestion, the coefficient for wait time would be underestimated if we did not take into account the endogeneity problem in the estimation. Moreover, past shuttle usage is correlated with unobserved customer-level characteristics. These unobservables lead to an endogenous selection of users into the low- and high-usage groups. Without considering this endogeneity issue, the coefficient for past usage will be overestimated in magnitude.

4.2 Instruments

Our approach to correct for the biases discussed in the previous section is to find valid instruments for the endogenous variables. We construct instrumental variables for the prices and cab wait time by utilizing the network structure of our data. Then, we select from a large set of valid instruments, the best set, following recent developments in the intersection of the causal inference and machine learning literatures. Second, we construct the instrumental variable for past shuttle usage by utilizing the timing of the introduction of shuttle routes.

Network Instruments The first step here is to define a set of network-based instruments following He et al. (2019). To explain the variation in prices for route jk, we look for exogenous characteristics of h that affect the popularity of j and k. We construct instrumental variables separately for origin j and destination k. Our proposed valid instruments are averages of the exogenous characteristics of all such feasible h that satisfy the relevance and the exclusion restriction criteria.

Lasso Instrument Selection Since the set of valid network instruments is large (1236), selecting the right ones is not trivial. Instead of handpicking some instruments or naively selecting the compete set, we use machine selection methods. We follow the method in Belloni et al. (2012) to select the best set of instruments in our setting. The method involves using a penalty function that penalizes the addition of more instruments to avoid a weak first stage in the IV estimation. The penalization is achieved through the square root Lasso estimator.

Shuttle Route Introduction Based Instrument for Shuttle Usage The instrument should provide an exogenous variation in customer i's past shuttle usage to recover the causal parameter of interest. We construct the instrument by utilizing the timing of the introduction of different shuttle routes in our data. Specifically, to explain the variation in customers' shuttle usage, we utilize the time of introduction of a route od_i attached to the customer i. Let T_{od_i} be the timing of the introduction of route od_i in the sample period and D_{od_i} the number of days from Jan 1 till T_{od_i}. Then, $\mathcal{Z}_i = 244 - D_{od_i}$ is a valid instrument for shuttle usage (Fig. 1).

The rationale for using the above instrument is as follows. The timing of opening up of shuttle operations on customer i's home route gives an exogenous shock to her shuttle usage. Thus, the length of time that the route od_i is active in the sample period affects customer i's shuttle usage. This is the relevance condition for the instrument. Also, the timing of the introduction of shuttle operations on a route is unlikely to be correlated with the unobserved customer characteristics ξ_i conditional on the market characteristics. This gives us the exclusion restriction condition.

4.3 Control Function Approach to Estimation

We use a control function approach with the instruments described in Sect. 4.2 to correct for the endogeneity in prices, wait time and past shuttle usage variables, following the method in Petrin and Train (2010). The control functions for the prices

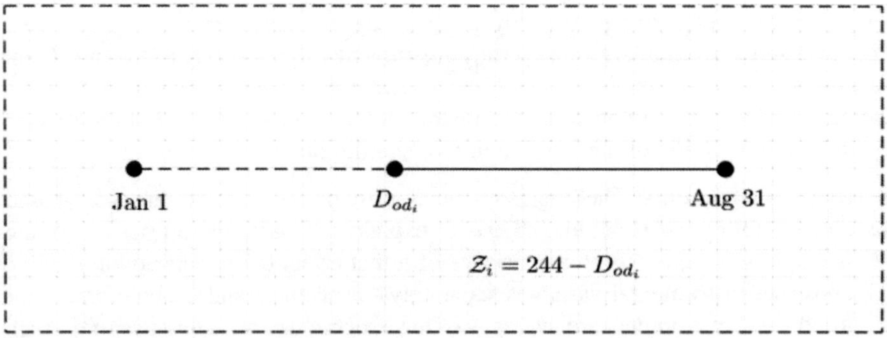

Fig. 1 Instrument for past shuttle usage of a customer. D_{od_i} represents the number of days from the start of the sample period to the introduction of home route od_i. The length of the solid line in days is the magnitude of the instrument

and wait time are specified below in the system of equations:

$$\begin{aligned} p_{jks} &= \alpha_{sp} + \kappa_{sp} Z_{sp}^{LASSO} + X_{jks}^{1'} \Lambda_{sp} + X_{jk}^{2'} \Phi_{sp} + \nu_{jks}^{p} \\ p_{jkc} &= \alpha_{sc} + \kappa_{cp} Z_{cp}^{LASSO} + X_{jkc}^{1'} \Lambda_{sp} + X_{jk}^{2'} \Phi_{cp} + \nu_{jkc}^{p} \\ w_{jc} &= \alpha_{wc} + \kappa_{cw} Z_{cw}^{LASSO} + X_{jkc}^{1'} \Lambda_{sp} + X_{jk}^{2'} \Phi_{cw} + \mu_{jc}^{w}. \end{aligned} \tag{4}$$

where, p_{jks}, p_{jkc} are average route-level prices for shuttle and cabs, and w_{jc} is the average cab wait time. $(Z_{sp}^{LASSO}, Z_{cp}^{LASSO}, Z_{cw}^{LASSO})$ are the Lasso selected sets of instrumental variables for the endogenous variables. X_{jk}^{2} is the same vector of exogenous market characteristics used in Eq. (1). Similarly, X_{jks}^{1} and X_{jkc}^{1} are the vectors of exogenous market-alternative level controls used in Eq. (1). $(\Lambda_{sp}, \Lambda_{cp}, \Lambda_{cw})$ and $(\Phi_{sp}, \Phi_{cp}, \Phi_{cw})$ are the associated coefficients. Using the control functions ν_{jkc}^{p}, ν_{jkc}^{p}, μ_{jc}^{p} and μ_{i}^{us} obtained from Eq. (4), the customer utility can be written as:

$$\begin{aligned} U_{ijkta} &= \alpha_a + \sum_r \beta_r p_{jka} d_{ir} + \gamma w_{jc} + \sum_r X_{jka}^{1'} \Theta_r d_{ir} + X_{jk}^{2'} \Delta + Q_{iw(t)}^{'} \Omega \\ &\quad + \mathrm{T}_{as(t)} + \lambda_1 \nu_{jkc}^{p} + \lambda_2 \nu_{jkc}^{p} + \lambda_3 \mu_{jc}^{p} + \lambda_4 \mu_i^{us} + \epsilon_{ijkta}. \end{aligned} \tag{5}$$

where λ_1, λ_2, λ_3 and λ_4 are the coefficients for the control functions. Then the customer's choice probability can be calculated as :

$$\begin{aligned} &P_{ijkta}(\beta_r, \gamma, \Theta_r, \Delta, \Omega, \lambda_1, \lambda_2, \lambda_3, \lambda_4) \\ &= \frac{exp(V_{ijkta} + \lambda_1 \nu_{jkc}^{p} + \lambda_2 \nu_{jkc}^{p} + \lambda_3 \mu_{jc}^{p} + \lambda_4 \mu_i^{us})}{1 + \Sigma_{a \in \{c,s\}} exp(V_{ijkta} + \lambda_1 \nu_{jkc}^{p} + \lambda_2 \nu_{jkc}^{p} + \lambda_3 \mu_{jc}^{p} + \lambda_4 \mu_i^{us}).} \end{aligned} \tag{6}$$

We recover the parameters $(\beta_r, \gamma, \Theta_r, \Delta, \Omega, \lambda_1, \lambda_2, \lambda_3, \lambda_4)$ by maximum likelihood estimation. The log likelihood is computed over all the choices observed in the data.

$$(\hat{\beta}_r, \hat{\gamma}, \hat{\Theta}_r, \hat{\Delta}, \hat{\Omega}, \hat{\lambda}_1, \hat{\lambda}_2, \hat{\lambda}_3, \hat{\lambda}_4) = \arg\max \mathcal{L}(((\beta_r, \gamma, \Theta_r, \Delta, \Omega, \lambda_1, \lambda_2, \lambda_3, \lambda_4)). \tag{7}$$

4.4 Results from the Choice Model

The estimated parameters from the choice model (second stage) are presented in Table 1. First, we note that the price coefficients β_r in the specification without using the IVs are −0.017 and −0.013 for groups 1 and 2, respectively. After using the control function approach, we recover β_r = −0.014 and −0.009. Hence, the

Table 1 Parameter estimates from the choice model. .s and .c are shuttle- and cab-specific coefficients. .1 and .2 are coefficients for the two customer usage groups (*,**,*** indicates statistical significance at 10%,5%,1% level)

Explanatory variable	Without IV	With IV
Shuttle intercept	−13.922*** (0.082)	−13.942*** (0.082)
Cab intercept	−12.887*** (0.080)	−13.032*** (0.080)
Price Paid.1	−0.017*** (0.0001)	−0.014*** (0.0001)
Price Paid.2	−0.013*** (0.0001)	−0.009*** (0.0001)
Wait	−0.129*** (0.001)	−0.135*** (0.001)
Time	−0.002*** (0.0001)	−0.004*** (0.0001)
Commute.1	−1.516*** (0.006)	−1.536*** (0.006)
Commute.2	−0.830*** (0.006)	−0.807*** (0.006)
Distance.1	−0.192*** (0.002)	−0.161*** (0.002)
Distance.2	−0.106*** (0.003)	−0.066*** (0.003)
Controlfunction shuttle price		−0.026*** (0.0002)
Controlfunction cab price		0.002*** (0.0002)
Controlfunction cab wait		0.069*** (0.005)
Morning.s	3.857*** (0.009)	3.879*** (0.009)
Morning.c	−13.754*** (0.082)	−13.766*** (0.081)
Cumulative Shuttle Usage.s	0.091*** (0.0002)	0.087*** (0.0002)
Controlfunction Usage.s		0.004*** (0.0001)
Cumulative Shuttle Usage.c	−3.923 (163.312)	−4.051 (165.234)
Controlfunction Usage.c		0.602 (1.321)
Recent Week Rides.s	0.948*** (0.002)	0.955*** (0.002)
Recent Week Rides.c	0.691*** (0.003)	0.680*** (0.003)
Market controls	Yes	Yes
Observations	1,323,413	1,323,413
McFadden R^2	0.580	0.585

price coefficients are adjusted down in magnitude—i.e., the price coefficient is overestimated without using the IVs. The control function for the shuttle price is significant and negative. The strong negative control function for the shuttle price indicates that the average price of the shuttle in a market is lower than what the observed market characteristics can explain. The control function for cab price is positive and much weaker in magnitude than that for the shuttle price. The positive sign indicates that the average price of cabs on a route is higher than as explained by the observed attributes. This is consistent with our discussion on the endogeneity of price in Sect. 4.1. Moreover, we find that there is substantial heterogeneity in the price coefficients across the two groups. The new users (group 1) are about 1.55 times more price-sensitive than experienced users (group 2). The coefficient for wait time γ in the model without IV is -0.129. The control function for cab wait time corrects for this bias expectedly. The corrected coefficient is -0.135. The coefficients for walking distance to the shuttle are negative and significant for both the user groups. The coefficient on walking distance captures the disutility incurred by the customer in walking to the shuttle pickup stop. Group 1 users are more sensitive to walking than group 2 users. The results allow us to estimate the monetary cost associated with the disutility of walking to the pickup stop. We estimate a disutility of ₹109 ($1.45) and ₹89 ($1.18), respectively, for walking 1 km to the shuttle stop to catch the shuttle for the two groups.

In addition to the operational levers, the customer's usage variables enter our model specification. First, the number of cumulative shuttle rides taken by a customer has a positive and significant effect on the probability of choosing the shuttle. The shuttle-specific coefficient of cumulative shuttle rides by a customer is 0.091 in the specification without IV. This shows that it is important to control for the usage variables when explaining the customer's choice. In the model with correction, we recover an estimate of 0.087. In addition to cumulative usage, the total number of rides on the platform in the recent week positively influences the probability of choosing both the shuttles and the cabs. The fixed effects for morning time slots are also significant. Our base category for the time slots is afternoon hours. As compared to the afternoon slot, people are more likely to choose shuttles and less likely to choose cabs in the morning. Finally, the pseudo R^2 for both specifications is around 0.58, which suggests that our rich model fits the data quite well.

5 Counterfactuals

In this section, we use our estimated model to conduct counterfactual analyses and provide prescriptive guidance to policy makers on congestion management in big cities. The question that we seek to answer is how to effectively increase the usage of pooled ride services and reduce the level of congestion on the roads. Our counterfactuals evaluate the impact of different policy interventions on customers' choices between private and pooled ride services and, therefore, on the number of vehicles on the road. Specifically, we examine three sets of strategies: (i) imposing congestion surcharges on private cabs and shuttles; (ii) providing discounts on shuttle service; and (iii) improving the service features of shuttle service. We call the first two sets of strategies the *price-based strategies* and the latter *operations-based strategies*. From a policy maker's point of view, it is challenging to evaluate the effectiveness of a strategy before implementing it. For example, it is very difficult to know a priori the right level of congestion surcharge to levy on the vehicles. This is where the strength of our model lies. Our estimated model allows us to measure customers' service choices when prices or service features are changed. Hence, using the estimated model, we can evaluate the relative efficacy of the different policies before implementation and, thus, provide the policy maker with a host of prescriptive solutions.

5.1 Applying Percentage Surcharges to a Congestion Zone

Local governments in large cities around the world, such as London, Singapore, and New York, have introduced congestion pricing policies to reduce congestion. In New York, for example, a surcharge is applied to all ride-hailing trips in a pre-determined congestion zone in Manhattan. In this counterfactual, we quantify the impact of imposing percentage congestion surcharges on cabs and shuttles on the number of vehicles on the road. Specifically, for all rides that cross the congestion zone—i.e., $j \in$ Zone 1 or $k \in$ Zone 1, irrespective of the time of the day, we calculate the customers' choices, the number of vehicles on the road, and the platform's revenue under policy $\mathscr{P}'$. We vary the level of the surcharge by applying different price multipliers, $(1 + \theta_c)$ and $(1 + \theta_s)$, to p_c and p_s for cab and shuttle services, respectively. Here, θ_c and θ_s are the percentage price surcharges. The corresponding θ_s for the shuttle ride is determined by the relationship:

$$\mathcal{m} p_c \theta_c = n_s p_s \theta_s. \tag{8}$$

where $\mathcal{m}$ is a multiplier and n_s is the number of seats on the shuttle. Specifically, in our simulations, we set $\mathcal{m}$ to 2, based on proposed recommendations in New York.[1] The number of seats in the shuttle, n_s, is set to 20, equal to the median number.

Figure 2 shows the effect of applying percentage congestion surcharges to shuttles and cabs. The x-axis is $100 \times (\theta_c)$. The status quo policy $\mathscr{P}_{Zone1}(p_c, p_s, X_s^1)$ is on the extreme left ($\theta_c = 0$). The *red line* in Fig. 2 shows the net percentage reduction in the number of total vehicles on the road calculated relative to the status quo level. The number of vehicles at any point is calculated as: # cab rides + $\frac{\text{\# shuttle rides}}{20}$. For this calculation, we assume that the outside option is public transportation which does not affect the total number of vehicles on the road. Although we do not observe the outside option of the customers, this calculation

[1] https://www.nytimes.com/2019/04/24/nyregion/what-is-congestion-pricing.html.

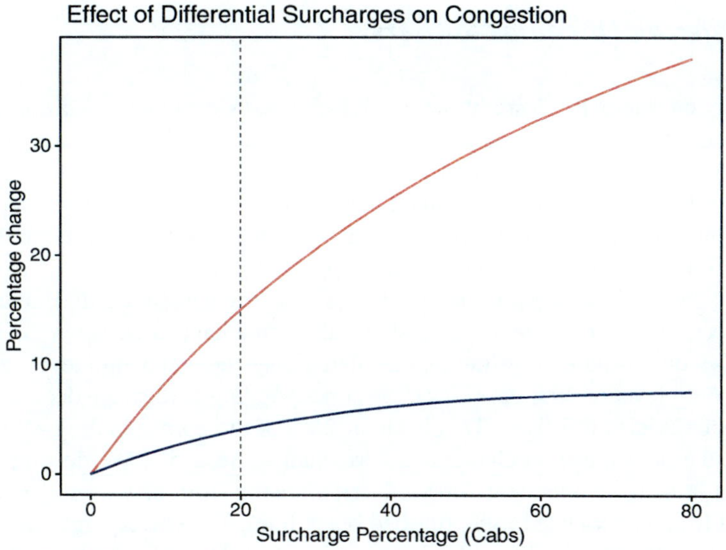

Fig. 2 Percentage reduction in total number of vehicles (red line) and percentage reduction due to pure substitution (blue line) after application of percentage surcharges. Vertical dashed line corresponds to $\theta_c = 0.2$

provides an upper bound to the magnitude of the impact of the surcharge policy on congestion reduction. We also quantify the reduction in vehicles due to cab customers substituting to the shuttle service only. The *blue* line shows the net percentage reduction due to the pure substitution between the cabs and the shuttles. For instance, a 20% surcharge on the cabs ($\theta_c = 0.2$) leads to a 15% net vehicle reduction on the road. The percentage vehicle reduction due to the pure substitution effect is around 4.04%. In this calculation, since we focus only on customers substituting from the cab service to the shuttle service without taking into account potential substitutions to the outside option which includes public transportation and other ride services, and that substituting to the outside option is unlikely to increase the number of vehicles on the road given the surcharge raises the price of riding in (or driving) smaller vehicles more than that of the big ones, the blue line provides a lower bound to the impact of the surcharge policy on congestion reduction.

We also disentangle the total reduction (blue line) into two components based on the customer segments that it arises from: (i) *new users* and (ii) *experienced users*. For $\theta_c = 0.2$, the effect from new users (solid black line) is close to 80% of the total congestion reduction. This suggests that customer heterogeneity is crucial in designing an effective congestion policy.

5.2 Offering Discounts to Users

In this counterfactual, we study the service choice of customers and quantify the decrease in the number of vehicles due to the substitution between the services—i.e., the lower bound of the impact on congestion reduction when discounts are offered to shuttle customers. Specifically, we change the shuttle price by applying a discount multiplier $\mathcal{B}$ while keeping the cab price the same as observed in the data. We vary $\mathcal{B}$ in increments of 0.05 over the support of [0.60,1].

Figure 3 shows the substitution effect after applying percentage discounts on the shuttle service. The *red line* in Fig. 3 shows the percentage decrease in the number of vehicles due to pure substitution, calculated over the status quo level ($\mathcal{B} = 0$). At a discount level of 20% ($\mathcal{B} = 0.7$), the corresponding percentage decrease in the number of vehicles is 1.04%. This is about 26% of the corresponding substitution effect in the percentage surcharge counterfactual in Sect. 5.1. We decompose this reduction into two components: the reduction arising from new users (solid black line) and from experienced users (dashed black line). The corresponding reduction from the two groups is 0.7% and 0.3% of the total number of vehicles on the road, respectively. Since the number of new and experienced users is the same, our finding suggests that targeting the new users when providing discounts is more than twice as effective as targeting the experienced users.

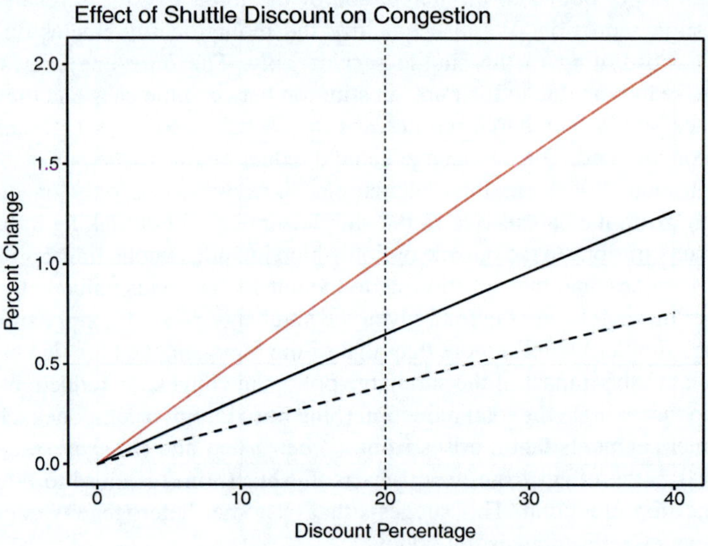

Fig. 3 Percentage reduction in the number of vehicles due to the pure substitution between the cabs and the shuttles (red line). The solid (dashed) black line is the contribution from the new (experienced) user group. Vertical dashed line corresponds to the discount percentage level of 20%

5.3 Improving Service Features

Using price-based strategies is an effective way to curb congestion on the road, but it comes with many drawbacks, such as deadweight loss to overall welfare due to its tax nature. An interesting alternative to achieve the same outcome is to use operations-based strategies, which have two advantages. First, they do not hurt customer welfare. Second, the strategies come "free" for the firm. The firm does not lose any surcharge revenue by employing these strategies. In this counterfactual, we apply multipliers $\mathcal{Q}$ to the service features and estimate the corresponding percentage reduction in the number of vehicles. We vary the multiplier in increments of 0.05 over the support of [0.65,0.95]. We do this exercise separately for the walking distance and the shuttle travel time features. As in the surcharge counterfactuals, we calculate both the total change in vehicles and the change in vehicles due to the substitution between the two services.

We report the corresponding effects in Table 2. We find that a 20% reduction in walking distance and travel time for shuttle rides leads to around a 1.46% and a 0.28% decrease in the number of vehicles on road due to the substitution between the two services. We can compare this decrease with the corresponding substitution in the price-based strategies. Considering the 20% surcharge scenario as the base case (see, Fig. 2), a 20% decrease in walking inconvenience can achieve around 35% of the total substitution achieved by the congestion surcharge. Similarly, a 20% decrease in the shuttle travel time can achieve 6.9% of the total substitution achieved by the congestion surcharge. In other words, a city could reap a significant portion of the benefits of congestion surcharges by utilizing the operational levers of the pooled ride service. The benefits could quickly *stack up* when the improvements in the various inconveniences are combined. Specifically, a 20% reduction in both shuttle ride time and walking inconvenience leads to a cumulative 1.51% vehicle reduction due to substitution (around 37.3% of the reduction achieved by the congestion surcharges in Sect. 5.1). From the policy maker's perspective, we find that improving the pooled ride service features proved to be an effective strategy to reduce congestion, while avoiding the drawbacks of the surcharge policies.

Table 2 Comparison of percentage reductions (total number of vehicles and substitution effect) when shuttle : travel time (left) and walking distances (right) are reduced

	Shuttle travel time inconvenience		Shuttle walking inconvenience	
Multiplier	Vehicle reduction	Substitution	Vehicle reduction	Substitution
0.95	0.05%	0.07%	0.07%	0.37%
0.90	0.10%	0.14%	0.15%	0.74%
0.85	0.15%	0.22%	0.25%	1.11%
0.80	0.21%	0.28%	0.36%	1.46%
0.75	0.26%	0.35%	0.48%	1.80%
0.70	0.31%	0.42%	0.62%	2.15%
0.65	0.36%	0.49%	0.79%	2.48%

6 Conclusion

In this paper, we study customer preferences of private and pooled transportation services and investigate how effective policies can be designed to incentivize customers to use pooled transportation to reduce congestion. Using detailed customer usage data from Ola's on-demand cab and fixed-route shuttle services in India, we estimate customer preferences of key service features using a discrete choice model. We account for the endogeneity of the service features, such as price and wait time, and of customers' past shuttle usage on the platform using the control function approach. We then conduct counterfactual analyses to evaluate the impact of congestion surcharge policies, discount policies, and improved pooled service features on the customers' choices and, therefore, the number of vehicles on the road. We find that, by changing operations levers such as pooled service features, instead of imposing a surcharge policy, cities can reduce a substantial amount of congestion without sacrificing consumer welfare. We also highlight the role of customer heterogeneity in improving the effectiveness of policy design. Our findings provide prescriptive recommendations to cities for designing effective policies for congestion management.

References

Ata, B., Barjesteh, N., & Kumar, S. (2019). Spatial pricing: An empirical analysis of taxi rides in neorking york city. Tech. rep., Working Paper.

Belloni, A., Chernozhukov, V., & Hansen, C. (2011). Lasso methods for gaussian instrumental variables models. arXiv:1012.1297v2

Belloni, A., Chen, D., Chernozhukov, V., & Hansen, C. (2012). Sparse models and methods for optimal instruments with an application to eminent domain. *Econometrica, 80*(6), 2369–2429.

Buchholz, N. (2020). Spatial equilibrium, search frictions and dynamic efficiency in the taxi industry. Working paper, Princeton.

Cohen, M., Fiszer, M.-D., & Kim, B.J. (2020a). Frustration-based promotions: Field experiments in ride-sharing. Working paper, SSRN.

Cohen, M., Fiszer, M.-D., Ratzon, A., & Sasson, R. (2020b). Incentivizing commuters to carpool: A large field experiment with waze. Working paper, SSRN.

Guajardo, J. A., Cohen, M. A., Kim, S.-H., & Netessine, S. (2012). Impact of performance-based contracting on product reliability: An empirical analysis. *Management Science, 58*(5), 961–979.

Han, L., Zhu, C., Wang, D.Z.W., Sun, H., Tan, Z., & Meng, M. (2019). Discrete-time dynamic road congestion pricing under stochastic user optimal principle. *Transportation Research Part E: Logistics and Transportation Review, 131*, 24–36.

He, P., Zheng, F., Belavina, E., Girotra, K. (2019). Customer preference and station network in the London bike share system. Working paper, Columbia Business School.

Kabra, A., Belavina, E., & Girotra, K. (2019). Bike-share systems: Accessibility and availability. *Management Science, 66*(9), 3799–4358.

Ostrovsky, M., & Schwarz, M. (2018). Carpooling and the economics of self-driving cars. Working Paper 24349, National Bureau of Economic Research.

Petrin, A., & Train, K. (2010). A control function approach to endogeneity in consumer choice models. *Journal of Marketing Research, 47*(1), 3–13.

Optimal Dispatch in Emergency Service System via Reinforcement Learning

Cheng Hua and Tauhid Zaman

1 Introduction

In the United States, medical responses by fire departments over the last four decades have increased by 367%, as reported by Evarts (2019). The reasons for the dramatic increase in medical calls include the aging population and health insurance that covers most of the ambulance costs (Boston Globe, Nov 29, 2015). Cities with tight budgets are short of response units to respond to the growing amount of medical calls in time. NBC10 in Philadelphia, on Feb 28, 2019, reported that two-thirds of the emergency medical calls had an ambulance response time of more than nine minutes. Thus, how to efficiently use the existing resources becomes an essential topic to decision-makers in emergency response departments.

In current practice, most cities dispatch ambulance units according to a fixed dispatch rule, which always dispatches the closest available unit. The dispatch policy is deemed a myopic policy as it only considers the current call and ignores the impact of dispatching a unit to future calls. In this paper, we model the ambulance dispatch problem as an average-cost Markov decision process (MDP). We propose an alternative MDP formulation for the problem using post-decision states that we show is mathematically equivalent to the original MDP formulation, but with a much smaller state space. Due to the curse of dimensionality, the applications of the formulations are restricted to small problems. To solve larger problems, we use temporal difference learning (TD-learning) with the post-decision states. We show

C. Hua (✉)
Antai College of Economics and Management, Shanghai Jiaotong University, Shanghai, China
e-mail: cheng.hua@sjtu.edu.cn

T. Zaman
School of Management, Yale University, New Haven, CT, USA
e-mail: tauhid.zaman@yale.edu

H. Yang et al. (eds.), *AI and Analytics for Public Health*, Springer Proceedings in Business and Economics, https://doi.org/10.1007/978-3-030-75166-1_3

that the TD-learning algorithm converges quickly, and the policies obtained from our method outperform the myopic policy.

The remainder of this paper is organized as follows. In Sect. 2, we provide a review of the relevant literature. In Sect. 3, we present the Markov decision process formulation. Section 4 presents the formulation using post-decision states which reduces the state space of the original formulation. In Sect. 5, we present the temporal difference learning algorithm that is based on the post-decision states and its theoretical properties, while in Sect. 6, we show the performance of the algorithm in numerical experiments. We conclude the paper in Sect. 7.

2 Literature Review

The optimal dispatch rule was first studied in Carter et al. (1972). The authors studied a simple system of two units. They provided a closed-form solution that determines the response areas to be served by each unit to minimize average response time. However, such a closed-form solution no longer exists in a system with more than two units and finding the optimal dispatch rule has been an important topic.

Jagtenberg et al. (2017a) studied whether dispatching the closest available unit is optimal in a dynamic ambulance dispatching problem based on a Markov decision process. The problem was discretized by time using one minute as the time interval. Two objectives were considered: mean response time and the fraction of calls with response time beyond a certain time threshold. The value iteration method was used to find the optimal solution. Jagtenberg et al. (2017b) provide an empirical bound for the gap between the existing solutions and the optimal solution.

Schmid (2012) followed the same formulation as introduced in Powell (2010) that uses approximate dynamic programming (ADP) with aggregation function to an ambulance relocation and dispatching problem to reduce the mean response time. A seminal paper by Maxwell et al. (2010) applied ADP to the ambulance redeployment problem, where they used Monte Carlo simulation with one-step bootstrap to estimate complex expectations and applied least squares regression with linear function approximation to learn the approximate value functions. Nasrollahzadeh et al. (2018) studied the problem of real-time ambulance dispatching and relocation, which is formulated as a stochastic dynamic program and solved using ADP. Their formulation and method are the same as proposed in Maxwell et al. (2010). The issues of this approach are that while most of the time they beat the benchmark policy, they usually never output the optimal policy. Also, it is not guaranteed that the learning method always converges, and finding useful basis functions for approximation is more of an art than science, which requires domain knowledge and testing.

Our paper is the first to model the emergency dispatch problem as an average-cost MDP, whose objective is more appropriate than discounted sum. We also show that the proposed TD-learning algorithm based on post-decision states is guaranteed to converge to the optimal solution.

3 Markov Decision Process Formulation

Consider a geographical region $R \subset \mathbb{R}^2$ that is served by a total of N emergency units, all of the same type, e.g., ambulance units. Calls arrive in region R according to a Poisson point process with an arrival intensity $\{\Lambda(x, y) : (x, y) \in R\}$ at location with coordinate (x, y). We partition the region R into J sub-regions and associate a center of mass with each sub-region $R_j \subset R$, which is also referred to as a demand *node*. Note that J can be as large as needed. Denote the total call rate in node j as $\lambda_j = \int_{R_j} \Lambda(x, y)dxdy$. The overall call rate in region R is denoted by $\lambda = \sum_j \lambda_j = \int_R \Lambda(x, y)dxdy$. We assume the mean service time follows a negative exponential distribution with rate μ_i for unit i. We assume that the service time includes turnout time, travel time, and time at the scene. The justification for this assumption is that travel time is usually a small portion of the total service time. With longer travel times, Jarvis (1975) mentioned a *mean service time calibration* to calibrate the service time to maintain the Markov property. Define t_{ij} as the average response time from the base of unit i to node j.

Let b_i represent the state of unit i, where $b_i = 0$ if the unit is available and $b_i = 1$ if the unit is busy. We denote the state space of all units as an N-dimensional vector $B = \{b_N \cdots b_1\} \in \mathscr{B}$, which is in a backward order similar to the representation of binary numbers. We define $B = b_i$ as the status of unit i. Note that $|\mathscr{B}| = 2^N$. If all units are busy when a call is received, we assume that it is handled by a unit outside of the system, such as a private ambulance company, or a unit from a neighboring jurisdiction, which is common mutual aid policy. This paper aims to find the optimal dispatch policy that minimizes the average response time of all served calls.

3.1 State Space

Define S as the state space S and $s \in S$ as the state of the system. We have $s = (j, B)$, which is a tuple of j and B that consists of the location of the current call and the state of all units (available or busy) at the time of the arrival. We denote $s(0) = j$ and $s(1) = B$ in state s. The entire state space has size $|S| = J \times 2^N$.

3.2 Action Space

When a call is received in the system, we decide on which unit to be dispatched to this call. An action in this problem is to dispatch a particular unit upon receiving a call, so the action space is given as $A = \{1, 2, \cdots, N\}$. Note that only an available unit may be dispatched. We define $A_s \subset A$ as the set of feasible actions at state s, where $A_s = \{i : B(i) = 0, i = \{1, 2, \cdots, N\}\}$. We define $a \in A_s$ as an action from the feasible action space.

3.3 Policy Space

We define the policy space as Π, the set of all feasible policies. A policy $\pi \in \Pi$ is a mapping from the state space to the action space, $\pi : S \rightarrow A$. Specifically, $\pi(s) = a, a \in A_s$. The optimal policy $\pi^* \in \Pi$ is the policy that minimizes the average cost over all time steps. Our goal is to find this optimal policy. We use a benchmark policy which sends the closest available unit, denoted by $\pi^m \in \Pi$. We have $\pi^m(s) = \arg\min_i t_{ij}, \forall i \in A_s, \forall s \in S$. Sending the closest available unit is myopic as it does not consider potential future calls. Saving the closest unit to the current call might greatly reduce the expected response time of a future call.

3.4 Costs

Define $c^\pi(s)$ as the cost of being in state s following policy π, which equals to the response time $c^\pi(s) = t_{ij}$ when the call location in state s is j, i.e. $s(0) = j$, and the policy dispatches unit i in state s, i.e. $\pi(s) = i$.

3.5 Transition Probabilities with Augmented Transitions

Define $p^\pi(s, s')$ as the transition probability from state $s = (j, B)$ to state $s' = (j', B')$ under policy π. In determining the transition state probability, we consider an augmented transition where a unit completes a service, and no dispatch action is needed. This is because the number of services completed between two arrivals is a random variable whose probability is complicated to compute. Introducing the augmented transition reduces the number of transition possibilities. Denote I_i as the vector of all 0's except for the ith element, which is 1. The transition rate $p^\pi(s, s')$ with augmented transition is given as

$$p^{\pi}(s,s') = \begin{cases} \frac{\lambda_{j'}}{\lambda+\sum_{k:B(k)=1}\mu_k+\mu_i}, & \text{if } s' = (j', B + I_i), \\ \frac{\mu_l}{\lambda+\sum_{k:B(k)=1}\mu_k+\mu_i}, & \text{if } s' = (\emptyset, B + I_i - I_l). \end{cases} \tag{1}$$

where the expression on the top corresponds to the transition from state s to state s' upon action $\pi(s) = i$ is taken and a new call arrives in node j', and the bottom expression corresponds to the augmented transition where no arrival occurs but a busy unit $l \in A/A_s$ completes its current service. No action is needed since there are no arriving calls in this transition.

3.6 Bellman's Equation

Define $V^{\pi} : S^{+} \mapsto \mathbb{R}$ as the value function for the MDP following policy π and the value of state s is $V^{\pi}(s)$, where S^{+} is the augmented state space that has dimension $|S^{+}| = (J+1)2^{N}$. Let μ^{π} be the average cost following policy π. The Bellman's equation for the average cost is

$$V^{\pi}(s) = c^{\pi}(s) - \frac{\mu^{\pi}}{2} + \sum_{s' \in S^{+}} p^{\pi}(s,s')V^{\pi}(s'), \ \forall s \in S^{+}. \tag{2}$$

Note that the 1/2 in the above equation is due to the existence of augmented transitions. A transition that is due to a service completion has zero cost and the number of service completions is always equal to the number of calls being served.

Define V^{π} as the vector of all state values, c^{π} as the vector of all state costs, P^{π} as the transition matrix, and e as the vector of all ones. The vector form of Bellman's equation is

$$V^{\pi} = c^{\pi} - \frac{\mu^{\pi} e}{2} + P^{\pi} V^{\pi}. \tag{3}$$

The solution to the above Bellman's equation is not unique. Instead, the set of all value functions takes the form $\{V^{\pi} + me \mid m \in \mathbb{R}\}$. Since shifting the value function by a constant value does not change the relative differences between state values, once we obtain a set of state values V^{π}, the policy can be updated as

$$\pi'(s) = \arg\min_{a \in A_s} t_{aj} + \sum_{s' \in S^{+}} p(s,s'|a)V^{\pi}(s'), \tag{4}$$

where $p^{\pi}(s,s'|a)$ is the one-step transition probability when taking action a instead of following the policy $\pi(s)$. This so-called policy iteration method is summarized in Algorithm 1.

Algorithm 1 Policy iteration method

```
Pick a random policy π_0. Set k = 0.
while π_k ≠ π_{k+1} do
    Compute the cost c^{π_k} and transition matrix P^{π_k}.
    Policy Evaluation: Solve the state values V^{π_k} from the Bellman's equation (3).
    Policy Improvement: For each state s ∈ S, update the actions of each state by
```

$$\pi_{k+1}(s) = \arg\min_{a \in A_s} t_{aj} + \sum_{s' \in S^+} p(s, s'|a) V^{\pi_k}(s').$$

```
    k = k + 1
end while
Output: Optimal Policy π* = π_k
```

4 Post-Decision State Formulation

The policy iteration method guarantees the convergence to the optimal dispatch policy that minimizes the average response time, requiring solving a linear system with $(J+1)2^N$ states repeatedly. In the section, by realizing the nature of the state transitions of the dispatch problem, we introduce the notion of post-decision states and use them as the new states in our problem. We show that the MDP formulation using post-decision states reduces the state space to 2^N, which also guarantees finding the optimal dispatch policy.

In the original formulation, a state is a tuple of call locations and all units' statuses $s = (j, B)$. A post-decision state s_x is a state that the system is in immediately after observing state s and taking action $a = \pi(s)$, before the next random information arrives into the system, which is the arrival of the next call location. Thus, given state s and unit i being dispatched, i.e., $a = \pi(s) = i$, the post-decision state s_x is $s_x = B + I_i$.

Note that by defining the post-decision state this way, we only need information about the statuses of all units. Define S_x as the post-decision state space; we have $|S_x| = 2^N$. Indeed, $S_x = \mathscr{B}$.

Lemma 1 *Let $p_x^\pi(s_x, s_x')$ be the corresponding transition probability from s_x to s_x'. We have*

$$p_x^\pi(s_x, s_x') = \begin{cases} \dfrac{\sum_{j \in \mathscr{R}_{l|s_x}^\pi} \lambda_j}{\lambda + \sum_{k:B(k)=1} \mu_k + \mu_i}, & \text{if } s_x' = s_x + I_l, \\ \dfrac{\mu_l}{\lambda + \sum_{k:B(k)=1} \mu_k + \mu_i}, & \text{if } s_x' = s_x - I_l, \end{cases} \tag{5}$$

where $\mathscr{R}_{l|s_x}^\pi$ is the set of demand nodes where policy π dispatches unit l, i.e.,

$$\mathscr{R}_{l|s_x}^\pi = \{j : \pi(s') = l, s' = (j, s_x)\}. \tag{6}$$

Proof For $s'_x = B + I_i - I_l$, where a transition happens when unit l completes its current service, the post-decision state transition is the same as the transition from s to the augmented state $s' = (\emptyset, B + I_i - I_l)$ as no call arrives and no action is needed for this state; thus $s' = s'_x$.

For $s'_x = B + I_i + I_l$, where a call arrives in post-decision state s_x, we need to capture the randomness of exogenous information, which is the location of call that arrives in s_x. We thus have

$$p_x^\pi(s_x, s'_x) = \sum_j p^\pi(s, s' = (j, s_x))\mathbb{1}_{\{\pi(s')=l\}} = \sum_j p^\pi(s, s')\mathbb{1}_{\{s'=(j,s_x)\}}\mathbb{1}_{\{\pi(s')=l\}}$$

$$= \sum_{j\in\mathscr{R}^\pi_{l|s_x}} p^\pi(s, s' = (j, s_x)) = \frac{\sum_{j\in\mathscr{R}^\pi_{l|s_x}} \lambda_j}{\lambda + \sum_{k:B(k)=1}\mu_k + \mu_i}.$$

□

Lemma 2 *The cost of post-decision state is* $c^\pi(s_x) = \frac{\sum_{l\in A_s}\sum_{j\in\mathscr{R}^\pi_{l|s_x}} \lambda_j t_{lj}}{\lambda+\sum_{k:B(k)=1}\mu_k+\mu_i}$.

Proof The cost of s_x is the expected one-step transition cost from s_x to s'_x under policy π. Let $c_l^\pi(s_x)$ be the expected cost of dispatching unit l in s_x. We have $c_l^\pi(s_x) = \frac{\sum_{j\in\mathscr{R}^\pi_{l|s_x}} \lambda_j t_{lj}}{\sum_{j\in\mathscr{R}^\pi_{l|s_x}} \lambda_j}$ if $l \in A_s$, and 0 otherwise. We thus have

$$c^\pi(s_x) = \sum_{l\in A_s} c_l^\pi(s_x)p_x^\pi(s_x, s_x + I_l) = \sum_{l\in A_s} \frac{\sum_{j\in\mathscr{R}^\pi_{l|s_x}} \lambda_j t_{lj}}{\sum_{j\in\mathscr{R}^\pi_{l|s_x}} \lambda_j}\frac{\sum_{j\in\mathscr{R}^\pi_{l|s_x}} \lambda_j}{\lambda + \sum_{k:B(k)=1}\mu_k + \mu_i}$$

$$= \frac{\sum_{l\in A_s}\sum_{j\in\mathscr{R}^\pi_{l|s_x}} \lambda_j t_{lj}}{\lambda + \sum_{k:B(k)=1}\mu_k + \mu_i}.$$

□

Note that all components defining $c^\pi(s_x)$ and $p_x^\pi(s_x, s'_x)$ are known, which are computed beforehand. Define $J^\pi : S_x \mapsto \mathbb{R}$ as the value function for the post-decision state space S_x. Let μ_x^π be the average cost following policy π in the post-decision state space. The Bellman's equation is

$$J^\pi(s_x) = c^\pi(s_x) - \frac{\mu_x^\pi}{2} + \sum_{s'_x\in S_x} p_x^\pi(s_x, s'_x)J^\pi(s'_x), \ \forall s_x \in S_x. \quad (7)$$

Let J^π, c_x^π and P_x^π be the corresponding vector representations. The vector form Bellman's equation around post-decision states is

$$J^\pi = c_x^\pi - \frac{\mu_x^\pi e}{2} + P_x^\pi J^\pi. \quad (8)$$

Algorithm 2 Policy iteration with post-decision states

1: Pick a random policy π_0. Set $k = 0$.
2: **while** $\pi_k \neq \pi_{k+1}$ **do**
3: Compute the cost $c_x^{\pi_k}$ and transition matrix $P_x^{\pi_k}$.
4: **Policy Evaluation:** Solve the state values J^{π_k} from the Bellman's equation (8).
5: **Policy Improvement:** For each state $s \in S$, update the actions of each state by

$$\pi_{k+1}(s) = \arg\min_{a \in A_s} t_{aj} + J^{\pi_k}(s_x = B + I_a).$$

6: $k = k + 1$
7: **end while**
8: **Output:** Optimal Policy $\pi^* = \pi_k$

Theorem 1 *The MDP formulation around post-decision states is equivalent to the original formulation. In particular*

(i) $\mu_x^\pi = \mu^\pi$;
(ii) For $s_x = B$, *let* $\Gamma = \lambda + \sum_{k:B(k)=1} \mu_k$, $s^{(j)} = (j, B)$ *and* $s^{[k]} = (\emptyset, B - I_k)$. *We have*

$$J^\pi(s_x) = \sum_j \frac{\lambda_j}{\Gamma} V^\pi\left(s^{(j)}\right) + \sum_{k:B(k)=1} \frac{\mu_k}{\Gamma} V^\pi\left(s^{[k]}\right). \tag{9}$$

The proof of the above theorem is obtained from expanding the value functions V^π in (9) by (2) and collecting terms. The details of the proof is shown in the full version of the paper online. Under this formulation, the new policy π' for state $s = (j, B)$ is updated as $\pi'(s) = \arg\min_{a \in A_s} t_{aj} + J^\pi(s_x = B + I_a)$. The new policy iteration around post-decision states is summarized in Algorithm 2.

5 Temporal Difference Learning with Post-Decision States

Let $\phi_{[p]} : S_x \mapsto \mathbb{R}$, $p = 1, 2 \cdots, P$, be the basis functions of post-decision states, and let $r = \{r_{[p]} : p = 1, 2 \cdots, P\}$ be the tunable parameters. The value function approximation is given by $\tilde{J}(s_x, r) = \sum_{p=1}^{P} r_{[p]}\phi_{[p]}(s_x)$. Let $\tilde{J}(r)$ be the vector of approximate state values of all states given parameter vector r and let Φ be an $2^N \times P$ matrix whose pth column is equal to the vector $\phi_{[p]}$ of all states in S^x. The vector form of the above equation is $\tilde{J}(r) = \Phi r$. Define $\{x_t \mid t = 0, 1, \ldots\}$ as the Markov chain on the post-decision state space S_x with transition matrix P_x^π.

Lemma 3 *The Markov chain corresponding to the state space* S_x *and transition matrix* P_x^π *is irreducible and has a unique stationary distribution.*

The proof of the above lemma is straightforward by noting that the Markov chain with post-decision state space forms a hypercube loss model whose property can be found in Larson (1974).

We define the temporal difference by $d_t = c(x_t) - \frac{\mu_t}{2} + \tilde{J}(x_{t+1}, r_t) - \tilde{J}(x_t, r_t)$, where $c(x_t) - \frac{\mu_t}{2} + \tilde{J}(x_{t+1}, r_t)$ is the differential cost function at state x_t based on the one-step bootstrap and $\tilde{J}(x_t, r_t)$ is the old approximate differential cost function at state x_t. Define r_t as the parameter vector at time t. We update the parameter vector r by $r_{t+1} = r_t + \gamma_t d_t \phi(x_t)$ and $\mu_{t+1} = (1 - \gamma_t)\mu_t + 2\gamma_t c(x_t)$. We let $\gamma_t = \frac{a}{a+t}$ where $a \geq 1$ is a hyper-parameter that controls the learning speed.

In this paper, we present a simple way of defining the basis function. We give an index i_x to each post-decision state $s_x = B$, where $i_x = \sum_i^N 2^i \mathbb{1}_{B(i)=1}$. We let the basis function $\phi_{[p]}(s_x)$ be

$$\phi_{[p]}(s_x) = \begin{cases} 1, & \text{if } p = i_x, \\ 0, & \text{if } p \neq i_x. \end{cases} \tag{10}$$

Algorithm 3 Policy iteration with TD-learning

1: Pick a random policy π_0. Set $k = 0$. Specify T and K.
2: **while** $k \leq K$ **do**
3: Set $t = 0$. Initialize r_0 and μ_0.
4: Starting from a random state x_0, generate a state trajectory $\{x_t \mid t = 0, 1, \ldots T\}$ corresponding to the Markov chain with state transition probability $P_x^{\pi_k}$ that is defined by the policy π_k.
5: **for** $t = 0$ to T **do**
6: Calculate the temporal difference d_t by

$$d_t = c(x_t) - \frac{\mu_t}{2} + \tilde{J}(x_{t+1}, r_t) - \tilde{J}(x_t, r_t).$$

7: Update the parameters by

$$r_{t+1} = r_t + \gamma_t d_t \phi(x_t),$$
$$\mu_{t+1} = (1 - \gamma_t)\mu_t + 2\gamma_t c(x_t).$$

8: **end for**
9: For each state $s \in S$, update the policy

$$\pi_{k+1}(s) = \arg\min_{a \in A_s} t_{aj} + \tilde{J}(s_x = B + I_a, r_T).$$

10: $k = k + 1$
11: **end while**
12: **Output:** Policy $\tilde{\pi}^* = \pi_K$

Theorem 2 *Algorithm 3 has the following three properties:*

(i) Converges with probability 1.

(ii) The limit of the sequence $\frac{\mu_t}{2}$ *at the kth iteration of the algorithm is the average cost* $\frac{\mu_x^{\pi_k}}{2}$*, i.e.,* $\lim_{t\to\infty}\mu_t = \mu_x^{\pi_k}$.

(iii) The limit of the sequence r_t *at the kth iteration of the algorithm, denoted by* r^{k*}*, is the unique solution of the equation* $T\left(\Phi r^{k*}\right) = \Phi r^{k*}$*, where* $T : \mathbb{R}^{2^N} \mapsto \mathbb{R}^{2^N}$ *is an operator defined by* $TJ = c_x^{\pi_k} - \frac{\mu_x^{\pi_k} e}{2} + P_x^{\pi_k} J$.

The proof of the above theorem follows from Tsitsiklis and Van Roy (1999) and Lemma 3, and the basis functions $\phi(s_x)$ being linearly independent for all states. It is also necessary that γ_t is positive, deterministic, and satisfies $\sum_{t=0}^{\infty}\gamma_t = \infty$ and $\sum_{t=0}^{\infty}\gamma_t^2 < \infty$. The details of the proof is shown in the full version of the paper. Theorem 2 guarantees that Algorithm 3 always returns the optimal policy if T is large enough. When T is moderately large, it is enough to obtain a policy close to optimal, as shown in the next section. Our algorithm avoids solving the complex Bellman's equation, which has exponential complexity. Once we calculate the cost vector c_x^{π} and transition matrix P_x^{π}, we easily obtain the temporal differences d_t by Monte Carlo simulation and evaluate the value functions that are needed for policy improvement in the policy iteration algorithm.

6 Numerical Results

In this section, we show the numerical results comparing the policy obtained from the TD-Learning method to the myopic policy that always dispatches the closest available unit for systems with $N = 5, 10$, and 15 units. We created an imaginary region which is partitioned into $J = 30$ demand nodes. We randomly locate units in the region and obtain the corresponding response times from each unit to each demand point. The policy from the proposed TD-Learning method with post-decision states is obtained by running the algorithm in 25 iterations. We perform a roll-out with 200,000 state transitions in each iteration and update the parameter vector r using the temporal differences d. We record the sample average response time in each iteration and the results are shown in Figs. 1a, b, and c, respectively.

We observe that the TD-Learning algorithm converges quickly in all cases as expected. We show the updates of the post-decision state values $\tilde{J}$ in one TD-Learning iteration for the case of $N = 5$ in Fig. 1d. The resulted policies in all three cases outperform the myopic policy that always dispatches the closest available units. We also observe that our algorithm obtains a superior policy reasonably quickly in about three iterations. In the case where $N = 15$, solve the Bellman's equation requires solving a system with $31 \times 2^{15} = 1{,}015{,}808$ states. In contrast, our method obtains a good policy in less than two minutes, and it applies virtually to systems of any sizes as guaranteed by its theoretical properties.

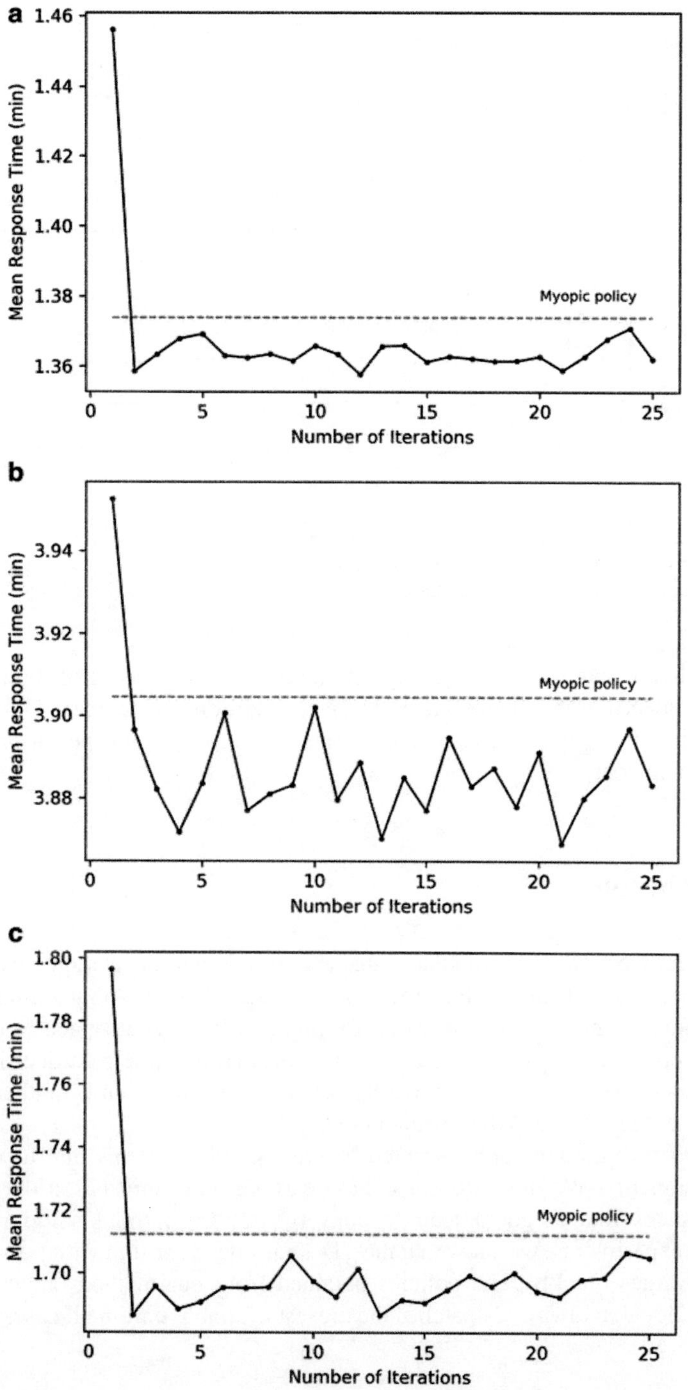

Fig. 1 (continued)

d

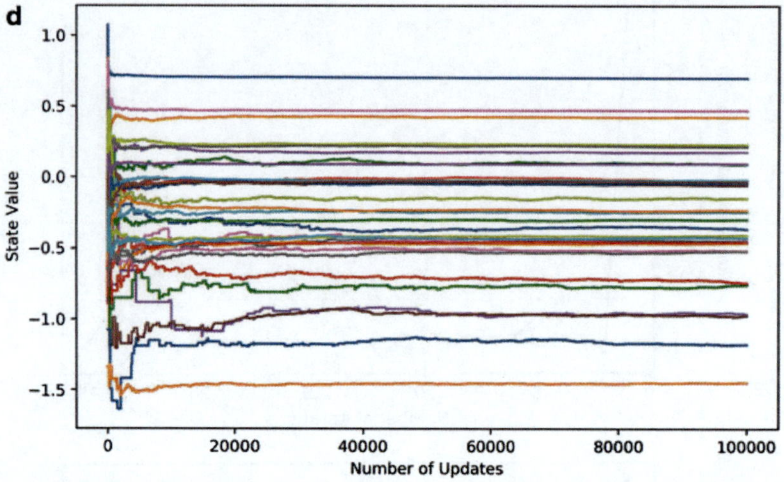

Fig. 1 Numerical results. (**a**) Mean response time comparison for $N = 5$ units. (**b**) Mean response time comparison for $N = 10$ units. (**c**) Mean response time comparison for $N = 15$ units. (**d**) State value updates in one TD-Learning iteration

The policies obtained from our algorithm result in an average of three seconds reduction in terms of response time with no additional resources. Our findings suggest that emergency response departments can improve their performance with minimal to no cost.

7 Conclusion

In this paper, we model the ambulance dispatch problem as an average-cost Markov decision process and aim to find the optimal dispatch policy that minimizes the mean response time. The regular MDP formulation has a state space of $(J + 1) \cdot 2^N$. We propose an alternative MDP formulation that uses the post-decision states and reduces the state space to 2^N. We show that this formulation is mathematically equivalent to the original MDP formulation.

The two formulations are restricted to only small problems due to the curse of dimensionality. We next present a TD-Learning algorithm based on the post-decision states that is guaranteed to converge to the optimal solution. In our numerical experiments, we show that the TD-Learning algorithm with post-decision states converges quickly. The policies obtained from our method outperform the myopic policy that always dispatches the closest available unit in all cases.

References

Carter, G. M., Chaiken, J. M., & Ignall, E. (1972). Response areas for two emergency units. *Operations Research, 20*(3), 571–594.

Evarts, B. (2019). Fire loss in the united states during 2018. *NFPA National Fire Protection Association, Quincy*.

Jagtenberg, C. J., Bhulai, S., & van der Mei, R. D. (2017a). Dynamic ambulance dispatching: is the closest-idle policy always optimal? *Healthcare Management Science, 20*(4), 517–531.

Jagtenberg, C. J., van den Berg, P. L., & van der Mei, R. D. (2017b). Benchmarking online dispatch algorithms for emergency medical services. *European Journal of Operational Research, 258*(2), 715–725.

Jarvis, J. P. (1975). *Optimization in stochastic service systems with distinguishable servers*. Ph.D. thesis, Massachusetts Institute of Technology.

Larson, R. C. (1974). A hypercube queuing model for facility location and redistricting in urban emergency services. *Computers & Operations Research, 1*(1), 67–95.

Maxwell, M.S., Restrepo, M., Henderson, S. G., & Topaloglu, H. (2010). Approximate dynamic programming for ambulance redeployment. *INFORMS Journal on Computing, 22*(2), 266–281.

Nasrollahzadeh, A. A., Khademi, A., & Mayorga, M. E. (2018). Real-time ambulance dispatching and relocation. *Manufacturing & Service Operations Management, 20*(3), 467–480.

Powell, W. B. (2010). Merging AI and OR to solve high-dimensional stochastic optimization problems using approximate dynamic programming. *INFORMS Journal on Computing, 22*(1), 2–17.

Schmid, V. (2012). Solving the dynamic ambulance relocation and dispatching problem using approximate dynamic programming. *European Journal of Operational Research, 219*(3), 611–621.

Tsitsiklis, J. N., & Van Roy, B. (1999). Average cost temporal-difference learning. *Automatica, 35*(11), 1799–1808.

Towards Understanding the Dynamics of COVID-19: An Approach Based on Polynomial Regression with Adaptive Sliding Windows

Yuxuan Xiu and Wai Kin (Victor) Chan

1 Introduction

Nowadays, COVID-19 is a major threat that all mankind needs to face. Up till now, some countries are approaching the end of the outbreak, thanks to vigorous responses to the epidemic. In other countries, however, the outbreak has not yet been brought under control, or there have been second waves of outbreak. In the current situation, the study of the dynamics of COVID-19 is of great importance. On the one hand, we can compare the impact of different control measures of the epidemic horizontally between different countries. On the other hand, we can refer to the dynamics of the epidemic in different countries to help predict possible future developments in countries where the epidemic has not yet ended.

Intuitively, the dynamics of each wave of COVID-19 can be classified into several stages, such as the early stage, the raising stage and the fading stage. However, the current criteria for partitioning COVID-19 time series are still subjective. Researchers lack quantitative standards to distinguish one stage from another. In other words, a mathematical definition of different stages is still needed. This paper aims at solving this problem by using polynomial regression with adaptive sliding windows as a mathematical standard to partition the COVID-19 time series into different stages.

Existing work on the dynamic analysis of COVID-19 can be broadly classified into the following four categories: epidemic models-based (e.g., SIR, SEIR) methods (Wangping et al., 2020; Calafiore et al., 2020), regression and autoregressive models (Ceylan, 2020; Singh et al., 2020), machine learning-based methods

Y. Xiu · W. K. (Victor) Chan (✉)
Shenzhen Environmental Science and New Energy Technology Engineering Laboratory, Tsinghua-Berkeley Shenzhen Institute, Tsinghua Shenzhen International Graduate School, Tsinghua University, Shenzhen, Guangdong, People's Republic of China
e-mail: chanw@sz.tsinghua.edu.cn

H. Yang et al. (eds.), *AI and Analytics for Public Health*, Springer Proceedings in Business and Economics, https://doi.org/10.1007/978-3-030-75166-1_4

(Kapoor et al., 2020), and hybrid models of the above three approaches (Yang et al., 2020). Methods based on epidemic models are well interpretable, that is, their parameters have practical meaning. However, the time-dependent parameters of the epidemic models are very difficult to estimate, due to the complexity of people's responses to COVID-19. Although multiple epidemic models with time-varying parameters have been proposed (Chen et al., 2020; Waqas et al., 2020), the determination of these time-varying parameters is still a difficult problem. Most epidemic models such as SIR and SEIR can not adaptively adjust their parameters. Therefore, when the control measures of COVID-19 change significantly, it is needed to assess and quantify the effect of these measures on the model parameters, which makes prediction rather difficult. Machine learning-based methods, on the other hand, can fit the data very well. In particular, machine learning-based methods can fuse and exploit data of multiple types and sources (e.g., population movement data) (Kapoor et al., 2020), in addition to the time series itself. But they are less interpretable, making it difficult to characterize the dynamics of the COVID-19 development. As a result, the usefulness of these methods is limited to forecasting. They can not quantify the impact of different responses on the development of the epidemic.

Regression and autoregressive models provide a good balance between interpretability and prediction accuracy. Thus, they are widely used in the study of COVID-19 time series. For example, Ceylan (2020) predicted total confirmed cases in Italy, Spain and France based on ARIMA model using data up to April 15, 2020. The optimal order of the ARIMA model is determined for each country. Similarly, Singh et al. (2020) exploited an advanced autoregressive integrated moving average (ARIMA) model to predict future trends in the 15 most affected countries at the time, forecasting confirmed cases, deaths, and recoveries in the following 2 months based on data up to April 24. Pandeya et al. (Gupta et al., 2020) compared the effectiveness of the SEIR model and polynomial regression for predicting the epidemic in India. Their results showed that the SEIR model had better prediction accuracy than polynomial regression.

These existing regression methods have a similar defect, that is, they all use the entire time series to fit the model. However, with the development of the epidemic, diverse responses are adopted, thus leading to changes in the inherent dynamics of the COVID-19 time series. Therefore, a more efficient approach would be to segment the COVID-19 time series according to different dynamic patterns and then fit a separate model for each segment.

In this paper, we adopt an approach based on polynomial regression with adaptive sliding windows to segment the COVID-19 time series. Such method has been proved to be able to efficiently extract the dynamic patterns of time series (Liu et al., 2020). This method uses the sliding windows with adaptive lengths to partition time series into segments. The dynamic pattern of each sliding window is defined as a fitted n-order polynomial. For a sliding window of length L, we consider the data within the sliding window to have the same dynamic pattern if the residual of the n-order polynomial regression is less than a predetermined threshold. We consecutively increase the length of the sliding window until the polynomial is insufficient to fit the newly added data. Subsequently, we set a new sliding window and repeat the above steps. Experimental results show that this method is able to adaptively segment the COVID-19 time series of different countries into segments that are highly intuitively interpretable.

Further, we analyze the segmentation results and define the similarity between segments using the dynamic time wrapping (DTW) distance (Berndt & Clifford, 1994). We visualize the similarity based on complex network analysis method. It can be observed that each wave of the epidemic outbreak can be broadly partitioned into three stages: the early outbreak, the rising stage, and the falling stage. Besides, significant similarities exist between the same stages of different waves and in different countries. In addition, some evidences suggest that the dynamics of the previous segment could provide useful information of the subsequent development of the epidemic.

2 Methodology

2.1 Data Sources

This paper studies the dynamics of COVID-19 through the time series of daily active cases. Our data is collected from https://www.worldometers.info/coronavirus/ at country level, from January 22, 2020 to September 2, 2020. In order to horizontally compare the development of the epidemic across countries, we normalize the number of the daily active cases by the total population of each country, which is collected from https://github.com/datasets/population.

Based on the above data, we first extract the dynamic patterns of different stages of COVID-19 in each country, then compare the dynamic patterns of 15 representative countries. These two steps are described as follows.

2.2 Extracting Dynamic Patterns of COVID-19 Time Series

For extracting the dynamic patterns, we adopt a similar approach to that proposed by Liu et al. (2020), whose original purpose is to study the self-similarity of the fluctuation behaviors of the WTI crude oil price. In this paper, we use a sliding window with adaptive length to partition the entire the time series of COVID-19 daily active cases into segments, fitting each segment with an n-order polynomial as shown in Eq. (1). In this paper, we choose $n = 3$.

$$y = \alpha_0 + \alpha_1 x + \alpha_2 x^2 + \cdots + \alpha_n x^n + \varepsilon \tag{1}$$

Since the time series within each sliding window is fitted using an n-order polynomial, the initial length of the sliding window is set to $n + 1$. Subsequently, we use Eqs. (2) and (3) as metrics for determining the length of the sliding window, where y_i is the real value of the number of COVID-19 active cases at the i-th day of the sliding window, and $\hat{y}_i$ is the corresponding predicted value by the model. The average number of the daily active cases inside the sliding window is expressed as $\overline{y}$. If both R^2 and $E_{\max}$ are greater than a predetermined threshold T, we increase the length of the window L by 1 day, that is, we move the front edge of the sliding window 1 day forward. We fit the data within the new sliding window and check whether the new fit satisfies $R^2 \geq 0.9$ and $E_{max} \geq 0.9$. We iteratively increase the length of the sliding window forward and perform polynomial regression until either R^2 or E_{max} is less than T. In this paper, we choose $T = 0.9$.

$$R^2 = 1 - \frac{\sum_{i=1}^{L} (y_i - \hat{y}_i)}{\sum_{i=1}^{L} (y_i - \overline{y})} \tag{2}$$

$$E_{\max} = 1 - \max\left(\frac{|y_i - \hat{y}_i|}{|y_i|}\right) \tag{3}$$

When the R^2 or E_{max} of a sliding window becomes less than T, we fix this sliding window to its current state and create a new sliding window. The new sliding window is created at the front edge of the old one, with the same initial length of $n + 1$. The same operations are performed consecutively until the time series is completely partitioned.

For each partition, its dynamic is described by a vector composed of coefficients of its polynomial regression, which is denoted as $[\alpha_1, \alpha_2, \cdots, \alpha_n]$. Notice that the intercept of the polynomial is not considered, since we are only interested in the variation of the time series in each partition, not in the specific values.

2.3 *Dealing with Structural Breakpoints of the Dynamics*

In this paper, we use *structural breakpoints* to refer to the data points that divide the time series into two fragments with different dynamics. In the method introduced above, structural breakpoints of the dynamics naturally appear at the ends of segments, and they participate in the polynomial regression. However, this operation is inappropriate, since the occurrence of a structural breakpoint implies that the dynamic pattern has changed. The structural breakpoint should be the beginning point of the next segment, rather than the end point of the previous segment. At the

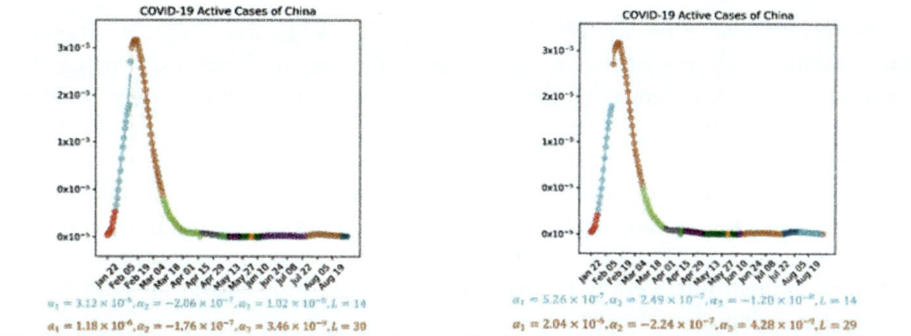

(a) Original partition and dynamics of the time series (b) Partition and dynamics after dealing with breakpoints

Fig. 1 An illustrative example for dealing with structural breakpoints. (**a**) Original partition and dynamics of the time series. (**b**) Partition and dynamics after dealing with breakpoints

same time, the structural breakpoint should not be involved in the regression of the previous segment. The reason is explained as follows.

The regression of n-order polynomials can be approximately regarded as an n-order Taylor expansion. Our decision on whether to create a new sliding window depends on whether the n-order Taylor expansion is sufficient to describe the dynamics within the current sliding window. In other words, we use the two criteria (i.e., $R^2 \geq 0.9$ and $E_{max} \geq 0.9$) to determine whether the n-order polynomial regression is a good fit to the data in the current sliding window. If not, a new sliding window will be created. Therefore, if the structural breakpoints are assigned into the previous sliding window, then dynamic pattern within that sliding window cannot be expressed by any n-order Taylor expansion. To ensure that the data within the current window can be well described by the n-order polynomial regression, structural breakpoints need to be assigned to the latter segment. In the rest part of this subsection, we demonstrate this idea through the case study of China.

In the time series analysis of COVID-19, the reason of the occurrence of a breakpoint, in addition to changes in COVID-19 dynamics, may also be an increase in detection capability or a change in statistical methods. Figure 1a illustrates this situation by a case study on China. In the CVOID-19 time series of China, there was an obvious jump up on February 12, dividing the already-occurring downward trend into two segments. This jump up was due to the fact that Hubei changed the statistical method of COVID-19, including clinically diagnosed cases as active cases. The overall downward trend of daily activity cases in China did not change. However, this outlier significantly affects the results of the polynomial regression of the previous segment, which can affect our judgment on the dynamics of the epidemic. More importantly, improvements in the detection capability and changes in the statistical methods can affect all the subsequent data after the structural breakpoint. In fact, these are two reasons that trigger changes in the dynamics of the COVID-19 time series. Therefore, assigning the structural breakpoints to the latter segments is a more proper operation.

In the regression shown in Fig. 1a, there is no special treatment on the structural breakpoints. From the second segment which is marked in blue, it can be clearly observed that the presence of the structural breakpoint in this sliding window results in an unsatisfactory regression result. In fact, the regression result is an approximately linear growth. It does not really reflect the dynamics of the time series within the window, which is actually in a fading stage. Considering that changes in detection capacity or statistical methods can affect both the breakpoints and their subsequent data, we assign breakpoints into the latter segments. Regression results obtained through this method is demonstrated in Fig. 1b, showing that we better extract the dynamics of the time series.

2.4 *Measuring the Similarity Among Segments*

After dividing the COVID-19 time series of each country into segments according to different dynamics, we further measure the similarity among these segments. Liu et al. (2020) measure the similarity between segments based on the Euclidean distance between the coefficient vectors of each segment's polynomial regression, as is shown in Eq. (4).

$$S_{ij} = \sqrt{\sum_{t=1}^{n} \left[\alpha_t^{(i)} - \alpha_t^{(j)}\right]^2} \tag{4}$$

where $\left[\alpha_1^{(i)}, \alpha_2^{(i)}, \cdots, \alpha_n^{(i)}\right]$ is the coefficient vectors of the polynomial regression of Segment i. Notice that the intercept is ignored.

However, we argue that this method may not be suitable for measuring the similarity between the segments of the COVID-19 time series. This is because there is a huge variation in the percentage of daily active cases per population among different countries. Therefore, we first normalized the time series of each country with its peak value, then use the dynamic time warping (DTW) distance (Berndt & Clifford, 1994) between each pair of segments as the similarity measure. The experimental results show that the DTW distance can effectively identify the similarity between the dynamics of each segment, thus matching the segments of the same stages of the outbreak in different countries.

3 Results

3.1 Partitioning COVID-19 Time Series

The trends of the COVID-19 epidemic can be broadly classified into three categories: rising, falling and equilibrium. Rising and falling mean that the number of active cases is significantly increasing or decreasing, while equilibrium means that the number of active cases does not change significantly. The equilibrium state includes the beginning and the end of the epidemic, as well as the dynamic equilibrium in the middle of the epidemic when the number of new infections and the number of cured cases are approximately equal.

In this paper, the COVID-19 time series (i.e., the percentage of active cases per population of each day) is partitioned into segments with different dynamics. Figure 1 shows the results of the partitioning of the time series for a representative set of 15 countries, where each fragment is labeled with a number. Overall, it can be observed that the COVID-19 time series are partitioned into a number of segments of different lengths. Especially, in the equilibrium state of the epidemic, the time series are partitioned into segments of short lengths, whereas the rising and falling trends are partitioned into long segments.

This result matches very well with our intuition. In both the rising and the falling state, the method adopted in this paper is able to nicely extract the corresponding trend. On the other hand, in the equilibrium state, when the number of active cases fluctuates, it is not possible to discern from the time series itself whether this is a random fluctuation (e.g., controllable imported cases) or a signal of the change in dynamics that indicates an upward or downward trend in the future. The approach adopted in this paper treats the data in the equilibrium state with caution, regarding each of the significant fluctuations as a signal of the change in dynamics. This results in the equilibrium state being split into small segments (Fig. 2).

Moreover, in many countries, a complete wave of an outbreak can be partitioned into four major segments: the beginning stage, the ascending stage, the descending stage, and the end stage. We can observe that the lengths of the beginning stages and end stages exceed the lengths of the short fragments during the equilibrium state, but are significantly smaller than the lengths of the ascending and descending stages. In addition, note that the ascending and descending stages are not segmented from the peak point, rather, these two stages are segmented from a time point before the time series reaches the peak. This is because we focus on the change in the dynamics of COVID-19, which is essentially caused by the epidemic being under control. In other words, the number of new daily infections gradually decreases and the number of cured cases gradually increases after the time point where the dynamics change, which leads to the number of daily active cases reaching the peak after a period of time.

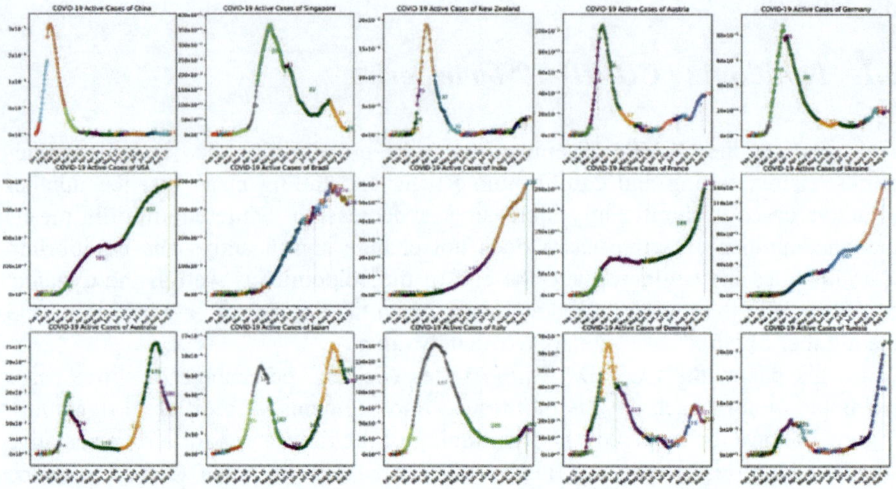

Fig. 2 Partitioning the COVID-19 time series according to different dynamics

3.2 Analyzing the Dynamics of the First Wave of the Outbreak

For the countries in Fig. 1, the development of the COVID-19 epidemic can be broadly classified into three categories: (1) the countries with only one peak where the epidemic is almost over, (2) the countries where the first wave of the epidemic is not yet under control, (3) the countries with two peaks where there is a second outbreak. We selected China, the United States, and Australia as representatives of each of the three categories. In this subsection, we compare the dynamics of the first wave of the COVID-19 outbreak in these three countries.

Table 1 selects some representative segments and shows the coefficients of the polynomial regression of these segments. Segment 0, Segment 100, and Segment 174 represent the dynamics of the initial stage of the outbreak in China, the USA and Australia, respectively. We can see that the polynomial coefficients of theses three segments are roughly proportional, suggesting that the same dynamics are present among the three countries at the beginning of the epidemic outbreak. However, the subsequent dynamics of the outbreak become very different. For China, the polynomial regression coefficients in Segment 1 already reflect the decreasing trend in the number of active cases, but the jump up on February 12 disrupts the dynamics we identified. For the U.S., the number of active cases in Segment 101 first decreases and then increases, indicating the occurrence of a second outbreak before the first wave of the epidemic ends. Segment 102 appears to be entering a downward trend, but the future is still unclear. For Australia, Segment 145 is similar to the Segment 1 of China, but the declining speed of Segment 176 is significantly slower than that of China's Segment 3, which may suggest the risk of a second outbreak.

Similar phenomenon can be observed from many other countries such as Japan, Italy and Denmark, where a slower declining speed of the first wave is followed

Table 1 Representative segments of the first wave of the outbreak in China, the U.S. and Australia

Country	Index of Segment	$[\alpha_1, \alpha_2, \alpha_3]$
China	0	$[7.27\times 10^{-8}, 5.87\times 10^{-8}, 4.03\times 10^{-9}]$
	1	$[5.26\times 10^{-7}, 2.49\times 10^{-7}, -1.20\times 10^{-8}]$
	2	$[2.04\times 10^{-6}, -2.24\times 10^{-7}, 4.28\times 10^{-9}]$
	3	$[-1.07\times 10^{-6}, 4.92\times 10^{-8}, -9.36\times 10^{-10}]$
USA	100	$[5.92\times 10^{-6}, 2.20\times 10^{-6}, 1.18\times 10^{-8}]$
	101	$[1.11\times 10^{-4}, -1.61\times 10^{-6}, 8.60\times 10^{-9}]$
	102	$[9.10\times 10^{-5}, -1.56\times 10^{-7}, -4.92\times 10^{-9}]$
Australia	174	$[5.86\times 10^{-7}, 1.25\times 10^{-7}, 1.24\times 10^{-8}]$
	175	$[9.25\times 10^{-6}, 1.03\times 10^{-6}, -6.91\times 10^{-8}]$
	176	$[-4.78\times 10^{-6}, 2.81\times 10^{-7}, -1.39\times 10^{-8}]$

by a second outbreak. By comparing the dynamics of the first wave of the COVID-19 outbreak across different countries, we propose an initial assumption that the dynamics at the end of the first wave may partially reflect the future development of the epidemic. A slower decline at the end of the first wave may suggest the risk of a second outbreak in the future.

3.3 Comparing the Dynamics of the First and Second Outbreak

In this subsection, we analyze a number of other representative segments. The additional segments are shown in Table 2. The main aim is to compare the similarities in epidemic dynamics across countries, as well as the similarities between the first and second outbreak.

Based on the method described in Sect. 2.4, we measure the similarity between the segments listed in Tables 1 and 2. We further normalize each column of the DTW matrix by the following Eq. (5).

$$S'_{ij} = \frac{S_{ij} - \frac{1}{n}\sum_{i=1}^{n} S_{ij}}{\frac{1}{n}\sum_{i=1}^{n}\left(S_{ij} - \frac{1}{n}\sum_{i=1}^{n} S_{ij}\right)} \tag{5}$$

Figure 3a demonstrates the matrix of DTW distance among those segments. We select the 15% percentile of all elements in the matrix as the threshold t. We drop the DTW distance greater than t to obtain the similarity network shown in Fig. 3b.

We further study the similarity network shown in Fig. 3b based on some complex network analysis methods. In this network, each node represents a corresponding segment and the edge (i,j) represents Segment i and Segment j are similar. The

Table 2 Additional representative segments of the first and second wave of the outbreak

Country	Index of Segment	$[\alpha_1, \alpha_2, \alpha_3]$
Australia: 2nd outbreak	178	$[-3.69\times 10^{-7}, 5.61\times 10^{-8}, 2.45\times 10^{-9}]$
	179	$[8.30\times 10^{-6}, 5.15\times 10^{-7}, -2.06\times 10^{-8}]$
	180	$[-1.52\times 10^{-5}, 1.04\times 10^{-6}, -3.75\times 10^{-8}]$
Japan: 1st outbreak	184	$[4.44\times 10^{-7}, -2.35\times 10^{-8}, 7.22\times 10^{-10}]$
	185	$[-5.04\times 10^{-7}, 2.48\times 10^{-7}, -4.40\times 10^{-9}]$
	186	$[5.10\times 10^{-6}, -2.13\times 10^{-7}, 1.59\times 10^{-9}]$
	187	$[-3.01\times 10^{-6}, -4.23\times 10^{-8}, 3.56\times 10^{-9}]$
Japan: 2nd outbreak	189	$[7.45\times 10^{-8}, 1.64\times 10^{-8}, 4.98\times 10^{-10}]$
	190	$[4.92\times 10^{-6}, 2.57\times 10^{-8}, -5.06\times 10^{-9}]$
	191	$[-2.50\times 10^{-6}, -1.58\times 10^{-7}, 1.66\times 10^{-8}]$

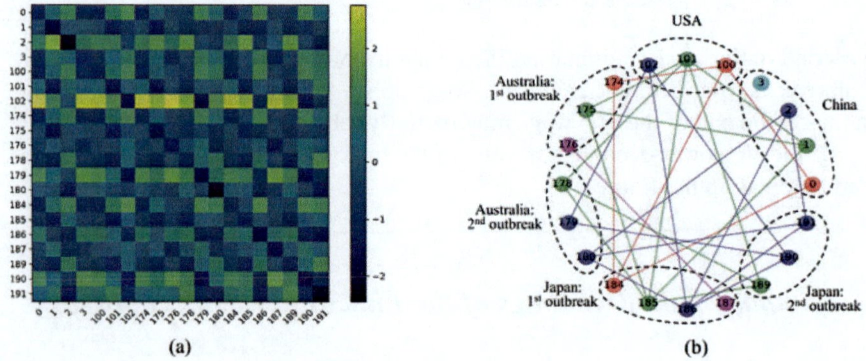

Fig. 3 Measuring the similarity among the representative segments. (**a**) The normalized DTW matrix. (**b**) The similarity network of the representative segments

similarity network is composed of six connected components. In Fig. 3b, the connectivity components are marked by different colors, with nodes belonging to the same connected component represented by the same color. After ignoring the outliers 3 and 199, it can be observed that the four major components (i.e., labeled red, blue, brown, and gray) represent roughly four different types of dynamics. The first three components represent the early outbreak, the raising, and the falling, respectively. Meanwhile, the segments that belong to the gray component are followed by second outbreaks.

Based on the above, some preliminary ideas can be suggested. Among the time series of COVID-19 daily active cases, some similarities can be observed in the dynamical structure. First, each wave of the outbreak can be roughly partitioned in to three stages: the early outbreak, the ascending stage, and the descending stage. In particular, the descending stage is characterized by the containment of the epidemic rather than a decrease in the number of daily active cases. Second, if we partition COVID-19 time series in to segments according to different dynamics, significant similarities can also be observed between the same stages in different countries and different waves of outbreak. Finally, the dynamics of the previous segment may implicitly inform the subsequent trends of the epidemic development, e.g., slower decline at the end of the first wave may suggest higher risk of a second outbreak in the future.

4 Conclusion

Understanding the dynamics of development of COVID-19 is crucial in the prediction and containment of the epidemic. This paper extracts the underlying dynamical structures of the time series of COVID-19 daily active cases, based on polynomial regression with adaptive sliding windows. Further, similarities among the partitioned COVID-19 time series of different countries are compared based on DTW distance and complex network analysis. We find that each wave of the outbreak can be roughly divided into three stages: the early outbreak, the ascending stage, and the descending stage. Significant similarities exist between the same stages of different waves in different countries. The dynamics of the previous stage may provide information of the subsequent development of the epidemic.

Acknowledgements This research was funded by the National Natural Science Foundation of China (Grant No. 71971127) and the Hylink Digital Solutions Co., Ltd. (120500002).

References

Berndt, D. J., & Clifford, J. (1994). Using dynamic time warping to find patterns in time series. In *Proceedings of the 3rd International Conference on Knowledge Discovery and Data Mining*, Seattle, WA.

Calafiore, G. C., Novara, C., & Possieri, C. (2020, March 31). A modified SIR model for the COVID-19 contagion in Italy. *arXiv:2003.14391* [physics].

Ceylan, Z. (2020). Estimation of COVID-19 prevalence in Italy, Spain, and France, (in en). *Science of the Total Environment, 729*, 138817.

Chen, Y.-C., Lu, P.-E., Chang, C.-S., & Liu, T.-H. (2020, April 28). A time-dependent SIR model for COVID-19 with undetectable infected persons. *arXiv:2003.00122* [cs, q-bio, stat].

Gupta, R., Pandey, G., Chaudhary, P., & Pal, S. K. (2020). SEIR and Regression Model based COVID-19 outbreak predictions in India. *Public and Global Health*, preprint 3 Apr 2020. Available: http://medrxiv.org/lookup/doi/10.1101/2020.04.01.20049825. Accessed on 22 Sept 2020 03:39:41.

Kapoor, A., et al. (2020, July 6). Examining COVID-19 forecasting using spatio-temporal graph neural networks. *arXiv:2007.03113* [cs].

Liu, S., Fang, W., Gao, X., Wang, Z., An, F., & Wen, S. (2020). Self-similar behaviors in the crude oil market. *Energy, 211*, 118682.

Singh, R. K., et al. (2020). Prediction of the COVID-19 pandemic for the top 15 affected countries: Advanced Autoregressive Integrated Moving Average (ARIMA) model, (in en). *JMIR Public Health and Surveillance, 6*(2), e19115.

Wangping, J., et al. (2020). Extended SIR prediction of the epidemics trend of COVID-19 in Italy and compared with Hunan, China. *Frontiers in Medicine, 7*, 169.

Waqas, M., Farooq, M., Ahmad, R., & Ahmad, A. (2020, May 10). Analysis and prediction of COVID-19 pandemic in Pakistan using time-dependent SIR model. *arXiv:2005.02353* [q-bio].

Yang, Z., et al. (2020). Modified SEIR and AI prediction of the epidemics trend of COVID-19 in China under public health interventions. *Journal of Thoracic Disease, 12*(3), 165–174.

Capturing the Deep Trend of Stock Market for a Big Profit

Robin Qiu, Jeffrey Gong, and Jason Qiu

1 Introduction

There are a variety of technical indicators in securities trading, including popular ones such as the Stochastic Oscillator, Relative Strength Index, Moving Average Convergence Divergence (MACD), Bollinger Band, and Money Flow Index. Leveraging technical indicators derived from historical securities trading data has been widely studied and applied in Wall Street. In conjunction with taking advantage of the advances of computing technology and data science, algorithmic trades in an automated high frequency trading system become a compelling choice by investors. In fact, about 90% of trading is done using automated trading approaches today. Obviously, investment firms and day traders have depended largely upon technical analysis based quantitative investing in securities trading (Gold, 2018).

In practice, quantitative investing that mainly involves data-driven mathematical modeling with the support of unceasingly increased computing powers continues to grow in popularity. There are a lot of individual investors; whereas stock and security investment portfolios are typically built and managed by fund managers and teams (Vezeris et al., 2018). We believe that, regardless of the fast advancement of artificial intelligence in the financial field, it is impossible to remove human intervention completely from automated trading systems in the process of securities

---

R. Qiu (✉)
Division of Engineering & Info Sci, Pennsylvania State University, Malvern, PA, USA
e-mail: robinqiu@psu.edu

J. Gong
Methacton School District, Eagleville, PA, USA

J. Qiu
Department of Computer Science, Duke University, Durham, NC, USA
e-mail: jason.qiu@duke.edu

H. Yang et al. (eds.), *AI and Analytics for Public Health*, Springer Proceedings in Business and Economics, https://doi.org/10.1007/978-3-030-75166-1_5

investment from beginning to end. This is particularly true when securities trading must be well aligned with the ever-changing requirement of a specific investment portfolio, including risk controls, short-term and long-term investment goals.

Despite the increasingly widespread application of securities technical analysis and algorithmic trading in the field, there is a big disagreement on their viability and effectiveness in literature (Gold, 2018). However, scholars and investors continue to heavily invest money, time, and energy to dig out the Holy Grail of technical indicators, trending patterns, trading algorithms and strategies based on data on hand. Intrigued by the overwhelming and rich information collected from the securities trading systems and current and past economic and social data, scholars and investors have developed different trading strategies by accounting for investors' psychological behaviors and their social interactions in finance, aimed at comprehensively exploring the financial market dynamics to maximize their investment returns (Hirshleifer, 2015; Lin et al., 2018).

It is well recognized that daily trading (e.g., investing in individual stocks through frequent trades on a daily basis) and long-term investment (e.g., capitalizing on holding mutual funds, index funds, or the like for investment appreciation over a period of time) execute different trading strategies. Regardless of the adoption of daily trading or long-term investment approaches, investors generally know that they carry out varying investment goals by taking on very different levels of investment risks. We understand that securities trading is well-known for its high risk and volatile nature. Consequently, most people who are non-professional traders are frequently intimidated by securities trading. In this study, we intend to develop a trading strategy that is easy to use, viable, and promising, ultimately leading to the development of people-centric investment service systems by leveraging the recent advances in AI and big data technologies (Qiu, 2009, 2014).

With a focus on the long-term investment in securities trading, we propose a trading strategy by enhancing MACD with Deep Learning. Our proposed trading strategy combines a variety of social, economic, and trading technical indicators to substantially improve the performance of securities trading. Table 1 shows the returns when $1 million are invested using various trading strategies. As shown, our proposed trading strategy performs well.

In the remaining part of this paper, we first present an enhanced MACD trading algorithm. Then, we show how to utilize deep learning to further enhance the proposed trading strategy, aimed at capturing the deep trend of stock market for improved returns. At last, by considering behavioral and social finance, we discuss how to explore systematically trading strategies to accommodate individuals' irrational trading behavior and the social and economic dynamics in the globalized world. At the end of this paper, a brief conclusion is provided.

Table 1 Returns of 1 million dollars investment by executing various trading strategies

Investment period	Buy and hold	MACD	Enhanced MACD	Enhanced MACD with DL
1/1/2018	1,000,000.00	1,000,000.00	1,000,000.00	1,000,000.00
12/31/2019	1,197,529.48	1,156,577.30	1,217,595.54	**1,261,842.92**

2 An Enhanced MACD Trading Algorithm

As we explore a trading framework for long-term investment, MACD is chosen to identify buy/sell signals. MACD measures the difference between two exponential moving averages (EMA) of closing prices using predefined time period lags to evaluate the market trend of an investment. In popular use, MACD is the difference between its 12-day EMA and 26-day EMA. A buy signal is triggered when the MACD crosses above a signal line that is the 9-day EMA of the adopted MACD, while a sell signal is triggered when the MACD crosses below the signal line (Fig. 1). Gold (2018) reveals that the returns of his experimental portfolio investment improves if MACD is used as an indicator to support the entry and exit conditions derived from Bollinger Bands, i.e. to signal an entry/exit when both Bollinger and MACD conditions are satisfied, if compared to an experiment using MACD alone.

At a given trading day t, the classic equations to implement the MACD trading algorithm or simply denoted as MACD (12, 26, 9) are formally defined as follows:

$$MACD_t = EMA_t \text{ (12-}day) \text{ } of \text{ } closing \text{ } price \text{ } - \text{ } EMA_t \text{ (26-}day) \text{ } of \text{ } closing \text{ } price \tag{1}$$

$$Buy\text{-}Sell\text{-}Trigger_t = MACD_t \text{ } - \text{ } EMA_t \text{ (9-}day) \text{ } of \text{ } MACD \tag{2}$$

Once the *Buy-Sell-Trigger* calculated from Eq. (2) changes its sign from negative to positive, it indicates a buying opportunity, i.e., a buy signal. On the contrary, when the *Buy-Sell-Trigger* calculated from Eq. (2) changes its sign from positive to negative, it creates a selling opportunity, i.e., a sell signal. As an example, we choose the second-most popular fund on the planet - the SPDR S&P 500 trust (i.e., SPY) to illustrate our study. SPY is an exchange-traded fund (ETF), which can be easily traded like a typical stock in the stock market. Figure 1 shows how the classic MACD is applied to SPY from 1/1//2018 to 12/31/2019. We use MACD indicators to test how this simple MACD trading algorithm performs. It seems that the classic MACD trading algorithm underperformed when compared to the simplest "buy-

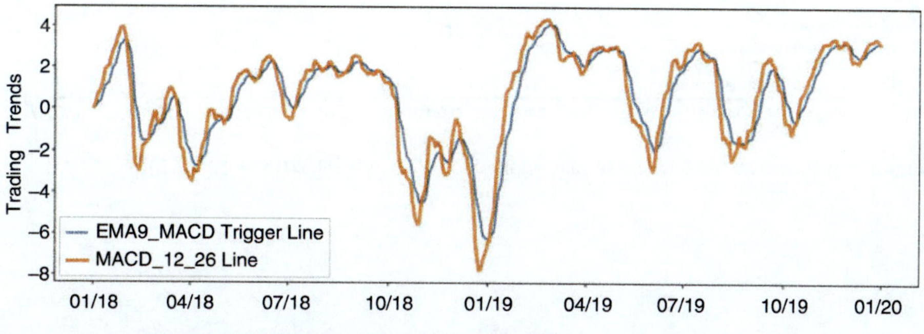

Fig. 1 MACD applied to SPY from 1/1/2018 to 12/31/2019

hold" trading strategy (please refer to Table 1 – the summary of all experiments in this study).

We add two more technical indicators, a daily price derivative and an extended MACD confirmation to evaluate the market momentum, to enhance the above-mentioned classic MACD trading algorithm. In addition to Eqs. (1) and (2), at a given trading day *t*, two more equations to enhance the MACD trading algorithm can be added as follows:

$$MACD2_t = EMA_t\,(6\text{-}day)\ of\ closing\ price\ -\ EMA_t\,(30\text{-}day)\ of\ closing\ price \tag{3}$$

$$Buy\text{-}Sell\text{-}Trigger2_t = MACD2_t\ -\ EMA_t\,(6\text{-}day)\ of\ MACD2 \tag{4}$$

$$Price_t^{'} = Close_t\ -\ Close_{t\text{-}1} \tag{5}$$

Essentially, we use a further extended measurement to determine the price trend of an investment to avoid frequent trading due to the increased market price fluctuations. To make EMAs more evenly distributed, we adjust Eq. (1) by reducing the parameters of 12-day and 26-day to 9-day and 18-day respectively, MACD (9, 18, 9). A price derivative is nothing but the true price moving direction, which is used to break the tie when there is a contradiction between two buy-sell signals. Figure 2 shows how the enhanced MACD^2 (9, 18, 9; 6, 30, 6; $Price_t^{'}$), i.e., MACD (9, 18, 9), MACD2 (6, 30, 6) and $Price_t^{'}$, is applied to SPY from 1/1//2018 to 12/31/2019. Table 1 shows an improved performance of trading SPY during the same 2-year period, resulting in a better return compared to the "buy-hold" trading strategy.

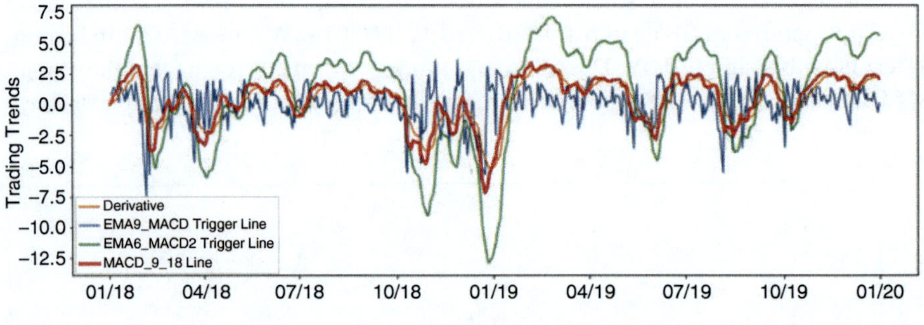

Fig. 2 An enhanced MACD or MACD^2 applied to SPY from 1/1/2018 to 12/31/2019

3 Deep Learning and Trading Strategies with Social and Economic Dynamics

3.1 Capturing the Deep Trend of Stock Market Using Deep Learning

From Sect. 2, we know that the enhanced $MACD^2$ can be adopted to improve the returns of securities investments. In Sect. 2, the dataset used in the evaluation is simply retrieved from Yahoo finance. It will be better if the future price of an investment can be well predicted, which helps confirm the trading decision when a trading signal is triggered. Many scholars have explored different approaches in predicting stock prices (Fister et al., 2019; Sang & Di Pierro, 2019). Deep learning such as recurrent neural networks, including Long Short-Term Memory (LSTM) and Gated Recurrent Units (GRU), has been widely explored in developing financial applications. When combined with technical indicators, applying LSTM in predicting securities momentums in the financial industry proves effective (Troiano et al., 2018).

To build a deep learning model for predicting the price trend of an investment, we adopt the LSTM machine learning algorithm. We retrieve the historical price data from 1/1/2000 to 12/31/2017 to create our dataset. The retrieved data are similar to the sample data shown in Table 2 (while excluding the Predicted Close and Average Close columns). The dataset with 4477 trading records in total is split using an 80/20 ratio – 3672 records for training and 805 records for testing. Table 3 summarizes the

Table 2 SPY sample data in the dataset used in the LSTM price prediction model

Date	High	Low	Open	Predicted close	Close	Volume	Average close
1/2/2018	268.81	267.4	267.84	265.68	268.77	86,655,700	268.10
1/3/2018	270.64	268.96	268.96	268.78	270.47	90,070,400	269.80
1/4/2018	272.16	270.54	271.2	270.23	271.61	80,636,400	271.35
1/5/2018	273.56	271.95	272.51	271.16	273.42	83,524,000	272.76
1/8/2018	274.1	272.98	273.31	273.26	273.92	57,319,200	273.54

Table 3 LSTM price prediction model architecture and parameters

Layer (type)	Output shape	Number of parameters
lstm_1 (LSTM)	(None, 50, 30)	4080
lstm_2 (LSTM)	(None, 50, 30)	7320
lstm_3 (LSTM)	(None, 30)	7320
dense_1 (Dense)	(None, 1)	31
model.compile(optimizer='adam', loss='mean_squared_error')		
model.fit(X_train, y_train, epochs=600, batch_size=32)		

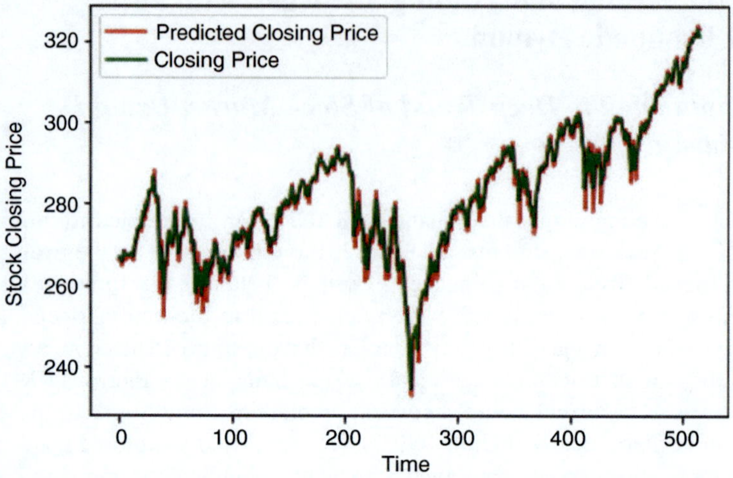

Fig. 3 SPY predicted closing price vs original closing price from 1/1/2018 to 12/31/2019

LSTM price prediction model in this study, highlighting the adopted architecture and parameters in the model training process.

Once the model gets trained and validated, we predict the closing price for SPY from 1/1/2018 to 12/31/2019. Figure 3 graphically presents the comparisons between SPY predicted closing prices and real closing prices over the prediction period. Figure 4 depicts how the trained model predicted the closing prices from 1/1/2018 to 12/31/2019. The percentage error of the predicted prices for 2 years of SPY trading averages −0.0016 (standard deviation = 0.0105). The predicted prices were then added into Table 2, which were used to replace the 'Close' price when computing the price derivative at a given day. Because of the available price of an investment on the next day, we reformulated Eq. (5) as follows:

$$PredPrice_t^{'} = PredictedClose_{t+1} - Close_t \quad (6)$$

Intuitively, we know that using this price trend derived from Eq. (6) to confirm a buy-sell decision in MACD^2 makes more sense than one derived from Eq. (5).

MACD^2 is then tested using Eq. (6) instead of Eq. (5). We also made some parameter adjustments in the enhanced MACD, i.e., MACD^2 (6, 18, 6; 3, 30, 3), to continue to test our proposed trading strategy. Figure 5 shows the MACD^2 (6, 18, 6; 3, 30, 3; $PredPrice_t^{'}$) model when applied to SPY from 1/1/2018 to 12/31/2019. As indicated in Table 1, we realized a further improved performance of trading SPY over the same evaluation period.

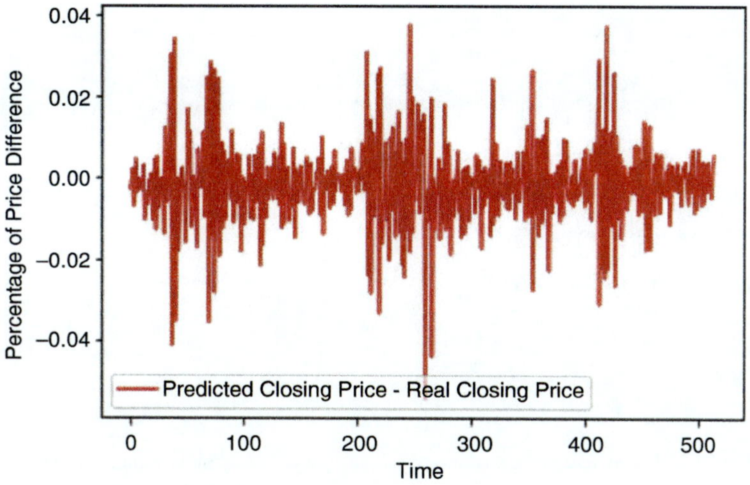

Fig. 4 SPY predicted closing price error percentage from 1/1/2018 to 12/31/2019

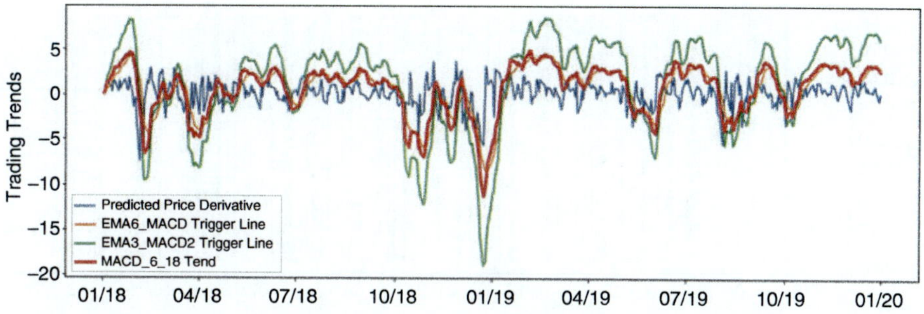

Fig. 5 Deep Trend based MACD^2 applied to SPY from 1/1/2018 to 12/31/2019

3.2 *Accounting for Behavioral and Social Finance to Continuously Enhance Trading Strategies*

Trading securities can be heavily influenced by emotions and instincts if not done automatically or at least on a team basis. As mentioned earlier, scholars and investors have developed different trading strategies by accounting for investors' psychological behaviors and their social interactions in finance, aimed at realizing better investment returns by accommodating well the uncertain and dynamic financial market (Hirshleifer, 2015; Lin et al., 2018). Incorporating the fear and greed model (Fig. 6) and a variety of economy barometers into trading strategies could be a great experiment (Liberto, 2019).

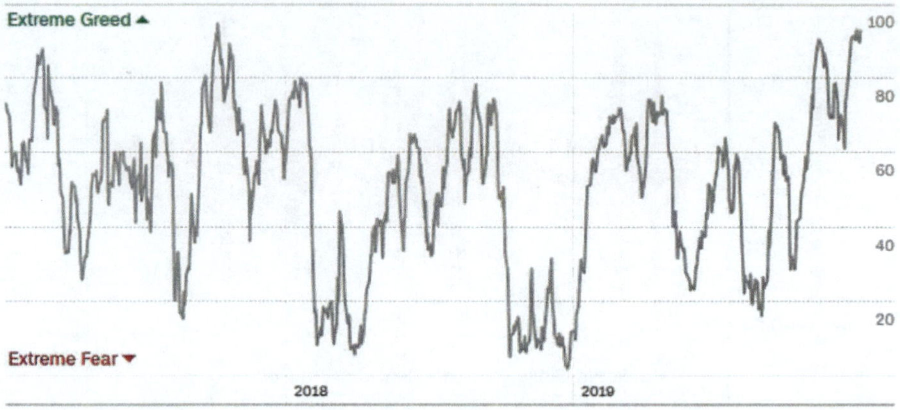

Fig. 6 The CNN money fear and greed index. (Copyright © https://money.cnn.com/data/fear-and-greed/)

Fig. 7 Trades based on Deep Trend MACD^2 applied to SPY from 1/1/2018 to 12/31/2019

Figure 7 shows how many trades were made when the deep trend based MACD^2 was applied to SPY over the two evaluation years. It will be interesting to see how various trading strategies can be developed by incorporating the fear and greed model and a variety of other economic barometers to meet investors' investment goals from time to time. Surely, viable and selective trading strategies could be further enhanced and developed when more social events and macro-economic data in general (Vargas et al., 2018; Long et al., 2019) can be considered in our study, which is in fact one of our ongoing research projects in the FinTech field.

The CNN money fear and greed index is based on the premise that great fear would result in securities trading well below their intrinsic values, while unbridled greed warrants irrational exuberance in trading, which has unduly escalated security values (Liberto, 2019). The CNN money fear and greed index can be enhanced by including more macro-economic data, such as the prime interest rate, the unemployment rate, the consumer sentiment index, and the producer price index

(PPI) in the USA. The enhanced index or customized indices may be great indicators and could be considered or built into our proposed MACD^2 trading strategy.

4 Conclusions

This paper briefly presented the preliminary outcomes of one of our FinTech research projects. The study is much limited given that more experiments using a variety of stocks or ETFs must be completed before we can claim that the proposed trading strategy works well for securities trading in general. As briefly discussed in Sect. 3.2, we are exploring more trading strategies by taking into consideration social events and macro-economic data in our ongoing study. Hopefully, in the near future we will be able to report a set of viable and promising trading strategies to meet both long-term and short-term investors' needs.

In summary, the proposed trading paradigm focusing on long-term investments can be easily scaled and transformed over time. Ultimately, the proposed trading paradigm can be developed as a public service system (Qiu, 2014) so that individual investors can access it freely to receive unbiased investment advice from time to time.

Disclaimer This project is purely for the purpose of research. Stock and security trading is risky. The authors have no responsibility for any loss if the proposed framework is adopted by individual investors or investment firms.

References

Fister, D., Mun, J. C., Jagrič, V., & Jagrič, T. (2019). Deep learning for stock market trading: A superior trading strategy? *Neural Network World, 29*(3), 151–171.

Gold, S. (2018). Stock market algorithmic trading: A test of Bollinger bands incorporating the squeeze effect and MACD conditions. *Journal of Applied Financial Research, 1*, 13–28.

Hirshleifer, D. (2015). Behavioral finance. *Annual Review of Financial Economics, 7*, 133–159.

Liberto, D. (2019). *Fear and greed index*. https://www.investopedia.com/terms/f/fear-and-greed-index.asp.

Lin, C. C., Chen, C. S., & Chen, A. P. (2018). Using intelligent computing and data stream mining for behavioral finance associated with market profile and financial physics. *Applied Soft Computing, 68*, 756–764.

Long, W., Lu, Z., & Cui, L. (2019). Deep learning-based feature engineering for stock price movement prediction. *Knowledge-Based Systems, 164*, 163–173.

Qiu, R. G. (2009). Computational thinking of service systems: Dynamics and adaptiveness modeling. *Service Science, 1*(1), 42–55.

Qiu, R. G. (2014). *Service science: The foundations of service engineering and management*. Wiley.

Sang, C., & Di Pierro, M. (2019). Improving trading technical analysis with TensorFlow Long Short-Term Memory (LSTM) Neural Network. *The Journal of Finance and Data Science, 5*(1), 1–11.

Troiano, L., Villa, E. M., & Loia, V. (2018). Replicating a trading strategy by means of LSTM for financial industry applications. *IEEE Transactions on Industrial Informatics, 14*(7), 3226–3234.

Vargas, M. R., dos Anjos, C. E., Bichara, G. L., & Evsukoff, A. G. (2018, July). Deep learning for stock market prediction using technical indicators and financial news articles. In *2018 International Joint Conference on Neural Networks (IJCNN)* (pp. 1–8). IEEE.

Vezeris, D., Kyrgos, T., & Schinas, C. (2018). Take profit and stop loss trading strategies comparison in combination with an MACD trading system. *Journal of Risk and Financial Management, 11*(3), 56.

Analysis on Competitiveness of Service Outsourcing Industry in Yangtze River Delta Region

Yanfeng Chu and Qunkai Peng

1 Introduction

After manufacturing outsourcing, service outsourcing refers to a new management model in which enterprises outsource their own non-professional business and complete their business on their behalf with the help of excellent teams of external professionals, thus enabling enterprises to focus on their core business. According to the different contents of service outsourcing, it can be divided into three categories: information technology outsourcing service (ITO), business process outsourcing service (BPO) and technical knowledge process outsourcing service (KPO).

In the past 10 years, China's service outsourcing industry has entered a rapid period of development. Driven by the "the belt and road initiative" strategy, the amount of service outsourcing contracts related to it reached 17.83 billion US dollars in 2015 alone, up 42.6% year-on-year, with an average annual growth rate of 40~60% in the past 5 years. China's service outsourcing industry has made remarkable achievements and is striving to become a major service outsourcing country.

The Yangtze River Delta region is one of China's economic cores, with dense population and industrial agglomeration. The development of its service outsourcing industry is unique and has attracted the attention of many experts and scholars. Xie Rongjian et al. (2017) used fuzzy analytic hierarchy process to study the competitiveness of service outsourcing industry in the Yangtze River Delta region, and built a two-level index evaluation system. It was concluded that the competitiveness of service outsourcing industry in the Yangtze River Delta was evaluated well in terms of human resources, transportation and communication

Y. Chu (✉) · Q. Peng
College of Economics and Management, Nanjing University of Aeronautics and Astronautics, Nanjing, Jiangsu, China
e-mail: yanfengc@nuaa.edu.cn

H. Yang et al. (eds.), *AI and Analytics for Public Health*, Springer Proceedings in Business and Economics, https://doi.org/10.1007/978-3-030-75166-1_6

facilities and economic environment. Xu Shan and Li Rongrou (2020) used factor analysis to measure the competitiveness of service outsourcing industry in major cities in Yangtze River Delta region, and found out the most critical factors that affect each region to undertake international service outsourcing. Dai Jun et al. (2015) has constructed a set of competitive ecological evaluation index system with 4 ecological dimensions and 31 consideration indicators, and made ecological measurement and situation analysis on Jiangsu's competitiveness in undertaking international service outsourcing of "the belt and road initiative". Through research, Di Changya and Xu Ying (2018) found that human capital, labor cost, infrastructure construction and supporting industry level played a positive role in promoting the development of service outsourcing competitiveness in Jiangsu Province. Shao and Chan (2011) pointed out the need to develop diversified international markets, establish outsourcing strategies, increase talent recruitment and training. Regarding the outsourcing strategy, Erna (2014) believes that the company must outsource to a company with a competitive advantage and continuously verify. He You-Shi and Qin Yong (2009) take the capacity of urban offshore software outsourcing as a research object and evaluate the capacity of offshore software outsourcing in Nanjing, Wuxi, Suzhou, and Changzhou by constructing a TOPSIS value function model. Regional competitiveness provides important guidance. At present, domestic and foreign experts and scholars have made some achievements in the research of the service outsourcing industry, but the research on the regional competitiveness of the service outsourcing industry in the Yangtze River Delta region is still relatively lacking. Therefore, this paper takes the service outsourcing industry in the Yangtze River Delta region as the research object, focuses on analyzing the competitiveness of service outsourcing industry in Jiangsu, Zhejiang and Shanghai provinces, finds out the difficulties and bottlenecks in the current industry development, and gives relevant suggestions for the future development of service outsourcing industry.

2 Building Evaluation Index System Based on the Diamond Model

2.1 Diamond Model Theory

Michael Porter believes that the competitiveness of a country or a region's industry mainly refers to the competitiveness reflected in the sales of the industry's products and the service provider of a specific industry (Shao-Wen & Wu-Chao, 2012). Factor conditions, demand conditions, related and supporting industries, corporate strategic structures, and peer industries The four factors of competition affect, the two variables of government and opportunity simultaneously affect the four factors, forming a diamond-like structural framework, called the diamond model (Fig. 1).

The diamond model theory provides a theoretical basis for analyzing the competitiveness of the service outsourcing industry. Based on this, considering

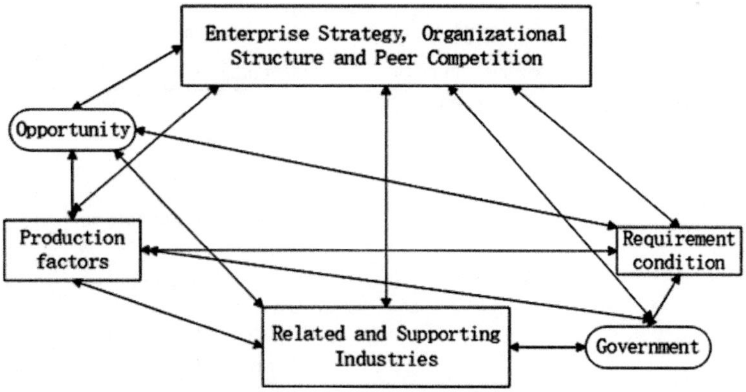

Fig. 1 Diamond model

the principles of comprehensiveness and completeness, Based on the existing research results and literature on service outsourcing at home and abroad, the competitiveness evaluation indexes of the service outsourcing industry are sorted out and summarized, as shown in Table 1.

2.2 Selection of Service Industry Competitiveness Indicators Based on Grey Correlation Degree

1. Improve the grey correlation model

Grey correlation degree analysis is a method to judge the correlation degree between each influencing factor and the given factor in a given system (Bing-Jun et al., 2005). The main idea of this method is to determine the correlation degree between each factor by experts scoring each index according to experience. The greater the final weight of each index, the higher the correlation degree, and the more important it is in the whole evaluation system. The specific steps are as follows:

In step 1, *m* experts are invited to score *n* indexes, i.e. weight assignment is performed to obtain an index weight matrix *D*, which can be expressed as

$$D = \begin{bmatrix} d_{11} & d_{12} & \cdots & d_{1m} \\ d_{21} & d_{22} & \cdots & d_{2m} \\ \vdots & \vdots & \vdots & \vdots \\ d_{n1} & d_{n2} & \cdots & d_{nm} \end{bmatrix} \quad (1)$$

Among them, d_{nm} is the empirical evaluation value of the *m-th* expert on the n-th index.

Table 1 Evaluation index of competitiveness of service outsourcing industry

	Primary indicator	Secondary index	Remarks
Regional Competitiveness of Service Outsourcing Industry	Element condition	Number of employees in the service outsourcing industry	X_1
		Assets of Service Outsourcing Enterprises	X_2
		Number of ports	X_3
	Requirement condition	Execution Amount of Offshore Service Outsourcing Contract	X_4
		The proportion of tertiary industry	X_5
		Foreign capital utilization	X_6
	Related and Supporting Industries	Number of ordinary colleges and universities	X_7
		Total freight volume	X_8
		Number of Internet Users	X_9
		Per capita GDP	X_{10}
	Enterprise strategic structure	Number of service outsourcing enterprises	X_{11}
		Operating Profit of Service Outsourcing Enterprises	X_{12}
	Government	Financial Allocation for Service Outsourcing Industry	X_{13}
	Opportunity	Number of service outsourcing demonstration cities	X_{14}
		International Service Outsourcing Contract Signing Amount	X_{15}

Step 2: Select the maximum scoring value from each column in the matrix D to become a reference weight vector D_O, which can be expressed as

$$D_O = (d_{O1}, d_{O2}, \ldots, d_{Om}) \tag{2}$$

Among them, d_{Om} is the weight determined for the *m-th* expert.

Step 3: Find each index vector, that is, the distance between each row D_i and D_O of D, that is

$$D_{Oi} = \sum_{k=1}^{m} (d_{Ok} - d_{ik})^2 \tag{3}$$

Where D_{0i} represents the distance between the index vector D_i and the reference weight vector D_0. d_{ik} indicates the weight given by the *k-th* expert to the *i-th* index.

Step 4: Calculate the weight of each index and normalize it, that is

$$\omega_i = 1/(1 + D_{0i}) \tag{4}$$

$$\overline{\omega}_i = \omega_i / \sum_{i=1}^{n} \omega_i \tag{5}$$

Among them, ω_i is the correlation degree between the *i-th* index and the maximum evaluation value of all exports, and $\overline{\omega}_i$ is the normalized *i-th* index weight.

2. Selection of competitiveness indicators for service outsourcing industry

Based on the grey relational grade analysis, three senior experts in the industry are invited to grade the index system shown in Table 1 by means of expert discussion, and the evaluation indexes are reasonably weighted and quantified by referring to relevant literature (Cao Tingting, 2014; Zhu Fulin, 2015; Chen Nana & He Zhixia, 2018), so as to avoid subjective assumptions caused by qualitative analysis and ensure the scientific and fair evaluation indexes. The weighting results are shown in the following table.

Further, according to Eqs. (3), (4) and (5) and the empirical matrix D shown in Table 2, the reference vector D_0 is formed by selecting the largest weight value

$$D_0 = (0.34, 0.3, 0.33)$$

At the same time, the calculation results of the corresponding index weight values are shown in Table 3.

From Table 3, we can see that the indicators that are highly correlated with the level of service outsourcing include: the amount of offshore service outsourcing contracts executed, the number of employees in the service outsourcing industry, the number of service outsourcing enterprises, and the assets of service outsourcing enterprises.

Table 2 Empowerment of competitiveness indicators of service outsourcing industry

	X_1	X_2	X_3	X_4	X_5	X_6	X_7	X_8	X_9	X_{10}	X_{11}	X_{12}	X_{13}	X_{14}	X_{15}
M_1	0.21	0.07	0.02	0.34	0.02	0.04	0.02	0.03	0.01	0.02	0.11	0.04	0.02	0.03	0.02
M_2	0.18	0.08	0.01	0.3	0.03	0.03	0.01	0.02	0.02	0.03	0.2	0.05	0.01	0.02	0.01
M_3	0.24	0.04	0.02	0.33	0.02	0.03	0.01	0.03	0.03	0.02	0.15	0.04	0.01	0.02	0.01

Table 3 Index weight value

Distance	Numerical	ω	Numerical	$\overline{\omega}$	Numerical
D_{01}	0.0394	ω_1	0.9621	$\overline{\omega}_1$	0.0779
D_{02}	0.2054	ω_2	0.8294	$\overline{\omega}_2$	0.0672
D_{03}	0.2826	ω_3	0.7797	$\overline{\omega}_3$	0.0631
D_{04}	0	ω_4	1	$\overline{\omega}_4$	0.0810
D_{05}	0.2714	ω_5	0.7865	$\overline{\omega}_5$	0.0637
D_{06}	0.2529	ω_6	0.7981	$\overline{\omega}_6$	0.0646
D_{07}	0.2889	ω_7	0.7759	$\overline{\omega}_7$	0.0628
D_{08}	0.2645	ω_8	0.7908	$\overline{\omega}_8$	0.0640
D_{09}	0.2773	ω_9	0.7829	$\overline{\omega}_9$	0.0634
D_{010}	0.2714	ω_{10}	0.7654	$\overline{\omega}_{10}$	0.0637
D_{011}	0.0953	ω_{11}	0.9130	$\overline{\omega}_{11}$	0.0739
D_{012}	0.2366	ω_{12}	0.8087	$\overline{\omega}_{12}$	0.0655
D_{013}	0.2889	ω_{13}	0.7759	$\overline{\omega}_{13}$	0.0628
D_{014}	0.2706	ω_{14}	0.7870	$\overline{\omega}_{14}$	0.0637
D_{015}	0.2889	ω_{15}	0.7759	$\overline{\omega}_{15}$	0.0628

3 Competitiveness Evaluation Based on Global Principal Component Analysis

3.1 *Global Principal Component Analysis*

Global Principal Component Analysis (GPCA) is based on the traditional principal component analysis method and integrates the idea of time series, which is suitable for evaluation from both vertical and horizontal aspects (Li Yangjie & Li Jing, 2020). The construction principle is as follows: Suppose K is a series of planar data tables arranged by time t: $K = \{X^t \in R^{n*p}, t = 1, 2, ..., T\}$. All data tables have sample points a_i with the same name and normalized variable indexes $X_1, X_2, ..., X_p$ in the same direction. It involves changes in time, so the sample group point at time t is expressed as: $a_i^t = \{a_i^t, i = 1, 2, \ldots, n\}$ and the global sample group point is $A_i = \bigcup_i T = 1\ A_i^t$. The center of gravity of the global data table can be expressed as g $= \sum_t T =_1 \sum_i n =_1 p_i^t a_i^t$. Where p_i^t is the weight of a_i^t and satisfied $\sum_t T =_1 \sum_i n =_1 p_i^t =1$. Therefore, $g^1, g^2, \ldots \ldots, g^T$ represents a time series at the overall level (Table 4).

3.2 *Empirical Analysis*

The test results show that the KMO value is 0.856, which is greater than 0.8, indicating that the constructed indexes have correlation and contain more common factors. Bartlett sphericity test value is 0.000, which is less than 1% significance

Table 4 KMO and Bartlett tests

KMO text		0.856
Bartlett sphericity test	Approximate chi-square	2250.086
	Variance	95
	Significant	0.000

Table 5 Global principal component eigenvalue and variance contribution rate

Principal component	Characteristic value	Contribution rate	Cumulative contribution rate
F_1	6.37	51.45%	51.45%
F_2	2.83	25.31%	76.76%
F_3	1.21	11.03%	87.79%

level, indicating that the indexes are independent of each other and pass the test, so the indexes and data can be analyzed by global principal component analysis.

After passing the test, three global principal components are extracted by principal component analysis according to the criterion that the characteristic value is greater than 1. The global principal component load matrix is established and rotated by the maximum variance method to calculate the eigenvalue, contribution rate and cumulative contribution rate of each global principal component.

As can be seen from Table 5, the cumulative variance contribution rate of the three global principal components F_1, F_2 and F_3 reached 87.79%, close to 90%, which shows that the changes of the values of these three principal components can almost replace the changes of the original nine indexes, which is sufficient to reflect the regional competitiveness of the service outsourcing industry.

By calculating the factor load of each index in the three principal components, it is found that: (1) The seven indicators X_4 (0.956), X_1 (0.925), X_{11} (0.902), X_2 (0.834), X_{12} (0.743), X_7 (0.731), X_{14} (0.722) have a larger load in the first global principal component, The production factors, demand structure, human resources and market scale that can represent the service outsourcing industry can be attributed to the industrial-strength factor. (2) In the second global principal component, the five indicators of X_{13} (0.856), X_8 (0.845), X_9 (0.807), X_6 (0.782), and X_5 (0.754) have a larger load, indicating government and related support Industry is an important factor in promoting the development of service outsourcing industry, and it can be attributed to the industry driving factor. (3) In the third global principal component, X_{15} (0.864), X_{10} (0.765), and X_3 (0.752) can be attributed to the industry potential factor.

3.3 Empirical Results

In this paper, the proportion of variance contribution rate of each global principal component to the total variance contribution rate is taken as the weight, and the comprehensive score of service outsourcing industry competitiveness is obtained by

Table 6 Comprehensive scores and rankings of service outsourcing industries in Jiangsu, Zhejiang, and Shanghai, 2013–2017

	2013		2014		2015		2016		2017	
	Score	Ranking	Score	Ranking	Score	Ranking	Score	Ranking	Score	Ranking
Jiangsu	0.847	1	0.956	1	1.155	1	1.346	1	1.612	1
Zhejiang	0.753	3	0.879	3	0.996	3	1.215	3	1.456	3
Shanghai	0.837	2	0.941	2	1.026	2	1.301	2	1.579	2

weighted calculation. It is expressed by F value. The higher the F value, the stronger the competitiveness of the service outsourcing industry, and vice versa. F_1, F_2 and F_3 respectively represent the scores of industrial-strength factor, industrial power factor and industrial potential factor.

Calculation steps: $F = 0.5145F_1 + 0.2531F_2 + 0.1103F_3.$

$$F_k = \sum_{i=1}^{9} \lambda_{ki} \times Y_i$$

F_k is the value of the *k-th* principal component; λ_{ki} is the factor load of the i index on the k principal component, and Y_i is the normalized value of the i index. The comprehensive scores and rankings of Jiangsu, Zhejiang, and Shanghai service outsourcing industries from 2013 to 2017 are calculated as shown in Table 6.

According to the comprehensive scores and rankings of service outsourcing industries of Jiangsu, Zhejiang, and Shanghai from 2013 to 2017 calculated in Table 6, it can be seen that the comprehensive scores of service outsourcing industries of the three provinces show an increasing trend in terms of time, which fully reflects the good development trend of service outsourcing industries in the Yangtze river delta region. Comparing the rankings of Jiangsu, Zhejiang, and Shanghai, it can be found that the comprehensive scores of service outsourcing industries of Jiangsu province have always been ahead of those of Zhejiang and Shanghai. In the increasingly fierce industrial competition, the service outsourcing industry in Jiangsu province has become the leader of the whole Yangtze River Delta region.

4 Summary and Thinking

4.1 Status and Challenges

1. Intensified internal and external competition

Although the service outsourcing industry in the Yangtze River Delta has a large number of enterprises and a large number of employees, it has a high degree of homogeneity, mainly concentrated in animation, software, finance, medicine, and other industries. The internal competition is increasingly fierce and lacks innovation and difference. India has an absolute advantage in human costs. According to relevant data, India's service outsourcing industry will save at least 50% of the cost for the world's multinational companies, which is an important reason why India's outsourcing industry is favored by the world's multinational companies.

2. Financing Difficulties for Service Outsourcing Enterprises

According to statistics, more than 90% of service outsourcing enterprises in the Yangtze River Delta region are emerging small and medium-sized enterprises. In small and medium-sized enterprises, due to their weak credibility, it is generally difficult to gain the trust of banks and investment institutions. Even if they can obtain financing, the amount of funds is very limited. Therefore, financing difficulty and expensive financing are common phenomena in the service outsourcing industry, which are also important factors restricting the further development and technological innovation of outsourcing enterprises.

3. Impact of Artificial Intelligence on Service Outsourcing Industry

With the continuous progress of Artificial Intelligence, it is not impossible for many traditional services in the service outsourcing industry to be replaced by intelligent robots in the future, which is also the inevitable trend of industrial transformation and upgrading. In the service outsourcing industry, the pace of development is fast, and the innovation ability of enterprises is crucial. Only a timely transformation can seize a piece of space in the fierce market competition.

4.2 Countermeasures and Suggestions

Because of the current development situation and challenges in the Yangtze River Delta region, this paper will give relevant suggestions from the aspects of talents, financing and industrial environment.

1. Strengthen talent training and introduction

Talent is the most critical factor in the outsourcing industry, which is different from other traditional resource-based industries. In the service outsourcing industry, talent represents the core competitiveness of enterprises. Therefore, colleges and universities should vigorously strengthen the training of outsourcing talents, establish and improve the professional skills training of service outsourcing. Besides, service outsourcing needs more practice and should increase the investment in the construction of service outsourcing practice bases to realize the seamless connection between college training and enterprise application. At the same time, the government should issue corresponding supporting policies, build a better development platform for talents, connect with the international community, strive to attract international talents based on preventing brain drain, and establish an international service outsourcing talent network.

2. Guarantee Funds, Form Industrial Scale and Build Brand Enterprises

Most of the service outsourcing enterprises in the Yangtze River Delta are small and medium-sized enterprises, which are small and scattered and lack influential leading enterprises. Therefore, it is difficult to contract large orders. The government should issue corresponding financial support policies to reduce the pressure on small and medium-sized enterprises and set up special funds to promote the development of small and medium-sized enterprises. For example, Wuxi has set up a special fund of 1.5 billion yuan, Suzhou and Hangzhou will invest a special fund of 100 million yuan every year to guarantee the financing of small and medium-sized outsourcing enterprises. Banks and financial institutions should also vigorously support the loans of enterprises, promote the formation of industrial-scale in the service outsourcing industry, change the current situation of operating independently, and strive to build brand enterprises with international competitiveness and improve their visibility.

3. Strengthen infrastructure construction and optimize industrial structure

The service outsourcing industry has very high requirements for the perfection of infrastructures, such as geographical conditions, logistics, and transportation, information network, labor force, etc., which will be directly linked to costs. The level of the cost will directly determine the development advantages of the industry. Experts point out that in the Yangtze River Delta region, service outsourcing enterprises in Nanjing, Shanghai, Hangzhou, and other cities have higher costs, while Suzhou and Changzhou have obvious advantages in costs. Therefore, the government can integrate enterprise resources, build a service outsourcing industry demonstration park, and improve the supporting construction of the industrial environment to reduce the costs of enterprises. At the same time, to optimize the industrial structure, enterprises must be innovative and constantly open up new markets to avoid internal homogeneous competition.

4. Vigorously support related and supporting industries

The development of service outsourcing industry is inseparable from related and supporting industries. Therefore, in the future industrial development, the government should vigorously support manufacturing, information technology, transportation, finance and other related supporting industries, improve the industrial layout, strengthen industrial cooperation, appropriately relax the industrial threshold, let more resources and service outsourcing industries develop, realize resource sharing among provinces and cities, build an internationally influential service outsourcing industrial cluster, and continuously enhance the comprehensive competitiveness of the service outsourcing industry in the Yangtze River Delta region.

5. Strengthen market supervision and improve relevant laws and regulations

Service outsourcing industry is a modern high-end service industry. With the rapid development of the industry, many problems and loopholes in market supervision have been exposed, the most obvious of which is the protection of intellectual property rights. Therefore, the government must improve the relevant laws and regulations, strengthen the protection of intellectual property rights, reduce property disputes, and establish a safe and healthy market environment, so that the service outsourcing industry can develop healthily and stably.

References

Bing-Jun, L., Si-Feng, L., & Bin, L. (2005). The hierarchic grey incidence analysis model with fixed weights and its application to a regional scientific-technical system. In *IEEE International Conference on Systems, Man & Cybernetics*. IEEE.

Cao Tingting. (2014). *Research on service outsourcing supplier selection method based on fuzzy analytic hierarchy process and grey correlation analysis*. Doctoral dissertation.

Chen Nana, & He Zhixia. (2018). Study on the international competitiveness of Hainan tourism service based on grey relational analysis. *Modern Economic Information, 21*, 488–489.

Dai Jun, Wu Hongzhen, Yan Shiqing, & Han Zhen. (2015). An empirical study on the competitiveness of China's 21 cities to undertake international service outsourcing. *Asia-Pacific Economy, 000*(005), 102–106.

Di Changya, & Xu Ying. (2018). Research on service outsourcing competitiveness and influencing factors in Jiangsu Province. *Market Weekly, 000*(012), 64–68.

Erna. (2014). *Suatu tinjauan mengenai penerapan strategi outsourcing dan strategi core competency studi kasus pada starbuck, Jurnal Akuntansi.*

Li Yangjie, & Li Jing. (2020). Dynamic evaluation of industrial ecology level in Yangtze River economic belt—Calculation based on global principal component analysis model. *Forestry Economy, 7.*

Shao, Q., & Chan, T. (2011). Strategies to improve Daqing service outsourcing competency. In *International Conference on Management & Service Science*. IEEE.

Shao-Wen, Z., & Wu-Chao, X. (2012). Empirical research on culture industry competitiveness evaluation in xi'an based on "diamond model" theory. *Journal of Jishou University*.

Xie Rongjian, Liang Liang, & Li Xiaodong. (2017). Evaluation of the competitiveness of service outsourcing industry in Yangtze River Delta based on fuzzy analytic hierarchy process. *Journal of Anhui Normal University (Humanities and Social Sciences Edition), 45*(001), 100–106.

Xu Shan, & Li Rongrou. (2020). Competitiveness and influencing factors of Yangtze River Delta undertaking international service outsourcing. *Journal of Hangzhou Dianzi University (Social Science Edition), 4.*

You-Shi, H., & Yong, Q. (2009). Comprehensive evaluation of city's undertake capability of offshore software outsourcing in Jiangsu Province. In *First IEEE International Conference on Information Science & Engineering*. IEEE Computer Society.

Zhu Fulin. (2015). Analysis of influencing factors of Indian service outsourcing competitiveness—An empirical study based on grey relational degree method. *World Economic Research*, (5 issues), 90–97.

OPBFT: Optimized Practical Byzantine Fault Tolerant Consensus Mechanism Model

Hui Wang, Wenan Tan, Jiakai Wu, and Pan Liu

1 Introduction

The blockchain derived from Bitcoin (Nakamoto, 2019) is a new type of decentralized protocol that integrates distributed data storage technology, P2P network transmission technology, consensus mechanism, encryption algorithms, programmable smart contract and other technologies (Yuan & Wang, 2016) to build a decentralized distributed system. Its decentralization, autonomy, immutability and anonymity make the blockchain not only play a huge role in digital assets such as digital currencies, but also receive widespread attention in areas such as financial services, public services, and social life. With the rapid development of the blockchain, its storage, consensus, supervision and security issues have become prominent. Among them, the consensus mechanism, as the core technology of the blockchain, is to solve the problem of node consistency in a distributed environment brought by the Byzantine General Problem (Lamport et al., 1982). The performance of the consensus mechanism directly affects the security and performance of the blockchain systems, restricting the wide-scale application of the blockchain technology. Thus the study of its related algorithms is particularly important.

H. Wang (✉) · J. Wu
College of Computer Science and Technology, Nanjing University of Aeronautics and Astronautics, Jiangsu, Nanjing, China

W. Tan
College of Computer Science and Technology, Nanjing University of Aeronautics and Astronautics, Jiangsu, Nanjing, China

School of Computer and Information Engineering, Shanghai Polytechnic University, Shanghai, China

P. Liu
Department of Information Technology, Shenzhen Easttop Supply Chain Management Co., Ltd, Shanghai, China

H. Yang et al. (eds.), *AI and Analytics for Public Health*, Springer Proceedings in Business and Economics, https://doi.org/10.1007/978-3-030-75166-1_7

The open network environment that the public blockchain system faces makes the design of its consensus mechanism fully consider security issues. Therefore, in order to address the security and efficiency of the consensus mechanism in the consortium blockchain, this paper introduces a reputation model and a degradation mechanism for malicious node based on the PBFT consensus algorithm, and utilizes an optimized consistency protocol to realize an improved blockchain consensus mechanism model.

The main contributions of this article are as follows:

1. This paper add a list of candidate nodes, which fill in the vacancy in time after the node has been revoked to participate in the consensus process due to its misdeeds.
2. The traditional consensus algorithm doesn't support the dynamic joining or exiting of nodes. The consensus nodes are defined in advance, and they take turns to package transactions to generate blocks. This paper establishes a reputation model to evaluate the credibility of each distributed node, and dynamically adjust its score and level through the behavior of the node, increasing the difficulty of node evildoing. This model is used in subsequent consensus rounds to update consensus nodes to ensure a high degree of credibility of the consensus nodes.
3. The optimized consistency protocol is used to replace the traditional consistency protocol process with time complexity as high as $O(N^2)$, which greatly reduces the network communication overhead and improves the consensus efficiency.
4. Although the traditional consensus consistency protocol has certain fault tolerance performance, it can't identify byzantine nodes. This paper introduces the detection and degradation mechanism of byzantine nodes, elaborate the methods of detecting byzantine nodes, find and remove malicious nodes in time, and shortens the period of malicious nodes being found.

The structure of this paper is as follows: Sect. 1 briefly introduces the research background and contribution of this article; Sect. 2 introduces the history and development of consensus algorithms; Sect. 3 focuses on the whole thought and detailed implementation of the OPBFT algorithm; Sect. 4 compares and analyzes the performance of the OPBFT algorithm and CBFT algorithm from the aspects of communication overhead, throughput, delay, and security; Sect. 5 summarizes the full text and proposes future work improvements.

2 Related Work

Consensus enables each node to reach an agreement in a decentralized, weakly trusted distributed system, and solves the problem of mutual trust between nodes on the basis of decentralization.

Satoshi Nakamoto adopted the Proof of Work (PoW) (Gervais et al., 2016) consensus mechanism in the Bitcoin system implemented in 2009. The core idea is that each node solve a difficult but easy to verify hash mathematical problem based on their computing power to competes with each other to obtain the accounting right, so as to ensure the consistency of the data and the security of the consensus. However, the waste of resources and the 10-mi transaction confirmation time caused by its strong consensus power have been widely criticized, and meet the general business needs difficultly (Yuan & Wang, 2016). The Proof of Stake (PoS) (Proof of stake, 2019) consensus algorithm was implemented in Peercoin released by Sunny King in 2012 for the first time. Its mining difficulty is determined by the stake held by the nodes rather than the computing power. Although the problem of wasting computing power in PoW has been reduced, it still needs to be mined in essence. The mechanism of coin age accumulation will gradually lead to a situation where "the rich are richer" (Houy, 2014).

The Delegated Proof of Stake (DPoS) (Schuh & Schieÿl, 2019) is based on PoW and PoS, and uses a node's stake as a ballot to select a certain number of representative nodes to generate and verify blocks by turns. Under this mechanism, there is no mining process that consumes computing power, which greatly reduces the number of the nodes directly participating in consensus, and can achieve consensus verification in seconds. However, it has brought along the following issues: inactive participation of nodes in voting, untimely removal of malicious nodes caused by too long voting cycles, and reliance on tokens making it difficult to apply to commercial applications.

The security of the consensus process is affected by a variety of factors, including the number of nodes, whether there are byzantine nodes, etc. (Bach et al., 2018). Byzantine nodes originate from the Byzantine General Problem. In a network without considering byzantine nodes, there are only non-malicious behaviors such as data packet loss and delay. Such nodes caused by non-malicious behavior attack stop working, which is called fail-stop failure (Schlichting & Schneider, 1983). Representative algorithms such as Paxos (Lamport, 2001) and Raft (Ongaro & Ousterhout, 2014) can achieve more efficient consensus. But in reality, it is difficult to build a system without byzantine nodes. Therefore, most consensus algorithms are based on the security model with byzantine nodes. Byzantine fault tolerance consensus algorithms need to rely on a weakly centralized consistency framework and higher communication overhead to provide instant consistency, thereby improving transaction processing throughput (Vukolic, 2015). Miguel Castro proposed the PBFT algorithm (Castro & Liskov, 1999) in 1999, which is an improvement on the Byzantine Fault Tolerance (BFT) algorithm. It reduces the algorithm complexity from an exponential level to a polynomial level, making it feasible in practical system applications, and providing $(n-1) / 3$ fault tolerance (where n is the total number of nodes) under the premise of liveness and safety. However, there are still some shortcomings. The static type of its network structure cannot dynamically detect the joining or exiting of nodes. The nodes rotate in turn according to their numbers to generate blocks. And the election method of primary node is arbitrary.

With the further development of blockchain technology, some new consensus algorithms are also emerging. In 2016, the Chinese blockchain community NEO proposed an improved byzantine fault-tolerant algorithm. This algorithm borrows from PoS design ideas based on PBFT and solves the problem of lack of final consistency between PoW and PoS, making the blockchain suitable for financial scenarios. In 2017, the dynamic authorization of byzantine fault tolerant (DDBFT) algorithm (Liu, 2017) was proposed. This algorithm applies DPoS to the PBFT algorithm, which makes PBFT dynamic. However, compared with the algorithm in this paper, DDBFT only uses the simple addition and subtraction operation to rank the consensus nodes, and can't accurately evaluate the node reputation. The consortium byzantine fault tolerant (CBFT) algorithm (Li, 2018) was proposed in 2018. It is based on PBFT algorithm, combined with block cache, block synchronization and signature, node change, with higher throughput and lower delay. However, compared with the algorithm in this paper, the essence of CBFT is still a three-phase protocol of PBFT, it will lead to great communication overhead with the increase of the number of nodes.

3 Analysis of OPBFT Algorithms

3.1 Whole Thought

During the implementation of the PBFT algorithm consensus protocol, a large amount of communication will occur between nodes. As the number of nodes increases, its communication overhead will increase exponentially, which will increase the bandwidth pressure and affect efficiency of the consensus algorithm. Therefore, this paper addresses the shortcomings of PBFT by adopting an optimized consensus protocol. The implementation of the consensus protocol is mainly divided into two cases: (1) there is no byzantine node in the consensus node, and the optimized consistency protocol is executed; (2) there are byzantine nodes in the consensus node, and the traditional consistency protocol of the PBFT algorithm is executed to ensure the fault tolerance of the algorithm.

According to the reputation model, the nodes in the network are divided into two categories. One type is consensus nodes, responsible for generating and verifying blocks, and participating in the consensus process. The other type is candidate nodes, which are used as a candidate set of consensus nodes to replace byzantine nodes in the consensus node set. They don't participate in the consensus process, but need to accept consensus results. In addition, each consensus node maintains a consensus node list (CNL) locally, which contains the current number of consensus nodes, node's number, node's public key, node's reputation value.

At the beginning of each round of consensus process, each consensus node locally initializes its own Digest Vectors (DVs), which are used to store the signed block header digest information sent by the node. The byzantine node detection and degradation mechanism is introduced into the algorithm. When a byzantine node appears in consensus nodes, a malicious node is detected and marked based on the received DVs information. Then the algorithm transfers to the traditional consensus protocol to complete the consensus process. After the conclusion of the consensus, the list of consensus nodes is updated and malicious nodes are removed according to the reputation model and the marked information.

3.2 Optimized Consistency Protocol

The consistency protocol of the PBFT algorithm needs to complete twice communications of nodes with a complexity of $O(N^2)$ during operation. In this paper, the Optimized Practical Byzantine Fault Tolerant (OPBFT) algorithm optimizes the consistency protocol in the absence of byzantine nodes, so that it can complete the consensus after completing the communications of nodes with O(N) complexity. Below N indicates the total number of consensus nodes, and f indicates the number of byzantine nodes.

An optimized consensus protocol proposed by (Fang et al., 2019), although it greatly reduces the problem of network communication overhead, it is not rigorous enough to identify byzantine nodes, and does not consider the situation where the primary node's evildoing. Therefore, combined with the improvement ideas of this paper, an optimized consistency protocol is designed as shown in Fig. 1.

The specific implementation process of the protocol is as follows:

1. Request phase: The client broadcasts <REQUEST, m, t, c> message to the entire network, where m represents the content that needs consensus, t represents the timestamp, and c represents the client that sent the message.
2. Pre-prepare phase: The primary node p receives the request from client c and generates a pre-prepare message << PRE-PREPARE, v, n, d, p, $E(d)_{p\_private}$ >, m>, and broadcasts to all consensus nodes, d is the digest information of message m, and $E(d)_{p\_private}$ means the digest information d is signed by the private key of the primary node p.
3. Feedback phase: The replica node verifies the received pre-prepare message, generates a feedback message <<BACK, v, n, d, i, $E(d)_{p\_private}$>, $Vote_{t/f}(m,i)$>

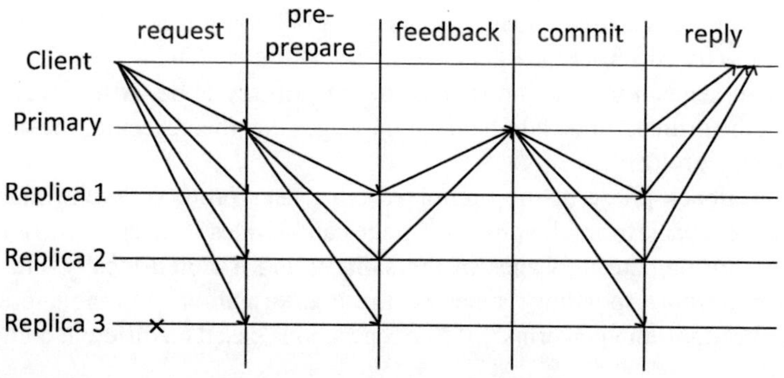

Fig. 1 Execution process of the optimized consistency protocol

after passing the inspection, and sends the message to the primary node, updates the local digest vector $DV_i[p] = E(d)_{p\_private}$. $Vote_{t/f}(m,i)$ is the vote on the message generated by node i.

4. Commit phase: When the primary node receives consistent feedback information from all other consensus nodes, it generates a commit message <<COMMIT, v, n, d, p>, Cert(m)> and broadcasts it to all nodes in the network. Cert(m) is the block certificate which contains all affirmative votes to verify the validity of the message. After all nodes receive the commit message, the transaction information is added to the local memory.

3.3 *Achievement of Consensus*

When the byzantine node is detected, the OPBFT algorithm still implements the traditional consensus protocol of the PBFT algorithm to reach a consensus. The consensus process mainly has three states: pre-prepare, prepare, and commit. The change of the consensus state depends on whether it has received at least $2f + 1$ confirmation message. The algorithm in this paper is slightly different. It is necessary to calculate the sum of the reputation values of the nodes that send confirmation information. Only when it is not less than 2/3 of the sum of the reputation values of all consensus nodes, as shown in Eq. (1), can we enter the next stage of consensus and improve the discourse rights of the node.

$$R = \frac{1}{N} * \left(2\left\lfloor \frac{N-1}{3} \right\rfloor + 1\right) * R_{total} \tag{1}$$

Where R_{total} represents the total reputation value of all consensus nodes.

The detailed process of OPBFT algorithm mainly includes the following steps:

1. Node initialization. Assume that there are S nodes, all nodes are numbered with $\{0,1,2, \ldots, | S | -1\}$, and the top N nodes with the highest reputation value are used as consensus nodes according to the reputation model $N = \{0, 1, 2, \cdots, | N | -1\}$, and the rest are candidate nodes $C = \{| N |, | N | + 1, \cdots, | S | -1\}$, where N must satisfy $N > 3 * f$.
2. The client sends a transaction request to the primary node. After receiving the request, the primary node numbers the messages and then executes the optimized consistency protocol.
3. In the feedback phase of the optimized consensus protocol, the primary node receives feedback from all consensus nodes and judges its accuracy, compares it with the pre-prepare messages of the same v and n stored locally, and judges whether the corresponding values of the d information with signature is the same. The algorithm performs different operations based on different comparison results.

(a) If the values are the same, the feedback information is correct. When all the feedback information received by the primary node is correct, the consensus node will generate block certificate continue to execute the optimized consensus protocol to complete the consensus process.
(b) If the values are different, it means that the information has been tampered with, that is, there are byzantine nodes in the consensus nodes. Algorithm will terminate the operation of the optimization consensus protocol, and instead implement the traditional consensus protocol to complete the consensus process to ensure system fault tolerance. At the same time, the byzantine nodes are marked according to the $DV_i[p]$ information, and they are downgraded according to the reputation model. Simultaneously, the entire consensus node set and the candidate node set are updated to ensure that consensus nodes are as honest as possible. In the next consensus process, the optimal consensus protocol continues to be implemented.

3.4 Node Behavior Distinguish and Reputation Model

As described above, the overall process of consensus reaching is introduced. How to detect the different abnormal behaviors of the consensus nodes is the key problem to be solved. Therefore, in the OPBFT algorithm, an additional data structure $DV_i[p]$ is used to record the block header digest information with node signatures, and the malicious behavior of the byzantine nodes is detected by comparing the value.

Node reputation model is a method to evaluate node credibility based on node behavior. The reputation-based model proposed in Lei et al. (2018), although it provides a more reliable guarantee for system security and liveness, but the model it builds changes linearly, which isn't enough to adapt to the complex changes of the actual situation. Therefore, in combination with the improved ideas of this paper, the reputation model is specifically constructed as follows: In the algorithm of this paper, the reputation value R is taken as a real number between 1 and 100, and the higher the value, the higher its credibility. For newly added nodes, the reputation value is initialized to 50. Here, let $R_i(t)$ be the reputation value of consensus node i in the t-th round of consensus, then $R_i(t + 1)$ can be discussed in the following two cases.

1. During the consensus process, the primary node is responsible for the generation of new blocks, and its abnormal behavior can be divided into the following two categories: (a) Due to blocking or intentional delay, the primary node doesn't pack block to send pre-prepare messages, resulting in no new blocks generated in this consensus. $DV_i[p] = 0$ can be used to detect this abnormal behavior and reduce its reputation value. (b) During the pre-prepare phase, the primary node sends inconsistent block information. In this case, the algorithm uses the public key of the primary node p to decrypt $DV_i[p]$. The result will be inconsistent with

the value of d stored locally. The primary node reputation value will be directly set to 0. Specifically, as shown in Eq. (2), where $0 < x < 1$.

$$R_i(t+1) = \begin{cases} \min(100, R_i(t) + \ln(R_i(t))), & \text{the proposed block is successfully appended to blockchain;} \\ \max(1, xR_i(t)), & \text{the node doesn't generate a new block} \\ 1, & \text{the node sends inconsistent messages to different nodes} \end{cases} \tag{2}$$

2. Similarly, there are two types of malicious behaviors of other nodes during the feedback phase: (a) The node does not send feedback information. At this case, it can be identified by detecting whether the feedback information $<E(d)_{p\_private}>_{i\_private}$ field is 0; (b) The node send tampered feedback. In this case, the node i is broadcasted to other nodes by its local $DV_i[p]$ information. The algorithm will not be able to successfully decrypt the $DV_i[p]$ with the public key of the primary node p, which indicates that the replica node i is a byzantine node. Specifically, as shown in Eq. (3), where $0 < x < 1$.

$$R_i(t+1) = \begin{cases} \min(100, R_i(t) + \ln(R_i(t))), & \text{the node isn't marked for abnormal behavior;} \\ \max(1, xR_i(t)), & \text{the node doesn't send feedback in time;} \\ 1, & \text{the node tampers and sends error messages;} \end{cases} \tag{3}$$

4 Simulation and Analysis of Results

In this paper, the PBFT algorithm is implemented through the Java programming language, and the OPBFT algorithm is obtained based on the improvement of the PBFT algorithm. The distributed consensus process is simulated using a local multi-node approach. Finally, the OPBFT algorithm and the CBFT algorithm are analyzed and compared, and it is found that the OPBFT algorithm has obvious improvements in data throughput and transaction latency.

4.1 Communication Overhead

In byzantine fault-tolerant consensus algorithms, the implementation of fault tolerance is based on data exchange. The data exchange process will bring communication overhead. Therefore, communication overhead is a key indicator to measure the efficiency of the algorithm. This paper compares the communication overhead of DDBFT algorithm, CBFT algorithm and OPBFT algorithm, and calculates the communication times required for a complete consensus by using the process of consensus algorithm.

By analyzing the communication process of DDBFT algorithm, it can be seen that the algorithm removes the final confirmation phase, transforms the three-phase protocol into a two-phase protocol, and also considers the flooding forwarding of the early transaction information. Its communication overhead is:

$$C_{DDBFT} = (n-1)^2 + (n-1) + (n-1)^2 = (2n-1)(n-1) \quad (4)$$

After analyzing the communication process of CBFT algorithm, it still adopts the three-phase protocol of PBFT algorithm. However, the general consensus node also creates and caches blocks in the pre-preparation stage. When the proposal block received from the primary node is the same as that created by itself, it can skip the block verification step and directly enter the voting stage. Otherwise, the proposed block will be cached after the verification, which reduces the cost of time compared with the PBFT algorithm to a certain extent.

The communication process of the OPBFT algorithm is analyzed. The algorithm is divided into two modes. One is the consensus process under the optimized consensus protocol. At this time, the consensus overhead is $3(n-1)$. The other is the same as the traditional consistency protocol, and its communication overhead is $2n(n-1)$. Assuming that the probability of the algorithm performing optimized consensus protocol is α, the probability of executing traditional consistency protocol is $1-\alpha$, so its communication overhead C_{OPBFT} is:

$$C_{OPBFT} = \alpha[3(n-1)] + (1-\alpha)[2n(n-1)], \alpha \in [0, 1] \quad (5)$$

It can be found from the analysis of formula (5) that the larger α is, the smaller the C is, which means that the longer the OPBFT algorithm executes the optimized consensus protocol, the smaller its communication overhead is. With the change of time, OPBFT can remove malicious nodes in time, ensure that the trustworthiness of consensus nodes is getting higher and higher, and the algorithm can run for a long time in the optimized consensus protocol mode.

4.2 Throughput

Throughput refers to the number of transactions completed in a unit of time, expressed as TPS (Transaction Per Second). The number of consensus nodes varies from 4 to 30, of which the number of byzantine nodes is $\left\lfloor \frac{N-1}{3} \right\rfloor$. As shown in Fig. 2a, the OPBFT algorithm's and CBFT algorithm's throughputs are shown. Figure 2b compares and analyzes the changes of the two algorithms' TPS of over time, the number of consensus nodes is fixed at 16.

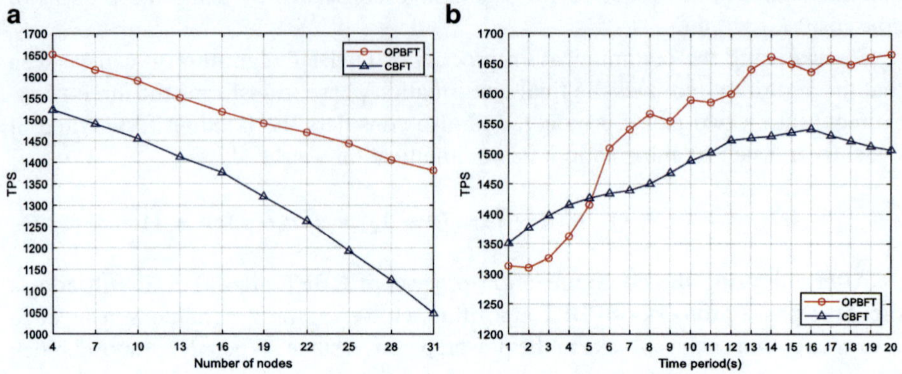

Fig. 2 (**a**) TPS comparison of the two algorithms; (**b**) The change of TPS over time

In Fig. 2a, as the number of nodes in the network increases, the throughput of both algorithms shows a downward trend. But overall, the throughput of the OPBFT algorithm is much higher than the throughput of the CBFT algorithm, and the average throughput has increased from 1320 TPS to 1511 TPS. In Fig. 2b, the early throughput of OPBFT is slightly lower than the CBFT. In the sixth period later, it overtook CBFT, and its TPS gradually stabilizes at about 1650. The main reason is that in the case of byzantine nodes, the OPBFT algorithm needs to complete the conversion of the consistency protocol and execute the traditional consistency protocol, which affects the throughput of the algorithm. As byzantine nodes are eliminated in the later period, the probability of algorithm executing optimization consistency is increasing, so the advantages of the later OPBFT algorithm over the CBFT algorithm are more and more obvious. On the contrary, the complexity of CBFT algorithm is $O(N^2)$, and the pressure on network bandwidth increases with the increase of the number of nodes, and even leads to network congestion, which greatly affects the throughput and consensus.

4.3 Delay

The delay in the blockchain indicates the time difference from the submission to the writing of the transaction. This article tests with different numbers of nodes. The following Fig. 3a shows the comparison of the transaction delay of the two algorithms, and the average value of multiple experiments is taken. Figure 3b shows the changes in the delay of the two algorithms over time.

It can be seen from Fig. 3a that with the increase of the number of nodes, the transaction delay of both algorithms gradually increases, But the delay of OPBFT algorithm is less affected by the number of nodes compared to the CBFT algorithm. Because the OPBFT algorithm needs to collect and store DVs information, it brings additional overhead. When there are byzantine nodes, it also needs to update the reputation value of consensus nodes by using the reputation model to complete the

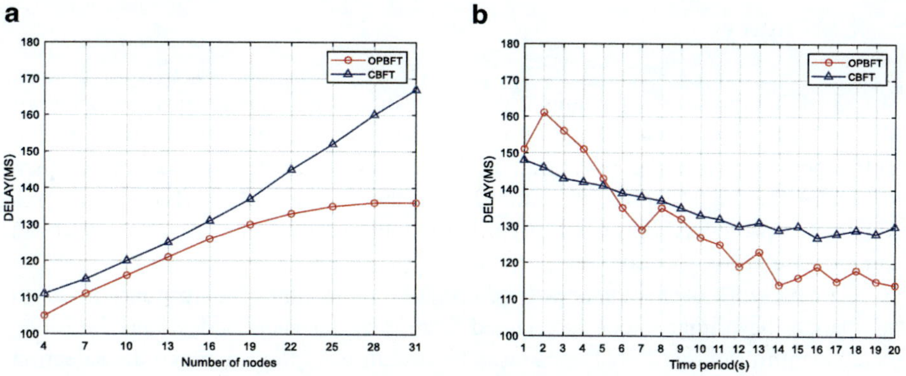

Fig. 3 (**a**) delay comparison of the two algorithms; (**b**) the change of delay over time

upgrade and degradation operations. Therefore, as shown in Fig. 3b, although the early delay of the OPBFT algorithm is slightly higher than the CBFT algorithm, soon with the improvement of the reliability of the consensus node set, the delay of the OPBFT algorithm is lower than the CBFT algorithm and gradually stabilizes at about 120 ms.

4.4 Security Analysis

Aiming at the design of the reputation model in this paper, the following security analysis is made on how the mechanism responds to potential threats. When a consensus node with a higher reputation value is attacked, according to the reputation model of this paper. Its byzantine behavior will be discovered in time, the reputation value will be sharply reduced, and finally it will be eliminated from the consensus node set. The reputation model uses a logarithmic function to limit the growth rate of the node's reputation value, which further increases the difficulty for a malicious node to want to become a consensus node again. Even if the malicious node disguise as an honest node to improve its own discourse rights, its reputation value won't increase as fast as honest nodes. Therefore, the proportion of honest nodes does not decrease over time.

As long as the number of byzantine nodes f does not exceed 1/3 of the total number of consensus nodes, the OPBFT algorithm can ensure the security and liveness of the blockchain system. During the system initialization phase, all nodes have the same reputation value. The PBFT algorithm is proved to guarantee the security and liveness of the system in (Castro et al., 2002). In the OPBFT algorithm, due to the construction of the reputation model, the proportion of honest nodes becomes higher and higher with the change of time, so the algorithm can still ensure the security and liveness of the system.

5 Summary

The OPBFT consensus mechanism proposed in this paper evaluates the behavior of each consensus node in the consensus process through a reputation model, and cooperates with the byzantine node detection and degradation mechanism to ensure the high degree of reliability of the system; Meanwhile, traditional consistency protocols' optimization further reduce system communication overhead and increase consensus efficiency.

The next work of this paper will be around the election of the primary node. In this paper, the primary node is selected from the consensus nodes, and the rotate mechanism in the traditional consensus algorithm is applied to make the selection of the primary node predictable. Increasing the randomness of the primary node selection will increase the difficulty of adversary attacks. Therefore, the design of the random selection method has a significant impact on the security of the consensus algorithm. In addition, with the rapid development of technologies such as machine learning and big data, the integration of cutting-edge technologies into the blockchain consensus mechanism will be of great benefit to analyze and predict node behavior in advance to discover byzantine nodes and resist conspiracy attacks.

Acknowledgements This work is partially supported by the National Natural Science Foundation of China (61672022, 61272036), Key Disciplines of Computer Science and Technology of Shanghai Polytechnic University (XXKZD1604).

References

Bach, L. M., Mihaljevic, B., & Zagar, M. (2018). Comparative analysis of blockchain consensus algorithms. In: *Proceedings of 2018 41st International Convention on Information and Communication Technology, Electronics and Microelectronics (MIPRO)* (pp. 1545–1550). Opatija: IEEE.

Castro, M., & Liskov, B. (1999). Practical byzantine fault tolerance. In: *Proceedings of the third Symposium on Operating Systems Design and Implementation*. New Orleans, Louisiana, USA (pp. 173–186). OSDI.

Castro, M., Liskov, B., & Pease, M. (2002). Practical byzantine fault tolerance and proactive recovery. *ACM Transactions on Computer Systems, 20*, 398–461.

Fang, W. W., Wang, Z. Y., Song, H. L., et al. (2019). An optimized PBFT consensus algorithm for blockchain [J]. *Journal of Beijing Jiaotong University, 43*(5) (方维维, 王子岳, 宋慧丽, et al. 一种面向区块链的优化PBFT共识算法[J]. 北京交通大学学报, *43*(5), 2019.).

Gervais, A., Karame, G. O., Karl, W., et al. (2016). On the security and performance of proof of work blockchains. In *[C]//ACM SIGSAC Conference* (pp. 3–16).

Houy, N. (2014). It will cost you nothing to 'kill' a proof-of-stake crypto-currency [J]. *Social Science Electronic Publishing, 34*(2), 1038–1044.

Lamport, L. (2001). Paxos made simple. *ACM SIGACT News, 32*(4), 18–25.

Lamport, L., Shostak, R., & Pease, M. (1982). The byzantine generals problem. *ACM Transactions on Programming Languages & Systems, 4*(3), 382–401.

Lei, K., Zhang, Q. C., Xu, L. M., et al. (2018). Reputation-based byzantine fault-tolerance for consortium blockchain. In *2018 IEEE 24th International Conference on Parallel and Distributed Systems (ICPADS)* (pp. 604–611). IEEE.

Li, J. F. (2018). *Research and application of blockchain consensus algorithm based on Byzantine Fault Tolerance mechanism* [D]. Zhengzhou University. (李剑锋. 基于拜占庭容错机制的区块链共识算法研究与应用[D]. 郑州大学, 2018.)

Liu, X. F. (2017). *Research on blockchain performance improvement based on Byzantine Fault Tolerance consensus algorithm based on dynamic authorization* [D]. Zhejiang University. (刘肖飞. 基于动态授权的拜占庭容错共识算法的区块链性能改进研究[D]. 浙江大学, 2017.)

Nakamoto, S. (2019, December 17). *Bitcoin: A peer-to-peer electronic cash system* [Online]. Available: https://bitcoin.org/bitcoin.pdf.

Ongaro, D., & Ousterhout, J. (2014). In search of an understandable consensus algorithm. In *Proceedings of Usenix Conference on Usenix Technical Conference* (pp. 305–319). Philadelphia: ACM.

Proof of stake [Online]. (2019, April 11). Available: https://en.bitcoin.it/wiki/ProofofStake.

Schlichting, R. D., & Schneider, F. B. (1983). Fail-stop processors: An approach to designing fault-tolerant computing systems [J]. *ACM Transactions on Computer Systems (TOCS), 1*(3), 222–238.

Schuh, F., & Schieÿl, S. (2019, November 29). *The BitShares Blockchain* [Online]. Available: https://github.com/bitshares-foundation/bitshares.foundation/blob/master/download/articles/BitSharesBlockchain.pdf.

Vukolic, M. (2015). The quest for scalable blockchain fabric: Proof-of-work vs. bft replication. *In International workshop on open problems in network security* (pp. 112–125).

Yuan, Y., & Wang, F. Y. (2016). Blockchain: The state of the art and future trends. *Acta Automatica Sinica, 42*(4), 481–494. (袁勇, 王飞跃. 区块链技术发展现状与展望. 自动化学报, *42*(4), 481–494, 2016.).

Entropy Weight-TOPSIS Method Considered Text Information with an Application in E-Commerce

Ailin Liang, Xueqin Huang, Tianyu Xie, Liangyan Tao, and Yeqing Guan

1 Introduction

As the rapid development of Internet economy, many companies are planning to publish and sell new products in the online market. However, how to design products and sell them online are the two major problems in the business process. As we all know, there are too much data that can be analyzed from Amazon, Taobao, JD.COM and other websites. These data have great commercial value. Through efficient collection, processing and analysis, we can help companies make more reasonable decisions on how to design products and how to sell online.

Some researchers take e-commerce products as the research object, based on the sentiment analysis of reviews (Liu Yulin & Tong Lirong, 2018; Li Huizong et al., 2019), putting forward the selection scheme and sales forecast of foreign trade e-commerce products, (Zhang Yue, 2019) studied online reviews of cold chain agricultural product e-commerce, and there are many other similar studies, such as (Yang Ruixin, 2017) on air conditioning and (Zhao Zhibin et al., 2017) on Chinese products. However, few studies focus on finding successful features from comments combined with other data, and help companies make decisions. Finding these features is very important, because for those companies who want to sell things online, there is a lot of wealth in data. This will help companies to make more correct decisions and gain more profits.

At present, there are many technical processes of generating data and information in real time. New technologies, such as big data and artificial intelligence, are the tools to analyze these data, which provide important value for companies. Sentiment analysis is a good way to analyze people's speech and individual behavior. It can

A. Liang (✉) · X. Huang · T. Xie · L. Tao · Y. Guan
College of Economics and Management, Nanjing University of Aeronautics and Astronautics, Nanjing, Jiangsu, China
e-mail: ailinliang@nuaa.edu.cn

H. Yang et al. (eds.), *AI and Analytics for Public Health*, Springer Proceedings in Business and Economics, https://doi.org/10.1007/978-3-030-75166-1_8

be used in many fields, such as predicting political election results (Paul et al., 2017), mental health care (Zhai et al., 2010), review analysis (Ye et al., 2005), and product analysis (Wang et al., 2018). Sentiment analysis mainly studies the emotional tendencies of texts from grammar, semantic rules and other aspects (Yanghui Rao et al., 2014). The texts from social network has the characteristics of few words, irregular grammar and noisy data, which increases the difficulty of sentiment analysis (Jun He et al., 2020). Therefore, a reasonable method is also needed to analyze the text in product reviews. Then, the sentiment analysis model based on LDA (Jun He et al., 2020; Wang Rui et al., 2020) is used to score each comment in the data set, and the success of each product is judged by considering the score. After obtaining the emotional score for each product, the data set needs preprocessing to facilitate further analysis. Actually, there are only nine useful data metrics: product_id, star_rating, helpfulvotes, total_votes, vine, verified_purchase, review_header, review_body and review_date. Based on Excel and Python as tools for this study, we can get other data characteristics through analysis and processing (see the subsequent analysis for details). The LDA-based sentiment analysis is used to score each comment, and then the score is used as an index to judge whether each product is successful, which also requires a reasonable weight matrix. There is a method of establishing weight matrix, namely entropy weight method. According to the basic principles of information theory, information is a measure of the degree of system order, and entropy is a measure of the degree of disorder. The smaller the information entropy of the index, the greater the information provided by the index. The greater the effect, the higher the weight. Therefore, we can use this method to calculate the weight of each index, which provides a basis for comprehensive evaluation of multiple indicators. Therefore, the method of weights determination based on entropy weight is objective and effective enough. Entropy method has been applied in many aspects, such as evaluation, decision-making and selection (Jian-qiang Zhang & Hai-hong Sun, 2019; Farhadinia, 2017; Yanmeng Zhang et al., 2017; Xiangxin Li et al., 2011) In 1981, C. L. David Henry Hwang and k. Yoon first proposed the method of ranking by similarity with ideal solution. TOPSIS is a method to calculate the relative distances between objects and the idealized target, and then sorting these objects based on distances. TOPSIS is a general multiple attribute decision making method. And entropy weight is always combined with TOPSIS method to tackle reality problems, such as supplier selection of home appliance industry supply chain (Yanmeng Zhang et al., 2017) and safety evaluation of coal mines (Xiangxin Li et al., 2011). Entropy method and TOPSIS are good at evaluating and find the best choice. On this basis, we can find ten successful products for people to analyze why these products sell well on the Internet.

As shown in Fig. 1, this study first collects data from Amazon, and we analyze the data, then delete unnecessary data and keep the data between 2012 and 2014, which are marked as "y" in vine, "n" in verified_purchase, and vice versa, and then sort the six data sets according to product_id. The index system is constructed based on correlation analysis. This system contains 4 important indexes composed of original measurements. Secondly, use LDA-based sentiment analysis with clean data, to quantify every emotional keyword and get the arithmetic mean as the emotional

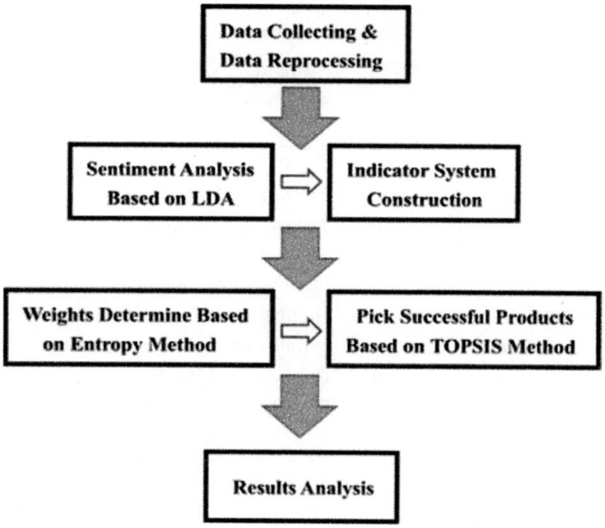

Fig. 1 Technical route

value of the whole review. Output score for each reviews_body. Then we pick out successful products based on the Entropy Weight-TOPSIS method. Through analyzing these top products, we can extract the successful features.

2 Data

2.1 Raw Data

In the online marketplace it created, Amazon provides customers with an opportunity to rate and review purchases. Individual ratings – called "**star ratings**" – allow purchasers to express their level of satisfaction with a product using a scale of 1 (low rated, low satisfaction) to 5 (highly rated, high satisfaction). Additionally, customers can submit text-based messages – called "**reviews**" – to express further comments and information about the product. Other customers can submit ratings on these reviews as being helpful or not – called a "**helpfulness rating**" – towards assisting their own product purchasing decision. We collect data on three products: a microwave oven, a baby pacifier, and a hair dryer. The three datasets provided include product user ratings and reviews extracted from the Amazon Customer Reviews Dataset through Amazon Simple Storage Service (Amazon S3). A detailed vocabulary of data tag definitions is as follows (Table 1):

Table 1 Data tags define vocabulary

Attributes	Meaning
customer_id (string)	Random identifier that can be used to aggregate reviews written by individual author.
review_id (string)	The unique ID of the review.
product_id (string)	The unique Product ID the review pertains to.
product_category (string)	The major consumer category for the product.
star_rating (int)	The 1–5 stars rating of the review given by people to rate a product with a number of stars.
helpful_votes (int)	Number of helpful votes
total_votes (int)	Number of total votes the review received.
vine (string)	Customers are invited to become Amazon Vine Voices based on the trust that they have earned in the Amazon community for writing accurate and insightful reviews. Amazon provides Amazon Vine members with free copies of products that have been submitted to the program by vendors. Amazon doesn't influence the opinions of Amazon Vine members, nor do they modify or edit reviews.
verified_purchase (string)	A "Y" indicates Amazon verified that the person writing the review purchased the product at Amazon and didn't receive the product at a deep discount.
review_body (string)	The review text.
review_date (bigint)	The date the review was written.

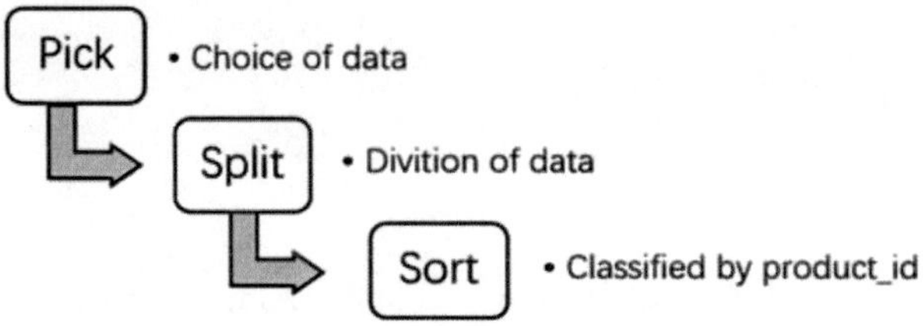

Fig. 2 Data preprocessing flow chart

2.2 *Data Preprocessing*

1. Import the three data set: hair_dryer.tsv, microwave.tsv and pacifier.tsv, into Excel sheet. Then observe them and analyze them by common sense. Choose the data of 2012, 2013 and 2014, because those deleted data (before 2012 and after 2015) are totally unreliable and redundant.
2. Split three datasets into six datasets. (Extract data measures "vine" and "verified_purchase", if "vine = Y", it means that the reviews are more accurate and insightful. Similarly, "verified_purchase = Y" indicates that the reviews are more objective. Therefore, the original data is divided into two parts according to "vine=Y & verified_purchase= N" or "vine=N & verified_purchase= Y".)
3. Sort all selected data by product_id & time(month) on python 3.7 (Fig. 2).

3 Modeling Preparation

3.1 Indicator System Construction

First, according to the previous analysis, we find that after mathematical processing, these flow data measures can be used as candidate indicators of the indicator system, and these four indicators are sorted by product _ id:

① $\overline{star_{rating}}$: The average of "star_rating" for each product. It's an important factor giving people reference, and the average can be more useful and meaningful, therefore we pick it into our indicator system.
② $\overline{\frac{\text{helpful}_{\text{votes}}}{\text{total}_{\text{votes}}}}$: The average of "helpful_votes/total_votes" for each product. It's an important factor to prove the authenticity and reliability of views, which displays the reputation of a product. (If total_votes $= 0$, $\frac{\text{helpful}_{\text{votes}}}{\text{total}_{\text{votes}}} = 0$)
③ $\overline{score}$: The Number which translated from "review_headline & review_body" based on sentiment analysis, shows the average product quality and consumer satisfaction for each product.
④ Sales: Sales volumes of each product according to product_id (Table 2).

3.2 LDA-Based Sentiment Analysis

1. Assumptions

① Consumers' emotions is closely related to product quality. ② Data is valid for completing emotion analysis.

2. The Foundation of LDA

LDA (Linear Discriminant Analysis) is a topic model, which can display the topic of each text in the form of probability distribution, and reverse the distribution of the topic based on a given document. After analyzing some texts to extract their topics (distribution), these texts and topics can be classified into clusters. At the same time, it is a typical bag-of-words model, that is, a text is composed of a group of words, and there is no sequential relationship between words. Establishing variables for LDA (Table 3):

Table 2 Indicators' symbols

Contents	Symbols
$\overline{star_{rating}}$	x_1
$\overline{\frac{\text{helpful}_{\text{votes}}}{\text{total}_{\text{votes}}}}$	x_2
$\overline{score}$	x_3
Sales	x_4

Table 3 Variables' define for LDA

Contents	Symbols
Keyword	K
Number of keywords in comments	n
Number of negative words in comments	m
Single keyword sentiment score	$e_i, i \in \{1, 2, \cdots, n\}$
Emotion score	E
Degree word sentiment value	level1(0.2 ~ 0.5)level2(0.4 ~ 0.7)level3(0.6 ~ 0.9)
	level4(1.2 ~ 1.6)level5(1.6 ~ 1.9)level6(2.0 ~ 2.5)
Negatives	D

The initial emotional value of positive word is 1, and the initial emotional value of negative word is − 1. We use this model and this idea to implement the text sentiment analysis of the views. The realization principle is as follows:

Step1: Loading emotional dictionary (divided into positive word/negative word/degree word/negative word) and word segmentation.

Step2: The emotional value of positive and negative words is accumulated by keyword matching. The emotional default value of positive and negative words are 1/−1.

$$K = \begin{cases} 1, word\ is\ positive \\ -1, word\ is\ negative \end{cases} \tag{1}$$

Step3: Retrieving and verifying the semantics of negatives D in text, and confirming the positive and negative again by multiple checks.

$$K' = -1^m K \tag{2}$$

Step4: Controlling the degree of repeated emphasis by a specific interval attenuation function. Determining the level of emphasis and multiplying K' with L (Degree word sentiment value) which is determined by the upper and lower limits of the interval and the attenuation coefficient.

$$e_i = K' \times L \tag{3}$$

Step5: Get an overall emotional value E of the text which is the arithmetic average of each single keyword's emotional value e_i, and the results of piecewise analysis.

$$E = \frac{1}{n}\sum_{i=1}^{n} e_i \tag{4}$$

3. Results (Run this model in python 3.7)

Output scores of each reviews_body. There are samples of results (Table 4):

Table 4 Sentiment scores on reviews for pacifiers

Reviews_body	Scores (E)
Sample1: Easy to clean, no weird smell, feels secure. These are pretty similar to Nuk pacifiers, but I like the material better. Nice pacifier.	2.67
Sample2: this is not worth it if you have any wind in your neighborhood. it does not have any weight to it and will simply blow down the street with any degree of wind. not worth it unless you get a 5 lb weight to hold it down.	−2.00
Sample3: very disappointed on this one. we need it for my husband to hold onto to enter the bath and it doesn't stick. it will lose suction at random times and you'll hear it crashing down into the tub. it's on smooth ceramic tiles as well. it should stay put. not happy with this purchase.	−2.89

4 Decision Making Based on Entropy Weight-TOPSIS Method

4.1 The Foundation of Model

To observe and confirm the characteristics of successful products by using multi-attribute decision-making method, it is necessary to establish a reliable index system. Then we choose entropy weight method to gain the weight matrix for indicators and TOPSIS method to find successful products. The modelling steps are as follows:

Step1: According to the indicator system and definitions, a decision matrix is obtained $X = (x_{ij})_{m \times n}$.

n: number of indicators, m: number of products, x_{ij} represents the j-th indicator value of the i-th evaluation object.

Step2: Normalize the raw matrix $Z = (z_{ij})_{m \times n}$ by column.

$$z_{ij} = \frac{x_{ij}}{\sqrt{\sum_{i=1}^{m} x_{ij}^2}} \ (1 \le i \le m, 1 \le j \le n) \tag{5}$$

Step3: Entropy can be expressed as the following formula by definition:

$$s\left(z_j\right) = \sum_{i=1}^{m} z_{ij} \ln z_{ij} \tag{6}$$

Step4: Calculate the entropy of each indicator.

$$S_j = -k \cdot s\left(z_j\right) \tag{7}$$

k is related to the number of samples m, and $k = \frac{1}{\ln m}$. The supplementary definition: If $z_{ij} = 0$, let $z_{ij} \ ln \ z_{ij} = 0$.

Step5: The Degree of difference of indicators.

$$G_j = 1 - S_j \tag{8}$$

Step6: Calculate the weight coefficient of each indicator.

$$c_j = \frac{G_j}{\sum_{i=1}^{n} G_i} \tag{9}$$

Step7: Because the importance of various indicators is different, the entropy weight of each indicator should be considered, and the normalized data is weighted to obtain the following weighted normalization matrix.

$$V = \left(v_{ij}\right)_{m \times n} = \begin{bmatrix} c_1 z_{11} & c_2 z_{12} & \cdots & c_n z_{1n} \\ c_1 z_{21} & c_2 z_{22} & \cdots & c_2 z_{2n} \\ \vdots & \vdots & \vdots & \vdots \\ c_1 z_{m1} & c_2 z_{m2} & \cdots & c_n z_{mn} \end{bmatrix} \tag{10}$$

Step8: Determine positive ideal solution and negative ideal solution.

$$V^+ = \left\{\left(\max_i v_{ij} | j \in J_1\right), \left(\min_i v_{ij} | j \in J_2\right)\right\} (i = 1, 2, \ldots, m) \tag{11}$$

$$V^- = \left\{\left(\min_i v_{ij} | j \in J_1\right), \left(\max_i v_{ij} | j \in J_2\right)\right\} (i = 1, 2, \ldots, m) \tag{12}$$

Among them J_1 is the benefit-type indicator set and J_2 is the cost-type indicator set.
Step9: Calculate distance.
The distances between the evaluated indicators and the positive ideal solution:

$$d_i^+ = \left[\sum_{j=1}^{n} \left(v_{ij} - v_j^+\right)^2\right]^{\frac{1}{2}} (i = 1, 2, \ldots, m) \tag{13}$$

The distance from the evaluated indicators to the negative ideal solution:

$$d_i^- = \left[\sum_{j=1}^{n} \left(v_{ij} - v_j^-\right)^2\right]^{\frac{1}{2}} (i = 1, 2, \ldots, m) \tag{14}$$

Step10: Calculate relative proximity. The relative proximity of each indicator to the ideal solution:

$$C_i = \frac{d_i^-}{d_i^+ + d_i^-} (i = 1, 2, \ldots, m) \tag{15}$$

Step11: Ranking relative proximity C_i.

According to the relative proximity values obtained above. The greater the relative closeness of the product, the more successful the product will be.

4.2 Results and Analysis (Run This Model in Python 3.7)

Six datasets have six weight matrixes, as shown in Table 5.

There is a sample result (picking the top 10 of pacifiers) in Table 6.

Firstly, $\boldsymbol{x_3}$ is the number which is translated from "review_headline & review_body" based on sentiment analysis. By sentiment analysis, the evaluation and qualitative analysis of products by consumers are transformed into quantitative indicators. Therefore, the index size of $\boldsymbol{x_3}$ represents the satisfaction degree of customers. The higher the score, the greater the customer's satisfaction and the more successful the product is. Therefore, the index of $\boldsymbol{x_3}$ can reflect the superiority of some characteristics of the product itself. For example, high-quality raw materials, aesthetic appearance, good customer experience, etc., greatly affect or determine the success of online sales of products.

In addition, $\boldsymbol{x_2}$ is the average value of "helpful_votes/total_votes" for each product. $\boldsymbol{x_2}$ represents the proportion of the number of consumers who voted that the

Table 5 The weights of datasets

Datasets	x_1	x_2	x_3	x_4
pacifier_vine	0.00857046	0.31428073	0.53113709	0.14601172
pacifier_verified_purchase	0.00920789	0.46741284	0.46477582	0.05860344
microwave_vine	0.00641744	0.01582893	0.41178463	0.56596899
microwave_verified_purchase	0.03262573	0.18562624	0.55308866	0.22865938
hairdryer_verified_purchase	0.01570945	0.27572281	0.61800476	0.09056299
hairdryer_vine	0.00728758	0.6814491	0.28472485	0.02653847

Table 6 Decision making on products

product_id	d_i^+	d_i^-	C_i	order
B0009XH6TG	0.136476	0.302574	0.689156	1
B00132ZG3U	0.136454	0.301603	0.688501	2
B003V264WW	0.137605	0.286589	0.675609	3
B00005O0MZ	0.147816	0.264795	0.641755	4
B000R80ZTQ	0.160629	0.224238	0.582637	5
B001QTW2FK	0.200724	0.158145	0.440677	6
B001UE7D2I	0.204837	0.155294	0.431216	7
B000A3I2X4	0.208026	0.153092	0.423938	8
B00APV7OWG	0.230470056	0.120650182	0.343615003	9
B0009XH6WI	0.238325699	0.114978564	0.325437806	10

product is useful to the total number of consumers who voted. The larger the index is, the greater the proportion of people who think the product is useful. Therefore, this index reflects the reputation and popularity of the product among consumers who purchase the product. This index reflects whether the product is welcomed by consumers more truly and reliably. Thus, this index can also reflect some important features of the product itself, which have the magic of attracting consumers and being favored by more consumers.

The weight ratio of $\boldsymbol{x}_2$ and $\boldsymbol{x}_3$ far exceeds the weight ratio of $\boldsymbol{x}_1$ and $\boldsymbol{x}_4$, which is up to 76.6326303%. This shows that consumers' preference information about products can be transformed into the data of these two indicators. Therefore, it is of more important theoretical and practical significance to analyze the characteristics of products based on $\boldsymbol{x}_2$ and $\boldsymbol{x}_3$ index data. Thus, for a company planning to design and sell online products, it should pay more attention to the information of consumers' favorite degree reflected by $\boldsymbol{x}_2$ and $\boldsymbol{x}_3$ indicators, as well as some characteristic information of the products themselves reflected based on this. According to the different characteristics of different products, it is easier to grasp the key features and contradictions of online sales products, design or optimize the corresponding products, and achieve the success of online sales. Moreover, It is also suggested that online sellers should pay attention to customer feedback information, use various measures to win consumers' favor, and establish good reputation, so as to increase the possibility of repurchase by old customers and attract new customers.

By analyzing the characteristics of the top ten products, we can provide some useful proposals for the company to making better decisions. Taking pacifiers as an example, this paper finds out the IDs of ten pacifiers by this method, and analyzes some characteristics of them as examples (Table 7).

For example, in terms of price, excluding the highest price and the lowest price, and calculating the average price of the remaining eight products, it can be found that if the price of the products is set at about 6.315$, the products are more likely to succeed in online sales. In terms of product materials, it can be found that 9/10 products are made from silica gel. Therefore, products made of this raw material are

Table 7 Product_id and some typical characteristics

product_id	Price	Material	Whether have free gift	...
B0009XH6TG	4.39$	Wacker silica gel	Yes	...
B00132ZG3U	5.99$	LSR food grade silica gel	No	...
B003V264WW	5.29$	Wacker silica gel	Yes	...
B00005O0MZ	4.39$	Nano silver liquid silica gel	Yes	...
B000R80ZTQ	9.99$	Food grade silica gel	Yes	...
B001QTW2FK	3.49$	Silastic	No	...
B001UE7D2I	16.39$	Food grade silica gel	Yes	...
B000A3I2X4	8.99$	Food grade silica gel	Yes	...
B00APV7OWG	5.99$	Nano silver liquid silica gel	Yes	...
B0009XH6WI	5.49$	Wacker silica gel	No	...

easier to sell successfully on the internet. As to whether there are free gifts or not, 7/10 products provide consumers with free gifts, so it is one of the successful ways to obtain online sales. There are more than three factors influencing the success of online sales. Therefore, when a company intends to open up the Internet market of a certain product, it can comprehensively evaluate the characteristics and analyze the advantages and disadvantages of the product, so as to make more scientific and effective decisions.

5 Conclusions

In our research, we use LDA sentiment analysis to convert online products reviews from text to numbers. Then, an index system with 4 indexes is constructed. Based on the entropy weight method, each index is given appropriate weight, and TOPSIS is used for comprehensive evaluation according to the characteristics of products, which provides quantitative basis for enterprises to make timely and reasonable product market decisions.

Through the above work, we find that x_2 and x_3 are very important and meaningful. Companies should pay more attention to x_2 and x_3 on the Internet to gain more trust from consumers. Different data has different values for different products. Managers should focus on different products, pay attention to different points, grasp the main contradictions, and analyze them, which can often grasp consumers' hearts.

On the one hand, this research is a way to enable managers to discover key points in big data faster and more accurately. On the other hand, it's a new and extensive method for TOPSIS. It enables TOPSIS to tackle indicator systems which contain text attributes.

Acknowledgements This research was supported by a project of the National Natural Science Foundation of China (72071111). At the same time, the authors would like to acknowledge the partial support of the project of Intelligence Introduction Base of the Ministry of Science and Technology (G20190010178).

References

Farhadinia, B. (2017). A multiple criteria decision making model with entropy weight in an interval-transformed hesitant fuzzy environment [J]. *Cognitive Computation, 9*(4).

Jian-Qiang Zhang & Hai-Hong Sun. (2019). Study on the evaluation index of cadet's physical training based on the entropy weight [C]. In *Proceedings of 2019 2nd International Conference on Informatics, Control and Automation (ICA 2019)* (pp. 88–92). Advanced Science and Industry Research Center: Science and Engineering Research Center.

Jun He, Hongyan Liu, Yiqing Zheng, Shu Tang, Wei He, & Xiaoyong Du. (2020). Bi-Labeled LDA: Inferring interest tags for non-famous users in social network [J]. *Data Science and Engineering, 5*(1).

Li Huizong, Yao Yao, Wang Xiangqian, et al. (2019). Sentiment analysis of online reviews of cold chain agricultural product e-commerce based on LDA [J]. *Journal of Nanyang Institute of Technology, 11*(2), 25–30.

Liu Yulin, & Tong Lirong. (2018). E-commerce online review data mining based on text sentiment analysis [J]. *Statistics and Information Forum, 33*(12), 119–124.

Paul, D., Li, F., Teja, M. K., Yu, X., & Frost, R. (2017). Compass: Spatio temporal sentiment analysis of US Election what Twitter says! In *Proceedings of the 23rd ACM SIGKDD International Conference on Knowledge Discovery and Data Mining, Halifax, Canada* (pp. 1585–1594).

Wang, F. F., Zhang, S. T., Zhang, J. L., et al. (2018). Research on the majority decision algorithm based on WeChat sentiment classification [J]. *Journal of Intelligent & Fuzzy Systems: Applications in Engineering and Technology, 35*(3), 2975–2984.

Wang Rui, Long Hua, Shao Yubin, & Du Qingzhi. (2020). Text feature extraction method based on labeled-LDA model [J]. *Electronic Measurement Technology, 43*(01), 141–146.

Xiangxin Li, Kongsen Wang, Liwen Liu, Jing Xin, Hongrui Yang, & Chengyao Gao. (2011). Application of the entropy weight and TOPSIS method in safety evaluation of coal mines [J]. *Procedia Engineering, 26*.

Yang Ruixin. (2017). *Emotion analysis of review data of e-commerce air conditioning products [D]*. Shanxi University.

Yanghui Rao, Jingsheng Lei, Liu Wenyin, et al. (2014). Building emotional dictionary for sentiment analysis of online news [J]. *World Wide Web, 17*(4), 723–742.

Yanmeng Zhang, Shengshi Zhou, & Di Wu. (2017). Research on supplier selection of home appliance industry supply chain based on entropy weight and TOPSIS method [C]. Research Institute of Management Science and Industrial Engineering. In *Proceedings of 2017 5th International Conference on Frontiers of Manufacturing Science and Measuring Technology (FMSMT 2017)* (pp. 672–676). Research Institute of Management Science and Industrial Engineering.

Ye, Q., Li, Y., & Zhang, Y. (2005). Semantic-oriented sentiment classification for Chinese product reviews: An experimental study of book and cell phone reviews. *Tsinghua Science and Technology, 10*(S1), 797–802.

Zhai, Z., Xu, H., & Jia, P. (2010). An empirical study of unsupervised sentiment classification of Chinese reviews. *Tsinghua Science and Technology, 15*(6), 702–708.

Zhang Yue. (2019). *Research on selection of foreign trade e-commerce based on comment sentiment analysis and sales forecast* [D]. Beijing Jiaotong University.

Zhao Zhibin, Liu Huan, Yao Lan, et al. (2017). Dimension mining and sentiment analysis of Chinese product reviews * [J]. *Computer Science and Exploration*.

Optimal Resource Allocation for Coverage Control of City Crimes

Rui Zhu, Faisal Aqlan, and Hui Yang

1 Introduction

As the complexity and challenge of the world grow, drug abuse, large population, shifting demographics, and advanced technologies bring more complicated criminal situations. One of the core responsibilities of governments is to protect citizens from crimes. Current law enforcement strategies can deal with a large scale of crimes. The decline of crimes in the United States is a positive sign for the safety of all citizens and this decline is credited to effective law enforcement strategies (Morris, 1997). However, due to the changing scale and nature of criminal activities, there would be insufficient people and support to scale up previous efforts achieved by traditional approaches (Hipp & Kane, 2017). As every second counts in the crisis, limited law-enforcement resources pose significant challenges to effectively and efficiently protect the city. Therefore, current situations of crimes call for new strategies of the law enforcement allocation to achieve faster responses, reduced costs, and highly efficient operations.

In the United States, crimes can be classified into two main categories, violent crimes and property crimes. Violent crimes consist of assault, robbery, rape, and murder (Freilich & Pridemore, 2007). Property crimes consist of burglary and theft. The crime rates corresponding to various types of crimes from 1981 to 2018 in

R. Zhu
Complex Systems Monitoring, Modeling and Control Laboratory, The Pennsylvania State University, University Park, PA, USA

F. Aqlan
Department of Industrial Engineering, The Behrend College, The Pennsylvania State University, Erie, PA, USA

H. Yang (✉)
Department of Industrial Engineering, Pennsylvania State University, University Park, PA, USA
e-mail: huy25@psu.edu

H. Yang et al. (eds.), *AI and Analytics for Public Health*, Springer Proceedings in Business and Economics, https://doi.org/10.1007/978-3-030-75166-1_9

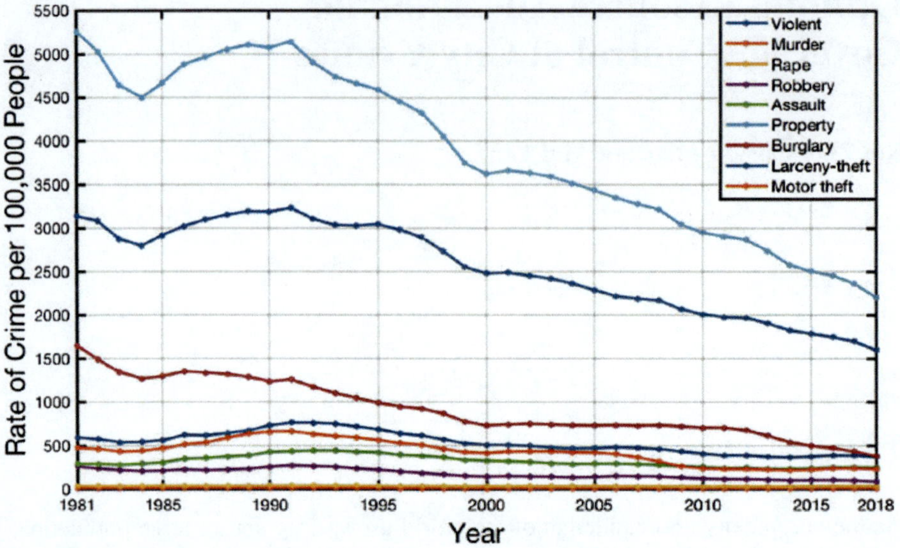

Fig. 1 Crimes in the United States from 1981 to 2018

the United States are shown in Fig. 1. The estimated number of violent crimes was 1,206,836 in 2018. Aggravated assault accounted for roughly 67% of the total violence. The estimated number of robberies was 282,061 and firearms were used in 38.5% of all robberies. The rough number of murders was 16,214 and firearms were used in 72.68% of murders. The estimated number of property crimes was 7,196,045 and resulted in losses of around $16.4 billion. The larceny-theft accounted for 72.5% of the property crimes (U. S. Department of Justice Federal Bureau of Investigation, 2018). Law enforcement officers call for better strategies that can help control and reduce crimes in the United States.

As the fourth largest city of Pennsylvania, Erie has an area of 19.37 square miles and a population of 101,786 people. The crime rate of Erie is relatively low compared to other large cities in New York, Ohio, and Pennsylvania. However, as shown in Fig. 2, Erie's crime rate is higher when compared to the national average, which is mainly because of economic factors that plague the city. Another reason for the high crime rate is the limited law-enforcement resources. In 2003, the police force had 214 officers. That number decreased to 161 officers in 2016. This resulted in Erie reaching its highest crime rate in 2008. Even in 2018, the crime rate in Erie was still higher than rates in 70.1% of U.S. cities (Crime rate in Erie, Pennsylvania (PA), 2020).

In this paper, we develop a new optimal learning algorithm to characterize multi-scale distributions of crimes and then determine an optimal policy for coverage control of city crimes. First, we categorize crimes into low, medium and high severity levels. Then, we characterize and model crime distributions for various severity levels. Second, we develop an optimal policy for coverage control to allocate limited resources of the law enforcement in areas of interest. Third, the model performance is measured based on average response time of an agent to

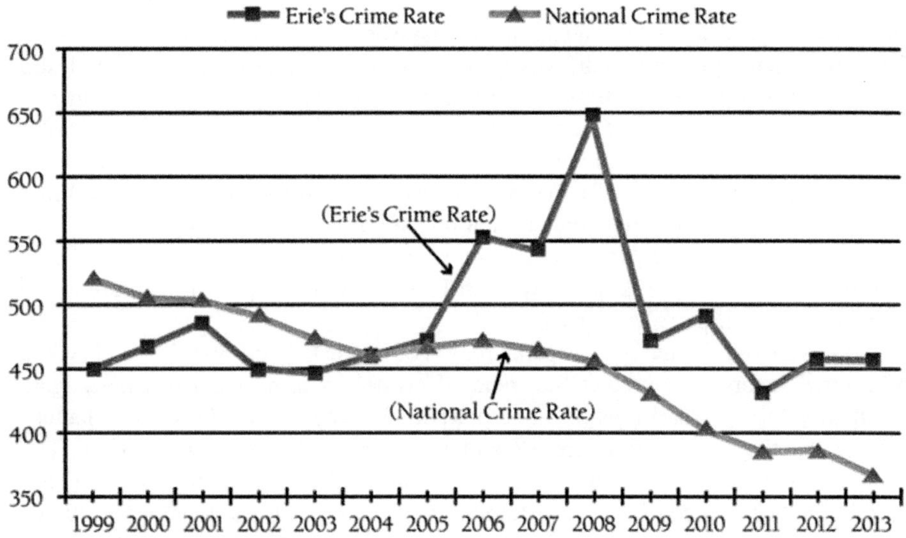

Fig. 2 Erie crimes vs. national average (Jefferson Educational Society, 2015)

reach crime scenes. Experimental results demonstrate that the proposed algorithm is effective and efficient in optimizing resource allocation for the coverage control of city crimes and shows a better performance in terms of average response time to crime scenes than zone-random patrol and uniform allocation. The rest of this paper is organized as follows: Section 2 introduces the state of the art in optimal resource allocation for city crimes. Section 3 presents the proposed methodology of optimal resource allocation for coverage control. Section 4 shows the experimental design and results, and Sect. 5 includes the conclusions arising from this investigation.

2 Research Background

Data analytics are widely studied by law enforcement officials and criminal investigators to analyze crime data. It is a challenging task to discover and understand the complex patterns of crimes from multiple perspectives. Risk Terrain Modeling (RTM) is one of the techniques to localize places where there is a high probability that a crime will occur. Also, RTM is able to identify risk factors of criminal events (Kocher & Leitner, 2015). A fuzzy association rule mining is developed by Buczak (Buczak & Uhrig, 1996) to investigate the underlying community crime pattern. As the crime data get big and complex, manual and visual analytics are shown to be limited in the ability to handle a large volume of data. A space-time and multivariate visualization system (VIS-STAMP) is then developed to identify hidden patterns of aggravated assault, robbery, burglary, etc. This approach is implemented to analyze the crime data of Philadelphia, Pennsylvania by Guo and Wu (Guo, 2009). Clustering techniques are commonly utilized to extract the crime pattern. Malleson et al. identify the area with significant crime rates by exploring crime clusters (Malleson & Andresen, 2016). Different features are given various weights in order to improve the accuracy of the model and remove outliers. The clustering model is leveraged to investigate how frequently crimes occur at different times. A weighting scheme is also developed to handle limitations of clustering techniques. Phillip et al. integrated crime data analytics along with socio-economic and socio-demographic factors to discover patterns that may contribute to the development of future crime occurrences (Phillips & Lee, 2012).

Crimes are distributed over a spatial region of interests (e.g., Erie, PA) and evolve over time. In order to predict the trend of crimes, it is necessary to investigate the spatiotemporal variations of crime patterns. In other words, how crime pattern changes in the region of interests? and what are the variations of crime data over different periods of time? Spatiotemporal modeling is conducive to better predict the occurrence of crimes. Although very little has been done to investigate spatiotemporal variations of crimes, spatiotemporal modeling has been developed and applied in many other disciplines, e.g., spatiotemporal modeling of electrical potentials on the body surface (Yao et al., 2018) and in the heart (Yang et al., 2013), and spatiotemporal modeling of service accessibility over a large geographic area (i.e., Georgia State, USA) for 13 years (Serban, 2011). The cluster confidence rate boosting has been utilized to identify the hierarchical structure of spatiotemporal patterns (Yu et al., 2016). Kotevska et al. have developed a model for temporal trends with multivariate spatiotemporal data streams to improve the performance of prediction (Kotevska et al., 2017).

In spite of rapid advancements in crime data analytics, very little has been done to utilize crime data for optimal allocation of law enforcement resources. As the complexity of crimes grows, current practice of law enforcement, which is simply increasing the number of resources to cover and control crimes, may not be effective. There is an urgent need to effectively and efficiently control crimes through optimal allocation of limited resources.

3 Research Methodology

This paper presents a new strategy for optimal allocation of law enforcement resources towards coverage control of city crimes. We develop an optimal learning algorithm to characterize multi-scale distributions of crimes and determine optimal coverage control of city crimes. First, we characterize and model distributions of city crimes within the area of interest. Second, the area of interest is partitioned into regions which optimally cover nonuniformly distributed crimes. Third, an optimal learning algorithm is developed to allow agents to asymptotically converge to centroids of corresponding regions.

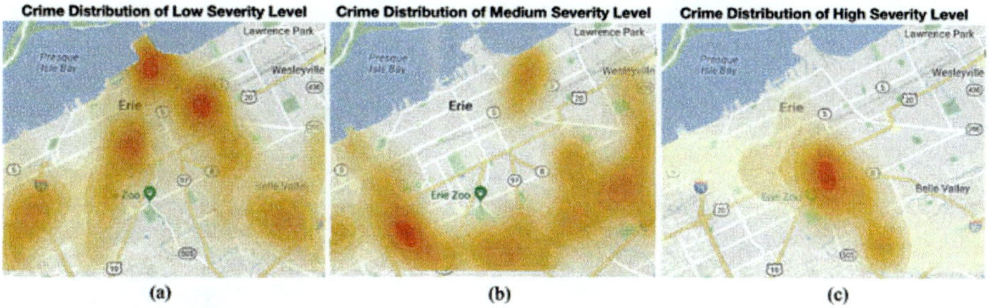

Fig. 3 Crime distribution: (**a**) Low severity; (**b**) Medium severity; (**c**) High severity

3.1 Crime Distribution Modeling

Crimes are categorized into three severity levels, namely low, medium, and high. The low severity level includes crimes such as theft, abandoned vehicle and burglary without force. Crimes of the medium severity level include drunkenness, disorderly conduct, burglary with force, etc. The high severity level involves missing person, assault, aggravated assault by prisoner, etc. Distributions of crimes at three severity levels are modeled to demonstrate the coverage of crimes in Erie city. Figure 3a shows the distribution of crimes at low severity level. The distribution of medium severity level is shown in sub-figure (b). Figure 3c demonstrates the distribution of crimes with high severity. The density of crimes varies from low to high while the color changes from light to dark. The proposed algorithm is implemented to cover and control crimes based on crime distributions with constrained resources.

3.2 Optimal Coverage Control

Let Ω be a convex with information density following Gaussian distribution as shown in Fig. 4a. If we only have one law enforcement agent to control the area Ω, intuitively, we will place the agent at the location with the highest density of crimes to minimize the information loss. The information loss for infinite points on the space Ω is formulated as

$$C(\theta_i) = \int_{\Omega} d(\|s - \theta_i\|)\,\sigma(s)ds \tag{1}$$

where $d(\|s - \theta_i\|)$ is a distance function between the agent θ_i and the location s on space Ω, $\sigma(s)$ is the information density at the location s. In this paper, the information density is referred to as the density of crimes at location s. The information loss increases along with the increment of function $d(\|s - \theta_i\|)$ and the information density $\sigma(s)$. Further, if we possess n agents and a complex distribution

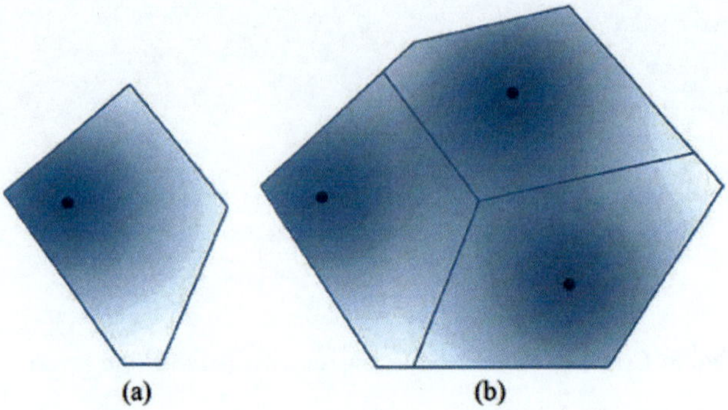

Fig. 4 Polygons with information densities following: (**a**) Gaussian distribution; (**b**) Complex distribution

on space Ω as shown in Fig. 4b, then we need to divide the space Ω into n regions and place one agent in each region. The objective is to minimize the total information loss through determining optimal tessellation and locations of agents, which is formulated as

$$C\left(\boldsymbol{\Theta}, \boldsymbol{\omega}\right) = \sum_{i=1}^{n} \int_{\omega_i} d\left(\|s - \theta_i\|\right) \sigma\left(s\right) ds \tag{2}$$

where $\boldsymbol{\Theta}$ is the agent set and $\boldsymbol{\Theta} = \{\theta_1, \theta_2, \ldots, \theta_n\}$, ω_i represents the region of space Ω, and $\boldsymbol{\omega} = \{\omega_1, \omega_2, \ldots, \omega_n\}$ is the tessellation, and n is the total number of agents.

Given a space Ω, Voronoi region V_i of agent θ_i is defined as

$$V_i = \left\{s \in \Omega : \|s - \theta_i\| \leq \left\|s - \theta_j\right\|, \forall j \neq i\right\} \tag{3}$$

where the set $\boldsymbol{V} = \{V_1, V_2, \ldots, V_n\}$ is the Voronoi tessellation of space Ω if $V_i \cap V_j = \varnothing$ for $i \neq j$. Each V_i in the set is referred to as Voronoi region corresponding to agent θ_i. Voronoi tessellation has different definitions using different distance functions. In this paper, we use the Euclidean norm to define Voronoi tessellation. The definition of Voronoi region ensures that all crime locations s within region V_i are the closest to agent θ_i. Therefore, the optimal tessellation for space Ω is the Voronoi tessellation.

Tessellation $\boldsymbol{\omega}$ in the objective function is then replaced with the Voronoi tessellation $\boldsymbol{V}$, and formulated as

$$C\left(\boldsymbol{\Theta}, \boldsymbol{V}\right) = \sum_{i=1}^{n} \int_{V_i} d\left(\|s - \theta_i\|\right) \sigma\left(s\right) ds \tag{4}$$

We compute the polar moment of inertia about the agent location as

$$J_{\theta_i} = \int_{V_i} \|s - \theta_i\|^2 \sigma\left(s\right) ds \tag{5}$$

If we set the function $d(\|s - \theta_i\|) = \|s - \theta_i\|^2$, then the objective function can be reformulated as

$$C(\Theta, V) = \sum_{i=1}^{n} \int_{V_i} \|s - \theta_i\|^2 \sigma(s) ds = \sum_{i=1}^{n} J_{\theta_i} \tag{6}$$

The mass and centroid of mass of each Voronoi region are formulated as

$$m(V_i) = \int_{V_i} \sigma(s) ds \tag{7}$$

$$c(V_i) = \frac{1}{m(V_i)} \int_{V_i} s \cdot \sigma(s) ds \tag{8}$$

We substitute J_{θ_i} in the objective function based on the parallel axis theorem and obtain

$$C(\Theta, V) = \sum_{i=1}^{n} J_{c(V_i)} + \sum_{i=1}^{n} m(V_i) \|\theta_i - c(V_i)\|^2 \tag{9}$$

From Eq. (9), it may be noted that the objective function is minimized when agent θ_i is located at $c(V_i)$. Therefore, optimal locations of law enforcement agents are determined as centroids of Voronoi regions.

3.3 Asymptotic Convergence of Optimal Allocation

We propose an optimal learning algorithm for agents to search for optimal locations. There are other existing methods, such as sequential sampling and Lloyd algorithm. However, sequential sampling algorithm is computationally expensive for a large space Ω (Du et al., 1999). Lloyd algorithm is a special case of the gradient flow algorithm with a step size equals to 1 (Du et al., 2006). The large step size poses the drawback that agent locations may never reach a steady state. The agent's movement follows

$$\theta_i(t+1) = \theta_i(t) - \alpha \cdot \frac{\partial C}{\partial \theta_i} \tag{10}$$

where $\frac{\partial C}{\partial \theta_i} = 2m(V_i)(\theta_i - c(V_i))$. Here, we assume the agent's movement obeys the first order dynamical behavior

$$\dot{\theta}_i = -\alpha_{step}(\theta_i - c(V_i)) \tag{11}$$

where α_{step} is the step size. Thus, agents converge asymptotically to their optimal locations by

$$\theta_i\,(t+1) = \theta_i\,(t) - \alpha_{step}\,(\theta_i\,(t) - c\,(V_i)) \tag{12}$$

4 Experimental Design and Results

We design a three-way layout experiment to test the performance of proposed algorithm as shown in Fig. 5. First, crimes are categorized into low, medium, and high severity levels and their crime distributions are modeled. Second, limited resources of the law enforcement are allocated with the proposed algorithm of optimal coverage control which includes settings of 8, 12, and 20 agents. Third, to evaluate the effectiveness and efficiency of the proposed algorithm, we measure the performance based on the response time of an agent to reach crime scenes and benchmark with both zone-random patrol and uniform allocation of law enforcement.

4.1 *Optimal Allocation of Law Enforcement Agents*

The proposed algorithm for optimal coverage control of city crimes is implemented for the three severity levels to optimally allocate law enforcement resources. We consider various settings of law enforcement agents, including 8-agent, 12-agent, and 20-agent allocations. Here, 12-agent allocation is utilized to demonstrate performance of the proposed algorithm.

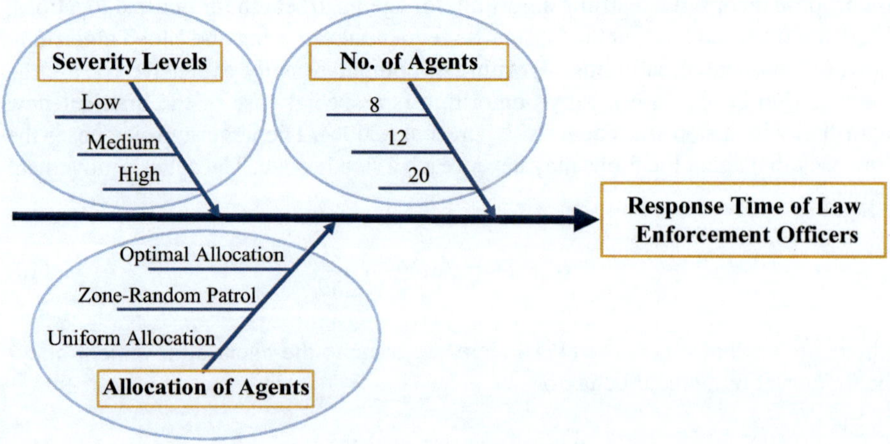

Fig. 5 The design of experiment for performance evaluation

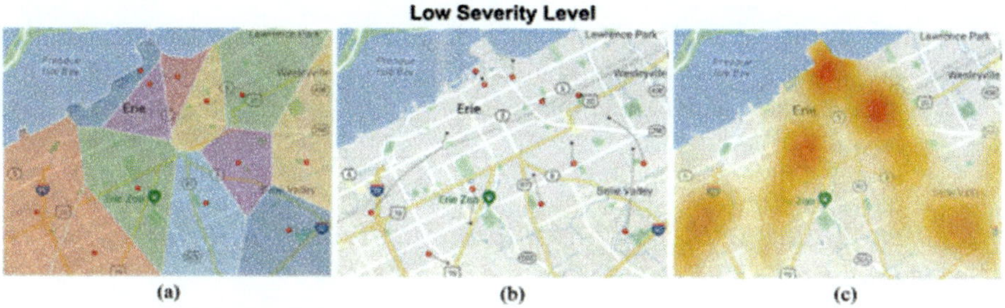

Fig. 6 Optimal allocation of 12 agents for the low severity level. (**a**) Optimal locations; (**b**) Routes from random starting locations to the optimal locations; (**c**) Crime distribution of low severity level

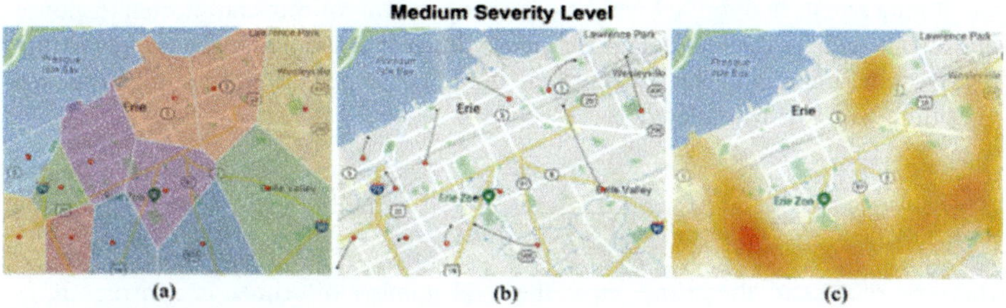

Fig. 7 Optimal allocation of 12 agents for the medium severity level. (**a**) Optimal locations; (**b**) Routes from random starting locations to the optimal locations; (**c**) Crime distribution of medium severity level

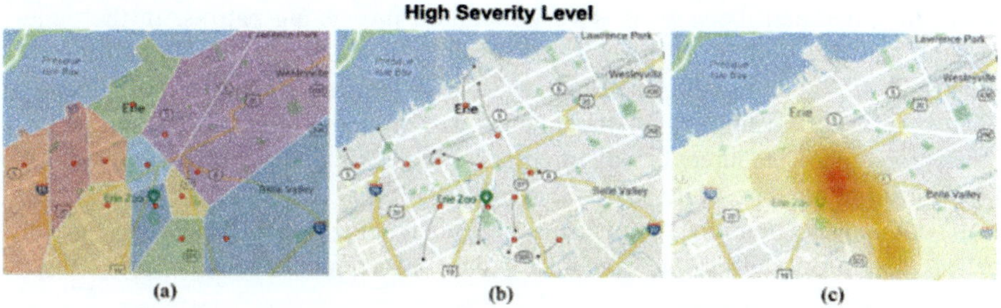

Fig. 8 Optimal allocation of 12 agents for the high severity level. (**a**) Optimal locations; (**b**) Routes from random starting locations to the optimal locations; (**c**) Crime distribution of high severity level

For crimes of the low severity level, Fig. 6a shows optimal locations of 12 agents. In Fig. 6b, routes of 12 agents from random starting locations (black dots) to optimal locations (larger red dots) are shown with black lines. Figure 7a b demonstrate the optimal allocation and routes of 12 agents for crimes at the medium severity level, respectively. The optimal allocation of 12 agents to cover and control high-severity crimes is demonstrated in Fig. 8a. Routes of 12 law enforcement agents to optimal locations are shown in Fig. 8b. Notably, optimal allocation of law enforcement agents varies depending on the distribution of crimes. Under a certain crime distribution as shown in sub-figure (c) of Figs. 6, 7 and 8, agents are likely to move towards the area with high crime density.

4.2 Benchmark with Zone-Random Patrol and Uniform Allocation

With the proposed algorithm for optimal coverage control of city crimes, law enforcement agents can provide an effective and efficient service to citizens. By positioning agents at optimal locations, the proposed algorithm can shorten response times from agents to crime scenes. Response time of an agent to reach crimes within its corresponding region is a key metric to quantify the performance of the proposed algorithm, which is formulated as

$$T_R = \frac{\sum_{i=1}^{n} \sum_{\gamma}^{m_i} \|c_\gamma - \theta_i\| \cdot t_R}{n} \tag{13}$$

where c_γ represents the crime, m_i is the total number of crimes in i th region, t_R is the time it takes for a law enforcement agent to travel a unit distance, T_R is the response time of one agent to reach all crimes within its corresponding region.

The performance of optimal resource allocation is benchmarked with zone-random patrol and uniform allocation of limited resources. We use zone-random patrol to benchmark because it is very common in existing patrols. In the zone-random patrol, Erie area is divided into 2 zones as shown in Fig. 9b and half of the agents are randomly allocated in each zone. All agents are uniformly allocated in the uniform allocation. We implement three strategies for various settings of law enforcement resources at all three severity levels and summarize response-time performances in Table 1. As shown in the table, response times of agents from optimal locations to crime scenes are much shorter than from zone-random patrol and uniform allocation. Specifically, law enforcement agents will give faster

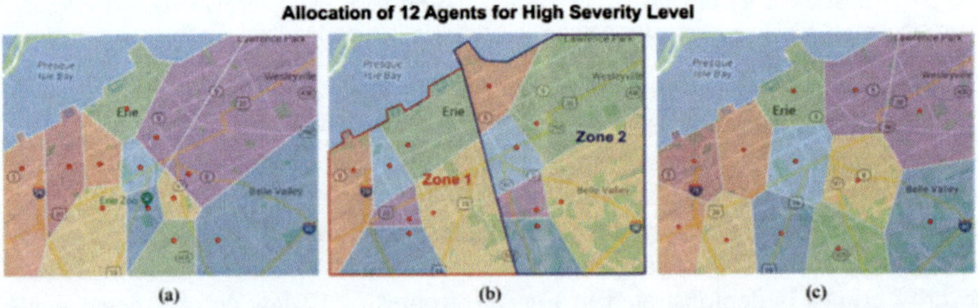

Fig. 9 Allocation of 12 agents for high severity level: (**a**) Optimal allocation; (**b**) Zone-random patrol; (**c**) Uniform allocation

Table 1 Response-time performances of optimal allocation, zone-random patrol and uniform allocation

Severity Level	8-agent Setting			12-agent Setting			20-agent Setting		
	Optimal	Zone-random Patrol	Uniform	Optimal	Zone-random Patrol	Uniform	Optimal	Zone-random Patrol	Uniform
Low	30.45	36.64	43.98	26.73	31.64	34.92	19.06	22.76	24.43
Medium	16.03	22.78	27.14	10.95	18.36	21.21	9.53	10.68	10.98
High	3.53	4.37	4.95	2.62	3.33	4.39	1.94	3.16	3.31

responses to crimes with the proposed algorithm. Further, the response time of an individual agent is shorter if we increase the number of law-enforcement agents.

5 Conclusions

As the scale and nature of crimes become more and more complex, resources of law enforcement become insufficient. Protecting citizens through effective and efficient strategies therefore becomes a challenging task, especially when the law enforcement resource is constrained. In this paper, we develop a new optimal learning algorithm to characterize multi-scale distributions of crimes and then determine an optimal policy for coverage control of city crimes. First, we categorize crimes into low, medium, and high severity levels. Then we characterize and model crime distributions for various severity levels. Second, we develop an optimal policy for coverage control to allocate limited resources of the law enforcement in the areas of interest. Third, the model performance is measured based on the response time of an agent to reach crime scenes. We evaluate and benchmark the performance of the optimal policy of coverage control with zone-random patrol and uniform allocation. Experimental results show that the proposed algorithm is effective and efficient in optimizing the allocation of limited law enforcement resources and demonstrate a better performance than zone-random patrol and uniform allocation of limited resources.

Acknowledgements This work is supported in part by the NSF grant (CMMI-1617148). The authors thank the Erie Police Department for sharing the crime information in this research. The author (HY) also thanks Harold and Inge Marcus Career Professorship for additional financial support.

References

Buczak, A. L., & Uhrig, R. E. (Sep. 1996). Hybrid fuzzy—Genetic technique for multisensor fusion. *Information Sciences (NY), 93*(3–4), 265–281.

"Crime rate in Erie, Pennsylvania (PA)." Accessed on Mar 27 2020. [Online]. Available: https://www.city-data.com/crime/crime-Erie-Pennsylvania.html.

Du, Q., Faber, V., & Gunzburger, M. (1999). Centroidal Voronoi tessellations: Applications and algorithms. *SIAM Review, 41*(4), 637–676.

Du, Q., Emelianenko, M., & Ju, L. (2006). Convergence of the Lloyd algorithm for computing centroidal Voronoi tessellations. *SIAM Journal on Numerical Analysis, 44*(1), 102–119.

Freilich, J. D., & Pridemore, W. A. (2007). Politics, culture, and political crime: Covariates of abortion clinic attacks in the United States. *Journal of Criminal Justice, 35*(3), 323–336.

Guo, D. (2009). Flow mapping and multivariate visualization of large spatial interaction data. *IEEE Transactions on Visualization and Computer Graphics, 15*(6), 1041–1048.

Hipp, J. R., & Kane, K. (2017). Cities and the larger context: What explains changing levels of crime? *Journal of Criminal Justice, 49*, 32–44.

Jefferson Educational Society. (2015). *Is Erie a safety city? Perception, realty, recommendation.*

Kocher, M., & Leitner, M. (2015). Forecasting of crime events applying risk terrain Modeling. *GI_Forum, 1*, 30–40.

Kotevska, O., Kusne, A. G., Samarov, D. V., Lbath, A., & Battou, A. (2017). Dynamic network model for smart city data-loss resilience case study: City-to-City network for crime analytics. *IEEE Access, 5*, 20524–20535.

Malleson, N., & Andresen, M. A. (2016). Exploring the impact of ambient population measures on London crime hotspots. *Journal of Criminal Justice, 46*, 52–63.

Morris, S. E. (1997). Crime and prevention: A Treasury viewpoint. *IEEE Spectrum, 34*(2), 38–39.

Phillips, P., & Lee, I. (2012). Mining co-distribution patterns for large crime datasets. *Expert Systems with Applications, 39*(14), 11556–11563.

Serban, N. (2011). A space-time varying coefficient model: The equity of service accessibility. *The Annals of Applied Statistics, 5*(3), 2024–2051.

U. S. Department of Justice Federal Bureau of Investigation. Crime in the United States, 2018.

Yang, H., Kan, C., Liu, G., & Chen, Y. (2013). Spatiotemporal differentiation of myocardial infarctions. *IEEE Transactions on Automation Science and Engineering, 10*(4), 938–947.

Yao, B., Zhu, R., & Yang, H. (2018). Characterizing the location and extent of myocardial infarctions with inverse ECG modeling and spatiotemporal regularization. *IEEE Journal of Biomedical and Health Informatics, 22*(5), 1445–1455.

Yu, C., Ding, W., Morabito, M., & Chen, P. (2016). Hierarchical spatio-temporal pattern discovery and predictive Modeling. *IEEE Transactions on Knowledge and Data Engineering, 28*(4), 979–993.

Application of Internet of Things (IoT) in Inventory Management for Perishable Produce

Jing Huang and Hongrui Liu

1 Introduction

Nowadays, people have an increasing demand on fresh produce for a healthier lifestyle. They shop either in local grocery stores or from online stores. In some states, like California, where farm industry is highly developed, there is always adequate supply of produce. That leads to a high competency between different grocery stores. While the costs of produce fluctuate slightly from one store to another, the quality of produce indeed determines how competitive the business is. Often times in a grocery store, items with the latest expiration date will be picked by customers first if items with different expiration dates were present and sold at the same price. While the freshest items are sold out, grocery stores confront the situation that the rest items are approaching the expiration date or expired already, which leads to high spoilage cost. The inventory management of produce is a very complex task. On one hand, customers desire variety of fresh produce at a reasonable price. On the other hand, different types of produce have different storage requirements and lifespans. The current labor-intensive inventory management costs a lot to ensure quality. Now with the development of high-tech tools, the conditions of the produce items can be tracked and monitored with the aid of IoT, and the inventory management policies can be adjusted timely and automatically to reduce cost.

The inventory management problems of perishable produce discussed in this paper include two aspects: inventory storage and sorting by produce status. IoT is an internet-based intelligent network aiming in tracking and sharing the real-time information about the products in supply chain through various technologies

J. Huang · H. Liu (✉)
Industrial and Systems Engineering, San Jose State University, San Jose, CA, USA
e-mail: hongrui.liu@sjsu.edu

H. Yang et al. (eds.), *AI and Analytics for Public Health*, Springer Proceedings in Business and Economics, https://doi.org/10.1007/978-3-030-75166-1_10

including Radio Frequency Identification (RFID), electronic identification (EID), advanced information tag, remote sensing (RS), and global positioning systems (GPS), etc. (Opara, 2003). With IoT, it is now possible to trace the status of perishable produce from the moment that it is produced to the point when it is received by the customers. Firstly, produce is tagged with its specific species, storage conditions, packaging date, freshness duration, and some other guides of operations on the packages. This tag is then scanned and uploaded to a cloud database. With the remote sensing installed in delivery and storage facilities, storage conditions are monitored. In case abnormality is detected, meaning, the sensed temperature does not meet the storage requirement of this produce, an instant feedback will be sent to a Wi-Fi enabled monitor to realign with desired storage conditions. Further analysis utilizing an ANN algorithm will be applied to the cloud database to sort the produce into different status categories. The sorted produce is then arranged on the shelf or picked for delivery according to the order of expiration dates. In addition, pricing strategies will be generated from an ANN algorithm to minimize the spoilage loss.

This article is organized as follows. In Sect. 2, we introduce recent studies about the applications of IoT in the supply chain for perishable produce. The design methodology is illustrated in Sect. 3. A case study for the inventory management of cherry utilizing the proposed methodology is given in Sect. 4. Conclusions and future work are discussed in Sect. 5. The working mechanism was illustrated to show how our proposal is realized.

2 Literature Review

For years, researchers have initiated various applications of IoT in the inventory management of perishable produce. Gao, et al. have applied IoT in designing a three-layer integrated agriculture system: a sensor layer to detect environmental information, a network layer to transmit data, and an application layer for data analysis and other supply chain services (Gao et al., 2015). John Livingston and Umamakeswari (2015) proposed an alternative way of locating the products with IP sensor node to make the transmission of information in supply chain more accurately and efficiently. Yan built a more advanced revenue model to quantify the benefits from the application of IoT in perishable products (Yan, 2017). A more advanced design and application of IoT-based warehouse management system was proposed to deal with the increasing complexity in selection of orders in Lee et al. (2018). The authors in this paper evaluated the operating expense with and without the use of IoT in the warehouse management and concluded that there is a total 50–55% cut in cost with the application of IoT. They also combined the IoT technologies with some order picking policies, which increased the efficiency in order fulfillment. The efforts made from above researchers have greatly contributed the development of IoT applications in supply chain and inventory management.

The inventory management for perishable produce can be further studied to address some special features in this category to improve their operation. In this paper, an application methodology of IoT is proposed to address specific real-life inventory problems for perishable produce.

3 Design Methodology

The design methodology of applying IoT in inventory management for perishable produce consists two parts: information tag and an intelligent sorting system. The information tag describes the attributes of the produce when it is packed. The information is continuously uploaded to a cloud database from the time it is packed to the time it is sold. The information in the cloud databased is passed to an ANN for analysis and shared with all sections in the supply chain, including the inventory section. The inventory storage conditions are tracked in real time with IoT technologies. The ANN model, as an analytic tool, will predict the status of the produce using the information in the cloud database to create sorting categories. Based on the sorting categories, a pricing strategy will be generated to minimize profit loss. After sales complete, all sales records will feedback to the cloud database to revise the sorting and pricing model that can improve the inventory management decision in the future. The working mechanism is illustrated in Fig. 1.

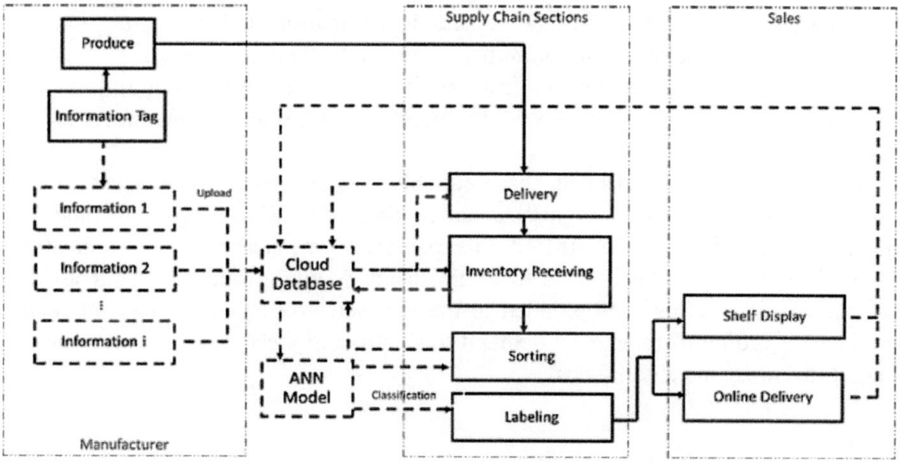

Fig. 1 Working mechanism of applying IoT in inventory management for perishable produce

3.1 Information Tag

Acquiring an information tag is the first step of the whole inventory management process. The information tag can be in the form of an RFID tag, which is made of special materials that can endure extreme conditions. (Opara, 2003). An RFID tag can be prestored with the following information:

(a) Produce name
(b) Produce species
(c) Manufacturer (address, phone number, and other contact information)
(d) Package size
(e) Fragility
(f) Storage temperature
(g) Storage humidity
(h) Lifespan (under proper storage conditions)

Produce name, species, and manufacturer information are the basic information that can be prestored in the RFID tag for tracing and sorting purpose. Package size is necessary for both inventory space management purpose and shelf/delivery arrangement. Fragility indicates the piling method. For example, leaf vegetables in plastic bags cannot be squeezed, while in clamshells, they are encouraged to be piled tightly to save storage spaces. Storage temperature and humidity are the key information to be recorded. Sensors can be used to periodically measure the environmental temperature and humidity (Ferrandez-Pastor et al., 2016). Monitors then compare these data with the recorded information and adjust the current environmental temperature and humidity accordingly. This ensures a fast response to the change of storage condition. Lifespan is a key constraint for determining the produce's turnover rate. It is also a reference for developing pricing strategies for near expiration produce.

When the produce is packed, these editable tags are generated with additional information: packing date, weight/size, and product code. With the packing date and lifespan, expiration date is estimated. The expiration date is an important indication of product status, which is used to establish the pricing strategies. After all the information in this tag are uploaded to the internet, grocery stores can order the produce according to forecasted demand to ensure a highest service level with the lowest inventory managerial cost.

3.2 Intelligent Sorting and Pricing System

The intelligent sorting system utilizes an ANN algorithm to train an inventory sorting model using the large amount of information of the produce that is uploaded to the internet. With this information, produce is sorted into several status categories. For example, if there are ten status categories defined, of which 10 is the freshest status, meaning that the produce in this category is just packed. On the contrary, 1 is the least fresh status, meaning that the whole package of the produce is spoiled. Based on historical sales records and the status of the produce, the ANN model predicts different levels of status of the produce, fits them into these status categories and generates a dynamic pricing strategy. For each category, a different discount is applied as an incentive to increase the sales.

This ANN model has two output measurements: service level and loss in profits. The service level is evaluated by the conforming rate of the produce in different status categories and its returning rate from customers. The conforming rate is calculated by comparing the predicted amount in each status category with the real status. To acquire the real status of the produce, the practical way is to randomly sample and inspect the produce. Since the returning rate is considered corelated with the conformity of produce status, it is no longer considered as one of the measurements in the service level. With the ANN model, the predicted loss in profits represents the total discounts. The real loss in profits includes the total discounts in sales, the refund cost, and the disposal costs. After each sales cycle, the sales records are submitted to the internet, and the errors are calculated in regarding to the difference between the predicted situation and the real situation. The errors come from two aspects: (1) the difference of the predicted status categories compared to the real status; (2) the difference between predicted loss in profits with the real loss in profits. With these errors, the weights of each information recorded are changed and the ANN model is adjusted, followed by a new discount strategy which assigns an updated price to each status category to optimize the sales with the lowest loss in profits.

With the information being processed, a sorting list is generated to help the warehouse staff arrange the shelf display and order picking. The produce displayed on the shelf with the same expiration date has the same price, while its price differs from the one with different expiration date. In addition, the updated pricing information goes to the internet-based information tag and sends an instruction to have the real price tags printed, so that the price of certain package of produce may be updated every day. The process of the intelligent sorting system is illustrated in Fig. 2.

4 Application of IOT in Inventory Management of Cherry

In this section, the design methodology discussed in the previous section is applied in the inventory management of cherry. The objective is to enhance the service level and minimize the loss in profits by introducing IoT in this case. The service level represents the conforming rate of status categories of cherry displayed on shelves or delivered to customers. The loss in profits consists of the total discounts of different status categories and the cost of disposed cherries. The service level and the loss in profits are estimated under three different scenarios:

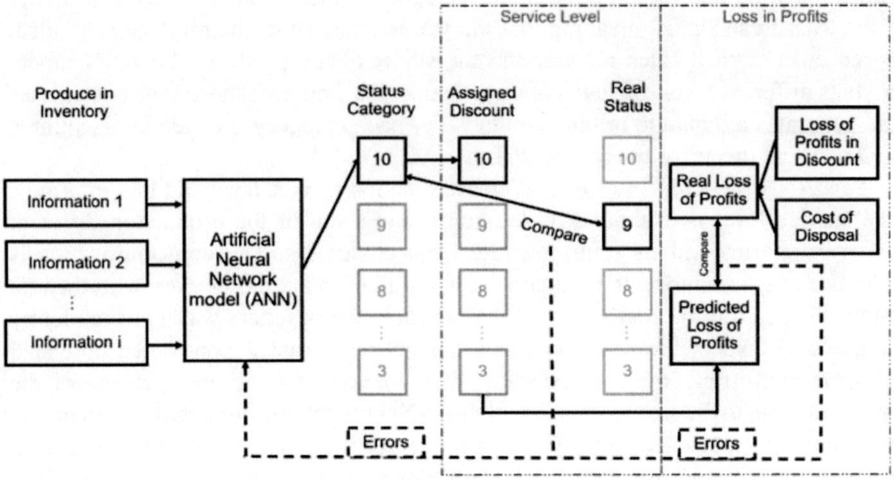

Fig. 2 Intelligent sorting and pricing system

(a) Traditional way in the inventory management of cherry.
(b) Applying IoT technology in the inventory management of cherry.
(c) Applying IoT technology and adjusted ANN model in the inventory management of cherry.

It should be mentioned that all three scenarios are based on the same time horizon. The information used in these scenarios are simulating the real cases to explain the process of applying IoT in inventory management of cherry. The following subparagraphs demonstrates the process of application with the problems described, assumptions, application of IoT in the inventory management of cherry, ANN model for sorting the status categories, and some further steps.

4.1 Problem Description

As known by all, cherry is a kind of perishable fruit usually with short lifespan and high disposal rate if it is stored improperly, and the cost of disposal is high. The Greens grocery aims to sell the freshest cherries directly delivered from the Best Cherry farm during the harvest season. The cost of cherry per package is \$3. The travel distance is about 3 h from the Greens to the Best Cherry, so they use a cold storage during the delivery. The temperature in the storage is fluctuating and it is difficult to monitor during the delivery. In order to lower the ordering cost, the Greens places each order with as many packages as possible according to estimated forecasted demand. However, the Greens encounters a high disposal rate and a loss in profits due to the limitation of storage conditions, shelf display, and fluctuated demand. Due to frequent entering the storage room, storage temperature is always higher than desired, and humidity is too high because of the high temperature, which speeds up the rate of decay among the cherries in the warehouse. The Greens piles half of the packages on the shelves for customers to pick up. The freshest packages were already sold out, while the rest with several decayed cherries are close to expiration date. Although a discount is applied on these near expiration date packages, the disposal rate is still high. The Greens are trying to introduce IoT technology in their inventory management of cherry and would like to extend the application to other produce if it helps raise the customer satisfaction and lower the loss in profits.

4.2 *Assumptions*

When talking about design methodology, the problem discussed are always the ideal cases. However, the reality is far more complex. We can have several assumptions to simplify the case and focus only on our problem of interests.

(a) The status of cherries may vary according to many factors, for example, the storage conditions. Since the three scenarios are based on the same time horizon, the cherries have consistent quality.
(b) Under the same storage conditions, the cherries have the same decay rate. With desired storage conditions, cherry has a lifespan of about 15 days.
(c) Due to the complexity of considering not only the material fee of IoT service like costs of RFID tags, internet accessible storage sensors and monitors, but also corresponding service fee, we assume that such kind of fee incurred by using IoT can be neglected, since the implicit benefits such as customer loyalty and reputation of the store cannot be quantified.
(d) During the sales, no out-of-stock will happen, meaning that nothing affects the service level except the conformity rate.
(e) In each scenario, when a discount of 50% or below is applied at the purchase, no return is allowed.
(f) If the cherries are in same status and the same discount is applied, the demand and return cases are assumed same for all scenarios. To keep the consistency for all scenarios, a demand matrix and a return cases matrix shown in Tables 1 and 2 are used to simulate real cases.

4.3 *Application of IOT in the Inventory Management of Cherry*

4.3.1 Scenario a. Traditional Way in the Inventory Management of Cherry

To lower the disposal rate, the Greens applies a 50% discount on cherries that are within 5 days of its expiration date to increase sales. Although there is non-refund

Table 1 The demand matrix for simulating the real cases

Discount\Status	10	9	8	7	6	5	4	3
1	150	130	110	80	50	20	0	0
0.9	160	140	120	90	63	32	10	5
0.8	170	150	130	100	76	45	20	10
0.7	180	160	140	110	90	58	30	16
0.6	–	170	150	120	100	70	40	22
0.5	–	–	160	130	110	80	50	28
0.4	–	–	–	140	120	90	60	34
0.3	–	–	–	–	130	100	70	40

Table 2 The return cases matrix for simulating the real cases

Discount\Status	10	9	8	7	6	5	4	3
1	2	3	5	7	10	14	20	28
0.9	1	2	4	6	8	11	16	23
0.8	0	1	3	4	6	8	12	18
0.7	0	0	1	2	4	6	9	14
0.6	0	0	0	1	2	4	7	10
0.5	–	–	–	–	–	–	–	–
0.4	–	–	–	–	–	–	–	–
0.3	–	–	–	–	–	–	–	–

policy for discounted cherries, customers who buy cherries at its original price may get a refund if they are not satisfied with the cherry. The cherries must be disposed after 15 days being displayed on shelf. The Greens ordered 1700 packages of cherry at a cost of \$3.00 per package for each shipment, with a retail price of \$10.00 per package. From Day 1 to Day 3, there is a high demand because the cherries are very fresh. However, from the fourth day, the demand starts to drop because the cherries do not look as fresh as they were several days ago. Until Day 10, there are still 560 packages left. From Day 11, the demand suddenly increases, since there is a 50% discount applied, which indicates that with 50% off the full price, cherries in this status are still acceptable. After that, it is hard to decide how much incentives to give to the customers to promote the sales of the rest packages while minimizing the loss in profits. On the other side, from Day 1 to Day 10, there is an increasing number of refund cases, because the quality of cherries drops while the price is the same. This causes the unsatisfaction from customers. The summary of sales and service level is shown in Tables 3 and 4.

Expired cases (e) stand for the number of packages left after 15 days sales and being disposed. The number of returned packages (r) are also counted in the total disposed quantity because they are not going to be recycled. For a single package, the costs of disposal consist of the cost of cherries (\$3.00) and the loss of profits (\$7.00) that they could have made. Total discount is also one of the loss in profits, and it depends on how much discount (c) it is applied on the cherries. Thus, the loss in profits can be calculated as follows:

Table 3 Daily sales – traditional way in the inventory management of cherry

Day	1	2	3	4	5	6	7	8	9	10	11	12	13	14	15
Demand (d)	150	150	150	130	130	110	110	80	80	50	80	50	50	28	28
Status category	10	10	10	10	10	10	10	10	10	10	5	5	5	5	5
Real status	10	10	10	9	9	8	8	7	7	6	5	4	4	3	3
Discount (c)	0	0	0	0	0	0	0	0	0	0	0.5	0.5	0.5	0.5	0.5
Price/pkg ($)	10	10	10	10	10	10	10	10	10	10	5	5	5	5	5
Returned case	2	2	2	3	3	5	5	7	7	10	0	0	0	0	0
Conformity	1	1	1	0	0	0	0	0	0	0	1	0	0	0	0

Table 4 Sales and service level summary – traditional way in the inventory management of cherry

Total Sales (n)	Expired (e)	Returned (r)	Total Disposed (e + r)	Costs of Disposal ($)	Total Discount ($)	Loss in Profits ($)	Service Level
1376	324	46	370	3700	1180	4880	0.27

$$Loss\ in\ Profits\ (\$) = (3 + 7) \times (e + r) + 10 \times \sum_{i=1}^{15} (1 - c_i)\, d_i$$

$$Service\ Level = \frac{Total\ Conformity\ days}{Total\ days.}$$

The Greens disposes 324 packages of cherry after 15 days, and they lost $4880.00 in the total profits. The total days of conformity of the status is 4, and the service level is 0.27.

4.3.2 Scenario b. Applying IoT Technology in the Inventory Management of Cherry

The Greens starts to cooperate with the Best Cherry farm to apply IoT technology in the whole supply chain. When the Greens places an order from the Best Cherry, they can immediately see the information online about the cherry as shown in Table 5.

After the cherries are harvested and packed, additional information about the packing date, weight and package number are uploaded as shown in Table 6.

The inventory system in the Greens is automatically updated with the information uploaded. Once the packages of cherry arrive, the storage temperature and humidity are ready. Since they have upgraded their thermostat system with an internet-based sensor and monitor, the storage room has relatively steady temperature and humidity. This helps improve the status of cherry and slow down the decay rate. With a better status in the same day compared to the traditional scenario, the demand

Table 5 Prestored information of ordered cherries shown online

Produce name	Species	Manufacturer	Package	Fragility	Storage condition		Lifespan (day)	Packed On	Weight	Pkg No.
					Temperature (°F)	Humidity (rh)				
Cherry	Bing	Best Cherry	Clear Bag	MF	30–31	90–95%	15			
Fragility	F	Fragile	Easily broken							
	MF	Medium fragile	No squeezing							
	N	Not fragile	Can be packed closely							

Table 6 Updated information of ordered cherries to store in the information tag

Produce name	Species	Manufacturer	Package	Fragility	Storage condition		Lifespan (day)	Packed On	Weight	Pkg No.
					Temperature (°F)	Humidity (rh)				
Cherry	Bing	Best Cherry	Clear Bag	MF	30–31	90–95%	15	May 16	2.18 lb	BC01
Fragility	F	Fragile	Easily broken							
	MF	Medium fragile	No squeezing							
	N	Not fragile	Can be packed closely							

Table 7 Daily sales – applying IoT technology in the inventory management of cherry

Day	1	2	3	4	5	6	7	8	9	10	11	12	13	14	15
Demand (*d*)	150	150	150	150	130	140	120	130	100	100	80	80	50	60	40
Status category	10	10	10	10	10	9	9	8	8	6	5	5	5	4	3
Real status	10	10	10	10	9	9	8	8	7	6	5	5	4	4	3
Discount (*c*)	0	0	0	0	0	0.9	0.9	0.8	0.8	0.6	0.5	0.5	0.5	0.4	0.3
Price/pkg ($)	10	10	10	10	10	9	9	8	8	6	5	5	5	4	3
Returned case	2	2	2	2	3	2	4	3	4	2	0	0	0	0	0
Conformity	1	1	1	1	0	1	0	1	0	1	1	1	0	1	1

Table 8 Sales and service level summary – applying IoT technology in the inventory management of cherry

Total sales (*n*)	Expired (e)	Returned (r)	Total disposed (e + r)	Costs of disposal ($)	Total discount ($)	Loss in profits ($)	Predicted loss in profits ($)	Service level
1630	70	26	96	960	2810	3770	2810	0.73

is increased. Based on the information, the ANN model predicts the status category and assigns the discounts accordingly. At the same time, the inventory managerial staff randomly samples some packages to check the real status of the remaining packages every day. After the deadline for returns, the sales and returns information is updated and feedback to the ANN model. Because of the error in status categories, the demand is lower than expected and there are some return cases in regarding to the quality of the cherries. This increases the cost in disposal, and as a result, the loss in profits is more than predicted. The ANN model then integrates these errors into the new model. An updated ANN model is then ready for the next sales circle. The complete sales and service level information is shown in Tables 7 and 8.

4.3.3 Scenario c. Applying IoT Technology and Adjusted ANN Model in the Inventory Management of Cherry

With adjusted ANN model, while the discounts are reconsidered based on the predicted status categories. We can see from Tables 9 and 10 that, although the predicted loss in profits are higher than the previous model, the real loss in profits is lower than that in Table 8. Due to more accurate prediction in the status categories, the customers can get the cherries which meet their expectation. This leads to a 100% service level. As a result, the sales are better in each day compared to scenario b, and less return cases happen in the following days. Finally, the real loss in profits in scenario c drops to $3710. In addition, a higher service level is achieved. The difference between predicted loss in profits with the real case becomes smaller as well. This is a demonstration of an improved ANN model to enhance the inventory management of the cherry, which eventually benefits the service level and revenue of the Greens.

Table 9 Daily sales – applying IoT technology and adjusted ANN model in the inventory management of cherry

Day	1	2	3	4	5	6	7	8	9	10	11	12	13	14	15
Demand (*d*)	150	150	150	150	140	140	130	130	110	100	80	80	60	60	40
Status category	10	10	10	10	9	9	8	8	7	6	5	5	4	4	3
Real status	10	10	10	10	9	9	8	8	7	6	5	5	4	4	3
Discount (*c*)	0	0	0	0	0.9	0.9	0.8	0.8	0.7	0.6	0.5	0.5	0.4	0.4	0.3
Price/pkg ($)	10	10	10	10	9	9	8	8	7	6	5	5	4	4	3
Returned case	2	2	2	2	5	3	2	3	3	2	0	0	0	0	0
Conformity	1	1	1	1	1	1	1	1	1	1	1	1	1	1	1

4.4 Pricing and Inventory Updates

After upgrading their facility to fit the IoT technologies, some operations in inventory management in the Greens become much easier than before. The Greens used to check the status of cherries every day and make the sales strategy accordingly. The staff there sometimes misclassified the new lots and the old lots, and the discounts were applied on the wrong packages. This frequently caused the loss in profits and, worse than all, the loss in customer's loyalty. With the IoT technologies, pricing is no longer an issue to the inventory management staff. They just get automatically printed price tags and stick them on the packages with a unique number. With each package sold or returned, the inventory record is updated. The probability of making mistakes in inventory management is minimized.

5 Conclusion

In this paper, an automatic sorting and pricing inventory methodology utilizing IoT technology is proposed for perishable produce. The proposed methodology is demonstrated through a case study, showing that the use of IoT technology and the ANN model helps in reducing spoilage cost and improving customer service level, which are critical in produce industry. In the case study, only a few factors that affect the sorting of perishable produce were considered and an ANN model did not demonstrate its potential. In real businesses when there are hundreds of produce categories, the information from harvesting to transportation and to storage is very complex. An ANN model can learn from a large amount of data aggregated by IoT and continuously improve its model output to minimize sorting and pricing error and cost. Moreover, the IoT, as an emerging technology, is far from reaching its full potential in supply chain management. Different models and algorithms that utilize the IoT data to improve supply chain operations need to be studied. In addition, the investment on the IoT technology may be significant to some businesses and

Table 10 Sales and service level summary – applying IoT technology and adjusted ANN model

Total sales (n)	Expired (e)	Returned (r)	Total disposed (e + r)	Costs of disposal ($)	Total discount ($)	Loss in profits ($)	Predicted loss in profits ($)	Service level
1670	30	22	52	520	3190	3710	3190	1

cannot justify the savings. We will apply our methodology to a large case with more produce categories and detailed cost parameters to obtain more practical solutions to the industry in the future.

References

Ferrandez-Pastor, F., Garcia-Chamizo, J., Nieto-Hidalgo, M., Mora-Pascual, J., & Mora-Martinez, J. (2016). Developing ubiquitous sensor network platform using Internet of Things: Application in precision agriculture. *Sensors, 16*, 1141.

Gao, S., Zhang, Y., Zhou, X., Xu, Y., Feng, B., & Zheng, L. (2015). Design of Internet of Things application and service detecting system in agriculture. *Journal of Shanghai Normal University (Natural Sciences)*, 51–59.

John Livingston, J., & Umamakeswari, A. (2015). Internet of things application using IP-enabled sensor node and web server. *Indian Journal of Science and Technology*, 207–212.

Lee, C., Lv, Y., Ng, K., Ho, W., & Choy, K. (2018). Design and application of Internet of things-based warehouse management system for smart logistics. *International Journal of Production Research, 56*(8), 2753–2768.

Opara, L. U. (2003). Traceability in agriculture and food supply chain: A review of basic concepts, technological implications, and future prospects. *Food, Agriculture & Environment, 1*(1), 101–106.

Yan, R. (2017). Optimization approach for increasing revenue of perishable product supply chain with the Internet of Things. *Industrial Management & Data Systems, 117*, 729–741.

Modified Risk Parity Portfolios to Limit Concentration on Low Risk Assets in Multi-Asset Portfolios

Fatemeh Amini, Atefeh Rajabalizadeh, Sarah M. Ryan, and Farshad Niayeshpour

1 Introduction

Portfolio managers are looking for new solutions to deliver a portfolio with maximal return and minimal risk. Since the introduction of Modern Portfolio Theory (MPT) by Markowitz (1952), researchers have proposed different models to construct an optimal portfolio according to various criteria (Kolm et al., 2014). These innovations continue because different models work well under different conditions.

Since the 2008 financial crisis, risk management and diversification have become more important. At that time researchers questioned traditional portfolio management approaches, demonstrated properties of the Risk Parity Portfolio (RPP), and compared it with two well-known portfolio management techniques, namely Mean-Variance Portfolio (MVP) and Equally Weighted Portfolio (EWP) (Maillard et al., 2010). They described RPP as a trade-off between these two approaches since it outperformed MVP in terms of diversification and beat the EWP in terms of individual asset risk. In other words, RPP concentrates on portfolio risk allocation rather than portfolio capital allocation. Additionally, RPP outperformed MVP during the crisis since it is not as sensitive to input parameters as MVP (Thiagarajan & Schachter, 2011; Chaves et al., 2011). RPP increases diversification and constructs a portfolio such that all assets contribute equally to the total portfolio risk, which leads to maximizing the highest return per unit of risk (Dalio et al., 2015). In other words, if an investor assumes the equal return contribution for all assets, the RPP has the highest Sharpe ratio (Chaves et al., 2011).

F. Amini · A. Rajabalizadeh · S. M. Ryan (✉)
Industrial and Manufacturing Systems Engineering, Iowa State University, Ames, IA, USA
e-mail: smryan@iastate.edu

F. Niayeshpour
Principal Financial Group, Des Moines, IA, USA

H. Yang et al. (eds.), *AI and Analytics for Public Health*, Springer Proceedings in Business and Economics, https://doi.org/10.1007/978-3-030-75166-1_11

A multi-asset portfolio constructed by risk parity optimization makes the total risk contribution of assets equal so that higher weights are assigned to less risky assets. By decreasing the portfolio risk due to downturn of a specific asset, it enables the portfolio manager to face unexpected circumstances. Moreover, RPP attempts to incorporate all assets in the portfolio while making the risk contribution equal, which increases diversification through distributing investment among several asset classes. However, RP tends to concentrate investment in the very low risk assets, whereas portfolio managers may want more diversification. Therefore, the purpose of this study is to survey and prototype different versions of the risk parity optimization model for a systematic multi-asset portfolio construction that not only focuses on assets' risk contribution but also on balancing the assets' weight allocation in the optimization process.

The "All Weather" fund, pioneered by Bridgewater Associates LP in 1996, was the first risk parity fund, but portfolio managers were skeptical about its performance. After discovering the importance of RPP in 2008, researchers introduced different formulation and computing methods to optimize the RPP. While the RPP model is roughly straightforward (Kazemi, 2012), yet there is no universally accepted formulation to optimize RPP. Maillard et al. (2010) proposed a convex formulation for RPP which was then modified by Bai et al. (2016). In this study, we have used the RP convex optimization model introduced in Bai et al. (2016). A full review of methods and application can be found in Qian (2016).

Considering the merits of implementing RPP, researchers are still improving its flexibility. One study suggested focusing on risk factors of each asset instead of equalizing the total risk (Bhansali, 2011). They mentioned that by focusing on risk factors, we can prevent the risk of investing in assets that are not likely to continue their historical performance. Moreover, a recent study recognized the high transaction cost of the RPP and mentioned that it is a risk-oriented model which does not consider other performance criteria (Wu et al., 2020). In this regard, we report Turnover (TO) of the RPP as well as other evaluation criteria to compare its performance with benchmarks. To increase the flexibility of RPP, we also have considered the desire of portfolio managers to modify the RPP by constraining the weights.

2 Methodology

To achieve the goal of this study, which is constructing a balanced portfolio in terms of asset risk contribution and weight allocation, first we introduce the basic risk parity (RP) portfolio optimization. Then two modified versions of RP are presented to prevent the portfolio from concentrating too much on a few low risk assets. The mathematical formulation of each proposed portfolio is described below. In this study, we assume that short selling is not allowed.

2.1 Risk Parity Portfolio

The formulation of the basic risk parity optimization is described in Eqs. (1, 2 and 3).

$$\min_w = \frac{1}{2}\, \boldsymbol{w}^T \boldsymbol{\Sigma}\,(\boldsymbol{w}) - \sum_{i=1}^{n} b_i \,\log w_i \tag{1}$$

$$\sum_{i=1}^{n} w_i = 1 \tag{2}$$

$$w_i \geq 0; \qquad \forall i \tag{3}$$

This is a convex optimization model for which Newton's method is guaranteed to find the stationary optimal point (Bai et al., 2016). In Eq. (1), w_i is the weight of asset i in the portfolio, n is the total number of assets, $\boldsymbol{w}_{n \times 1}$ is the weight vector of all assets, $\boldsymbol{\Sigma}_{n \times n}$ is the covariance matrix of assets' rates of return, and b_i is the desired relative risk contribution of asset i to the risk portfolio calculated by Eq. (4), where TRC_i is the total risk contribution of asset i and is defined as Eq. (5).

$$b_i = \frac{TRC_i}{\sum_i TRC_i} \tag{4}$$

$$TRC_i = \frac{w_i (\boldsymbol{\Sigma} \mathrm{w})_i}{\sigma_P} \tag{5}$$

Here, σ_P represents the standard deviation of the portfolio's return and is calculated in Eq. (6).

$$\sigma_P = \sqrt{\boldsymbol{w}^T \boldsymbol{\Sigma}\,(\boldsymbol{w})} \tag{6}$$

At a stationary point for the convex objective function in Eq. (1), the risk contributions are equal and risk parity is achieved indirectly by minimizing this function. Equation (2) makes sure that funds are fully invested. And, $\mathrm{TRC_i} = \frac{\mathrm{w_i}(\Sigma \mathrm{w_i})}{\sigma_\mathrm{P}}$ Eq. (3) enforces the assumption that short selling is not allowed. This formulation (Eq. 1) results in weight concentration on low-risk assets. Because managers may not like how much RP emphasizes these assets, we impose bounds in the following proposed portfolios to prevent too much concentration.

2.2 *Uniformly Bounded (UB) Risk Parity Portfolio*

In this portfolio, constant values have been chosen as the lower bound (L) and upper bound (U) for all assets' weights in all asset classes. These bounds are judiciously chosen after analysing the results of the basic risk parity model to channel the portfolio towards allocating weights to high risk assets as well as low risk ones. Therefore, UB is achieved by substituting Eq. (7) for Eq. (3) in the optimization problem.

$$L \leq w_i \leq U; \qquad \forall i \tag{7}$$

2.3 *Differentially Bounded (DB) Risk Parity Portfolio*

In this portfolio, different boundaries have been chosen for assets' weights in each asset class, separately. These bounds also originated from the results of the UB portfolio that try to distribute the weights among assets as uniformly as possible while concentrating on the objective function. To construct this portfolio, a unique lower and upper bound are set for assets within a class. Therefore, we modified the optimization problem of RP by substituting Eq. (8) for Eq. (3).

$$L_j \leq w_j \leq U_j; \qquad \forall j \in asset\ class\ k \tag{8}$$

3 Results

The daily closing price data of assets falling into three broad classes, namely, commodities (CO), equities (EQ), and fixed income (FI) are considered in this study. Asset (sub)classes and the ETF proxies associated with each are described in Appendix. The dataset includes the daily closing prices of 20 ETFs from 07/01/2013 to 12/31/2019. All steps in the simulation process have been implemented in *R*, using the *riskParityPortfolio* package (Griveau-Billion et al., 2019). We evaluated the performance of the proposed portfolios along with three benchmarks regarding the in-sample (the whole dataset is used for model implementation and performance analysis) and out-of-sample performance (different segments of the data are used for model implementation and performance analysis). Three benchmarks considered in this study are, 60/40 (60% EQ, 40% FI), 50/40/10 (50% FI, 40% EQ, 10% CO), and Equally-Weighted (EW) portfolio. In the 60/40 and 50/40/10 benchmarks, the weights are equal within each class. In UB, we set $L = 0.02$, $U = 0.1$, and in DB, we set $L_j = 0.02$, $U_j = 0.06$; $\forall\ i \in FI$ and $L_j = 0.04$, $U_j = 0.1$; $\forall\ i \in EQ, CO$. In the in-sample performance analysis, we experimented with different time frames over which to estimate the covariance matrix. Moreover, we assumed that high-frequency trades are not allowed; i.e., trades in small time windows like seconds and minutes are not considered. Thus, we consider other time frames for estimation such as monthly, 3-monthly, 6-monthly, 9-monthly, and annual. The returns over longer time frames are calculated based on the approximate compounding return formula and the covariance matrix is also calculated separately for each time frame (Luenberger, 1998).

In the out-of-sample performance analysis, we considered 1 month as the rebalancing period (Maillard et al., 2010) and simulated the process over 2 years. The training window consists of 54 months from 07/01/2013 to 12/31/2017. The 2 year simulation window starts at 01/01/2018 and ends on 12/31/2019. Both training and simulation windows are rolling for 1 month in each rebalancing repetition.

3.1 In-Sample Performance

To obtain insight into the motivation for introducing modified risk parity portfolios, we compared the weight allocation and relative risk contribution of assets when using parameters estimated on a daily basis. Similar patterns are seen for other estimation time frames as well.

As can be seen in Fig. 1, weights are distributed more uniformly among assets in all classes when moving from RP to DB, which confirms the contribution of this study.

Figure 2 demonstrates the relative risk contribution of assets in each risk parity portfolio. It can be seen that in the RP, all assets contribute equally to the risk of portfolio. However, in UB and DB the risk contributions of assets are not

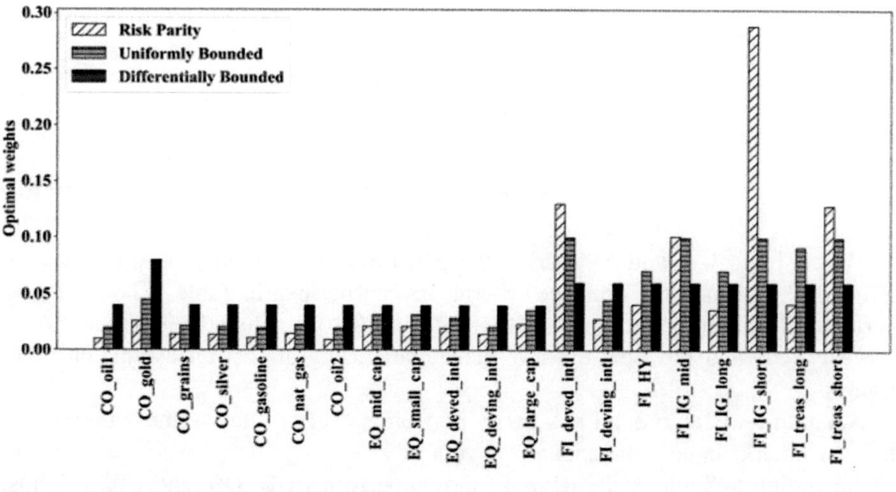

Fig. 1 Weight distribution of different portfolios

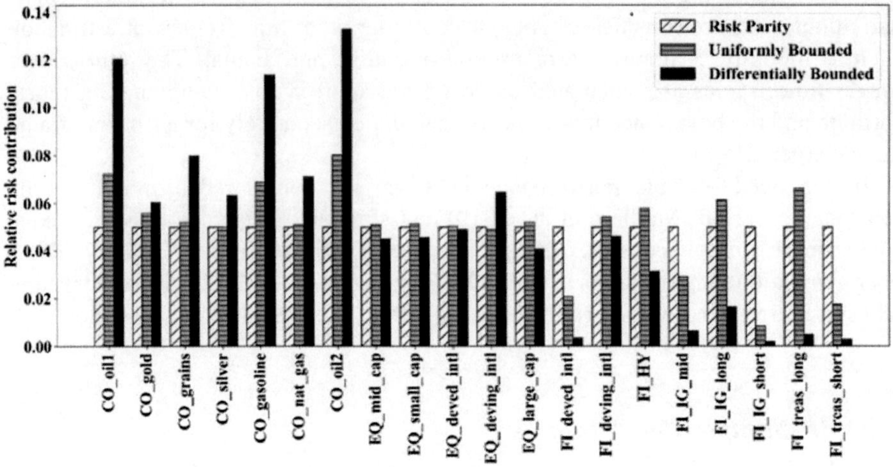

Fig. 2 Relative risk contribution of different portfolios

uniformly distributed. The reason behind this unbalanced risk contribution is the new constraints imposed on the optimization problem of RP, which make the objective function worse and result in unequal risk contribution.

To compare the performance of the different portfolios in different time frames, we calculated their in-sample performances (when the whole dataset is used) based on Annualized Return (AR), Annualized Volatility (AV), Annualized Sharpe Ratio (ASR), Maximum Drawdown (MDD), and percentage of Positive Rolling Return (PRR). MDD and PRR are calculated using Eq. (9) and Eq. (10), respectively where $W(t)$ in Eq. (9) is the portfolio price at time $t \in [0, \tau]$, τ denotes the time frame and N_t in Eq. (10) is the number of days in $[0, \tau]$ in which the portfolio return is positive.

$$MDD = \frac{\left(\max_{t\in[0,\tau]} W(t) - \min_{t\in[0,\tau]} W(t)\right)}{\min_{t\in[0,\tau]} W(t)} * 100 \tag{9}$$

$$PRR = \frac{N_t}{\tau} * 100 \tag{10}$$

Tables 1, 2, 3, 4 and 5 describe the performance of alternative portfolios in terms of the defined performance criteria. As can be seen in Table 1, none of risk parity portfolios compete with the 60/40 and 50/40/10 benchmarks in terms of AR; however, UB and DB beat the EW benchmark using almost all estimation time frames.

According to Table 2, all risk parity portfolios over all time frames outperform the benchmarks in terms of annualized volatility.

According to Table 3, the RP portfolio outperforms UB, DB, and EW in terms of ASR over all time frames; however, it fails to beat the 60/40 and 50/40/10 benchmarks.

Table 1 Annualized return (%) of alternative portfolios over different time frames

Model	Daily	Monthly	3-monthly	6-monthly	9-monthly	Annual
RP	2.256	2.348	2.501	2.945	2.588	1.648
UB	2.594	2.806	3.015	3.649	3.343	2.885
DB	2.668	2.878	3.094	3.722	3.410	2.985
EW	2.656	2.656	2.656	2.656	2.656	2.656
60/40	7.199	7.199	7.199	7.199	7.199	7.199
50/40/10	5.387	5.387	5.387	5.387	5.387	5.387

Table 2 Annualized volatility (%) of alternative portfolios over different time frames

Model	Daily	Monthly	3-monthly	6-monthly	9-monthly	Annual
RP	4.490	4.613	4.792	5.154	4.711	3.749
UB	5.746	5.886	5.957	6.492	6.340	6.179
DB	6.913	6.994	6.939	7.810	7.045	7.641
EW	7.977	7.977	7.977	7.977	7.977	7.977
60/40	11.042	11.042	11.042	11.042	11.042	11.042
50/40/10	9.266	9.266	9.266	9.266	9.266	9.266

Table 3 Annualized Sharpe ratio of alternative portfolios over different time frames

Model	Daily	Monthly	3-monthly	6-monthly	9-monthly	Annual
RP	0.503	0.509	0.522	0.571	0.549	0.440
UB	0.452	0.477	0.506	0.562	0.527	0.467
DB	0.386	0.412	0.446	0.476	0.484	0.390
EW	0.333	0.333	0.333	0.333	0.333	0.333
60/40	0.651	0.652	0.652	0.651	0.651	0.652
50/40/10	0.581	0.581	0.581	0.581	0.581	0.581

Table 4 Maximum drawdown (%) of alternative portfolios over different time frames

Model	Daily	Monthly	3-monthly	6-monthly	9-monthly	Annual
RP	–	6.305	8.462	9.311	10.848	12.459
UB	–	7.956	10.856	11.844	13.291	15.562
DB	–	9.098	12.425	13.632	14.420	17.370
EW	–	11.346	15.574	16.392	15.701	18.854
60/40	–	16.323	21.678	23.574	21.678	25.484
50/40/10	–	13.794	18.492	20.058	18.492	21.380

It can be seen in Table 4 that RP has the lowest MDD over all time frames in comparison with other portfolios. As the data is collected on a daily basis, MDD is not defined on daily basis since it requires a time interval to be calculated. The differences among percentages of PRR is negligible over different time frames among alternative portfolios according to Table 5. This value lies in the interval (52.8, 54.3) for all portfolios.

Table 5 Positive rolling returns (%) of alternative portfolios over different time frames

Model	Daily	Monthly	3- monthly	6-monthly	9-monthly	Annual
RP	54.2	54.2	54.1	54.4	54.0	53.0
UB	53.8	53.9	54.0	54.6	54.3	54.0
DB	52.9	52.7	53.1	53.8	53.1	53.1
EW	52.8	52.8	52.8	52.8	52.8	52.8
60/40	54.3	54.3	54.3	54.3	54.3	54.3
50/40/10	54.1	54.1	54.1	54.1	54.1	54.1

Table 6 Out-of-sample performance of alternative portfolios on basis of 1 month rolling window

Model	AR (%)	AV (%)	ASR	MDD (%)	PRR (%)	TO (%)
RP	−0.540	5.557	−0.096	9.836	55.666	21.574
UB	1.882	6.708	0.280	12.714	55.864	8.979
DB	5.389	7.706	0.699	13.315	55.864	5.643
EW	4.490	8.957	0.501	14.083	55.268	–
60/40	4.658	12.536	0.371	19.077	55.069	–
50/40/10	4.296	10.684	0.402	16.707	54.274	–

3.2 *Out-of-Sample Performance*

All portfolios are compared in terms of out-of-sample performance criteria (Peterson et al., 2015) and shown in Table 6. The RP portfolio results in negative AR and consequently negative ASR; however, it has the lowest AV and lowest MDD over the rebalancing period. DB outperforms alternative portfolios in terms of ASR and has the lowest Turnover (TO). The TO calculation is shown in Eq. (11). It is not defined for the benchmarks since portfolio rebalancing is implemented only for risk parity portfolios.

$$TO = \frac{1}{D_y} \sum_{t=1}^{D_y} \sum_{i=1}^{n} \left(\left| w_{i,t+1} - w_{i,t+} \right| \right) * x_y * 100 \tag{11}$$

Here, D_y is the number of days in all test windows, x_y is the number of time slots of rolling window in a year, and $w_{i,\,t+1}$ and $w_{i,\,t+}$ are the weight of asset i in the portfolio after and before rebalancing, respectively.

Finally, the comparison of investment growth of portfolios over the most recent 2 years (i.e., the simulation window) has been analyzed as shown in Fig. 3. It is assumed that $1000 is invested in each portfolio on the first day of the rebalancing period. As can be seen in Fig. 3, all portfolios show similar trends of investment growth over time. There is no significant difference between the behavior of portfolios over the year 2018, but DB starts to surpass other portfolios in 2019 and ends up with highest investment growth at the end of the simulation window. It should be noted that RP has the lowest fluctuation in investment growth. This fact

Fig. 3 Investment growth of portfolios during the rebalancing period

can be confirmed in the last days of 2018, when almost all portfolios experienced a drop; however, RP demonstrated the lowest decrease of investment value. This performance suggests that RP is an appropriate model for investors who are not interested in high return in a short amount of time and prefer, instead, to capture a steady stream of returns while not exposing the fund to high risks.

4 Conclusion

In this paper, we applied portfolio manager sentiment on relative asset weights to construct two modified version of RP portfolios for a wide range of assets. The first one (UB) applies the same bounds for all asset weights in all classes and the second one (DB) specifies different bounds for assets by class. The risk parity portfolio along with two modified version of it were implemented on ETF data representing 20 assets in three different classes. The performance of these three risk parity portfolios and three well-known benchmarks are compared in terms of multiple performance criteria including annualized return, annualized volatility, annualized Sharpe ratio, maximum drawdown, and percentage of positive rolling return on multiple time frames. Based on the in-sample results, RP outperforms UB, DB, and benchmarks in terms of annualized volatility, maximum drawdown, and annualized Sharpe Ratio; however, DB beats UB, RP and some benchmarks in terms of annualized return. Out-of-sample results based on a 1-month rebalancing period demonstrated that DB has the best performance in terms of annualized return, annualized Sharpe ratio, and turnover, but it also has the highest annualized volatility. In terms of maximum drawdown, DB outperforms RP and UB portfolios, while it cannot beat the benchmarks. Finally, UB and DB have the highest PRR. The out-of-sample results suggest that an investor might increase return and decrease risk by tailoring their investment approach to the current volatility regime – for example, using a traditional approach like 50/40/10 in low volatility periods and switching to RP in volatile ones. On the other hand, because DB appears to perform relatively well throughout the simulation horizon, it could be seen as a promising "all-weather" approach.

Appendix (Table 7)

Full name of all ETFs that have been used in this paper is summerized in Table 7.

Table 7 ETFs dictionary

Asset class	Sub-class	Symbol	Name
EQ	small_cap	SLY	SPDR S&P 600 Small Cap ETF
	mid_cap	MDY	SPDR S&P MIDCAP 400 ETF Trust
	large_cap	SPY	SPDR S&P 500 ETF Trust
	deved_intl	SPDW	SPDR Portfolio Developed World ex-US ETF
	deving_intl	SPEM	SPDR Portfolio Emerging Markets ETF
FI	IG_short	SPSB	SPDR Portfolio Short Term Corporate Bond ETF
	IG_mid	SPIB	SPDR Portfolio Intermediate Term Corporate Bond ETF
	IG_long	SPLB	SPDR Portfolio Long Term Corporate Bond ETF
	HY	JNK	SPDR Bloomberg Barclays High Yield Bond ETF
	deving_intl	EBND	SPDR Bloomberg Barclays Emerging Markets Local Bond ETF
	deved_intl	BNDX	Vanguard Total International Bond Index Fund ETF Shares
	treas_short	SPTS	SPDR Portfolio Short Term Treasury ETF
	treas_long	SPTL	SPDR Portfolio Long Term Treasury ETF
CO	silver	SLV	iShares Silver Trust
	gold	GLD	SPDR Gold Shares
	oil1	DBO	Invesco DB Oil Fund index fund
	oil2	USO	United States Oil Fund, LP
	gasoline	UGA	United States Gasoline Fund, LP
	nat_gas	UNG	United States Natural Gas Fund, LP
	grains	JJG	iPath Series B Bloomberg Grains Subindex Total Return ETN

References

Bai, X., Scheinberg, K., & Tutuncu, R. (2016). Least-squares approach to risk parity in portfolio selection. *Quantitative Finance, 16*(3), 357–376.

Bhansali, V. (2011). Beyond risk parity. *Journal of Investing, 20*(1), 10,137–10,147.

Chaves, D., Hsu, J., Li, F., & Shakernia, O. (2011). Risk parity portfolio vs. other asset allocation heuristic portfolios. *Journal of Investing, 20*(1), 108.
Dalio, R., Prince, B., & Jensen, G. (2015). *Our thoughts about risk parity and all weather*. Bridgewater Associates. https://www.scribd.com/document/283103005/Our-Thoughts-About-Risk-Parity-and-All-Weather.
Griveau-Billion, T., Richard, J., & Roncalli, T. (2019). *Package riskParityPortfolio*. https://cran.r-project.org/web/packages/riskParityPortfolio/riskParityPortfolio.pdf.
Kazemi, H. (2012). An introduction to risk parity. *Alternative Investment Analyst Review, 1.*
Kolm, P. N., Tütüncü, R., & Fabozzi, F. J. (2014). 60 years of portfolio optimization: Practical challenges and current trends. *European Journal of Operational Research, 234*(2), 356–371.
Luenberger, D. G. (1998). *Investment science*. Oxford University Press.
Maillard, S., Roncalli, T., & Teiletche, J. (2010). The properties of equally weighted risk contribution portfolios. *Journal of Portfolio Management, 36*(4), 60.
Markowitz, H. (1952). The utility of wealth. *Journal of Political Economy, 60*(2), 151–158.
Peterson, B., Carl, P., Boudt, K., Bennett, R., Yollin, G., & Martin, R. D. (2015). *Package PortfolioAnalytics*. https://cran.r-project.org/web/packages/PortfolioAnalytics/PortfolioAnalytics.pdf.
Qian, E. E. (2016). *Risk parity fundamentals* / Edward E. Qian, PhD, CFA. Boca Raton.
Thiagarajan, S., & Schachter, B. (2011). Risk parity: Rewards, risks, and research opportunities. *Journal of Investing, 20*(1), 8,79–8,89.
Wu, L., Feng, Y., & Palomar, D. P. (2020). General sparse risk parity portfolio design via successive convex optimization. *Signal Processing, 170.*

A Data Analysis Method for Estimating Balking Behavior in Bike-Sharing Systems

Aditya Ahire and Ashkan Negahban

1 Introduction and Background

Understanding customer behavior can have significant implications for any business. In particular, bike-sharing systems can benefit substantially from better understanding of rider behavior as it would enhance demand estimation. Virtually all critical short- and long-term decisions related to the design and operation of these systems rely on accurate demand estimates. The main decision-making problems in this context include number of stations and their location, station size (number of docks), fleet size (number of bikes), pre-balancing and rebalancing operations, subscription options/pricing and per trip charges. The more accurate the demand estimates, the more appropriate these decisions. As a result, demand analysis is one of the main research areas in the bike-sharing literature.

1.1 Previous Work on Bike-Sharing Demand Analysis

Existing studies on bike-sharing demand analysis can be classified into the following four categories:

- **Sole bike-sharing demand analysis:** These papers solely analyze usage data without considering other factors (e.g., socio-demographics) or other modes of transportation (taxi, rideshare, subway, etc.) (Bordagaray et al., 2016; Vogel et al., 2011; Oppermann et al., 2018; Bargar et al., 2014; Come et al., 2014; Rudloff & Lackner, 2014).

A. Ahire · A. Negahban (✉)
School of Graduate Professional Studies, The Pennsylvania State University, Malvern, PA, USA
e-mail: anegahban@psu.edu

H. Yang et al. (eds.), *AI and Analytics for Public Health*, Springer Proceedings in Business and Economics, https://doi.org/10.1007/978-3-030-75166-1_12

- **Multi-factorial demand analysis:** The papers in this category analyse the demand for bike-sharing systems in conjunction with other factors such as weather, socio-demographic and socio-economic factors, leisure travel and infrastructure (bicycle tracks, etc). For a sample list of these studies, see (Caulfield et al., 2016; El-Assi et al., 2017; Singhvi et al., 2015; Tran et al., 2015).
- **Multi-modal demand analysis:** The papers in this category analyse the demand for bike-sharing systems in conjunction with the demand for other modes of transportation, namely taxi, train/subway, and bus system. For a sample of these studies, see (Singhvi et al., 2015; Tran et al., 2015).
- **Demand censoring:** Data on successful bike pickups censor part of the demand from customers that were unable to pick up a bike due to bike unavailability. These studies propose data cleaning/filtering (O'Mahony & Shmoys, 2015), non-parametric (Albiński et al., 2018), and simulation-based inference (Negahban, 2019) methods to address the censoring problem.

1.2 Contribution of This Paper

Existing demand analysis studies primarily focus on *aggregate* demand patterns at the station, region, or city level. To the best of the authors' knowledge, there is no paper on *individual-level* analysis of balking behavior. This paper contributes to the bike-sharing literature by proposing a data analysis method for inferring the balking threshold and timing of balking decision for individual customers from system-generated data on observed bike pickup times (readily available for virtually any bike-sharing system). Since individuals' true balking behavior is unknown and unobservable, we use simulation to mimic customer behavior and generate synthetic data that are similar to those generated by real-world bike-sharing systems. We then apply our method on the simulated data and assess its efficacy by comparing the estimates with the input parameters used in the simulation model to generate the data.

The remainder of the paper is organized as follows. Section 2 describes the estimation problem. The discrete-event simulation model, simulated data, and the proposed data analysis method are presented in Sect. 3. Section 4 summarizes the results. Finally, conclusions and future research opportunities are discussed in Sect. 5.

2 Problem Description: Estimation of Balking Threshold and Timing of Balking Decision

Our goal is to derive insights on users' balking behavior by estimating their balking threshold (in terms of bike availability) and timing of balking decision. In this section, we frame the estimation problem investigated in this paper.

2.1 The Research Subjects and Their Balking Behavior

We consider a user that regularly picks up a bike from a station according to a relatively fixed schedule. For example, consider a user that has picked up a bike from the same station at around 7:40 AM on many weekdays (say, to ride to work). We let T_A be the random variable representing the user's intended arrival time at the station (or intended pickup time). We assume the user (hereafter referred to as the *subject subscriber*) checks bike availability sometime before her intended pickup time by checking the station status via the service provider's mobile app or website. We use a random variable T_S to represent the time the subject subscriber checks the station status ($T_S \leq T_A$). We also let random variable T_W represent the elapsed time between checking bike availability and arrival at the station (if the subject subscriber decides to pick up a bike). For example, T_W may represent the time it takes the subject subscriber to walk to the bike station. Therefore, we have $T_S + T_W = T_A$. Note that when $T_W = 0$, then $T_S = T_A$, which basically indicates that the subject subscriber does not check bike availability in advance of arrival at the station.

We also assume that the subject subscriber has a balking threshold (represented by random variable B_T) so that she balks if there are fewer than B_T bikes available at the station at T_S when she checks the station status. In other words, based on her past experience, the subject subscriber expects that the station will be out of bikes by her arrival time (T_A) if there are fewer than B_T bikes available at the station T_W minutes before T_A. Our goal is to estimate the balking threshold (B_T) and time of balking or checking station status (T_S) – and consequently T_W – solely from the observed pickup times and bike availability data, even though the system-generated usage data do not provide any direct information on the events that take place that lead to different possible outcomes. These outcomes are discussed next.

2.2 Possible Scenarios

We can group the realizations of the interval of interest (say, 7:00 AM – 8:00 AM on weekdays) as follows:

- **Group 1 – Days with successful pickup:** Days that the subject subscriber picked up a bike from the station according to her regular schedule (say, around

7:40 AM). Two conditions were met on these days: (a) the station had more than B_T bikes available at time T_S when the subject subscriber checked the station status; and, (b) there was at least one bike available when the subject subscriber arrived at the station at T_A. These days can be directly identified from the system-generated data based on the subject subscriber's pickup time.

- **Group 2 – Days without pickup**: This group includes the days that the subject subscriber did not pick up a bike from the station according to her regular schedule. There are four possibilities/subgroups:
  - **Subgroup 2.1 – No pickup due to insufficient bike availability at T_S:** Days that the subject subscriber balked at T_S because bike availability was less than her balking threshold (B_T) when she checked the station status. These days cannot be identified from usage data since T_S and B_T are unobservable.
  - **Subgroup 2.2 – No pickup due to empty station at T_A:** It is possible that the station had more than B_T bikes available at time T_S when the subject subscriber checked the station status, but then ran out of bikes by the time the subject subscriber arrived at the station at T_A. This group includes such days, which are not identifiable from the system usage data since T_S and B_T are unobservable. Moreover, T_A for these days is also unknown due to censoring. In other words, there is no way to tell whether the subject subscriber actually visited the bike station and failed to pick up a bike due to outage.
  - **Subgroup 2.3 – No pickup due to bad weather:** It has been shown that weather conditions have a significant effect on bike sharing demand (El-Assi et al., 2017). We specifically include days with bad weather conditions as a separate subgroup since we can use the available weather data to distinguish these days (which can include rainy, snowy, and extremely cold or hot days).
  - **Subgroup 2.4 – No pickup due to other reasons:** There are many other possible reasons that may result in the subject subscriber not using the bike-sharing system even on days with sufficient bike availability and pleasant weather conditions. These reasons include but are not limited to illness, random delays and time constraints (say, due to oversleeping), customer mood (say, the subject subscriber may feel lazy to ride a bike to work), being on a work-related or personal travel, random changes in customer's schedule (say, the subject subscriber decided to work from home for the entire or part of the day), and other conflicting commitments (say, a doctor's appointment). Information on these days is also unavailable.

3 Methodology

We employ discrete-event simulation to generate synthetic data and assess the proposed data analysis method. There are two major reasons behind using simulation instead of real-world data:

I. Public data on bike-sharing systems do not include any identifier for users (say, real or censored subscriber ID). While this is mainly for privacy concerns and to prevent tracking individual users, this would prevent us from identifying appropriate subject subscribers to use in our study.
II. Individuals' true balking behavior is unobservable, prohibiting validation of the resulting estimates. Simulation allows us to verify and assess the accuracy of our data analysis method by comparing the resulting estimates with the input distributions for balking threshold (B_T) and time of checking station status (T_S) used in the simulation to generate the synthetic data in the first place.

3.1 The Discrete-Event Simulation Model

We assume that the nonstationary bike demand for a station can be approximated by a piecewise-constant rate function with a series of smaller intervals (say, hourly) where demand is assumed to be stationary, and independently and identically distributed (IID). A schematic example is provided in Fig. 1. This is a common approach for modeling and generating nonstationary stochastic processes in the simulation literature (Morgan et al., 2016) and bike sharing studies (Negahban, 2019; Patel et al., 2019). There are several methods and tools that can help identify an appropriate piecewise-constant rate function, namely the change-point analysis from (Chen & Gupta, 2011) and visual assessment tools (Vincent, 1998; Ansari et al., 2014; Negahban et al., 2016). In this paper, we consider a situation where these intervals are already determined and the goal is to estimate the balking threshold and timing of balking decision for a particular subject subscriber in one of these intervals (say, from 7:00 AM to 8:00 AM on weekdays).

We develop a discrete-event simulation model in Simio (Smith et al., 2018) to mimic the operation of the bike station during the time window of interest. Table 1 summarizes the parameters of the simulation model. Each replication represents a realization of bike availability trajectories during the interval of interest. In all

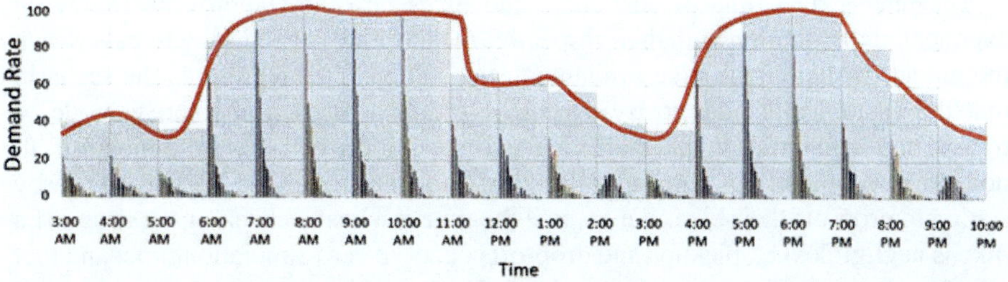

Fig. 1 A piecewise-constant rate function with hourly intervals for approximating the underlying nonstationary bike demand for a station represented by the continuous solid line. The HistoRIA tool developed in (Ansari et al., 2014) is used to generate the piecewise-constant rates as well as the histogram of inter-arrival times within each 1-h block

Table 1 Input parameters of the simulation model of the bike station

Parameter	Description	Value/Range
Input parameters related to the bike station		
N_D	Number of docks at the station	45
$N_B(0)$	Initial number of bikes at station at t = 0	Uniform (5, 30)
CIAT	Customer inter-arrival time	Exponential (0.7) minute
BIAT	Bike inter-arrival time	Exponential (1) minute
P_{BW}	Probability of bad weather conditions	0.15
Input parameters related to subject subscriber (assumed to be unknown)		
T_S	Time of checking station status	Triangular (28, 30, 32)
B_T	Balking threshold	Discrete uniform (10, 11)
T_W	Elapsed time between T_S and arrival time at station (T_A)	Triangular (9, 10, 11)
P_{OR}	Probability of other reasons for not using a bike	0.1

replications, the number of docks at the station is 45 and remains unchanged during the 1-h interval. However, the initial bike inventory at the beginning of the interval on any given day is a random variable and follows a uniform distribution as shown in Table 1. We consider a "busy" station, where the demand rate for bikes is higher than the demand rate for docks (i.e., bike drop-off rate), meaning that the station is likely to have low bike availability during the interval of interest if the initial number of bikes at the beginning of the time interval is small. There are three types of entities in the simulation:

- *Bicycle* entities, which represent riders that attempt to drop-off a bike at the station. Bicycle entities are generated according to the *BIAT* distribution.
- *Subject Subscriber*, which is a marked entity under study, hence there is only one instance of this entity type in any simulation run. This is explained in more detail later in this subsection.
- *Customer* entities represent individuals (other than Subject Subscriber) that attempt to pick up a bike from the station. Customer entities are generated according to the *CIAT* distribution.

The inter-arrival time of customers and bikes into the station both follow an exponential distribution with their respective mean value. Customers will balk (leave the model) if there is no bike available at the station. This represents the scenario that the customer decides to try a nearby station or use an alternative mode of transportation. Similarly, bikes arriving into the station will leave the model if all docks are occupied. This represents the scenario that the rider decides to try a nearby station to drop-off their bike. We assume the time it takes to check-out or drop-off a bike is negligible, i.e., pickups and drop-offs occur in zero simulation time, and that all bikes/docks are functional in the simulation model. For real-world applications, this can be adjusted to account for broken bikes and docks.

To focus on the estimation problem, we generate a *marked* customer sometime (say, around 7:30 AM) during the simulation of the interval of interest. This marked entity represents our subject subscriber and the time it is inserted in the simulation

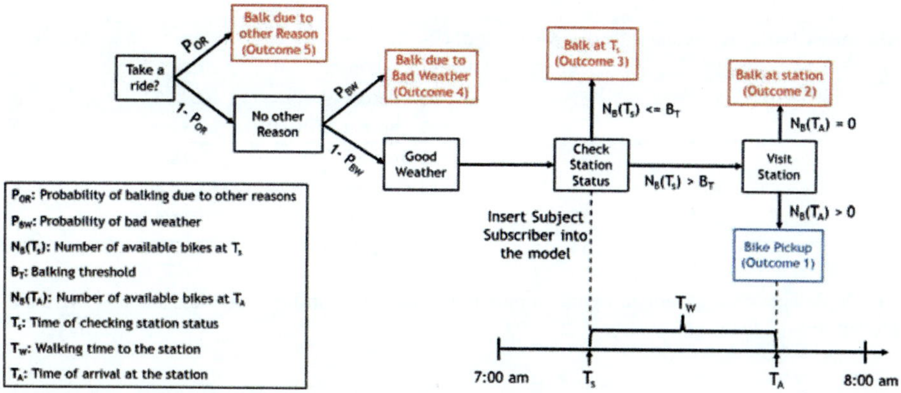

Fig. 2 Different outcomes for the subject subscriber in the simulation

run represents T_S at which she checks bike availability at the station. For example, this can represent a subscriber that picks up a bike at around 7:40 AM on most weekdays to go to work and checks the station status at around 7:30 AM. There are five possible outcomes for the Subject Subscriber entity, which directly correspond to the possible scenarios discussed in Sect. 2.2. The outcomes are summarized in Fig. 2 and can be described as follows:

- Decide to visit the station if $N_B(T_S) > B_T$, where $N_B(T_S)$ denotes the number of bikes available at the station at T_S. Once the marked customer arrives at the bike station at $T_A = T_S + T_W$, there are two possible outcomes:
  - **Outcome 1:** The marked customer will pick up a bike if there is any available.
  - **Outcome 2:** Balk if there is no bike available at the station.
- Decide not to visit the station:
  - **Outcome 3:** Balk after checking station status at T_S due to insufficient bike availability if $N_B(T_S) \leq B_T$.
  - **Outcome 4:** Balk due to bad weather conditions, which occurs with a probability of P_{BW}.
  - **Outcome 5:** Balk due to other reasons, which occurs with a probability of P_{OR}.

It is worth noting that the observations related to the marked customer across n simulation replications will be IID, so standard statistical methods are still applicable. Moreover, inserting arrivals generally carries an inherent risk of distorting the underlying arrival process (in this case, *CIAT* ~ Exponential (1) minute). However, this effect is negligible in our case as we only insert a single marked customer during a 1-h simulation period.

Dayindex	Pickdrop	Availablebikes	Availabledocks	Objects	Time	Bad Weather
1	1	8	37	Customer	07:38:17	0
1	1	7	38	Customer	07:38:27	0
1	1	6	39	Subject_Subscriber	07:38:29	0
9	2	17	28	Bicycle	07:16:12	1
9	1	16	29	Customer	07:16:28	1
9	2	17	28	Bicycle	07:16:42	1
9	1	16	29	Customer	07:17:31	1

Fig. 3 Sample data generated by the simulation model. The dashed line indicates that there are hidden rows in between

3.2 Synthetic Data Generated by the Simulation Model

We consider B_T and T_S distributions used in the simulation to be unknown and unobservable (as in reality). We strive to estimate B_T and T_S solely from the pickup time and bike availability data generated by the simulation, which mimic real-world data readily available on bike-sharing systems. We simulate 300 realizations of the interval of interest. Figure 3 shows a snapshot of the simulated data. The "Day index" is the index for the realization (replication), hence varies from 1 to 300. The "Pick/drop" column indicates whether the row corresponds to a pickup or drop-off event, indicated by 1 and 2, respectively. The number of "Available bikes" and "Available docks" indicate the value just after the respective pickup/drop-off event. The "Object" column indicates the type of entity corresponding to the event. The "Time" column indicates the time stamp when the corresponding event occurred. Finally, the "Bad Weather" column represents the weather condition for that day (0 = good weather, 1 = bad weather). Before performing the estimation analysis, data pre-processing is performed by converting time stamps to real values. For example, by moving the time origin to 7:00 AM, a time stamp of 7:10:30 AM is converted to 10.5 min.

3.3 The Proposed Heuristic Data Analysis Method

Table 2 presents the notations used in this section. To motivate the proposed method, we consider a simplified version of the problem where the true (unknown) values for balking threshold, time of checking station status, and arrival time at the station, respectively denoted by B_T^*, T_s^*, and T_A^*, are deterministic and constant over all days included in the analysis. Without loss of generality, we set $P_{BW} = 0$. For any B_T and T_S, we define two conditional probabilities:

$$\pi_{B_T,T_s}^{P} = \Pr\{N_B(T_s) \geq B_T | E\} = \frac{\Pr\{N_B(T_s) \geq B_T \cap E\}}{\Pr\{E\}}$$

Table 2 Notations related to the proposed data analysis method

Notation	Description
E	The event that the subject subscriber picks up a bike
R	The event that the subject subscriber does not use the system due to other reasons
$N_B(t)$	Number of bikes available at the station at time t
s^P	Number of days that subject subscriber picked up a bike
s^{NP}	Number of days that subject subscriber did not pick up a bike
$n^P_{B_T,T_s}$	Number of days that subject subscriber picked up a bike and $N_B(T_S) \geq B_T$
$n^{NP}_{B_T,T_s}$	Number of days that subject subscriber did not pick up a bike and $N_B(T_S) \geq B_T$
$\pi^P_{B_T,T_s}$	Proportion of days that subject subscriber picked up a bike and $N_B(T_S) \geq B_T$
$\pi^{NP}_{B_T,T_s}$	Proportion of days that subject subscriber did not pick up a bike and $N_B(T_S) \geq B_T$

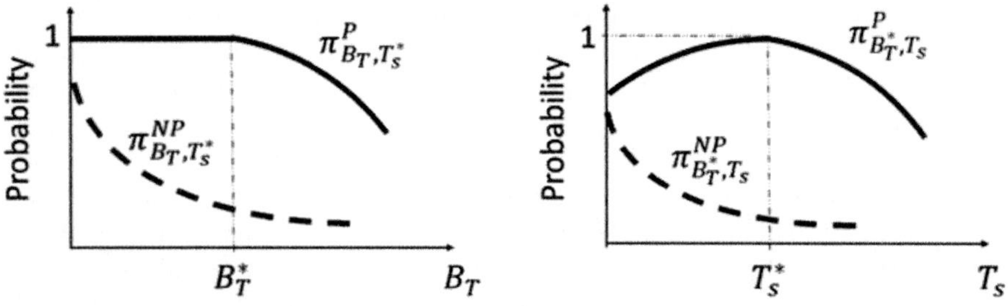

Fig. 4 Possible behavior of the two conditional probabilities around the correct estimates B_T^* and T_s^*. In the left figure, we have $T_s = T_s^*$, and in the figure on the right, $B_T = B_T^*$. The proof for concavity/convexity of these functions is deferred to future research

$$= \frac{\Pr\left\{N_B\left(T_s\right) \geq B_T \cap N_B\left(T_s^*\right) \geq B_T^* \cap N_B\left(T_A^*\right) > 0 \cap R'\right\}}{\Pr\left\{E\right\}},$$

$$\pi^{NP}_{B_T,T_s} = \Pr\left\{N_B\left(T_s\right) \geq B_T | E'\right\} = \frac{\Pr\{N_B(T_s) \geq B_T \cap E'\}}{\Pr\{E'\}}$$

$$= \frac{\Pr\{N_B(T_s) \geq B_T \cap [R \cup (N_B(T_s^*) < B_T^* \cap R') \cup (N_B(T_s^*) \geq B_T^* \cap N_B(T_A^*) = 0 \cap R')]\}}{\Pr\{E'\}}.$$

Clearly, for ($B_T \leq B_T^*$, $T_s = T_s^*$), the first conditional probability will take its maximum possible value of 1. A simple assessment of these two conditional probabilities at B_T^* and T_s^* (i.e., correct estimates) suggests the possibility that the magnitude of their difference is maximized at ($B_T = B_T^*$, $T_s = T_s^*$) as schematically shown in Fig. 4.

Motivated by the above, we propose the heuristic method shown in Fig. 5 to solve the general case where B_T^*, T_s^*, and T_A^* are random variables. We first average the observed pickup times to compute the subject subscriber's expected arrival time at station, T_A. The subject subscriber checks the station status sometime before or at

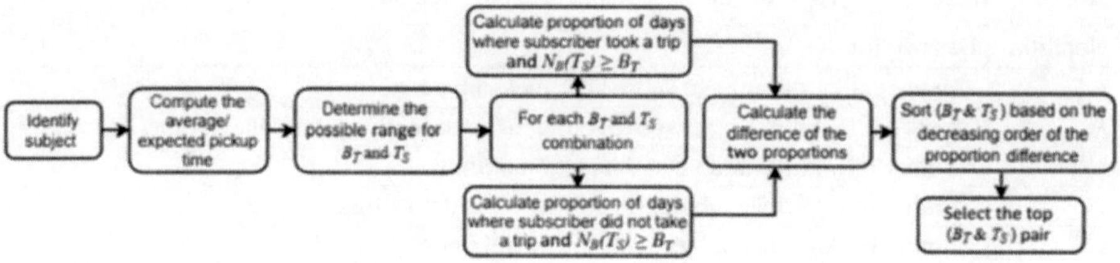

Fig. 5 General steps in the proposed data analysis method for estimating balking threshold and time of checking station status

T_A, hence the T_A value estimated in this fashion is the maximum possible average value for T_S (recall that $T_A = T_S + T_W$). We then start the first loop of the estimation algorithm by setting T_S equal to its upper bound T_A. In other words, we initially assume $T_S = T_A$. We then sequentially decrement the assumed T_S in each subsequent iteration of the search. In the analysis presented in this paper, we decrement T_S by 1 min at a time, although a smaller increment can be used for higher precision. It is unlikely for a customer to make balking decisions a long time before their intended pickup time (say, half an hour before) as bike availability can change drastically by the time they get to the station. The minimum possible value for T_S can be set to the adjusted time origin ($t = 0$) or an arbitrary small value between 0 and the estimated T_A. Here, we set the lower bound for T_S to zero to guarantee that the search covers the true unknown T_S.

Under each potential T_S value, we run a second loop by varying B_T within its possible range. Of course, the minimum possible value for B_T is 0. The maximum B_T value included in the search can be set to an arbitrary large value. In the analysis presented in this paper, we use a conservative maximum B_T value of 20. Although, it is highly unlikely for a customer to balk if the station has 20 bikes available when she checks bike availability (given good weather and no other reason not to ride a bike). By setting this upper bound to 20, we guarantee that our search covers the true unknown B_T. For each potential (T_S, B_T) pair included in the search, we compute two proportions:

$$\pi^{P}_{B_T,T_s} = \frac{n^{P}_{B_T,T_s}}{s^{P}} \text{ and } \pi^{NP}_{B_T,T_s} = \frac{n^{NP}_{B_T,T_s}}{s^{NP}}.$$

We then compute the difference of these two proportions under each potential (T_S, B_T) pair included in the search and sort the searched (T_S, B_T) combinations based on a decreasing order of this difference. The (T_S, B_T) pair for which the difference is maximum would be our estimate of the time of checking station status and balking threshold.

4 Implementation and Results

Figure 6 shows the calculations for some of the (T_S, B_T) pairs searched. Out of the 300 simulated realizations of the interval of interest (7:00 AM to 8:00 AM), there were 68 realizations that the subject subscriber picked up a bike ($s^P = 68$) and 232 that she did not pick up a bike ($s^{NP} = 232$). The average of the observed 68 pickup times was 39.39 min. Therefore, in the first loop of the algorithm, we vary possible T_S values from 39.39 to 0.39 in 1-min increments. In the second loop, we vary B_T from 1 to 20. Consider the first row in Fig. 6 corresponding to $T_S = 0.39$ and $B_T = 1$. There was at least one bike at the station at time 0.39 in 178 days out of the 232 days that the subject subscriber did not pick up a bike ($n^{NP}_{B_T=1,T_s=0.39} = 178$). Similarly, there was at least one bike at the station at time 0.39 in 55 days out of the 68 days that the subject subscriber picked up a bike ($n^{P}_{B_T=1,T_s=0.39} = 55$).

In the last step, we sort (T_S, B_T) combinations based on a decreasing order of their proportion difference. As shown in Fig. 7, we observe that $(T_S = 30, B_T = 11)$ and $(T_S = 30, B_T = 10)$ have the two largest proportion difference values, hence would be our top point estimates of T_S and B_T. Based on Table 1, we know these estimates are correct since they match the mean of the T_S and B_T distributions used in the simulation model to generate the data in the first place. It is important to note that we tested the efficacy of the proposed method in 15 other simulated cases with different parameter configurations and were able to estimate the correct balking threshold and time of checking station status in all cases. However, space limitations preclude the inclusion of all results in this paper.

B_T	T_s	$n^{NP}_{B_T,T_s}$	$n^{P}_{B_T,T_s}$	s^{NP}	s^P	$\pi^{NP}_{B_T,T_s}$	$\pi^{P}_{B_T,T_s}$	$\pi^{P}_{B_T,T_s} - \pi^{NP}_{B_T,T_s}$
1	0.39	178	55	232	68	0.76724	0.80882	0.04158
2	0.39	177	55	232	68	0.76293	0.80882	0.04589
3	0.39	176	55	232	68	0.75862	0.80882	0.05020
12	12.39	89	57	232	68	0.38362	0.83824	0.45461
13	12.39	82	57	232	68	0.35345	0.83824	0.48479
14	12.39	74	56	232	68	0.31897	0.82353	0.50456
13	24.39	25	50	232	68	0.10776	0.73529	0.62754
14	24.39	19	47	232	68	0.08190	0.69118	0.60928
15	24.39	11	41	232	68	0.04741	0.60294	0.55553

Fig. 6 Sample calculations for different (T_S, B_T) pairs included in the search process. The dashed lines indicate hidden rows

B_T	T_s	$n^{NP}_{B_T,T_s}$	$n^{P}_{B_T,T_s}$	s^{NP}	s^{P}	$\pi^{NP}_{B_T,T_s}$	$\pi^{P}_{B_T,T_s}$	$\lvert\pi^{P}_{B_T,T_s} - \pi^{NP}_{B_T,T_s}\rvert$
10	29.39	21	66	232	68	0.090517241	0.970588235	0.880070994
11	30.39	10	62	232	68	0.043103448	0.911764706	0.868661258
10	30.39	19	64	232	68	0.081896552	0.941176471	0.859279919
9	29.39	33	68	232	68	0.142241379	1	0.857758621
9	30.39	30	65	232	68	0.129310345	0.955882353	0.826572008
10	28.39	30	65	232	68	0.129310345	0.955882353	0.826572008
9	31.39	27	64	232	68	0.11637931	0.941176471	0.82479716
11	29.39	13	59	232	68	0.056034483	0.867647059	0.811612576
8	29.39	44	68	232	68	0.189655172	1	0.810344828

Fig. 7 Final step of the proposed method. Sorted (T_S, B_T) pairs based on a decreasing order of proportion difference

5 Conclusions and Future Research

We propose a simple yet effective heuristic data analysis method for estimating balking behavior of bike-sharing users that visit a station according to a somewhat fixed schedule on a regular basis (e.g., a subscriber that takes a bike to work most weekday mornings). We assume such users check the station status sometime before their intended pickup time to decide whether or not to visit the station based on bike availability at that time. Our heuristic method aims to estimate the time the user checks the station status and their balking threshold in terms of bike availability. We tested and confirmed the efficacy of the proposed method using several simulated scenarios.

An immediate extension of this work involves analytical proof for conditions that determine the concavity or convexity of the two conditional probabilities used in our heuristic method. Another important extension involves validation via real-world data. There are two main obstacles that make validation challenging for researchers: (a) due to privacy concerns and to prevent tracking individuals, service providers do not provide any identifier for subscribers in the data that they make public. Our algorithm needs this information to identify subjects and compute their expected pickup time. Service providers, however, have access to such data; and, (b) collecting information on individual's balking behavior requires requesting information directly from the subjects (say, via a survey or interview).

References

Albiński, S., Fontaine, P., & Minner, S. (2018). Performance analysis of a hybrid bike sharing system: A service-level-based approach under censored demand observations. *Transportation Research Part E: Logistics and Transportation Review, 116*, 59–69.

Ansari, M., Negahban, A., Megahed, F. M., & Smith, J. S. (2014). HistoRIA: A new tool for simulation input analysis In *Proceedings of the 2014 Winter Simulation Conference* (pp. 2702–2713).

Bargar, A., Gupta, A., Gupta, S., & Ma, D. (2014). Interactive visual analytics for multi-city bikeshare data analysis. In *Proceedings of the 3rd Urbcomp*.

Bordagaray, M., dell'Olio, L., Fonzone, A., & Ibeas, A. (2016). Capturing the conditions that introduce systematic variation in bike sharing travel behavior using data mining techniques. *Transportation Research Part C: Emerging Technologies, 71*, 231–248.

Caulfield, B., O'Mahony, M., Brazil, W., & Weldon, P. (2016). Examining usage patterns of a bike-sharing scheme in a medium sized city. *Transportation Research Part A, 100*(2017), 152–116.

Chen, J., & Gupta, A. K. (2011). *Parametric statistical change point analysis: With applications to genetics, medicine, and finance* (2nd ed.).

Come, E., Randriamanamihaga, N. A., Oukhellou, L., & Aknin, P. (2014). Spatio-temporal analysis of dynamic origin-destination data using latent dirichlet allocation: Application to vélib' bike sharing system of Paris. In *TRB 93rd Annual meeting*, France, 19p.

El-Assi, W., Salah Mahmoud, M., & Nurul Habib, K. (2017). Effects of built environment and weather on bike sharing demand: A station level analysis of commercial bike sharing in Toronto. *Transportation, 44*, 589–613.

Morgan, L. E., Titman, A. C., Worthington, D. J., & Nelson, B. L. (2016). Input uncertainty quantification for simulation models with piecewise-constant non-stationary Poisson arrival processes. In *Proceedings of the 2016 Winter Simulation Conference* (pp. 370–381).

Negahban, A. (2019). Simulation-based estimation of the real demand in bike-sharing systems in the presence of censoring. *European Journal of Operational Research, 277*, 317–332.

Negahban, A., Ansari, M., & Smith, J. S. (2016). ADD-MORE: Automated dynamic display of measures of risk and error. In *Proceedings of the 2016 Winter Simulation Conference* (pp. 977–988).

O'Mahony, E., & Shmoys, D. B. (2015). Data analysis and optimization for (citi)bike sharing. In *Proceedings of the 29th Conference on Artificial Intelligence (AAAI'15)* (pp. 687–694).

Oppermann, M., Möller, T., & Sedlmair, M. (2018). Bike sharing atlas: Visual analysis of bike-sharing networks. *International Journal of Transportation, 6*(1), 1–14.

Patel, S. J., Qiu, R., & Negahban, A. (2019). Incentive-based rebalancing of bike-sharing systems. In H. Yang & R. Qiu (Eds.), *Advances in service science* (pp. 21–30). Springer.

Rudloff, C., & Lackner, B. (2014). Modeling demand for bikesharing systems: Neighboring stations as source of demand and reason for structural breaks. *Transportation Research Record: Journal of the Transportation Research Board, 2430*, 1–11.

Singhvi, D., Singhvi, S., Frazier, P. I., Henderson, S. G., O' Mahony, E., Shmoys, D. B., & Woodard, D. B. (2015). Predicting bike usage for New York City's bike sharing system. In *Computational sustainability: Papers from the 2015 AAAI Workshop* (pp. 110–114).

Smith, J. S., Sturrock, D. T., & Kelton, W. D. (2018). *Simio and simulation: Modeling, analysis, applications* (5th ed.). Simio LLC.

Tran, T. D., Ovtracht, N., & D'arcier, B. F. (2015). Modeling bike sharing system using built environment factors. In *Procedia CIRP, Elsevier, 2015, 7th Industrial Product-Service Systems Conference - PSS, industry transformation for sustainability and business, 30* (pp. 293–298).

Vincent, S. (1998). Input data analysis. In J. Banks (Ed.), *Handbook of simulation* (pp. 55–90).

Vogel, P., Greiser, T., & Mattfeld, D. C. (2011). Understanding bike-sharing systems using data mining: Exploring activity patterns. *Procedia – Social and Behavioral Sciences, Elsevier 2011, 20*, 514–523.

The Impact of Scalability on Advisory and Service Delivery Efforts of Nonprofits

Priyank Arora, Morvarid Rahmani, and Karthik Ramachandran

1 Introduction

There are over 5000 nonprofit organizations (NPOs) in the United States that provide services related to mental health and crisis intervention, civil rights and advocacy, and employment search and training (National Center for Charitable Statistics, 2019). These mission-driven NPOs face a complex combination of challenges in serving their clients: First, since their clients often vary greatly in terms of their needs (Drucker, 1995; Hasenfeld, 2009), NPOs might be drawn to offer a variety of services that enable different pathways to wellness (Sawhill & Williamson, 2001; Ebrahim & Rangan, 2014). Second, since these NPOs are not revenue-generating and rely on external funding from government and private donors, they operate under a scarcity of resources (Feng & Shanthikumar, 2016). Finally, their clients are often unable to articulate their needs as they are unaware of the true causes of their situation (Holdsworth & Tiyce, 2013) or have endured traumatic experiences resulting in symptoms of PTSD, low self-esteem, or anxiety (Stewart et al., 2004). As such, clients may seek and receive services that are not best-suited to their needs. While mismatched clients continue to consume resources, an NPO's efforts to serve them produce limited social impact. As a result, many NPOs in this domain serve in an interpretive role by providing advisory support to their clients to help them receive the most appropriate services (Emanuel & Emanuel, 1992). However, because such guidance/advisory support does not create a direct impact and also requires resources (funds) that can also be used for other

P. Arora (✉)
Isenberg School of Management, University of Massachusetts Amherst, Amherst, MA, USA
e-mail: parora@isenberg.umass.edu

M. Rahmani · K. Ramachandran
Scheller College of Business, Georgia Institute of Technology, Atlanta, GA, USA

H. Yang et al. (eds.), *AI and Analytics for Public Health*, Springer Proceedings in Business and Economics, https://doi.org/10.1007/978-3-030-75166-1_13

impact-creating activities, it creates a service design dilemma for these NPOs. Despite growing evidence documenting challenges faced by such NPOs, their operational issues have received limited attention from the academic community (Berenguer & Shen, 2019; Besiou & Van Wassenhove, 2020). In this study, we address the following questions: For NPOs that serve distressed individuals, *(i) what are the optimal investments in advisory and service delivery activities that creates the most social impact? and (ii) how does the degree of scalability of services affect these optimal investments?*

2 Related Literature

There are many distinctive objectives for design of services depending on the context. These objectives include reducing customer wait time and system congestion in call centers and hospitals (e.g., Shumsky & Pinker, 2003; Lee et al., 2012), optimizing the sequence of service encounter in entertainment industries (Das Gupta et al., 2015), and maximizing the quality of services delivered in healthcare and legal consulting (Anand et al., 2011; Tong & Rajagopalan, 2014). In this paper, we focus on the service design of NPOs toward maximizing service quality, which in our context is equivalent to generating a higher social impact.

Providers can directly control the perceived quality of their services by carefully choosing the level of resources (Green et al., 2013; Lu & Lu, 2017). However, unilaterally increasing efforts at a service step may not be optimal in some scenarios (Bellos & Kavadias, 2021). Specifically in customer-intensive services, Anand et al. (2011) show that there is a trade-off between offering a deep experience (requiring slowness) and offering a fast and congestion-free service (requiring speed). In for-profit settings, Soteriou and Hadjinicola (1999) and Soteriou and Chase (2000) study resource allocation toward improving service quality, but in the context where the stages of service provision are independent and their qualities are additive (e.g., patient satisfaction during visits to a medical clinic). An important distinction in the context that motivates our study is the interdependence between the provider's efforts in different service stages (i.e., advisory and service delivery efforts). That is, while the NPO's advisory and service delivery activities are complementary in generating social impact, one activity cannot be improved without adversely affecting other activities given the scarcity of resources. We therefore propose an optimal service design for NPOs whose activities are interdependent.

All NPOs must overcome several hurdles in their quest to serve important social needs, and the operational nature of these hurdles may vary from one context to another (see Feng and Shanthikumar (2016) for a detailed review of the challenges faced by NPOs). For NPOs, the complexity of service design and improving quality (impact) arises from the scarcity of resources (Lien et al., 2014) and scalability of their services (Bradach, 2003; Hurst, 2012). Some NPOs have resorted to managing these challenges by allocating a portion of their service capacity to revenue-generating consumers (de Véricourt & Lobo, 2009). We contribute to this literature on non-profit service design by identifying another source of complexity for NPOs: the loss of social impact due to mismatches between the services clients receive and their true needs. We consider a new issue that has not received much attention in the literature: clients may be unable to identify services that suit their needs, which can lead to lower overall impact generated by the NPO's efforts.

3 Model

To answer our research questions, we develop an analytical model, in which an NPO that has a limited amount of resources, denoted by $S > 0$, has to decide on how to invest that in various client-facing activities (i.e., advisory and service delivery efforts) to maximize its overall social impact. In order to capture the differences between clients' needs and the services offered by the NPO, we consider a simple setting with two client types, denoted by $i \in \{a, b\}$, and two service types, denoted by $j \in \{a, b\}$. The A-type (B-type) service is best suited to the service needs of a-type (b-type) clients. However, clients may seek the services that are not best suited to their needs.

Clients We denote by $p \in (0, 1)$ the proportion of a-type clients and by $1 - p$ the proportion of b-type clients. The NPO might have a greater impact by investing the same amount of resources in serving one type of clients than the other; this, for example, may be due to differences in economic impact between the needs of clients. We define $I_i \in R^+$ as a measure for the social impact that the NPO creates by investing a unit of its resources in providing the best suited service to clients of type i for $i \in \{a, b\}$. Without loss of generality, we consider $I_a \geq I_b$. Consequently, we define $k \doteq I_a/I_b \geq 1$, which we refer to as the *impact factor*.

Mismatch Because clients might not be able to articulate the root causes of their needs, they may seek services that are not best suited to their needs. We denote by $\delta_{ij} \in [0, 1]$ the degree of loss of impact due to mismatches between clients' needs and services they receive for $i \in \{a, b\}$ and $j \in \{A, B\}$. For instance, when a-type clients receive the NPO's B-type service (which is not best-suited to their needs), the social impact that the NPO creates by investing a unit of its resources in such service encounters is $\delta_{aB} \cdot I_a \leq I_a$. For simplicity of exposition, we consider $\delta_{aB} = \delta_{bA} = \delta = [0, 1]$, and refer to it as the degree of *loss of impact* due to mismatches. When there is no mismatch, $\delta_{aA} = \delta_{bB} = 1$. The parameter δ can be interpreted as the degree of similarity in clients' needs. For example, if the two types of clients have similar needs (δ is high), the loss of impact due to mismatches is low.

Advisory Effort In order to reduce the loss of impact from service mismatches, the NPO can provide guidance to their clients on choosing the most appropriate services for their needs. This could be in the form of hiring and training employees to design and conduct extended in-take interviews and professional tests of skills, improving intake processes and technology (e.g., software, web-resources), or administering health and behavioral examinations. We denote by $\theta(e_G)$ the proportion of clients who would receive correct services when the NPO invests e_G in its advisory effort, where $\theta(e_G)$ increases in e_G.

Service Delivery Efforts The NPO can increase its impact by investing more resources into the delivery of its services. We denote by $e_A \geq 0$ and $e_B \geq 0$ the NPO's efforts in providing the *A*- and *B*-type services, respectively. These efforts could be in the form of hiring and training employees for delivery of a particular type of service, contracting with specialists (e.g., lawyers and tutors), or investing in infrastructure for service delivery (e.g., shelters and temporary housing). We model the impact generated by the NPO when it exerts e_j towards the *j*-type service, $j \in \{A, B\}$, and delivers it to the *i*-type clients, $i \in \{a, b\}$, as $I_{ij} = \delta_{ij}. I_i \cdot (e_j)^{\gamma}$. We refer to $\gamma \in (0, 1]$ as the *scalability* level of the NPO's services, which we explain next.

Scalability The parameter γ captures returns to scale of the NPO's service delivery efforts. When $\gamma = 1$, the impact generated by the NPO riscs at a constant rate with any increase in service delivery efforts. However, when $\gamma < 1$, the marginal impact created by the NPO decreases with an increase in service delivery efforts. The scalability of the NPO's services may be limited by several practical constraints (Bradach, 2003; Forti & Andrew, 2014). The NPO transforms effort into impact by connecting clients to several sources, such as partners, governments, and volunteers (Wong, 2015). Thus, any bottleneck in accessing these sources could limit the scalability of the NPO's services (Hurst, 2012). For instance, legal services might be provided to clients through a combination of in-house administrative work and pro-bono legal experts. While the NPO can increase in-house staffing by spending more resources, it gets progressively more difficult for it to enhance legal expertise.

Service Design Problem The NPO aims to maximize the total expected social impact generated through its activities. For a given level of advisory effort (e_G), the proportion of a-type clients that receive the *B*-type service is $(1 - \theta(e_G))\,p$, and the proportion of b-type clients that receive the *A*-type service is $(1 - \theta(e_G))(1 - p)$. Accordingly, we obtain the total expected impact (*TEI*) that the NPO delivers as follows:

$$
\begin{aligned}
TEI\ (eG, eA, eB) \doteq p\,\theta\,(e_G)\ (kI_b \cdot (e_A)^{\gamma}\ ) + p\ (1 - \theta\,(e_G))\ (\delta k I_b \cdot (e_B)^{\gamma}) \\
+ (1 - p)\,\theta\,(e_G)\,(I_b \cdot (e_B)^{\gamma}\ ) + (1 - \theta\,(e_G))\,(1 - p)\,(\delta I_b \cdot (e_A)^{\gamma})\,.
\end{aligned}
$$

The first and third terms in equation above correspond to the NPO's impact for serving clients who receive the best-suited service for their needs. The second and fourth terms in equation above correspond to the NPO's reduced impact for the two cases of mismatch (accordingly, these terms contain δ). The NPO chooses the optimal investments for its advisory effort (e_G^*) and service delivery efforts (e_A^*, e_B^*) to maximize its total expected impact using its limited resources (S). The NPO's optimization problem captures the following key and central trade-off: While increasing advisory effort increases the likelihood of clients receiving the appropriate services, it comes at the cost of limiting the NPO's service delivery efforts. Note that advisory and service delivery efforts are complementary in the objective function, but they are drawn from the same pool of resources.

4 Results

Our analysis generates the following first-order managerial insights for NPOs that serve clients in distress: *First*, although the NPO may have a tendency to provide several types of services to cater to different client types, we show that it can be sub-optimal. Specifically, when an NPO's services are scalable (i.e., $\gamma = 1$), the NPO should offer only the service type that generates a higher overall impact. However, when the NPO's services are non-scalable (i.e., $\gamma < 1$), the NPO can generate higher social impact if it balances its investments toward both types of services, as opposed to investing all resources in only one type of service. *Second*, we find that when the NPO (scalable or non-scalable) is severely resource-constrained, it is optimal to offer only basic guidance to its clients; instead, all its resources should be directed toward service delivery activities. In contrast, when the NPO has sufficient amount of resources, it is optimal to spread resources between both advisory and service delivery activities. Both these insights are illustrated in Fig. 1. In addition, our analysis reveals that the optimal advisory effort (when non-zero) should be higher when different types of clients are not evenly mixed in the population, or when mismatches lead to higher loss of impact.

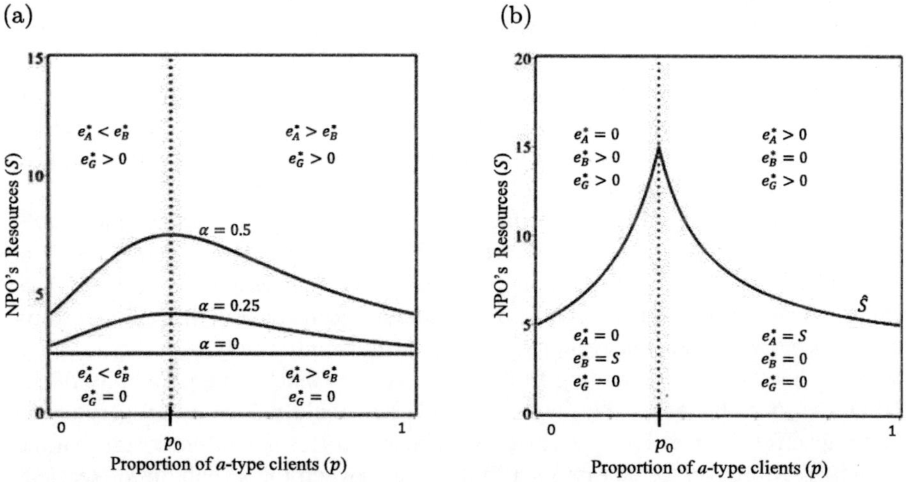

Fig. 1 NPO's Optimal Service Design. (**a**) Non-Scalable NPO (with $\gamma < 1$). (**b**) Scalable NPO (with $\gamma = 1$). *Note:* Parameters: $k = 2$, $\theta\ (eG) = 0.5 + 0.05\ e_G$, $S = 10$, $p = 0.35$, and $\delta \geq 0$ (left) and $\delta = 0.5$ (right)

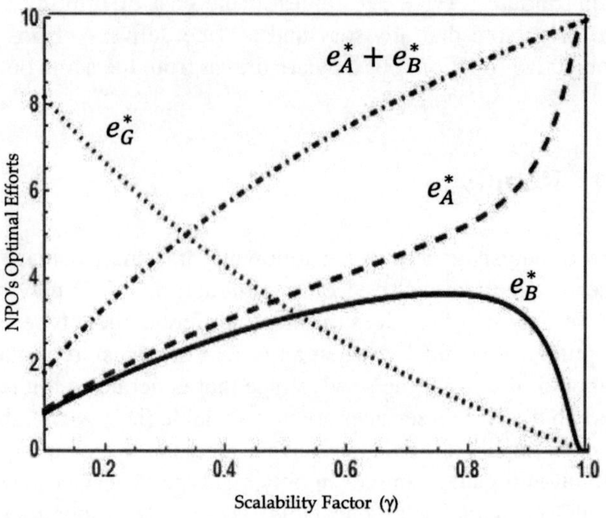

Fig. 2 Effect of Scalability Factor on the NPO's Optimal Efforts. *Note:* Parameters are the same as in Fig. 1 with $\delta = 0$

An important practical implication of these findings is as follows: When services are scalable, it is optimal for the NPO to specialize in only one type of service delivery. This implies that the NPO should not attempt to provide "everything for everyone;" instead, they should determine the type of a service to offer based on the impact factor (k) and the client mix (p). In particular, the NPO should focus on the service type that generates the greatest overall impact (depending on $pk \lesseqgtr 1 - p$). Further, irrespective of the salability of services, the NPO should invest in advisory effort only when it has sufficient amount of resources. However, it is important to note that this threshold amount of resources is smaller when the degree of scalability (γ) is lower. Naturally, the lack of scalability imposes a limit on the impact that the NPO can generate by its service delivery efforts. In such a scenario, the NPO can obtain a greater impact by increasing its advisory effort (which reduces mismatches), than through its service delivery efforts (which have decreasing marginal returns to scale).

Figure 2 illustrates our results on the impact of degree of scalability on the NPO's optimal service delivery efforts. We find that the optimal service delivery efforts toward each type of service are more balanced when services are less scalable (γ is small); however, as services become more scalable (γ increases), the ratio of efforts becomes more skewed and eventually the NPO offers only one type of service when $\gamma \to 1$. Another notable finding from this analysis is that, as the scalability of the NPO's services increases (i.e., $\gamma \to 1$), the NPO should reduce its delivery of the less impactful service. This prioritization primarily arises due to the scarcity of the resources. Moreover, as can be seen in the figure, when the scalability of services is lower than a threshold, the NPO should invest more in its advisory effort, and its advisory effort should even exceed its total service delivery effort.

Additionally, we use numerical examples to understand how our results may apply in practice for an NPO's service design decisions. To do so, we estimate model parameters based on practitioner reports on domestic violence in the U.S. and our conversation with managers at a Houston, Texas-based NPO that empowers survivors of domestic abuse. Summarily, our numerical illustrations show that in designing their services, NPOs should take into account the scalability of their services as well as the loss of impact from mismatches. Although obtaining exact estimates of these parameters may be difficult in practice, NPOs can benchmark themselves with respect to peer organizations, and also observe directional trends in these situational factors. For instance, as an NPO gets more mature, it may become more efficient in delivering its services and build improved access to external resources (via expanding its network and building trust), which implies higher scalability. Similarly, the loss of impact from mismatches may decline over time as the NPO implements client management routines and recovery procedures. As parameters such as scalability and loss of impact evolve, our findings can help NPOs decide on how to invest their resources in various activities, such as hiring employees, expanding infrastructure, and training volunteers.

5 Conclusion

This paper studies the optimal service design of non-profit organizations (NPOs) that serve distressed individuals. Based on our experience and involvement with several NPOs, we realized these organizations operate under a complex combination of challenges such as limited funding, heterogeneity in clients' needs, the limited scalability of their services, and mismatches between clients' needs and services provided. Our analysis has revealed two rules of thumb nonprofits should consider in designing their services by allocating funds between their advisory and service delivery activities. (i) Nonprofits can generate more social impact by offering a smaller subset of services. This is not a comfortable thought for mission-driven nonprofits that don't want to turn away a client in trouble. However, it's important to recognize that when an NPO's services are scalable, it is vital to focus on a few services in order to create a higher social impact. (ii) When more funds are available, the first investment should be in providing guidance to clients about the appropriate services rather than increasing the breadth of offered services.

Improvement in the operations of NPOs can significantly reduce economic and social burdens on the society. We hope that this paper opens new ways and further interests in studying operational complexities of nonprofit service providers.

References

Anand, K. S., Pac, M. F., & Veeraraghavan, S. (2011). Quality-speed conundrum: Trade-offs in customer-intensive services. *Management Science, 57*(1), 40–56.

Bellos, I., & Kavadias, S. (2021). Service design for a holistic customer experience: A process framework. *Management Science, 67*(3), 1718–1736.

Berenguer, G., Shen, Z.-J. (2019). Challenges and strategies in managing nonprofit operations: An operations management perspective. *Manufacturing & Service Operations Management, 22*(5), 888–905.

Besiou, M., & Van Wassenhove, L. N. (2020). Humanitarian operations: A world of opportunity for relevant and impactful research. *Manufacturing & Service Operations Management, 22*(1), 135–145.

Bradach, J. L. (2003). Going to scale. *Stanford Social Innovation Review, 1*(1), 19–25.

Das Gupta, A., Karmarkar, U. S., & Roels, G. (2015). The design of experiential services with acclimation and memory decay: Optimal sequence and duration. *Management Science, 62*(5), 1278–1296.

de Véricourt, F., & Lobo, M. S. (2009). Resource and revenue management in nonprofit operations. *Operations Research, 57*(5), 1114–1128.

Drucker, P. F. (1995). *Managing the non-profit organization: Practices and principles*. Taylor & Francis.

Ebrahim, A., & Rangan, V. K. (2014). What impact? A framework for measuring the scale and scope of social performance. *California Management Review, 56*(3), 118–141.

Emanuel, E. J., & Emanuel, L. L. (1992). Four models of the physician-patient relationship. *Journal of the American Medical Association, 267*(16), 2221–2226.

Feng, Q., & Shanthikumar, J. G. (2016). Not-for-profit operations management. In S. Gupta & M. Starr (Eds.), *The Routledge companion for production and operations management, Ch. 29*. Taylor & Francis.

Forti, M., & Andrew, Y. (2014). Social good = scale x impact (who knew?). *Stanford Social Innovation Review*.

Green, L. V., Savin, S., & Savva, N. (2013). "Nursevendor problem": Personnel staffing in the presence of endogenous absenteeism. *Management Science, 59*(10), 2237–2256.

Hasenfeld, Y. (2009). *Human services as complex organizations*. Sage Publications.

Holdsworth, L., & Tiyce, M. (2013). Untangling the complex needs of people experiencing gambling problems and homelessness. *International Journal of Mental Health and Addiction, 11*(2), 186–198.

Hurst, A. (2012). Demystifying scaling. *Stanford Social Innovation Review*. http://bit.ly/2G5g66D. Last accessed on 9 Apr 2019.

Lee, H.-H., Pinker, E. J., & Shumsky, R. A. (2012). Outsourcing a two-level service process. *Management Science, 58*(8), 1569–1584.

Lien, R. W., Iravani, S. M., & Smilowitz, K. R. (2014). Sequential resource allocation for nonprofit operations. *Operations Research, 62*(2), 301–317.

Lu, S. F., & Lu, L. X. (2017). Do mandatory overtime laws improve quality? Staffing decisions and operational flexibility of nursing homes. *Management Science, 63*(11), 3566–3585.

National Center for Charitable Statistics. (2019). *The nonprofit sector in brief*. https://nccs.urban.org/project/nonprofit-sector-brief. Last accessed on 10 July 2020.

Sawhill, J., & Williamson, D. (2001). Measuring what matters in nonprofits. *McKinsey Quarterly*. https://mck.co/2H08p2h. Last accessed on 10 July 2020.

Shumsky, R. A., & Pinker, E. J. (2003). Gatekeepers and referrals in services. *Management Science, 49*(7), 839–856.

Soteriou, A. C., & Chase, R. B. (2000). A robust optimization approach for improving service quality. *Manufacturing & Service Operations Management, 2*(3), 264–286.

Soteriou, A. C., & Hadjinicola, G. C. (1999). Resource allocation to improve service quality perceptions in multistage service systems. *Production and Operations Management, 8*(3), 221–239.

Stewart, A. J., Steiman, M., Cauce, A. M., Cochran, B. N., Whitbeck, L. B., & Hoyt, D. R. (2004). Victimization and posttraumatic stress disorder among homeless adolescents. *Journal of the American Academy of Child & Adolescent Psychiatry, 43*(3), 325–331.

Tong, C., & Rajagopalan, S. (2014). Pricing and operational performance in discretionary services. *Production and Operations Management, 23*(4), 689–703.

Wong, C. (2015). Scale and sustainability – What's a funder to do? *Nonprofit Quarterly*. http://nonprofitquarterly.org/2015/12/16/scale-and-sustainability-whats-a-funder-to-do. Last accessed on 5 Feb 2018.

Green Location-Routing Problem with Delivery Options

Mengtong Wang, Lixin Miao, and Canrong Zhang

1 Introduction

Due to the discrete temporal and spatial distribution of customer demands that results in a high delivery failure and a low vehicle utilization, the last-mile delivery has become a time-consuming and uncertain process. As a result, parcel logistics companies are exploring and implementing innovative tools such as drones and pavement-based droids to optimize last-mile deliveries. However, such solutions are difficult to be widely adopted due to significant public acceptability and regulatory barriers (https://www.nic.org.uk/publications/better-delivery-the-challenge-for-freight/). Recently, lockers for self-service collection and return of parcels have received positive feedback from both customers and industries by improving the user experience for the former and providing scale benefits for the latter (Vakulenko et al., 2017). Furthermore, contactless delivery has become a new hotspot in the context of COVID-19, bringing new opportunities for the development of lockers. Therefore, a crucial issue for a parcel logistics company is to redesign its pick-up and delivery operations to include lockers as an option.

The first author is a student and we would like to compete for the Best Student Paper award.

M. Wang
Department of Industrial Engineering, Tsinghua University, Beijing, China

Division of Logistics and Transportation, Shenzhen International Graduate School, Tsinghua University, Shenzhen, China

L. Miao · C. Zhang (✉)
Division of Logistics and Transportation, Shenzhen International Graduate School, Tsinghua University, Shenzhen, China
e-mail: crzhang@sz.tsinghua.edu.cn

H. Yang et al. (eds.), *AI and Analytics for Public Health*, Springer Proceedings in Business and Economics, https://doi.org/10.1007/978-3-030-75166-1_14

In addition, there is a rising awareness of the need for alleviating the environmental consequences of logistics operations by deploying low-emission vehicles, e.g., electric vehicles (EVs) (Shen et al., 2019). For example, FedEx has expanded the size of its EV fleets to minimize environmental impacts (http://csr.fedex.com/pdf/FedEx_GCR_FINAL_4.17.19_144dpi.pdf). EVs are now sufficient to meet the needs of short- and medium-distance transportation, and do not need to be charged during the trip. Consequently, there are two delivery options that parcel logistics companies can choose to serve customers. The first delivery option is the direct-to-customer delivery where EVs transport parcels to customers' homes or workplaces. Another delivery option is the direct-to-locker delivery where EVs transport parcels to lockers and customers pick up them at their own convenience.

In this paper, we first address a green location-routing problem with delivery options (GLRP-DO) where a parcel logistics company needs to simultaneously determine the location of lockers and the routing plans for EVs from a single depot. Routes for the replenishment of lockers and routes for the delivery of parcels to customers are separated due to several practical considerations. Hence, we consider two types of EV fleets with different load capacities and battery driving ranges, where large EVs are dedicated to replenishing lockers by providing the round-trip service and small EVs are dedicated to servicing the customers. A locker can serve customers within its pre-specified coverage range and has an accommodation capacity limitation. Each customer must be served by either a locker or a small EV. The goal is to minimize the total cost from the perspective of a parcel logistics company, including the opening cost and handling cost of lockers, as well as the routing costs of EVs. Figure 1 shows a schematic example of the GLRP-DO distribution system.

Despite receiving considerable attention in the last-mile distribution system, research on lockers appears to be scarce. The most related problem is the multi-depot two-echelon vehicle routing problem with delivery options in (Zhou et al., 2018). In this problem, the delivery option for each customer is pre-set by giving

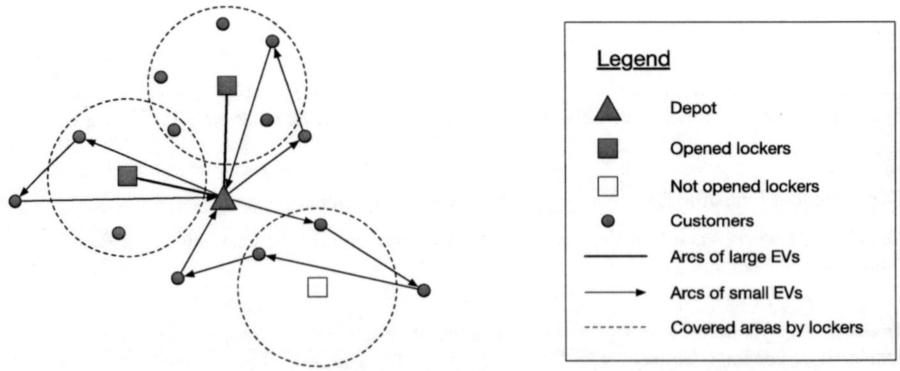

Fig. 1 A schematic example of the GLRP-DO distribution system

a node set including three cases (only served by direct delivery, only served by pick-up facilities, served by either-or). Another related problem is a simultaneous facility location and vehicle routing problem in (Veenstra et al., 2018). In contrast to our problem where a patient within a service range may not be assigned to the corresponding locker (probably due to the limitation of its accommodation capacity), a patient in (Veenstra et al., 2018), is forced to be served by an open locker if it is within the coverage distance of the locker, leading to an un-capacitated locker. Then, a recent study extended the above research to a two-echelon system and developed an efficient adaptive large neighbourhood search heuristic to provide high-quality solutions for large-sized instances (Enthoven et al., 2020).

Different from the above literature, some studies have investigated a similar locker network with the objective of maximizing the company's overall profit. It is not necessary to satisfy all customer demands in these studies. Deutsch and Golany (2018) designed a parcel locker network as a solution to the last-mile distribution system, and used discounts in the delivery cost for customers who choose the locker service. Hosseini et al. (2019) developed a generalization of the capacitated location-routing problem from the perspective of a company engaged in collecting used products from customer zones to maximize its overall profit. A financial incentive is defined to help determine the quantity of used products which are returned by customers. In this problem, the idea of collection centers is similar to our lockers. However, our study focuses on minimizing the total cost and satisfying the total customer demand. Moreover, there is no limit on the accommodation capacity of collection centers and the driving range of vehicles in Hosseini et al. (2019).

According to the above review, although the above studies have mentioned lockers or delivery options in various contexts, they have not totally regarded delivery options as decision variables and have not considered both coverage ranges of lockers and battery driving ranges of EVs. Furthermore, so far, there has been no research to develop exact algorithms for similar problems, but this is very important to obtain available benchmark solutions for larger instances. To address practical issues and fill the research gap, we use an integrated modelling approach to assist in the planning process to deploy a new last-mile distribution system with two delivery options (i.e., the GLRP-DO) and propose an effective branch-and-price (B&P) algorithm that can solve to optimality instances with moderately larger size.

2 Model Formulation

We apply a Danzig-Wolfe decomposition (Dantzig & Wolfe, 1960), to formulate the GLRP-DO as a set partitioning formulation and treat it as a master problem (MP) that links the columns generated through pricing subproblems. In this study, two types of pricing subproblems are proposed for providing feasible columns. The first type of pricing subproblems is the locker coverage service subproblem (LCSP) that generates a pattern of service customers with negative reduced cost for each locker, and the second type of pricing subproblems is the shortest path problem with battery driving range constraints (SPPBDRC) that helps generate a negative reduced-cost small EV route.

2.1 Problem Statement

In this paper, the GLRP-DO is defined in a graph $G = (V, A)$, where $V = \{0\} \cup V_l \cup V_c$ is the set of vertices, $\{0\}$ represents the depot, V_l is the set of potential lockers, and V_c is the customer set. Let the arc set $A = \{(i,j) : i,j \in V, (i,j) \notin V_l \times V_l\}$, which comprises the arcs connecting the depot to the customers and lockers, and those connecting pairs of customers. Associated with a locker $l \in V_l$ are, the fixed open cost f_l (per day), the handling cost a_l (per parcel), the accommodation capacity Q_l and the coverage range r_l. For replenishing lockers, the set of large EVs K^0 with load capacity Q_e^0 and battery driving range B^0 is available at the depot, and provides the round-trip service due to the high volume transported between the depot and lockers. For serving customers, the set of small EVs K with load capacity Q_e and battery driving range B is available at the depot and provides the direct delivery service. The distance, the travel cost of arc $(i,j) \in A$ and the charging consumption rate are given as d_{ij}, c_{ij} and h, respectively. It is assumed that the distance and travel cost observe the triangle inequality. Each customer $i \in V_c$ has a known and deterministic demand q_i, and can be served by a small EV or an open locker. We also assume that no customer demand is greater than the capacity of small EVs and lockers, and no accommodation capacity of lockers is greater than the load capacity of large EVs. Therefore, only one round-trip service is needed for each open locker. Let ϕ be a factor that represents the economies of scale of the round-trip service, then $\overline{f_l} = f_l + \phi\,(c_{0l} + c_{l0})$ can be treated as the fixed cost of locker l, consisting of the opening cost f_l of locker l and the routing cost $\phi(c_{0l} + c_{l0})$ of large EVs.

2.2 Master Problem

We redefine the set V_l as $\{l |\, d_{0l} + d_{l0} \leq B^0, l \in V_l\}$ to ensure that lockers can be visited by large EVs. Let P be the set of all feasible patterns. Each pattern $p \in P_l$ means a set of customers served by locker l within its coverage range and accommodation capacity, and $\cup_{l \in V_l} P_l = P$. Let R be the set of all feasible routes. Each route $r \in R$ starts from the depot, visits one or several customers in V_c, and ends at the depot. Moreover, each route does not violate the load and battery capacities of small EVs by construction. Let $\alpha_{ip} \in \{0, 1\}$ be a binary parameter that equals 1 if customer i is assigned to pattern p, and 0 otherwise. Let $\beta_{ir} \in \{0, 1\}$ be a binary parameter that equals 1 if customer i is visited by route r, and 0 otherwise. The costs of each pattern $p \in P$ and route $r \in R$ are c_p and c_r, respectively. Then, let v_l be a binary variable that takes value 1 if locker l is opened, and 0 otherwise. Let z_p be a binary variable that takes value 1 if pattern $p \in P$ belongs to the solution, and 0 otherwise. Let x_r be a binary variable that equals 1 if route $r \in R$ belongs to the solution, and 0 otherwise. Finally, the MP is formulated as follows:

$$\min_{v,z,x} \sum_{l \in V_l} \overline{f}_l v_l + \sum_{p \in P} c_p z_p + \sum_{r \in R} c_r x_r \tag{1}$$

$$\text{s.t.} \quad \sum_{p \in P} \alpha_{ip} z_p + \sum_{r \in R} \beta_{ir} x_r = 1 \qquad \forall i \in V_c \tag{2}$$

$$\sum_{p \in P_l} z_p = v_l \qquad \forall l \in V_l \tag{3}$$

$$v_l \in \{0, 1\} \qquad \forall l \in V_l \tag{4}$$

$$z_p \in \{0, 1\} \qquad \forall p \in P \tag{5}$$

$$x_r \in \{0, 1\} \qquad \forall r \in R \tag{6}$$

The objective function (1) minimizes the total cost including the fixed cost of opened lockers (the opening cost of opened lockers and the large EV routing cost for these lockers), the handling cost of opened lockers, and the small EV routing cost for customers. Constraint (2) ensures that each customer is served by exactly one route or one pattern, thereby achieving a partitioning scheme for customers. Constraint (3) guarantees that at most one pattern will be chosen for each locker. To introduce the subsequent two types of pricing subproblems, let $\boldsymbol{\mu}$ and $\boldsymbol{\tau}$ be the dual variables of the constraints (2) and (3), respectively. Constraints (4), (5) and (6) specify the domains of the decision variables.

2.3 The Locker Coverage Service Subproblem

Let z_i^l be a binary variable that takes value 1 iff customer i is served by locker l. Then all pricing subproblems for each locker are identical except for the dual price τ_l considered in the objective function. Hence, the LCSP- l is simply dedicated to each open locker l and is presented as follows:

$$\min_{z} \sum_{i \in V_c} (a_l q_i - \mu_i) z_i^l - \tau_l \tag{7}$$

$$\text{s.t.} \quad d_{il} z_i^l \leq r_l \qquad \forall i \in V_c \tag{8}$$

$$\sum_{i \in V_c} z_i^l = Q_l \tag{9}$$

$$z_i^l \in \{0, 1\} \qquad \forall i \in V_c \tag{10}$$

In this formulation, the objective function (7) is to minimize the reduced cost of a pattern for locker l. Constraints (8) and (9) are implemented to define the coverage range and the accommodation capacity constraints of locker l, respectively. The binary requirement of the solution is ensured by constraint (10). The LCSP- l is essentially a variant of knapsack problem that can be solved in pseudo-polynomial time. Considering that commercial MIP solvers can easily solve instances of the knapsack problem with thousands of variables (Poss, 2013). We call the CPLEX solver to find the exact solution of the LCSP-l.

2.4 The Shortest Path Problem with Battery Driving Range Constraints

Let x_{ij} be equal to 1 iff a small EV traverses arc (i,j). b_i represents the remaining battery of a small EV when it arrives at node $i \in \{0\} \cup V_c$. Then, the SPPBDRC is presented as follows:

$$\min_x \sum_{i \in \{0\} \cup V_c} \sum_{j \in \{0\} \cup V_c, j \neq i} \bar{c}_{ij} x_{ij} \tag{11}$$

$$\text{s.t.} \sum_{j \in V_c} x_{0j} = 1 \tag{12}$$

$$\sum_{j \in V_c} x_{j0} = 1 \tag{13}$$

$$\sum_{j \in \{0\} \cup V_c, j \neq i} x_{ji} = \sum_{j \in \{0\} \cup V_c, j \neq i} x_{ij} \qquad \forall i \in V_c \tag{14}$$

$$\sum\nolimits_{i \in V_c} \sum_{j \in \{0\} \cup V_c, j \neq i} q_i x_{ij} \leq Q_e \tag{15}$$

$$b_i \leq B - h * d_{0i} x_{0i} \qquad \forall i \in V_c \tag{16}$$

$$b_j \leq b_i - h * d_{ij} x_{ij} + B\left(1 - x_{ij}\right) \qquad \forall i \in V_c, \forall j \in \{0\} \cup V_c, j \neq i \tag{17}$$

$$x_{ij} \in \{0, 1\} \qquad \forall i \in V_c, \forall j \in \{0\} \cup V_c, j \neq i \tag{18}$$

$$b_i \geq 0 \qquad \forall i \in \{0\} \cup V_c \tag{19}$$

The objective function (11) minimizes the reduced cost of the constructed route. Constraints (12) and (13) are associated with the routing decision, and constraint (14) ensures the flow balance. Constraint (15) relates to the total small EV load capacity. Constraints (16) and (17) enforce sub-tour elimination using the cumulative battery capacity consumed upon visiting a node for each small EV. Constraints (18) and (19) define the domains of decision variables. Note that the new arc cost $\overline{c}_{ij}$ is defined as $\overline{c}_{ij} = c_{ij} - \mu_i, \forall i \in V_c, \forall j \in \{0\} \cup V_c, j \neq i$ and $\overline{c}_{0j} = c_{0j}, \forall j \in V_c$.

It can be shown that the SPPBDRC is modeled as an elementary shortest path problem with resource constraints which is NP-hard (Dror, 1994), but it can be solved by dynamic programming in pseudo-polynomial time. Therefore, we employ the label setting algorithm to generate the optimal route of the SPPBDRC. Labels are attached to each node to identify the state of the resources (reduced cost, load capacity and battery capacity) when a corresponding feasible path is found from the depot to the present node. In addition, we extend labels and accelerate the solution procedure by adopting the method proposed by Feillet et al. (2004).

3 Methodology

First, the linearly relaxed version of MP (LMP) is solved to obtain dual values for defining reduced costs. Considering that a restricted LMP (LRMP) involves a small subset of columns at each branch node, then the column generation is called to identify the subsets of patterns and routes with negative reduced costs. These patterns and routes as new columns are added to the LMP and resolved iteratively. This procedure is repeated until no new column exists and then we can obtain the optimal solution of LMP. If it is fractional, some branching rules are applied hierarchically in the B&P algorithm and the detail is described later. The best-first strategy is implemented to explore the branch-and-bound tree, which guarantees that the child node with the best lower bound will be explored first.

3.1 Branching Rules

In this study, two types of pricing subproblems are proposed for constructing feasible columns. In order to create adequate branching rules that are compatible with these pricing subproblems, we consider four-layer hierarchical branching rules in the proposed B&P algorithm.

The first branching rule branches on the location variable v_l. Fixing $v_l = 1$ enforces the use of locker l with its LCSP- l solved, and fixing $v_l = 0$ is achieved by imposing a value of $z_p = 0$ for all of the patterns $p \in P_l$ in the RMP.

The second branching rule restricts the service assignment for each customer. We define $\rho_i = \sum_{p \in P} \alpha_{ip} z_p$ as the value of choosing the locker service for customer i. We branch on the value of ρ_{i^*} that is most fractional (whose fractional part is closest to 0.5). Fixing $\rho_{i^*} = 1$ means that customer i^* is only visited by the locker service, and can be achieved by deleting all routes that visit customer i^* and no longer allowing small EVs to visit customer i^* in its corresponding subproblems. Fixing $\rho_{i^*} = 0$ ensures that customer i^* cannot be satisfied by the locker service. We delete all patterns that visit customer i^* and prohibit any locker to serve customer i^* in its pricing subproblems.

The third branching rule branches on the arc (i,j). Let R_{ij} be the set of all routes that visit the arc (i,j), and we select the arc (i^*,j^*) such that $\varphi_{i^*j^*} = \sum_{r \in R_{i^*j^*}} x_r$ is most fractional. Then, we impose $\varphi_{i^*j^*} = 0$ by removing the arc (i^*,j^*) from the network for small EVs and dropping all routes that include arc (i^*,j^*) in one branch. In the other branch, we impose $\varphi_{i^*j^*} = 1$ by removing all arcs (i',j^*) and (i^*,j') such that $i' \neq i^*$ and $j' \neq j^*$, and dropping all route-related columns that do not satisfy this constraint.

The fourth branching rule restricts the locker assignment for each customer. We define $\theta_{il} = \sum_{p \in P_l} \alpha_{ip} z_p$ as the value of choosing locker l for serving customer i. We branch on the value of $\theta_{i^*j^*}$ that is most fractional. Fixing $\theta_{i^*j^*} = 1$ means that customer i^* is only visited by locker l^*, and can be achieved by deleting all patterns from locker $l \neq l^*$ that visit customer i^* and no longer allowing locker $l \neq l^*$ to visit customer i^* in its corresponding subproblems. Fixing $\theta_{i^*j^*} = 0$ ensures that customer i^* cannot be satisfied by locker l^*. We delete all patterns from locker l^* that visit customer i^* and prohibit locker l^* to serve customer i^* in its pricing subproblems.

3.2 A Tight Upper Bound

Through numerical experiments, we find that in most cases, the best lower bound has reached the optimal value after the first three steps of branching rules. It means that the best routes have been found but the best patterns haven't been identified. To reduce the number of branching, we propose an acceleration technique with the aim of finding a better feasible solution and thus improving the performance of our B&P algorithm.

If a solution of RLMP satisfies the first three branching rules, we can obtain a feasible small EV routing plan and its cost $z_{EV} = \sum_{r \in R} c_r x_r$. In addition, the set of lockers to open V_l' and the set of customers served by lockers V_c' can also be acquired. Then, we develop the following locker assignment problem (LAP) to find a feasible solution for assigning remaining customers to the opened lockers.

$$z_{Locker} = \min_{v,z} \sum\nolimits_{l\in V_l'} \overline{f}_l v_l + \sum_{i\in V_c'} \sum\nolimits_{l\in V_l'} a_l q_i z_i^l \tag{20}$$

$$\text{s.t.} \quad \sum_{l\in V_l'} z_i^l = 1 \qquad \forall i \in V_c' \tag{21}$$

$$z_i^l d_{il} \le r_l v_l \qquad \forall i \in V_c', \forall l \in V_l' \tag{22}$$

$$\sum_{i\in V_c'} q_i z_i^l \le Q_l v_l \qquad \forall l \in V_l' \tag{23}$$

$$v_l \in \{0, 1\} \qquad \forall l \in V_l' \tag{24}$$

$$z_i^l \in \{0, 1\} \qquad \forall i \in V_c', \forall l \in V_l' \tag{25}$$

Once the first three branching rules are met and the fourth branching rule is violated, we first solve the LAP to obtain a new upper bound $z' = z_{Locker} + z_{EV}$ to update the best upper bound, rather than implementing the fourth branching rule. In the computational experiments section, we will test that the above acceleration strategy to observe its impact on the reduction of running time.

4 Computational Experiments

We conduct numerical experiments with two aims. First, we assess the performance of our B&P algorithm in comparison with the original MIP model using the branch-and-cut algorithm implemented via CPLEX with version 12.7. Second, the sensitivity analysis and managerial insights are given to consider the impact of delivery options of the GLRP-DO. All experiments are coded by JAVA and run on a computer with a 16GB RAM and 4.0 GHz CPU.

4.1 Description of Problem Instances

For the GLRP-DO, we construct our instances by extending three well-known sets of benchmark instances (Perboli et al., 2011), namely "Set 2" and "Set 3". Moreover, some new information is added and summarized in Table 1, which includes the name of sets (Set), the number of instances (#), the numbers of customers (N_c), lockers (N_l) and EVs (N_e), and other values of related parameters. Note that we assume the parameter settings of EVs are sufficient to ensure the feasibility of testing instances.

Table 1 Characteristics of the GLRP-DO instances

Set	#	N_c	N_l	N_e	f_l	ϕ	a_l	Q_l	Q_e^0	R	B^0	B	h
2/3	6	21	3	16	60	0.1	0.001	6000	3000	30	180	120	1
	6	21	4	16	60	0.1	0.001	4500	3000	30	180	120	1
	6	50	5	23	100	0.1	0.01	90	45	50	300	200	1
	6	50	6	23	100	0.1	0.01	75	45	50	300	200	1

4.2 Computational Performance of the B&P Algorithm

Partial results of performance comparison between our B&P algorithm and CPLEX are shown in Table 2. The first and second columns show the name and the set of potential lockers of each instance, respectively. For CPLEX, Columns 3–5, respectively, represent the objective value (optimal solutions or best upper bound found) obtained with CPLEX (Best), the solver optimality gap within 3600 seconds (Gap(%)), and the computing time of CPLEX (T(seconds)). The remaining columns relate to the B&P algorithm. Columns 6–9 report the following: (i) the optimal objective value or the upper bound obtained within 3600 seconds (Best); (ii) the optimality gap at termination or within 3600 seconds (Gap(%)); (iii) total CPU time of the B&P algorithm without the tight upper bound described in Sect. 3.2 (T_{no}(seconds)); (iv) total CPU time of the B&P algorithm with the tight upper bound (T(seconds)).

The results show that the superior performance of the B&P algorithm over the B&B/C algorithm implemented with CPLEX mainly comes from the following two aspects: (i) The proposed tight upper bound presented in previous section is beneficial for accelerating the B&P procedure. For large-size instances, the substantial CPU time savings achieved by this acceleration technique are about 30% on average for the instances in Set 2, and about 12% on average for the instances in Set 3. For medium-size instances, the performance of the tight upper bound is not outstanding, or even worse. The reason may be that this upper bound cannot provide a tighter upper bound but increases the redundant upper bound calculation. (ii) The proposed B&P algorithm is both effective and efficient in solving the GLRP-DO. In fact, CPLEX failed to solve the GLRP-DO for 50% of instances to provable optimality within 3600 seconds. In contrast, the proposed B&P algorithm solved all testing instances to optimality in an average of 5 seconds for instances with three lockers and 21 customers, an average of 18 seconds for instances with four lockers and 21 customers, an average of 186 seconds for instances with five lockers and 50 customers, and an average of 572 seconds for instances with six lockers and 50 customers.

Table 2 Comparative results for CPLEX vs. B&P algorithm

	Description	CPLEX			B&P algorithm			
Inst.	Lockers	Best	Gap (%)	T (seconds)	Best	Gap (%)	T_{no} (seconds)	T (seconds)
Set 2								
En51s5-1	2,4,5,17,46	841.61	1.21	3623.45	841.61	0.00	194.83	163.24
En51s5-2	6,12,27,32,37	843.83	1.38	3625.53	843.83	0.00	219.72	159.10
En51s5-3	7,11,19,27,47	835.19	1.33	3630.36	835.19	0.00	305.66	202.33
Average			1.31	3626.45		0.00	240.07	174.89
En51s6-2	6,12,27,32,37,45	951.17	3.07	3614.81	945.46	0.00	325.15	271.91
En51s6-4	3,4,8,17,39,46	953.12	3.25	3621.27	946.95	0.00	1000.23	743.30
En51s6-5	2,13,22,27,32,37	954.74	3.28	3624.18	946.03	0.00	396.15	327.12
Average				3620.08			573.84	447.44
Set 3								
En51s5-1	12,18,39,41,43	841.94	1.53	3622.92	841.94	0.00	274.58	188.78
En51s5-2	13,19,40,41,42	844.97	1.39	3623.22	844.97	0.00	292.15	186.50
En51s5-3	13,40,41,42,44	848.95	1.22	3617.99	848.95	0.00	198.19	155.89
Average				3621.37		0.00	254.97	177.06
En51s6-1	12,18,21,39,41,43	951.66	3.03	3608.38	945.41	0.00	423.17	399.12
En51s6-2	13,19,20,40,41,42	955.79	3.24	3618.43	948.33	0.00	514.64	447.49
En51s6-4	16,22,24,28,41,43	959.05	3.78	3618.56	947.26	0.00	407.94	355.05
Average				3615.12		0.00	448.58	400.55

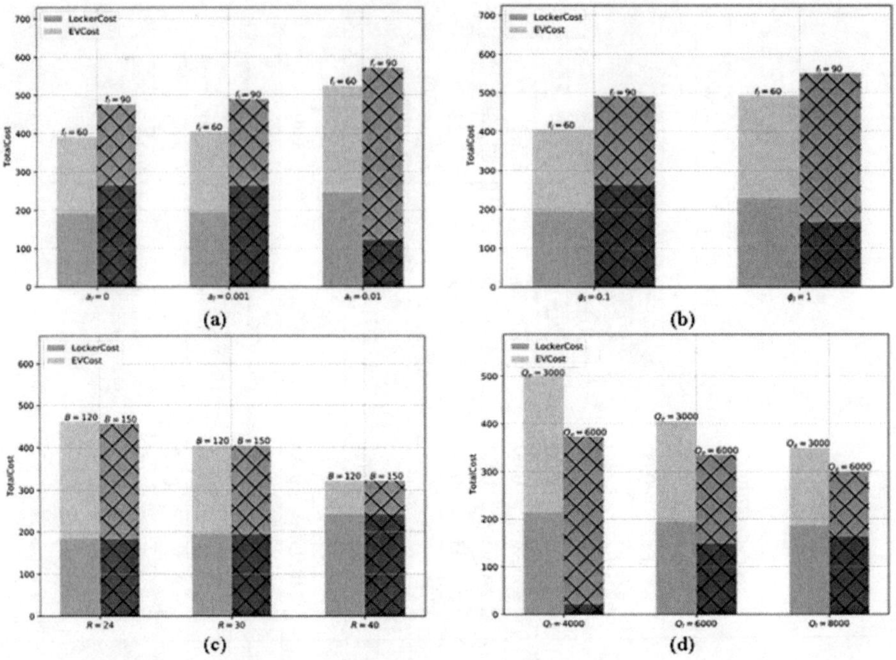

Fig. 2 Sensitivity analysis of the GLRP-DO

4.3 Sensitivity Analysis for the GLRP-DO

In this section, we investigate three sets of model parameters that may affect the solutions of the GLRP-DO. The first one is the sensitivity of the result to the cost of lockers including the opening cost, the handling cost and economies of scale factor. The second one is the impact of the coverage range of lockers and the battery driving range of small EVs, and the third one is the effect of the accommodation capacity of lockers and the load capacity of small EVs.

Examples of such lockers that provide self-service options can include lockers and self-built service stations. As a result, the costs of these lockers and the realization of economies of scale may be very different. Figure 2a and 2b contrast the composition of the optimal total cost under different opening cost f_l, holding cost a_l and economies of scale factor ϕ settings. The results show that when the opening cost is high, as the holding cost or economies of scale factor increases, the total cost increases, where the cost of locker service decreases and the cost of small EVs increases. In addition, we observe the GLRP-DO system is more sensitive to the handling cost than the opening cost.

Given a depot, a set of customers and a set of potential lockers can be opened, the changes in the coverage range R and the battery driving range B may result in different delivery solutions for serving all the customers. Figure 2c illustrates that with the increase in the coverage range of lockers, the total cost decreases, and the delivery proportion of lockers increases. Furthermore, when the battery driving range of EVs reaches a certain level, it will not have much impact on the choice of delivery options and cost.

Indeed, different types of lockers or EVs can provide different accommodation/load capacities. Thus, planning an efficient GLRP-DO system requires examining the impact of the accommodation capacity Q_l of lockers and the load capacity Q_e of small EVs. As can be seen in Fig. 2d, the accommodation capacity and load capacity can provide better competitive advantages for their corresponding delivery options (lockers and EVs). Moreover, the solutions of the GLRP-DO are more sensitive to the accommodation capacity than to the load capacity.

5 Summary

In this paper, we introduce the GLRP-DO, a practical last-mile delivery problem, that can deal with the presence of delivery options (lockers or direct delivery) and the application of EVs. We develop an effective B&P algorithm for the GLRP-DO, where two types of pricing subproblems are solved exactly and some useful acceleration techniques are proposed. Due in part to the tighter upper bound, the computing time of the algorithm is reduced by 20% on average for large-size testing instances. Furthermore, the B&P algorithm greatly outperforms the commercial solver CPLEX over all testing instances. Our computational study also illustrates how the GLRP-DO can support parcel logistics companies to make better decisions in the relevant context. First, it is very important to improve the utilization rate of locker accommodation capacity as much as possible if companies consider including lockers as a delivery option, as the comparative advantage of using lockers depends largely on the economies of scale. In addition, experimental results demonstrate that the GLRP-DO system is more sensitive to the handling cost than to the opening cost of lockers. Second, EVs are suitable for such hybrid delivery systems and mixed-fleet policies for the management of EVs are profitable in such a delivery network. Consequently, interesting extensions on this research consist of involving customer participation in decision-making process to maximize the utilization rate of the opened lockers, and investigating the GLRP-DO with special aspects such as multi-trips or customer time windows.

Acknowledgements This work was supported by the National Key R&D Program of China under grant No. 2018AAA0101705, and the National Natural Science Foundation of China under grants 71771130 and 71872092.

References

Dantzig, G. B., & Wolfe, P. (1960). Decomposition principle for linear programs. *Operations Research, 8*(1), 101–111.

Deutsch, Y., & Golany, B. (2018). A parcel locker network as a solution to the logistics last mile problem. *International Journal of Production Research, 56*(2), 251–261.

Dror, M. (1994). Note on the complexity of the shortest path models for column generation in VRPTW. *Operations Research, 42*(5), 977–978.

Enthoven, D. L. J. U., Jargalsaikhan, B., Roodbergen, K. J., et al. (2020). The two-echelon vehicle routing problem with covering options: City logistics with cargo bikes and parcel lockers. *Computers & Operations Research, 118*.

Feillet, D., Dejax, P., Gendreau, M., et al. (2004). An exact algorithm for the elementary shortest path problem with resource constraints: Application to some vehicle routing problems. *Networks, 44*(3), 216–229.

Hosseini, M. B., Dehghanian, F., & Salar, M. (2019). Selective capacitated location-routing problem with incentive-dependent returns in designing used products collection network. *European Journal of Operational Research, 272*(2), 655–673.

Perboli, G., Tadei, R., Vigo, D., et al. (2011). The two-echelon capacitated vehicle routing problem: Models and math-based heuristics. *Transportation Science, 45*(3), 364–380.

Poss, M. (2013). Robust combinatorial optimization with variable budgeted uncertainty. *A Quarterly Journal of Operations Research, 11*(1), 75–92.

Shen, Z. M., Feng, B., Mao, C., et al. (2019). Optimization models for electric vehicle service operations: A literature review. *Transportation Research Part B-methodological, 128*, 462–477.

Vakulenko, Y., Hellstrom, D., Hjort, K., et al. (2017). What's in the parcel locker? Exploring customer value in e-commerce last mile delivery. *Journal of Business Research, 88*, 421–427.

Veenstra, M., Roodbergen, K. J., Coelho, L. C., et al. (2018). A simultaneous facility location and vehicle routing problem arising in health care logistics in the Netherlands. *European Journal of Operational Research, 268*(2), 703–715.

Zhou, L., Baldacci, R., Vigo, D., et al. (2018). A multi-depot two-echelon vehicle routing problem with delivery options arising in the last mile distribution. *European Journal of Operational Research, 265*(2), 765–778.

Molecular Bioactivity Prediction of HDAC1: Based on Deep Neural Nets

Miaomiao Chen, Shan Li, Yu Ding, Hongwei Jin, and Jie Xia

1 Introduction

Pharmaceutical chemistry has always been a hot branch in the field of pharmaceutical research (Xu, 2019). With the advancement of drug synthesis technology, the number of medicinal chemical molecules has exploded (Chaurasia et al., 2016; Matteelli et al., 2017). Adopting information research methods into prediction of on-target and off-target activities can help to narrow the scope of the experiment and greatly reduce the manpower and material resources of searching for new medicinal molecules, as well as the elimination rate in clinical trials (Zhang et al., 2016; Zhu et al., 2019).

With the advent of 'Big data' era, deep neural networks (DNNs) have achieved remarkable performance in the fields of image recognition, speech recognition, and natural language processing. DNNs is a full connected neural network composed of an input layer, an output layer and several hidden layers, with a number of neurons in each layer. There is no connection between neurons in the same layer while each neuron in the Nth layer is all connected with those in the (N-1)th. And the outputs of neurons from the (N-1)th layer are the inputs of those from the Nth. Figure 1 shows a DNN model with one hidden layer. See that each connection is mapped to one weight. Neurons are the basic unit of neural network. Each neuron

M. Chen · S. Li (✉) · Y. Ding
College of Economics and Management, Nanjing University of Aeronautics and Astronautics, Nanjing, China
e-mail: lishan@nuaa.edu.cn

H. Jin
School of Pharmaceutical Sciences, Peking University, Beijing, China

J. Xia
Department of New Drug Research and Development, Institute of Materia Medica, Chinese Academy of Medical Sciences and Peking Union Medical College, Beijing, China

H. Yang et al. (eds.), *AI and Analytics for Public Health*, Springer Proceedings in Business and Economics, https://doi.org/10.1007/978-3-030-75166-1_15

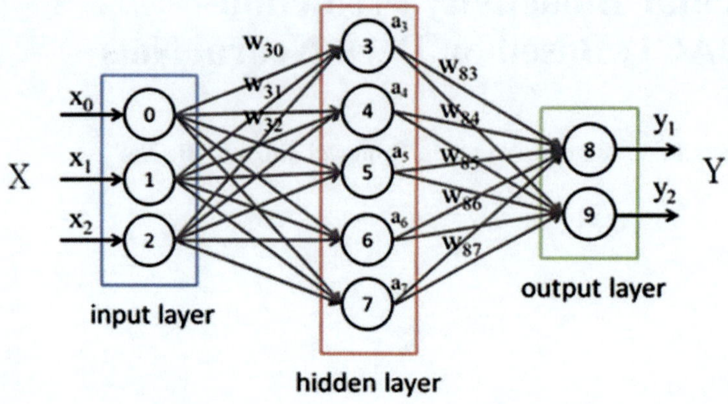

Fig. 1 Structure diagram of DNN

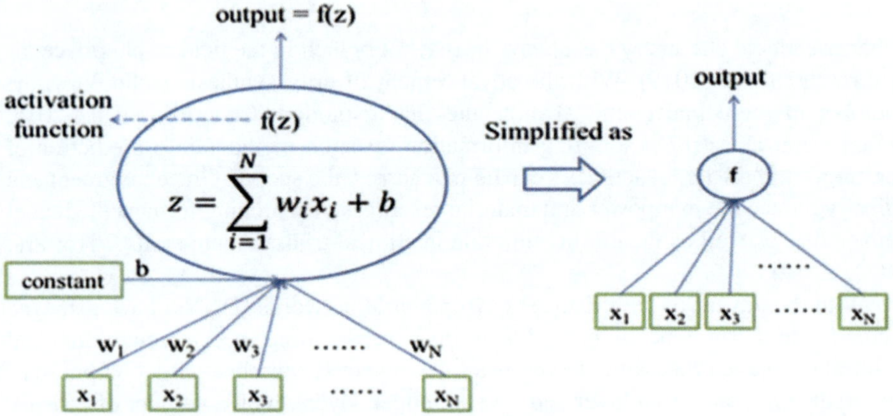

Fig. 2 Detailed and simplified diagrams of how neurons work

has a corresponding activation function, which can realize nonlinear transformation from the sum of inputs under different weights into the output of the neuron. Figure 2 displays the detailed and simplified operation of neurons. After initial standardization of the selected features, one or more output values are obtained through the calculation of neurons in the input layer and hidden layers, which we called the forward propagation of the neural network. Weights are parameters initialized as random number. The HDAC1 applied for the study is a dataset with labels of activity. Training using datasets with labels of pattern is called supervised learning, in which a loss function in the output layer will be defined to calculate the deviation between the true label values and the results of forward propagation. And then back propagation (BP) is used to update the weights until deviation minimization (Lecun et al., 2014). Both forward and back propagation are not visible. That's why deep learning is called "black box" algorithm.

Quantitative structure-activity relationship (QSAR) is a very commonly used technique in pharmaceutical industry for predicting molecular bioactivity (Worachartcheewan et al., 2009). Actually, DNNs has so far been applied for building a QSAR model for pharmaceutical molecules and achieved remarkable improvement compared to traditional approaches (Goh et al., 2017; Pereira et al., 2016). In the Kaggle competition sponsored by Merck in 2012, Dahl and Hinton et al. proposed a method with DNNs based on 15 QSAR datasets, which consistently outperformed RF by a margin (mean R2 averaged from 0.42 to 0.49) for the first time in the past 10 years. Later, a team from Merck applied the model to the other 15 extended datasets for comprehensive evaluation and analysis, which helped to develop a generic DNN model with fixed parameters for QSAR (Ma et al., 2015). Different from the single output of the aforementioned models, Dahl et al. (Dahl et al., 2014) proposed a multitask neural networks model based on the database called PubChem, developing a shared and learnable feature extraction pipeline for multitask model, which enabled DNNs to process thousands of associated feature inputs. And it was shown that the predictive performance of DNNs had an obvious advantage over traditional machine learning methods. In the same year, Hochreiter et al. (Mayr et al., 2016) applied the multitask DNNs trained with ECFP4 fingerprints for a larger database called ChEMBL. The AUC value of this model was 0.83, above all the other trained machine learning models. Then Panda Group and Google (Ramsundar et al., 2015) submitted a large-scale study result on arxiv, which displayed that multitask DNNs outperformed other traditional machine learning methods like logistic regression and RF but only based on specific data sets. In addition, Korotcov et al. (2017) pointed out the shortcomings in previous studies: the number of other machine learning methods compared with DNNs is deficient; the type of datasets used for experiment is limited; the metrics used for measuring performance are lacking. So they compared 6 machine learning methods including DNNs using FCFP6 fingerprints based on 8 varying datasets that were applicable to pharmaceutical research. And measured by 7 different metrics, DNNs offered obvious improvement in performance.

In view of the researches above, we propose DNNs as a method to develop a QSAR model for molecular bioactivity prediction of HADC1, a new but essential tool for drug discovery, with less relative research based on deep learning methods (Chen et al., 2015; Tan et al., 2009). Besides, we also propose training separate DNNs using 3 different input features based on the same dataset, which provides a reference for feature selection when constructing QSAR of inhibitor drugs like HDAC1 based on DNNs.

2 Experimental Section

2.1 Data Set

The data HDAC1 we used in this study came from ChEMBL, an online database of active small molecules developed by the European Bioinformatics Institute. There were 5572 raw data with properties including SMILES, IC50 value, confidence score, rotatable bonds (RBs), molecular weight (MW) etc. SMILES was used for computing molecular properties and IC50 value was used for discriminating active and inactive molecules. So firstly we removed the data without SMILES or IC50 value and those duplicated. Data with a confidence score below 4 were removed as well. Additionally, the statistical results of present marketed drugs show that molecular structure with MW not below 600 or RBs above 20 has no ideal druggability. After removing this part of data, 3812 pieces of data remained. Depending on whether the IC50 value was above 1uM or not, they were labeled as 2499 active molecules and 1313 inactive molecules. Eventually the whole dataset was randomly divided into training set and testing set in ratio of 7:3.

2.2 Feature Competition and Extraction

Three main properties containing molecular attribute descriptors, MACCS fingerprints (166 bins data set) and ESTATE fingerprints (79 bins data set) (Yu-Huan & Ke-Jiang, 2009) were computed from SMILES of HDAC1. As for the molecular attribute descriptors, we used a third-party library based on python—ChemoPy to compute four sets of feature descriptors, which described physicochemical properties and structures of molecules from diverse perspectives. They were respectively constitutional descriptors with 40 segment attributes, Basak descriptors with 21 segment attributes, Burden descriptors with 64 segment attributes and MOE-type descriptors with 60 segment attributes, summing up to 175 molecular attribute descriptors (Fan et al., 2017). Since information might be redundant across these attribute descriptors and for reducing the complexity of training, we used several statistical methods to extract features. Firstly we used the Eq. (1) to calculate the Person CC of molecular attribute descriptors and their corresponding IC50 values:

$$\rho_{XY} = \frac{Cov(X, Y)}{\sqrt{D(X)}\sqrt{D(Y)}} = \frac{E((X - EX)(Y - EY))}{\sqrt{D(X)}\sqrt{D(Y)}} \tag{1}$$

The less-relevant attribute descriptors whose CC values were below 0.05 were removed. Then we calculated the CC values between two attribute descriptors. In the pairs whose results were above 0.9, the descriptors with less relevance to IC50 were eliminated, so as to do the dimension reduction. Next we used SVM-RFE based

Table 1 36 Selected molecular attribute descriptors

Groups	Descriptors
Basak	IC6, SIC6, IC3, CIC0, IC2, SICO
Burden	bcutm2, bcutm3, bcutm4, bcutm5,bcutm8, bcutm10
	bcutp1, bcutp2, bcutp3, bcutp9, bcutp10, bcutp11, bcutp12, bcutv13, bcutv14, bcutv16
	bcute1, bcute3, bcute10, bcute8
Constitutional	PC2, nring, nnitro, nhev, naro
MOE-type	EstateVSA3, PEOEVSA2, PEOEVSA8
	MRVSA2, slogPVSA7

on five-fold cross validation to calculate the importance of the remaining attribute descriptors. After removing those with little importance, we finally saved 36 best attribute descriptors (see Table 1).

2.3 DNNs Trained with Molecular Attribute Descriptors

Apart from the parameters automatically updated by Deep Neural Networks algorithm itself, those set by researchers depending on experience are called hyper-parameters, such as number of hidden layers, number of neurons, activation function and loss function. In this study, we used Keras with Theano backend to develop a basic DNNs model at first. Then starting from the number of hidden layers, did the hyper-parameters optimization by a step based on ten-fold cross validation on training set. The optimization process of DNNs trained with 36 selected attribute descriptors was displayed as follows. The Function Seed was used to keep initial random number same for each training. As for the initial number of nodes in the hidden layers, one common strategy turns to set it as the mean of number of input layer and output layer, which are 19.

Number of hidden layers Except for ReLU as activation function in hidden layer, sigmoid as activation function in output layer, batch_size = 5 and epoch = 100, other hyper-parameters were set as default. After comparing the prediction accuracy on testing set of DNNs with 1–4 hidden layers (Fig. 3), also with considering the computational complexity, we used DNNs with 2 hidden layers.

Grid search hyper-parameters settings for improving prediction accuracy We used grid search method from Scikit-learn (Bergstra et al., 2011). Set those hyper-parameters mutable variables and could be optimized in the scope: number of nodes (50, 100, 200, 300, 512, 1024), batch size (3, 4, 5, 6, 7, 8, 9, 10), number of epoch (50, 100, 150, 200, 300, 400), network weight initialization (zero, uniform, normal, lecun_uniform, lecun_normal, glorot_uniform, glorot_normal, he_uniform, he_normal), optimization algorithm (adam, adamax, Adadelta, rmsprop), activation function (softmax, softsign, relu, tanh, sigmoid), learning rate (0.001, 0.01, 0.02,

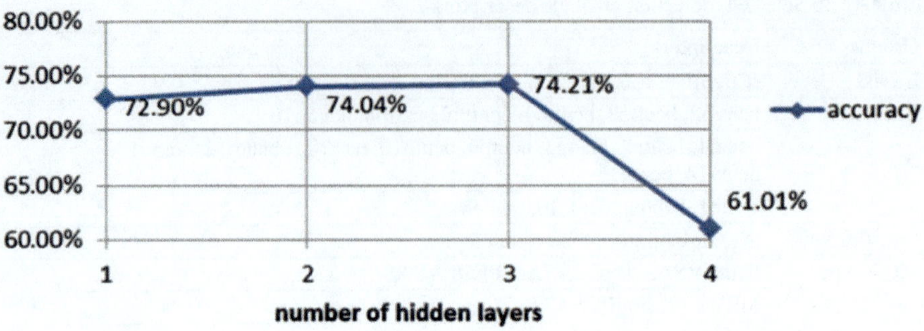

Fig. 3 Prediction accuracy of DNN models with 1–4 hidden layers

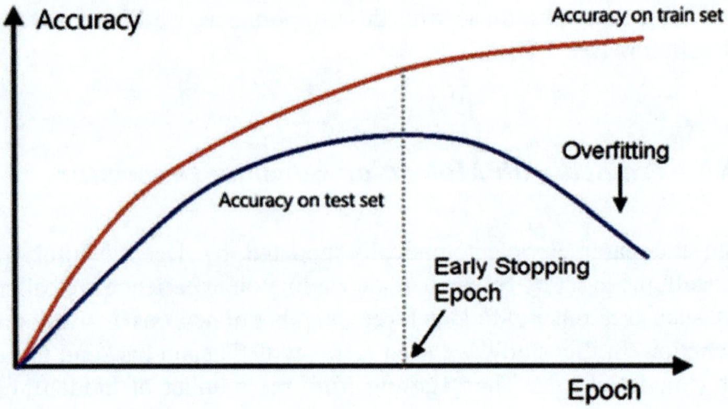

Fig. 4 Principle of using EarlyStopping

0.1, 0.2). The results showed that when number of neurons = 300, batch size = 4, number of epochs = 300, weight initializer = he_normal, optimizer = adamax, learning rate = 0.01, activation function of the 1st hidden layer = softsign, and of the 2nd hidden layer = ReLU, the precision accuracy on training set was up to 95%, while on testing set was only approximately 80%.

Dropout, L2 norm and EarlyStopping for overfitting When the accuracy on training set is much higher than that on testing set, overfitting may occur during the training. Two basic approaches to avoid DNN model overfitting are used in training including L2 norm and drop out regularizations for all hidden layers. Proposed by Hinton (Hinton et al., 2006), Dropout helps turn several weights or outputs of hidden layers to zero so that interdependency between neurons can be reduced (Srivastava et al., 2014). And the principle of L2 norm is to add a penalty to the loss function that is proportional to the weight of the model. Besides, too many training epochs can also cause overfitting (Fig. 4). The global optimum may have occurred before the training epochs reach 300. Hence EarlyStopping was used to early terminate the training when prediction accuracy on testing set didn't improve in the last 50 epochs.

After using the grid search to set dropout rate = 0.3, L2 regularization = 0.01, as well as the EarlyStopping, the model's precision accuracy on testing set was improved to 83.74%.

2.4 *DNNs Trained with Molecular Fingerprints*

After computing from SMILES by ChemoPy, we got a data set of MACCS fingerprints which has 166 dimensions and a data set of ESTATE fingerprints which had 79 dimensions. The value of each dimension was either 0 or 1. Thus except for the number of nodes in the input and hidden layers, other two DNNs separately trained with MACCS fingerprints and ESTATE fingerprints as attribute descriptors were built by the same method. The results showed that the accuracies on test set of MACCS and ESTATE were 85.14% and 82.61%.

3 Metrics

Apart from precision accuracy on test set, we also use other 6 metrics to evaluate the performance of three trained DNNs from diverse perspectives, including precision, recall, specificity (SP), sensitivity (SE), F1-Score, ROC curve and the area under it (AUC). The abbreviations from confusion matrix (Table 2) are used to define these metrics. See that TP is the number of true positives, FP is the number of false negatives, FN is the number of true negatives.

SP measures the prediction accuracy on inactive molecules while SE inversely measures that on active molecules: SP = TN/(TN + FP), SE = TP/(TP + FN). They are two common metrics for molecular bioactivity prediction experiments. Model precision is the probability a predicted true label is indeed true and is defined: Precision = TP/(TP + FP). Similarly, model recall can be thought of percentage of true class labels correctly identified by the model as true and is defined: Recall = TP/(TP + FN). The F1-Score is simply the harmonic mean of the recall and precision: F1 = 1/(1/P + 1/R) = (2*P*R)/(P + R). Accuracy we have calculated is another measure of the overall model classification performance and is the percentage of correctly identified labels out of the entire population: Accuracy = (TP + TN)/(TP + FP + TN + FN). The ROC curve can be computed by plotting the recall vs the false positive rate (FPR) at various decision thresholds T, where FPR = FP/(FP + TP). In this study, all constructed models are capable of assigning a probability estimate of a sample belonging to the true class. Thus, we can construct an ROC curve (Fig. 5) by measuring the recall and FPR performance

Table 2 Confusion matrix

	Actual positive	Actual negative
Predicted positive	TP	FP
Predicted negative	FN	TN

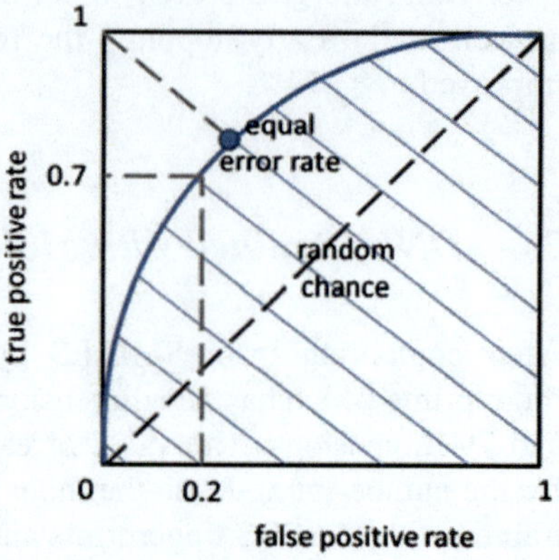

Fig. 5 The interpretation of ROC curve

Table 3 The parameter settings and accuracy of 3 DNNs

Fingerprint	Attribute descriptors	MACCS	ESTATE
Number of hidden layer	2	1	1
Number of neuron	300	300	300
Batch_size	4	4	4
Kernel_initializer	he_normal	he_normal	he_normal
Activation function	softsign+ReLU	ReLU	ReLU
Learning rate	0.01	0.01	0.001
Optimizer	Adamax	Adamax	Adamax
Dropout rate	0.01	0.005	0
Accuracy on test set	**0.8374**	**0.8514**	**0.8261**

when we considered a sample with a probability estimate > T as being true for various intervals between 0 and 1. The AUC can be constructed from this plot and can be thought of as the ability of the model to separate classes, where 1 denotes perfect separation and 0.5 is random classification.

4 Results and Analytics

Although three characteristic properties were all computed from SMILES, they had different number of dimensions and thus represent diverse structure information. Table 3 displays the exact hyper-parameters settings of three models after grid search in the same scope. It was found that, DNNs trained with different structural descriptors calculated from a same dataset respectively got their best performance with different number of hidden layers, learning rate and Dropout rate. However, there was no obvious pattern of input data dimensions and these three hyper-parameters, which was probably due to the same and not large number of input data. The results of hyper-parameters settings provided little reference for setting original grid search scope.

Table 4 Performance evaluated by different metrics of 3 DNN models

Metrics	Attribute descriptors	MACCS	ESTATE
Accuracy	83.74%	85.14%	82.61%
SP	84.24%	88.38%	84.16%
SE/Recall	82.45%	78.81%	78.87%
Precision	66.92%	77.61%	67.43%
F1-Score	73.88%	78.21%	72.70%
AUC	0.89	0.90	0.87

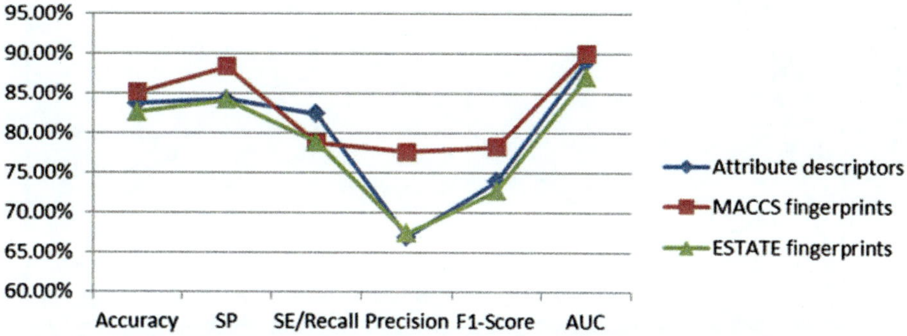

Fig. 6 Line chart of performance evaluated by different metrics of 3 DNN models

Then we focused on the model evaluation index. Apart from the accuracy on test set, we used metrics package from Scikit-learn to calculate the other Precision, Recall (SE), SP, F1-Score of three DNNs and used pyplot to plot their ROC curves. Table 4 displays the results and we used the line chart (Fig. 6) to show the comparison more visually. It is found that on the metrics of Accuracy, SP, F1-Score and AUC, the order of model performance from high to low is: MACCS fingerprints > molecular attribute descriptors > ESTATE fingerprints, which consistently shows the best all-around model robustness of DNN model trained with MACCS fingerprints. For the SE/Recall, the order is: molecular attribute descriptors > ESTATE fingerprints > MACCS fingerprints; as on the Precision, the order is: MACCS fingerprints > ESTATE fingerprints > molecular attribute descriptors. The difference on SP and SE tells us that DNN model trained with MACCS fingerprints has an advantage on the identification of inactive molecules, while DNN model trained with attribute descriptors can do better on active molecules.

We did not only focus on one descriptor but calculated three different kinds of characteristic properties from SMILES of HDAC1 and developed three diverse DNN models, as well as used 7 metrics to evaluate their performance from different perspectives. Although the results provide a reference idea of feature selection for different application requirements of QSAR model based on DNNs, it lacks a good explanation.

With regard to the ROC curve, results of three DNN models were plotted together in Fig. 7. The AUC values also show that, DNN trained with MACCS fingerprints has the best overall model classification performance. However, if we set different

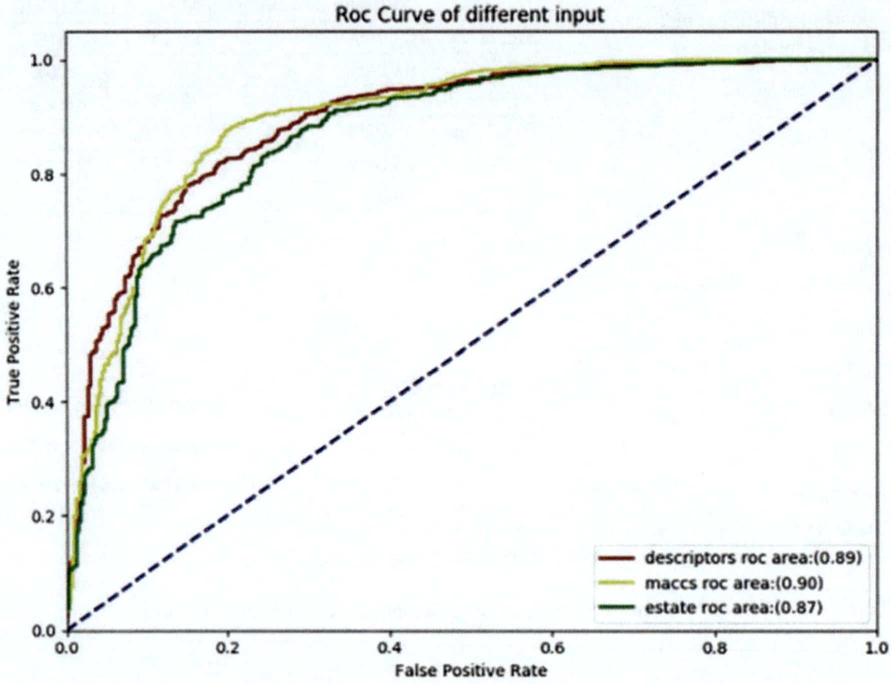

Fig. 7 ROC curves of 3 DNN models

thresholds to distinguish between active and inactive molecules, there are situations when DNN model trained with attribute descriptors perform better than that of MACCS fingerprints. Meanwhile, we can see that when the threshold is close to 0 or 1, there is no significant difference in the prediction performance of the three DNN models.

5 Conclusion

In this study, we have demonstrated that DNNs can be used as a practical QSAR method for HDAC1, while whose relevant researches are mostly using traditional approaches. During the training of models, we systematically optimized 8 main hyper-parameters which may influence the model performance to some degree and thus we got a good precision performance on test set (the best AUC is up to 0.9). Nevertheless, when faced with DNN models with more hidden layers or neurons, the gird search method for parameter optimization may show its weakness. And actually it is also one of the points which limit the scope of parameter searches. Another optimization method called random search has been proved fast and convenient in other fields. Therefore, future studies may apply this method to search the optimal parameters in a larger scope and thus achieve a better prediction performance.

Acknowledgements This work was supported in part by the Fundamental Research Funds for the Central Universities (NO.NJ2019023), National Natural Science Foundation of China (Grant No. 81603027 and 81973238), the National Science and Technology Major Projects for Major New Drugs Innovation and Development (Grant No. 2018ZX09711001-012), the Humanities and Social Sciences Funds of Ministry of Education (No.15YJC630122).

References

Bergstra, J. S., Bardenet, R., Bengio, Y., & Kegl, B. (2011). Algorithms for hyper-parameter optimization. *Advances in Neural Information Processing Systems, 25*.

Chaurasia, P. K., Bharati, S. L., & Sarma, C. (2016). Laccases in pharmaceutical chemistry: A comprehensive appraisal. *Mini-Reviews in Organic Chemistry, 13*(999), 1.

Chen, Y., Du, J., Zhao, Y. T., et al. (2015). Histone deacetylase (HDAC) inhibition improves myocardial function and prevents cardiac remodeling in diabetic mice. *Cardiovascular Diabetology, 14*(1), 14–99.

Dahl, G. E., Jaitly, N., & Salakhutdinov, R. (2014). Multi-task neural networks for QSAR predictions. *Computer Science*.

Fan, D., Yang, H., Li, F., et al. (2017). In silico prediction of chemical genotoxicity using machine learning methods and structural alerts. *Toxicology Research, 10*, 1039.

Goh, G. B., Hodas, N. O., & Vishnu, A. (2017). Deep learning for computational chemistry. *Journal of Computational Chemistry, 38*(16), 1291–1307.

Hinton, G. E., Osindero, S., & Teh, Y. W. (2006). A fast learning algorithm for deep belief nets. *Neural Computation, 18*(7), 1527–1554.

Korotcov, A., Tkachenko, V., Russo, D. P., et al. (2017). Comparison of deep learning with multiple machine learning methods and metrics using diverse drug discovery data sets. *Molecular Pharmaceutics, 14*(12).

Lecun, Y., Boser, B., Denker, J. S., et al. (2014). Backpropagation applied to handwritten zip code recognition. *Neural Computation, 1*(4), 541–551.

Ma, J., Sheridan, R. P., Liaw, A., Dahl, G. E., & Svetnik, V. J. (2015). Deep neural nets as a method for quantitative structure−activity relationships. *Journal of Chemical Information and Modeling, 55*, 263.

Matteelli, A., Carvalho, A. C., Dooley, K. E., et al. (2017). TMC207: The first compound of a new class of potent anti-tuberculosis drugs. *Future Microbiology, 5*(6), 849–858.

Mayr, A., Klambauer, G., Unterthiner, T., et al. (2016). DeepTox: Toxicity prediction using deep learning. *Frontiers in Environmental Science*.

Pereira, J. C., Caffarena, E. R., & Santos, C. D. (2016). Boosting docking-based virtual screening with deep learning. *Journal of Chemical Information and Modeling, 56*(12), 2495.

Ramsundar, B., Kearnes, S., Riley, P., et al. (2015). Massively multitask networks for drug discovery. *Computer Science*.

Srivastava, N., Hinton, G. E., Krizhevsky, A., Sutskever, I., & Salakhutdinov, R. (2014). Dropout: A simple way to prevent neural networks from overfitting. *Journal of Machine Learning Research, 15*, 1929–1958.

Tan, Y., Huang, W., & Yu, N. (2009). Structure-activity relationships of histone deacetylase inhibitors. *Acta Pharmaceutica Sinica, 44*(10), 1072–1083.

Worachartcheewan, A., Nantasenamat, C., Naenna, T., et al. (2009). Modeling the activity of furin inhibitors using artificial neural network. *European Journal of Medicinal Chemistry, 44*(4), 1664–1673.

Xu, Q. (2019). Application of pharmacological chemistry in the teaching of pharmacology. *Technology Wind, 07*, 37.

Yu-Huan, T., & Ke-Jiang, L. (2009). Application of 2D molecular fingerprints based on molecular similarity in visual screening. *Journal of China Pharmaceutical University, 40*(2), 178–184.

Zhang, R., Cao, X., Liu, Y., et al. (2016). A new method for identifying compounds by luminescent response profiles on a cataluminescence based sensor. *Analytical Chemistry, 83*(23), 8975.

Zhu, L., Huang, L., Wang, A., Li, Q., Guo, J., Wang, L., & Zhang, G. (2019). The evaluation of an immunoperoxidase assay applicable in antiviral drug screening. *Biologicals: Journal of the International Association of Biological Standardization, 57*.

Risk Assessment Indicators for Technology Enterprises: From the Perspective of Complex Networks

Runjie Xu, Nan Ye, Qianru Tao, and Shuo Zhang

1 Introduction

In recent years, the technology industry has adopted different business models from the traditional industry to increase market share and pursue future excess returns as far as possible (Rometty, 2006; Johnson et al., 2008). These companies have built their strength through free services, cash subsidies and the acquisition of new companies. The technology companies' business model assumes that value is created not only by producers but also by customers and other members of the ecosystem. From this perspective, enterprises in the Internet industry only need to make strategic investments and acquisitions in a series of fields to acquire businesses and users in this field, and through these businesses to lay out infrastructure construction and serve existing users, so as to strengthen their position in the digital economy. This business model concept challenges the assumptions of traditional value creation and value acquisition theories (Shuhidan et al., 2016), and further promotes more technology enterprises to gain monopoly status through similar frequent acquisitions and investments. However, the continuous investment and acquisition of these technology enterprises does not take profit as the first purpose, but depends on whether the field can provide infrastructure, technology, service or product supporting services for its core business development, which leads to the fact that the acquired technology-based companies are often unable to make profits now or even in the future (Carpenter et al., 2003).

The risks in the technology industry are mainly concentrated in law, equity, technology, management, etc. (Etges et al., 2017; Trott, 2012). For example, S Romanosky et al. (2019) considered data leakage and security incidents caused by

R. Xu (✉) · N. Ye · Q. Tao · S. Zhang
College of Economics and Management, Nanjing University of Aeronautics and Astronautics, Nanjing, China

H. Yang et al. (eds.), *AI and Analytics for Public Health*, Springer Proceedings in Business and Economics, https://doi.org/10.1007/978-3-030-75166-1_16

technical problems of technology companies, which led to legal lawsuits. Therefore, they considered the way of network construction to study loss insurance. Tu et al. (2014) studied the risks implied in supply chain management of technology enterprises and found that supply chain management ability was closely related to innovation ability. Teece et al. (2016) used traditional economic methods to explain the uncertain risks of scientific and technological innovation enterprises. There are also domestic research on the related risks of science and technology enterprises.

In a large number of risk researches on technology companies, the data used are mostly from income statement, cash flow statement, balance sheet and other data in financial statements. But clearly, these quantitative content is not the only factor to measure the enterprise risk, the information in the financial reports of enterprises not only includes the disclosure of the objectivity of digital information, also including the enterprise risk assessment of board, management and analyst, so the evaluation cannot be measured only by objective of digital information. Our target is to combine these abstract narratives with the quantitative information in the financial statements, to explore the hidden relationships therein, and to explain the unique risk characteristics of technology enterprises. Specifically, the political Outlines, risk descriptions and strategic developments in financial reports are often communicated to investors through written statements that are difficult to quantify in risk studies and have not been taken into account in previous studies.

Therefore, this study using the results of the abstract text messages to build risk network of enterprise technology, by focusing on analysis of financial report for risk, strategy, such as content description, we put this kind of non Euclidean domain in the form of data processing for containing the network relations, the network contains a risk of infection probability, the probability of infection and the total risk associated. In the past, a large number of enterprise risk studies were mainly based on stock price, daily return rate and other securities market information (Sakamoto & Vodenska, 2017; Xu et al., 2017).

Therefore, the analysis of abstract text information in financial statements to build risk network is the first contribution of this paper. The second contribution of this paper is to extract the most significant risk factors of science and technology enterprises through the analysis of network characteristics, and construct the RLC risk measurement index based on these risks. In order to prove the effectiveness of the indicators, this paper selects the financial reporting data of global top100 technology enterprises to verify the correlation with risk measurement indicators (RLC). The experiment shows that the net income of technology enterprises is significantly negatively correlated with RLC risk index, that is to say, the higher the risk level factor of technology enterprises is, the lower the net income of enterprises will be. In addition, the model can predict the future net income level of enterprises through the risk level coefficient, and the risk level factor has a significant negative correlation with the stock price. Through this study, the management level, government regulatory departments and legislative organs of technology enterprises can better understand the nature of the correlation between development and risks of technology enterprises, and take actions to manage and avoid the risks of technology enterprises according to the actual needs of macroeconomic development.

2 Risk Network Construction

Based on the classic method of complex networks (Albert & Barabási, 2002; Boccaletti et al., 2006), define the risk of science and technology enterprises association network consists of two basic elements: node v e and the associated way, it will be deemed v v collections, as a collection of e, e and e of each side (association) has a v e a pair of point (I, j) and the matching, then the whole network can use the symbol G = (v, e). If G has n nodes, it is denoted as V = (1, 2 ..., n), and suppose.

The network can be completely described by the matrix $G = (g_{ij})_{n \times n}$. At the same time, because the complex system is a typical directed network, it can be judged that $G = (g_{ij})_{n \times n}$ is an asymmetric matrix.

For the degree k_i of node v_i in the network, it is defined as the number of edges connected to the node: $k_i = \sum_{j=1}^{n} g_{ij}$. Namely, the connection probability between a new node and the original node.

In the same way, the probability of edges connected by a node is: $P(k_i)^{\rightarrow} = \frac{k_i^{\rightarrow}}{\sum_j k_j}$, the value represents the infection intensity of a node. From scientific and technological enterprise internal risk evolution mechanism analysis, suppose the system unstable state is only one risk source *i* before, if the risk source *i* can lead to other risks occurred one after another, the influence of the risk source *i* ability stronger, namely $P(k_i)^{\rightarrow}$ value is higher. This indicates that the more likely a risk factor is to play a risk-induced role in the system, the more attention should be paid to it. On the contrary, a low value indicates that a node has less influence on the outside world.

However, the true contagion capacity of risk sources must also take into account the vulnerability of other nodes in the whole risk system. Therefore, the probability of nodes being connected is also required: $P(k_i)^{\leftarrow} = \frac{k_i^{\leftarrow}}{\sum_j k_j}$, which represents the sensitivity of a node to attack. The larger the value is, the more vulnerable the node is to infection by other risks, and conversely, the less susceptible it is to infection by outsiders. Referring to previous studies (Xu et al., 2020), the core mechanism of network construction in this paper is to determine whether there is a direct induction relationship between different risk factors. If there is a direct infection, then the directed circuit of both sides of the infection will be established.

Liu (2012) defined emotional analysis as the research field of analyzing people's views, emotions, evaluations, attitudes and emotions on entities such as products, services, individuals, organizations, events, issues, topics and their attributes. In fact, the environmental and linguistic differences between the author and the reader make emotional analysis an extremely difficult task.

This paper focuses on the evaluation of corporate risk, development and future expectations by the board of directors, management and analysts in corporate financial statements. These texts are mixed with political overview, risk description, development strategy and other contents. According to the method of literature induction, the statement of "pointing nature" about risks in the financial report was sorted out (as shown in Table 1), relevant nodes were extracted, and a network was

Table 1 Sample node extraction

Original financial statement (case)	Semantic analysis
Breaches of our cybersecurity measures could result in unauthorized access to our systems, misappropriation of information or data, deletion or modification of user information, or a denial-of-service or other interruption to our business operations.	Business risk – network security vulnerability
	Operational risk – data privacy risk
	Legal risk – consumer complaint risk
Our revenue growth also depends on our ability to continue to grow our core businesses as well as businesses we have acquired.	Business risk – critical business service capabilities
	Investment risk – acquisition, investment, alliance risk
We may also face protectionist policies that could, among other things, hinder our ability to execute our business strategies and put us at a competitive disadvantage relative to domestic companies in other jurisdictions.	Legal risk – constraint risk
	Operational risk – international business capability risk
If we are not able to continue to innovate or if we fail to adapt to changes in our industry, our business, financial condition and results of operations would be materially and adversely affected.	Investment risk – innovation and industry change risk
We face risks relating to our acquisitions, investments and alliances.	Investment risk – acquisition, investment, alliance risk

constructed according to the relationship between nodes, with a total of 660 nodes and 840 edges. All the nodes in the network are from the risks mentioned in the financial report, and the connections in the network are from the language expressed by the analyst. The extraction process is shown in Fig. 1.

Since most technology companies choose to go public in the United States, the financial report data is referred to (https://www.sec.gov). The semantic analysis of this paper relies on Alibaba's 2019 financial statements. According to the results of semantic analysis, the network diagram (Fig. 2) is constructed, and corresponding annotation is made for significant risk factors. In addition, for convenience of analysis, this paper ranks the risk amount (Degree), risk contagion amount (Outdegree) and risk infected amount (Indegree).

Based on the analysis of the above three networks, we summarize the important risk nodes under the three conditions of connectivity, connectivity, and connectivity as legal risk, business risk, investment risk, and operational risk.

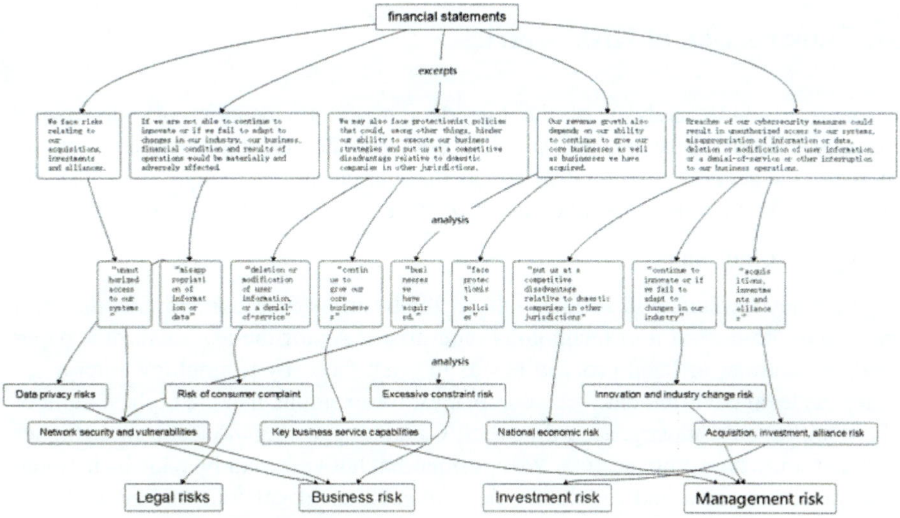

Fig. 1 Node extraction process diagram

Fig. 2 Risk relational, risk relational of connectivity and Risk relationship of outliers

3 Construction of Risk Metrics

Through semantic analysis of the text data in the financial report, this paper constructs the risk network, and sorts the degree of connectivity, connectivity and degree of the network, and finally extracts the four categories of legal risk, business risk, investment risk and operational risk as the significant risk sources.

1. Legal risk indicators

Too many lawsuits will seriously affect the business development, income level and market competition of technology enterprises. As for the legal risk, this paper takes the number of legal proceedings as the risk faced by technology enterprises under the legal environment. Tech companies, for example, had the highest number of lawsuits of any industry, according to SEC data, suggesting the sector faces higher legal exposure than other sectors. We consider that laws are mainly made by national judicial departments and regulatory departments, so in general, this type of risk belongs to the category of external risks. Therefore, this paper adopts the annual litigation growth rate of the technology industry to reflect the external risks faced by enterprises. To be specific, the higher the rate of litigation in the industry, the greater the risk faced by the enterprise.

2. Business risk indicators

Business risks include business problems caused by technical problems in science and technology enterprises. For example, the main business of science and technology enterprises will cease to be serviced due to technical problems, which will have a great impact on science and technology enterprises. Business risks also include the decline of market share in the process of business development due to competition, industry changes and other reasons. Combining these two points, we use the growth rate of operating income to reflect the business risk. When using this indicator, we consider that the decline in business growth rate may be due to the smaller growth space caused by the size of the enterprise itself. However, the actual technology industry often chooses to expand new businesses to avoid the "ceiling" problem of a certain business. Therefore, the decline in business growth rate can still indicate the potential risks faced by enterprises.

3. Investment risk indicators

Technology companies are using strategic investments and acquisitions in a range of sectors to strengthen their leadership in the digital economy. These investments and acquisitions are not for profit in the first place, but depend on whether the field is relevant to the current primary business or whether it provides the infrastructure, technology, services or products for its business development. However, such strategic investments and acquisitions continue to adversely affect the financial performance of Alibaba's technology businesses. For example, buy companies with low margins or losses that may not make a profit at all in the future. To this end, we use the ratio of net cash flow generated by investment activities to capital to reflect the investment risk of enterprises. The bigger this index is, the greater the risk of investment and merger will be.

4. Indicators of operational risk

Business risk refers to the change of market value caused by the change of production and operation in external environment of science and technology enterprises, which affects the change of future cash flow of science and technology enterprises. The management risk of the technology industry on the one hand, mainly comes from general complex and changeful market environment, on the other hand, in general, the cycle of the product is through the start-up stage, growth stage, mature stage and decline stage, but the products of science and technology cycle is compactness, initial risk big, the investment is more, other companies are scrambling to imitate and high speed of knowledge update makes the products in the recession time is shorter, thus entered a new round of cycle. Therefore, this paper USES the ratio of net cash flow generated from operating activities to total capital to reflect the operating risk of an enterprise. This index reflects the amount of each capital invested in operating activities. The bigger the index is, the greater the operating risk is.

In addition, the objective of this paper is to build a comprehensive index to reflect the risk quantification situation of science and technology enterprises, so we use dimension reduction method to reasonably constitute the unified index of these four types of risks to objectively reflect the real risk situation of enterprises. In order to make the comprehensive index lose as little information as possible in the original variables, so as to achieve the purpose of comprehensive analysis of the collected data, principal component analysis (PCA) is adopted to achieve dimensionality reduction of the four risk dimensions. Suppose there are n samples, and each sample has p variables, thus an $n \times p$ matrix is formed:

$$X = \begin{bmatrix} x_{11} & \cdots & x_{1p} \\ \vdots & \ddots & \vdots \\ x_{n1} & \cdots & x_{np} \end{bmatrix}$$

The original variables are $x_1, x_2 \cdots, x_p$, and the comprehensive index is set as q_1, $q_2 \cdots, q_k$ $(p \leq k)$, then

$$\begin{cases} q_1 = l_{11}x_1 + l_{12}x_2 + \cdots + l_{1p}x_p \\ q_2 = l_{21}x_1 + l_{22}x_2 + \cdots + l_{2p}x_p \\ \quad \cdots\cdots \\ q_k = l_{k1}x_1 + l_{k2}x_2 + \cdots + l_{kp}x_p. \end{cases}$$

l_{ij} shall satisfy the following conditions:

Table 2 KMO and Bit spherical inspection

KMO	0.582	
Bartlett	Approx. Chi-Square	240.738
	df	6
	Sig.	0

Table 3 Common factor variance

	Initial	Extract
Litigation growth rate	1	0.725
Increase rate of main business revenue	1	0.459
Investment efficiency	1	0.469
Operational efficiency	1	0.624

1. q_i is not related to q_j, where $i \neq j$ and $i, j = 1, 2 \cdots, k$
2. q_1 is the largest deviation of all linear combinations of $x_1, x_2 \cdots, x_p$, q_2 is the largest deviation of all linear combinations of $x_1, x_2 \cdots, x_p$ independent of q_1, and by the same way, q_k is the largest deviation of all linear combinations of x_1, $x_2 \cdots, x_p$ independent of $q_1, q_2 \cdots, q_{k-1}$.

Then the comprehensive variable $q_1, q_2 \cdots, q_k$ is called the principal component variable of the original variable $x_1, x_2 \cdots, x_p$.

Determine the load l_{ij} of each synthetic variable q_i on all original variables $x_1, x_2 \cdots, x_p$, and get the score of each synthetic variable. Then, according to the relevant literature, the variance contribution rate of each index was calculated to calculate the comprehensive score of each enterprise so as to measure the risk of each enterprise:

$$RLC = \frac{\omega_1}{\sum_{i=1}^{k} \omega_i} q_1 + \frac{\omega_2}{\sum_{i=1}^{k} \omega_i} q_2 \cdots + \frac{\omega_k}{\sum_{i=1}^{k} \omega_i} q_k \tag{1}$$

The ω_i is the variance contribution rate of the comprehensive variable q_i.

The premise of principal component analysis is to preprocess the data and conduct correlation test to determine whether this method can be used for analysis. According to Bartlett test, P value is less than 0.05 and KMO value is greater than 0.5, which basically meets the requirements. Principal component analysis can be performed, and the test results are shown in Table 2. In addition, Table 3 shows the common degree data of all variables and the information extraction of the original variable by the new factor. The common degree value of all the original variables is higher than 0.4, which means that the original variable has a strong correlation with the new factor and the factor can effectively extract the information.

In this paper, according to the extraction principle of eigenvalues greater than 1, variance contribution rate and cumulative variance contribution rate of initial common factor eigenvalues are solved according to principal component analysis, and the number of common factors is determined by variance analysis. Table 4 shows that the eigenvalues of the first two factors are greater than 1, so the principal components of the first two factors are extracted for evaluation as comprehensive factors.

Table 4 Population variance explained by factors

Element	Initial eigenvalues			Extract the sum of squares and load			Sum of squares of rotation loaded		
	Total	% of the variance	Cumulative %	Total	% of the variance	Cumulative %	Total	% of the variance	Cumulative %
1	1.261	31.535	31.535	1.261	31.535	31.535	1.233	30.823	30.823
2	1.015	25.374	56.909	1.015	25.374	56.909	1.043	26.086	56.909
3	0.977	24.414	81.323						
4	0.747	18.677	100.00						

Table 5 Component score coefficient

	Factor1	Factor2	Factor3	Factor4
F1	0.217	0.248	0.559	−0.628
F2	0.809	−0.538	0.119	0.074

Table 5 shows the component score coefficient matrix. Through this matrix, the scores of two comprehensive factors can be calculated, namely:

$$\begin{cases} F_1{=}0.217x_1{+}0.28x_2{+}0.559x_3{-}0.628x_4 \\ F_2{=}0.809x_1{+}0.538x_2{+}0.119x_3{-}0.074x_4 \end{cases} \quad (2)\text{–}(3)$$

Where x_1 is the growth rate of litigation, x_2 is the growth rate of operating income, x_3 is the net cash flow generated from investment activities/total assets, and x_4 is the net cash flow generated from operating activities/total assets.

The new two factors reflect the enterprise risk level from different aspects, and it is difficult to make a comprehensive evaluation by using a single common factor. Therefore, the comprehensive score is considered to be calculated according to the variance contribution rate corresponding to each common factor as the weight, that is, the required risk factor coefficient (RLC) is obtained:

$$RLC = \frac{30.823}{56.909} \times F_1 + \frac{26.086}{56.909} \times F_2 = 0.489x_1 - 0.113x_2 + 0.353x_3 - 0.31x_4 \quad (4)$$

4 Empirical Test of Risk Metrics

In this paper, risk level factors are obtained based on risk network construction, and risk measurement index (RLC) is obtained accordingly. In order to verify the scientific nature of this index, this paper analyzes and studies the correlation between the development level of science and technology enterprises and RLC. The data takes the global top100 technology companies by market capitalization as samples, and adopts the financial statements and stock price data from 2005 to 2019. The data are obtained from the Wind database and the open data set published by the us securities commission.

In terms of variables, the development level of an enterprise adopts net income, which reflects the total profit of an enterprise, that is, the income or income balance after deducting business costs, taxes and other expenses from the total income, which can directly reflect the development status of an enterprise. The control variable is the logarithmic form of assets and the r&d investment level of the enterprise (R&D expense/asset). The r&d investment level will inevitably affect the development of the enterprise. The success of r&d investment may bring qualitative development to the enterprise, while the failure will bring capital loss to the enterprise, which will have an impact on the development of the enterprise.

Table 6 Regression results

Variable	Net income
RLC	-1.647^{***} (−2.734)
Ln(assets)	24.837^{***} (17.245)
R&D/Assets	101.261^{***} (2.568)
Constant	-559.604^{***} (−15.836)
Observations	1170
Adjusted R-squared	0.294
RLC/RLC	1
Stock Price/RLC	-0.081^{***}

Note: ***, ** and * mean significant at the level of 1%, 5% and 10%, respectively. The value of t is in parentheses

Therefore, the following model is constructed:

$$\text{Net Income} = \alpha + \beta \text{RLC} + \gamma \text{R}\&D/\text{Assets} + \delta \text{Ln}\,(\text{Assets}) \tag{5}$$

Table 6 reflects the results of regression and correlation analysis. The experiment shows that the net income of enterprises is significantly negatively correlated with RLC, that is, the higher the risk level factor of enterprises is, the lower the net income of enterprises will be. Through this model, we can predict the net income level of enterprises through the risk level coefficient, and the risk level factor has a significant negative correlation with the stock price.

Table 6 is the results of regression analysis and correlation analysis. Through regression results, the relationship between RLC and the net income of technology enterprises can be obtained:

$$\text{Net Income} = -\,1.65 \times \text{RLC} + 101.26 \times \text{R}\&D/\text{Assets} + 24.84 \times \text{Ln}\,(\text{Assets}) - 559.6 \tag{6}$$

According to the experiment, the net income of science and technology enterprises is significantly negatively correlated with RLC, that is, the higher the risk level factor of enterprises is, the lower the net income of enterprises is, which is consistent with the generally recognized relationship between risk and return. Moreover, it can be seen from this model that the risk level coefficient can accurately predict the net income level of science and technology enterprises, and the risk level factor has a significant negative correlation with the stock price, so this index is scientific to a certain extent.

The rapid development of science and technology enterprises is bound to be accompanied by the acquisition and investment of other enterprises, so as to enhance their comprehensive strength, but at the same time, it will also bring many hidden dangers, such as labor disputes, legal compliance and other problems. One of these pitfalls, especially for foreign technology firms, can bankrupt a technology firm rather than allow it to expand rapidly. Of course, due to the nature of science and technology enterprises being eliminated without development, domestic science and technology enterprises are bound to choose the world's advanced technologies for their own development. Therefore, domestic science and technology enterprises have to acquire or cooperate with foreign science and technology enterprises. Hidden dangers such as legal compliance and copyright will have a huge negative impact on enterprises. It is worth noting that the acquisition and investment behaviors of science and technology enterprises centering on the development of science and technology do not necessarily promote the growth of their net income. While focusing on innovation and development, science and technology enterprises cannot ignore the hidden dangers in the process.

5 Conclusion

Based on science and technology in the earnings of Euclidean domain data, considering to build the network transmission of infection, infection and risk total three network, combined with the feature of the network to complete the core of many risk factors for the enterprise internal risk source filtering, extracting legal risk, business risk, investment risk, management risk four types. After that, this paper quantifies the four indicators, constructs the risk measurement index (RLC), and calculates the risk quantification results of science and technology enterprises.

In order to prove the scientific nature of the index, this paper selects the net income and stock price of the global top100 technology enterprises to verify the correlation between risk measurement index (RLC). The experiment proves that RLC has a significant negative correlation with corporate net income and stock price, and RLC can predict corporate net income and stock price. In addition, although science and technology enterprises have a high requirement for innovation ability, innovation is not a decisive factor, and the overall risk of science and technology enterprises plays a decisive role in inhibiting the development of enterprises.

Although the technological innovation of technology enterprises may bring huge improvement to the enterprise's net income and stock price, technology enterprises often ignore the risk of copyright and labor disputes that follow the innovative technology, and the inaccurate positioning of mass demand may encounter unpredictable business risks. If technology enterprises only pursue the development of their main business, but blindly invest and acquire, they will misjudge the overall life cycle and thus bring more operational risks. This is still a serious problem in the development process of the world's top technology enterprises. These risks cannot be ignored because of the huge profits brought by a few successful scientific and technological innovations. Therefore, while improving our competitiveness in the industry through scientific and technological innovation, we should also pay attention to the control ability of risks, so as to ensure the healthy and steady development of the scientific and technological industry.

Acknowledgments This work is supported by the National Social Science Foundation of China (17BGL055) and Innovation Project Fund of NUAA (2019EC01, 2019EC09, 2020CX009040).

References

Albert, R., & Barabási, A. L. (2002). Statistical mechanics of complex networks [J]. *Reviews of Modern Physics, 74*(1), 47.

Boccaletti, S., Latora, V., Moreno, Y., et al. (2006). Complex networks: Structure and dynamics [J]. *Physics Reports, 424*(4–5), 175–308.

Carpenter, M. A., Pollock, T. G., & Leary, M. M. (2003). Testing a model of reasoned risk-taking: Governance, the experience of principals and agents, and global strategy in high-technology IPO firms [J]. *Strategic Management Journal, 24*(9), 803–820.

Etges, A. P. B. d. S., Souza, J. S. d., & Kliemann Neto, F. J. (2017). Risk management for companies focused on innovation processes [J]. *Production*, 27.

Johnson, M. W., Christensen, C. M., & Kagermann, H. (2008). Reinventing your business model [J]. *Harvard Business Review, 86*(12), 57–68.

Liu, B. (2012). Sentiment analysis and opinion mining [J]. *Synthesis Lectures on Human Language Technologies, 5*(1), 1–167.

Romanosky, S., et al. (2019). Content analysis of cyber insurance policies: How do carriers price cyber risk? [J]. *Journal of Cybersecurity, 5*(1), tyz002.

Rometty, V. G. (2006). *Expanding the innovation horizon: The global CEO study 2006* [R]. IBM Business Service.

Sakamoto, Y., & Vodenska, I. (2017). Erratum to "systemic risk and structural changes in a bipartite bank network: A new perspective on the Japanese banking crisis of the 1990s" [J]. *Journal of Complex Networks, 5*(3), 512–512.

Shuhidan, S. M., et al. (2016). Market orientation within technological companies: Risk based approach [C]. In *2016 3rd International Conference on Computer and Information Sciences (ICCOINS)* (pp. 43–48). IEEE.

Teece, D., Peteraf, M., & Leih, S. (2016). Dynamic capabilities and organizational agility: Risk, uncertainty, and strategy in the innovation economy [J]. *California Management Review, 58*(4), 13–35.

Trott, P. J. (2012). *Gestão da inovação e desenvolvimento de novos produtos* [M]. .

Tu, C., Hwang, S., & Wong, J. (2014). How does cooperation affect innovation in micro-enterprises? [J]. *Management Decision, 52*(8), 1390–1409.

Xu, R., Wong, W.-K., Chen, G., & Huang, S. (2017). Topological characteristics of the Hong Kong stock market: A test-based P-threshold approach to understanding network complexity [J]. *Scientific Reports, 7*, 41379.

Xu, R., Mi, C., Mierzwiak, R., & Meng, R. (2020). Complex network construction of Internet finance risk. *Physica A: Statistical Mechanics and its Applications, 540*, 122930.

Subsidy Design for Personal Protective Equipments (PPEs) Adoption

Ailing Xu, Qiao-Chu He, and Ying-ju Chen

1 Introduction

Since December 2019, the COVID-19 outbreak has spread in over 100 countries and regions at a stunning pace. To prevent humanitarian health hazards such as COVID-19, people are strongly suggested to purchase and use Personal Protective Equipments (PPEs) for self-protection. However, the fraction of the population who refused to comply with the PPEs is high (and also much higher in some regions than others). In this paper, we focus on an empirically tested behavioral explanation for the compliance obstacle (a lack of self-control) based on the *present-bias effect*, which means the trend to give a higher valuation to a present reward but a lower valuation to a future reward (O'Donoghue & Rabin, 2006). Since the utility of PPEs is realized in the future, a consumer may postpone his purchase decision but finally abandon his purchase plan in the future period due to this present-bias effect. The key take-away we focus on is that advance selling can be beneficial to the consumers as a *commitment device* (Bryan et al., 2010). However, the effect of advance selling may be limited, especially for consumers with low valuation, and

A. Xu
School of Business and Management, The Hong Kong University of Science and Technology, Clear Water Bay, Kowloon, Hong Kong
e-mail: axuaj@connet.ust.hk

Q.-C. He (✉)
Faculty of Business Administration, Southern University of Science and Technology, Shenzhen, People's Republic of China
e-mail: heqc@sustech.edu.cn

Y.-j. Chen
School of Business and Management & School of Engineering, The Hong Kong University of Science and Technology, Clear Water Bay, Kowloon, Hong Kong
e-mail: imchen@ust.hk

H. Yang et al. (eds.), *AI and Analytics for Public Health*, Springer Proceedings in Business and Economics, https://doi.org/10.1007/978-3-030-75166-1_17

can only encourage a part of consumers to purchase PPEs. Advance selling alone cannot fully address the compliance obstacles in PPEs.

To encourage more aggregate adoption in PPEs, we discuss a combination of advance selling and a subsidy policy. Our research question is how to design policy instruments, combining advance selling and subsidy programs, to resolve healthcare compliance obstacles and boost aggregate adoptions in PPEs? We propose a model consisting of a manufacturer of PPEs, consumers who suffer from *present-bias* and have heterogeneity in the awareness of their present-bias and a donor who provides a subsidy to boost the adoption of PPEs. We restrict ourselves to the manufacturer subsidy policy, where the subsidy is provided only to the manufacturer. The customer subsidy refers to the opposite one. It is because the consumer subsidy will be undercut by the manufacturer by charging a higher price, which makes the consumer subsidy less cost-effective than the manufacturer subsidy. In particular, when the budget is tight, the manufacturer subsidy should be only provided in the spot-period, because the spot-period subsidy reduces the effects of advance selling. In contrast, subsidy in both periods should be provided, when the budget constraint is relaxed. The results provide actionable insights towards overcoming healthcare compliance obstacles.

2 Literature Review

Our paper is related to the product/technology adoption puzzle. The first explanation of this puzzle is the lack of information. Conley and Udry (2010) study the effect of social learning in the knowledge diffusion and adoption of fertilizer in agricultural industry. Alternatively, Duflo et al. (2011) use present-bias to explain Kenya farmers' procrastination in fertilizer purchase. Present- bias is an important factor in explaining compliance obstacles. Following O'Donoghue and Rabin (2006), we model this effect using quasi-hyperbolic discounting framework.

Our paper is also related to the subsidy policy design. Burkart et al. (2016) point out the importance of carefully designed funding systems in the success of humanitarian organizations. Cohen and Dupas (2010) find affirmative evidence of subsidy by a randomized field experiment in Kenya, wherein malarial insecticide-treated nets are sold to pregnant women at some randomized prices. And our paper is closely related to Yu et al. (2018), which shows that manufacturer subsidy is more cost-effective than consumer subsidy.

3 Model

We consider a three-period model in which a manufacturer produces and sells PPEs to consumers. Let $t = 0$ denote the advance-period, $t = 1$ denote the spot-period and $t = 2$ denote the period when all the decisions are made and utilities are realized.

Consumers The population size of consumers (she) who get informed about PPEs is Λ_0 and Λ_1 in two periods, respectively. The utility of consuming PPEs is θV, which will be realized at $t = 2$. $V > 0$ refers to the intrinsic and deterministic value of the PPEs, while θ refers to a random valuation for the PPEs, which is privately observed by the consumer at $t = 1$. θ follows an uniform distribution on [0, 1].

Quasi-hyperbolic discounting Following O'Donoghue and Rabin (2001), we model the present-bias effect using quasi-hyperbolic discounting framework. A consumer' expected payoff at $t = 0$ is $u_0 = v_0 + \beta \sum_{t=1,2} \delta^t v_t$, where $\delta \in (0, 1]$ is the discount factor between two periods, and $\beta \in (0, 1]$ is the *present-bias* factor between the current period and future periods. Furthermore, consumers at $t = 0$ are heterogeneous in the awareness (unawareness) about her present-bias. Only a portion γ of customers are aware of their present-bias and know true β, while a portion $1 - \gamma$ are unaware and thus think their present-bias factor is 1 at $t = 0$.

Manufacturer The manufacturer (he) controls the production channel and dictates the prices. The manufacturer sets static prices P_0 and P_1 in the advance- and spot-period to maximize his profit as follows:

$$\max_{P_0} \pi_0 (P_0) = (P_0 - c)\, q_0,$$

$$\max_{P_1} \pi_1 (P_1) = \alpha (P_1 - c)\, q_1 + \pi_0 (P_0),$$

wherein q_0 and q_1 are adoption quantities in two periods, c is the unit production cost and the manufacturer's discount factor α is different from the consumers' in general.

Donor We also assume a donor who aims to incentivize compliance behaviors. The subsidy design is determined *a priori*, i.e., $t = -1$. The donor provides subsidy to the manufacturer in the form of $\mu_0 c$ and $\mu_1 c$ in advance- and spot-period, wherein $\mu_0, \mu_1 \in [0, 1]$, and to consumers in the form of $\lambda_0 P_0$ and $\lambda_1 P_1$, where λ_0, $\lambda_1 \in [0, 1]$. The subsidy program is constrained by an exogenous dollar amount B. The donor's decision problem is as follows:

$$\max_{\mu_0,\mu_1;\lambda_0,\lambda_1} \mathrm{Q} = q_0 + q_1$$

$$\text{s.t.}\, q_0 \,(\mu_0 c + \lambda_0 P_0) + \alpha\, q_1 \,(\mu_1 c + \lambda_1 P_1) \le B$$

4 Analysis

To start with consumer behavior analysis, consumers are strategic, which implies that consumers make decisions by payoff calculations and comparison among different periods. For a clear demonstration, we do not consider the subsidy at this stage. A consumer's expected payoff is $u_1 = \beta\delta\theta V - P_1$, if she makes the purchase in period 1. Hence, she makes the purchase in period 1 if $\theta \geq \frac{P_1}{\beta\delta V}$. And the analysis is similar for period 0. A consumer will make purchase decisions to maximize her expected payoff.

Then we can characterize the manufacturer's pricing strategies as follows:

Proposition 1 *Three pricing strategies are sustained in equilibrium*[1]*:*

- *Equilibrium-D, "discount advance selling": All consumers who arrive in period* 0 *make purchases in period 0 (pooling equilibrium), denoted by the superscript D;*
- *Equilibrium-P, "premium advance selling": Among those who arrive in period 0, sophisticated consumers make purchases in period 0, while naive consumers do not (separating equilibrium), denoted by the superscript P;*
- *Equilibrium-N, "no advance selling": No consumers who arrive in the period* 0 *participate in the advance-selling market (pooling equilibrium), denoted by the superscript N.*

Furthermore, we have $Q^D > Q^P > Q^N$*, for the given subsidy.*

We define three different pricing strategies for the manufacturer. The divergent purchasing behaviors are driven by the heterogeneity in consumers' sophistication. Furthermore, since the total adoption quantities increase when more consumers make purchase in the advance period. It indicates that the advance pricing strategy is an effective instrument to stimulate the adoption quantity.

After characterizing consumer behaviors and manufacturer selling strategies, we start by investigating the donor's problem. We consider simplifying the general form of the subsidy program, as it is complicated to solve. We compare the manufacturer subsidy and the consumer subsidy and get:

Lemma 1 *The optimal subsidy design requires* $\lambda_0^* = \lambda_1^* = 0$.

Lemma 1 indicates that the donor should provide the subsidy only to the manufacturer. When given equal μ_1 or λ_1 (μ_0 or λ_0), it costs more to subsidize consumers than to subsidize the manufacturer. Moreover, any consumer subsidy will be undercut by the monopolistic manufacturer who is able to charge a higher price, which increases the cost for the donor to subsidize consumers. Hence, it is

[1]The terminology for different advance pricing strategies follows classic literature, e.g., Xie and Shugan (2001). The term "premium advance selling" speaks relatively to "discount advance selling", and does not imply a high price in absolute terms.

optimal for the donor to offer the subsidy only to the manufacturer. In the following discussion, it is sufficient to restrict ourselves to manufacturer subsidy policy.

Since the adoption quantity in the advance-period is dependent on the fraction of naive consumers, the advance-period subsidy will have no effect on the adoption quantity under any given selling strategy. Hence, we first pursue the analysis by investigating the spot-period subsidy effects on manufacturer profits. And we have:

Lemma 2

1. *Under the no advance pricing strategy, we have* $\frac{\partial \pi^N}{\partial \mu_1} > 0$.
2. *Under the premium advance pricing strategy, if and only if* $\frac{\Lambda_1}{\Lambda_0} > \frac{\beta^2\delta[c\mu_1-c+(2-\beta)\delta V]}{2\alpha(c\mu_1-c+\beta\delta V)}$*, we have* $\frac{\partial \pi^D}{\partial \mu_1} > 0$.
3. *Under the discount advance pricing strategy, if and only if* $\frac{\Lambda_1}{\Lambda_0} > 2\alpha + \frac{\gamma\delta[c(\mu_1-1)(2\beta-1)+\beta\delta V]}{2\alpha(c\mu_1-c+\beta\delta V)}$*, we have* $\frac{\partial \pi^P}{\partial \mu_1} > 0$.

Interestingly, Lemma 2 indicates that the spot-period subsidy can reduce the manufacturer's profits when combining with advance selling strategies. The result is due to the inter-temporal cannibalization effect. The profit loss is generated when the increase in the spot-period subsidy forces the manufacturer to reduce the advance-period price and shift the sales from the advance period to the spot period. Thereby, only when the population ratio is relatively large, the profit increase in the spot-period outweighs the profit loss in the advance-period such that the spot-period subsidy benefits the manufacturer.

Finally, we go back to the donor's problem in Equation [equation-donor] and consider the optimal combination of subsidies in two periods.

Proposition 2 *Given the government budget B, there exists two thresholds* $\underline{B}$ *and* $\overline{B}$ *such that the donor provides subsidy only in the spot-period when* $B \leq \underline{B}$ *and distributes the subsidy in both advance- and spot-period when* $B \geq \overline{B}$*. When* $\underline{B} < B < \overline{B}$*, either policy is possible.*

Proposition 2 indicates that the optimal subsidy policy depends on the budget. Intuitively, the subsidy distributed in the advance-period will induce advance selling and thus a higher adoption quantity, but it incurs additional cost for the donor. When the budget is tight, it is not financially feasible to induce advance pricing strategies, so the donor should subsidize the manufacturer only in the spot-period. Conversely, when the budget gets relaxed, the donor has incentives to provide subsidy in the advance-period to induce advance selling. However, when the budget is intermediate, the optimal subsidy policy depends on the subsidy cost to induce advance selling and the advance-period subsidy is determined by profit loss for the manufacturer to choose advance selling. When the profit loss is small under advance pricing strategies, the donor will distribute the subsidy in both periods. Our results provide policy guidelines for designing such subsidy programs.

5 Conclusion

To prevent humanitarian health hazards such as COVID-19, we propose a stylized model with present-bias to understand a lack of compliance in Personal Protective Equipments (PPEs). Furthermore, we investigate the optimal subsidy policy when incorporating a donor. Advance selling strategies are effective in incentivizing the adoption quantity, but the increase in the spot-period subsidy discourages the manufacturer to adopt advance selling strategies. Finally, when the subsidy program is budget-constrained, the donor should provide subsidy only in the spot-period. In contrast, when the budget constraint is relaxed, subsidies should be provided in both periods. Our research is pioneering work in understanding and mitigating the adoption of preventive measures to prevent humanitarian health hazards. Future research is needed to further operationalize our actionable insights.

References

Bryan, G., Karlan, D., & Nelson, S. (2010). Commitment devices. *Annual Review of Economics, 2*(1), 671–698.

Burkart, C., Besiou, M., & Wakolbinger, T. (2016). The funding humanitarian supply chain interface. *Surveys in Operations Research and Management Science, 21*(2), 31–45.

Cohen, J., & Dupas, P. (2010). Free distribution or cost-sharing? Evidence from a randomized malaria prevention experiment. *Quarterly Journal of Economics, 125*(1).

Conley, T. G., & Udry, C. R. (2010). Learning about a new technology: Pineapple in Ghana. *American Economic Review, 100*(1), 35–69.

Duflo, E., Kremer, M., & Robinson, J. (2011). Nudging farmers to use fertilizer: Theory and experimental evidence from Kenya. *American Economic Review, 101*, 2350–2390.

O'Donoghue, T., & Rabin, M. (2001). Choice and procrastination. *Quarterly Journal of Economics*, 121–160.

O'Donoghue, T., & Rabin, M. (2006). Optimal sin taxes. *Journal of Public Economics, 90*(1011), 1825–1849.

Xie, J., & Shugan, S. M. (2001). Electronic tickets, smart cards, and online prepayments: When and how to advance sell. *Marketing Science, 20*(3), 219–243.

Yu, J. J., Tang, C. S., & Shen, Z.-J. M. (2018). Improving consumer welfare and manufacturer profit via government subsidy programs: Subsidizing consumers or manufacturers? *Manufacturing & Service Operations Management, 20*(4), 752–766.

Early Detection of Rumors Based on BERT Model

Li Yuechen, Qian Lingfei, and Ma Jing

1 Introduction

With the continuous development of the Internet, the network data generated by users on Chinese social media grows exponentially every day, which also promotes the generation and spread of Internet rumors. Users can post any information directly on the microblog, and the users who read the microblog information do not have enough ability to identify the authenticity of the information. At present, social platforms mainly rely on manual audit and rumor refutation platform to detect rumors, which is not only inefficient, but also consumes a lot of human costs and time, which can not meet the needs of real-time rumor detection.

The effect of traditional model on rumor detection is relatively backward, and it has a great adverse effect when it is detected. Early detection of rumors is carried out at the beginning of rumors, which is more in line with the needs of current network environment supervision, and can effectively avoid the social harm caused by rumor spreading.

2 Related Work

The earliest research on rumor detection on social platforms originated from twitter. Scholars built models by using rumor related features (Gist, 1951). Later research models mainly focus on traditional classification methods and deep neural network.

L. Yuechen (✉) · Q. Lingfei · M. Jing
College of Economics and Management, Nanjing University of Aeronautics and Astronautics, Nanjing, Jiangning, China

H. Yang et al. (eds.), *AI and Analytics for Public Health*, Springer Proceedings in Business and Economics, https://doi.org/10.1007/978-3-030-75166-1_18

Traditional classification methods focus on the shallow features of rumors. Castillo et al. (2011) used four types of features, extracted 15 kinds of features from information, users, topics and dissemination, and used decision tree algorithm to realize rumor detection. Previous research is based on the data of twitter platform. Yang et al. (2012) conducted rumor detection based on Sina Weibo background for the first time, introduced user location features and client-side features, and verified the effectiveness of the features through corresponding quantitative experiments. However, previous studies did not consider the impact of the daily cycle and the external impact cycle of rumors. Kwon et al. (2013) proposed a detection model based on time cycle, using temporal features, structural features and linguistic features to improve the effect of rumor detection.

Although artificial feature construction has achieved certain results in rumor detection, it has some disadvantages such as laborious work and poor generalization ability. In recent years, with the continuous development of deep learning, scholars began to try to use deep learning model for rumor detection research. Ma et al. (2016) proposed to use RNN (recurrent neural network) model combined with time series information to realize rumor detection. Deep learning model can achieve better accuracy than machine learning model. Chen et al. (2018) introduced attention mechanism on the basis of traditional RNN model, so as to achieve effective extraction of text features of twitter, and achieved good detection results.

Through the analysis of the existing rumor detection models, it is concluded that machine learning and deep learning models have achieved good detection results in the field of automatic online rumor detection. In addition, the rumor detection model based on deep learning model can achieve better results than the traditional machine learning model. In the early stage of microblog publishing, a large number of microblog comments and forwarding information can not be obtained, and most of the existing research models use microblog comments, forwarding and other features to achieve effective rumor detection. Therefore, it is a challenge for the research of rumor detection that only limited features can be used to detect rumors in the early stage.

3 Method

In recent years, the neural network training method based on the pre training model and fine tuning the vertical task has achieved good results in many natural language processing projects. The Bert (bidirectional encoder representations from transformers) model (Devlin et al., 2018) is a new method of pre training language representation. Its essence is to fine tune the target data set on the pre trained model under the large context environment, so as to represent the target input as a fixed feature vector for classification, The feature representation of Bert can be directly used as the word embedding feature of the task. The model migrates from the general domain to the target domain, and then performs the specific classification task.

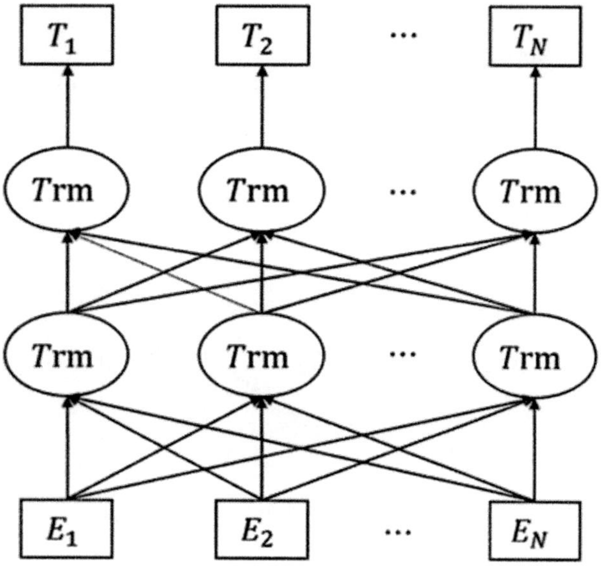

Fig. 1 The structure of BERT

The main structure of the model is bidirectional transformers encoder, which mainly uses attention mechanism to model sentences. The structure of the model is as follows:

E in Fig. 1, $E_1, E_2, \ldots, E_N$ represents every word in the input text, and a "TRM" refers to a transformer encoder. Through the double-layer transformer encoder, the model can learn the information of the front and back sides of each word, so as to obtain a more comprehensive word vector representation.

Transformer encoder (Yu et al., 2018) is composed of encoder and decoder. Its core idea is attention mechanism. The encoder converts variable length input sequence into fixed length vector, and decoder decodes the fixed length vector into variable length output sequence.

The encoder part is the main structure of transformer, as shown in Fig. 2 below:

As can be seen from the figure, the encoder is composed of six identical layers, each of which has two sublayers. The first sub layer is the multi head attention mechanism layer, and the second sub layer is a simple fully connected forward neural network. The residual network structure is used to connect the two sublayers, followed by a regularization layer.

The input of encoder is the word embedding representation of the whole sentence. In natural language processing, the word order information of text is also very important. However, the self attention mechanism can not extract the word order features. Therefore, before entering the self attention layer, the corresponding position information of each word is added to the word embedding representation of the input layer.

In addition to the word vector, the input of the Bert model includes two parts: text vector and position vector. Text vector is automatically learned in the process

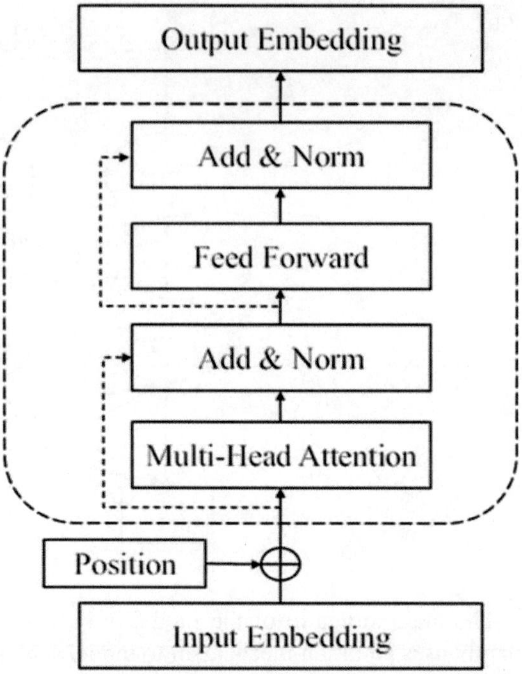

Fig. 2 The encoder part of transformer

of model training, which is used to represent the global semantic information, and is integrated with the semantic information of each word. Because the semantic information carried by the words / words in different positions of the text is different, a different position vector is added to the words in different positions to distinguish them.

The main module of the encoder is the self attention layer (Kim, 2014). Its idea is to calculate the relationship between each word and all words in a sentence, and adjust the weight of each word by using the relationship to obtain a new expression. The representation also includes the relationship with other words on the basis of the semantics of the word itself, which can realize the distinction of polysemy of a word.

The input of multi head self attention layer is a query matrix Q, a key matrix K and a value matrix V composed of word vectors:

$$Attention\,(Q, K, V) = softmax\left(\frac{QK^T}{\sqrt{d_k}}\right)V \tag{1}$$

The above formula calculates the weight of each key in K according to the Q matrix, and the vector dimension of each key in K is expressed as d_k. When the vector dimension is too high, the calculation result is too large_ K performs the square root operation to scale the weight. After the weight matrix of the key is obtained, it is multiplied by the value matrix to get the final calculation result of each key.

Softmax function is used to map the output real number field of linear model to the effective real number space of [0,1] representing probability distribution. Here, the softmax function is used for normalization, and the specific calculation process is as follows:

$$Softmax\left(z_1, z_2, \cdots, z_N\right) = \frac{1}{\sum_i^N e^{z_i}}\left(e^{z_1}, e^{z_2}, \cdots, e^{z_N}\right) \tag{2}$$

The matrix is normalized by row, and the row vector elements are compressed in equal proportion. The sum of compressed vector elements is 1. After a series of calculations above, the attention vector of each word in the original input statement is obtained. Here, the vector integrates the word information of other positions, and arranges them in rows to get a matrix, which is the final attention value.

In order to calculate attention more comprehensively, transformer introduces multi attention mechanism. Firstly, different linear mappings are performed on the input, and then the scaling point product attention of the mapping results is calculated. Each calculation result is called a head, and the attention matrix obtained by multiple operations is horizontally spliced, and then multiplied by a weight matrix to compress into a matrix. The specific calculation formula as follows:

$$\begin{aligned} MultiHead\left(Q, K, V\right) &= Concat\left(head_1, head_2, \cdots, head_n\right) W^o \\ head_i &= Attention\left(QW_i^Q, KW_i^K, VW_i^{V\cdot}\right) \end{aligned} \tag{3}$$

W_i^Q, W_i^K, W_i^V represents the three weight matrices corresponding to the i-th header. The concat function splices the calculation results of multiple headers. W^O is the weight matrix used in splicing.

In the add & norm layer, the results of attention layer are normalized, and the idea of residual connection is used to avoid the degradation problem caused by too deep network layer. The formula is as follows:

$$LN\left(x_i\right) = \alpha \times \frac{x_i - \mu_L}{\sqrt{\sigma_L^2 + \varepsilon}} + \beta \tag{4}$$

The feed forward layer is a fully connected network including two layers of full connection calculation and a relu activation function. The calculation formula is as follows:

$$FNN(x) = \max\left(0, W_1 \cdot x + b_1\right) \cdot W_2 + b_2 \tag{5}$$

The output of the whole model is obtained by normalizing the output of the fully connected network and calculating the residual connection.

The output of the Bert model has two kinds of vectors: character level and sentence level. Character level means that each character of the input text corresponds to a vector representation, and the sentence level is the vector of the left most [CLS] symbol of the model output, which can represent the semantics of the whole sentence.

The Bert model automatically adds [CLS] and [SEP] symbols at the beginning and end of each sentence. After model calculation, each character will get the corresponding vector representation. This paper uses the vector corresponding to the [CLS] symbol in the output. Compared with the traditional text representation method, there is no need to do feature extraction and feature vector stitching.

Finally, the calculated feature matrix is transferred into the full connection layer, and all local features are combined into global features by softmax calculation, which is used to calculate the score of each category. Represents the calculation process of the prediction probability result corresponding to the output result sequence.

$$P(y|x) = \frac{e^{s(x,y)}}{\sum_{y \in Y} e^{s(x,y)}} \quad (6)$$

4 Experimental Setup and Results

In this section, we describe our experimental setup. The dataset and Evaluation methodology is described in Sects. 4.1 and 4.2. The experiments that we conduct to test the performance of our model is given in Sect. 4.3.

4.1 Dataset

In this experiment, the data is from the microblog data set published in the 2016 Ma (Ma et al., 2016) literature, including 2351 normal microblogs and 2313 rumor microblogs, with a total of 4664 microblog events. The original microblog data contains more data. In order to prove that the model can effectively detect microblog rumors, only the first microblog text is selected.

4.2 Evaluation Methodology

The problem studied in this paper belongs to the classification problem. The most commonly used evaluation indexes are accuracy, precision, recall and F1 score, and their definitions are as follows:

Table 1 Results of comparative experiments

Model	Accuracy	Precision	Recall	F1-score
CNN	0.7051	0.7073	0.7103	0.7082
RNN	0.6924	0.6935	0.6919	0.6893
LSTM	0.7263	0.7261	0.7206	0.7253
BERT	0.9261	0.9223	0.9284	0.9253

$$Accuracy = \frac{TP + TN}{TP + TN + FN + TN} \tag{7}$$

$$Precision = \frac{TP}{TP + FP} \tag{8}$$

$$Recall = \frac{TP}{TP + FN} \tag{9}$$

$$F1 - score = \frac{2 \times Precision \times Recall}{Precision + Recall} \tag{10}$$

4.3 Experiments

In order to verify the superiority of our model in rumor early detection, we compare the effect of early rumor detection with existing detection models. The specific experimental results are shown in Table 1:

The experimental results show that the BERT model has stronger semantic feature extraction ability than the traditional model, and is far better than the traditional neural network model in accuracy, precision, recall and F1 score. By comparing the experimental results, we can see that the rumor detection model proposed in this paper can effectively improve the timeliness of rumor detection, and better realize the early detection of microblog rumors.

The training is carried out on a 4G memory GPU. After the 40th round of training, the model has achieved a high accuracy, and the training time is within 1 h. With the increase of data volume, the training time of the model will also increase correspondingly. However, the detection time of the trained model for a single microblog will not increase. It will only be affected by the text length of the microblog, and the detection results can be obtained in a very short time, which can meet the real-time requirements in the actual detection.

The above experimental results can prove that the proposed microblog rumor early detection model can effectively improve the timeliness of detection and achieve effective early detection of rumors. The early rumor detection model based on Bert can also achieve better detection effect than other detection models.

5 Conclusions

Aiming at the problems of more required features and less timeliness in rumor detection, this paper tries to realize the early detection of rumors spread in microblog without relying on other information such as the comments and forwarding of microblog, so as to detect the rumors in the early stage. The results show that the detection effect of the proposed method is better than the traditional machine learning algorithm and the detection algorithm based on deep learning, and through a number of indicators, it is proved that this model can achieve better results in early rumor detection task.

The innovation of this paper is to use the Bert pre training model, which has achieved good results in the field of natural language processing recently, to obtain better deep semantic features, so that only using the microblog text data can realize the early detection of microblog rumors.

At the same time, the detection model in this paper has some limitations and shortcomings, such as the number of microblog data sets to be strengthened, how to add more early features to the model to improve the generalization ability of the model. On the other hand, there are various forms of Internet rumors, most of which contain images, videos and other information. How to combine these information into rumor detection model is the focus of future rumor detection research.

References

Castillo, C., Mendoza, M., & Poblete, B. (2011). Information credibility on twitter [C]. In *Proceedings of 20th International Conference on world wide web Hyderabad* (pp. 675–684). ACM.

Chen, T., Li, X., Yin, H., et al. (2018). Call attention to rumors: Deep attention based recurrent neural networks for early rumor detection [C]. In *Proceedings of the 2018 Pacific-Asia Conference on Knowledge Discovery and Data Mining* (pp. 40–52).

Devlin, J., Chang, M. W., Lee, K., et al. (2018). Bert: Pre-training of deep bidirectional transformers for language understanding [J]. *arXiv preprintarXiv:1810.04805.*

Gist. (1951). Rumor and public opinion [J]. *American Journal of Sociology, 57*(2), 159–167.

Kim, Y. (2014). Convolutional neural networks for sentence classification [J]. *arXiv:1408.5882v2.*

Kwon, S., Cha, M., Jung, K., Chen, W., & Wang, Y. (2013, December). Prominent features of rumor propagation in online social media. In *2013 IEEE 13th International Conference on Data Mining* (pp. 1103–1108). IEEE.

Ma, J., Gao, W., Mitra, P., et al. (2016). Detecting rumors from microblogs with recurrent neural networks [C]. In *Proceedings of the Twenty-Fifth International Joint Conference on Artificial Intelligence* (pp. 3818–3824). IJCAI.

Yang, F., Liu, Y., Yu, X., & Yang, M. (2012, August). Automatic detection of rumor on Sina Weibo. In *Proceedings of the ACM SIGKDD Workshop on Mining Data Semantics* (pp. 13). ACM.

Yu, A. W., et al. (2018). Qanet: Combining local convolution with global self-attention for reading comprehension. *arXiv preprint arXiv: 1804.09541.*

Research on the Cause of Personal Accidents in Electric Power Production Based on Capacity Load Model

Penglei Li, Chuanmin Mi, and Jie Xu

1 Introduction

Accident-causing theory is a basic theory that guides safety management. It is of great significance for enterprises to fully understand the causes of accidents and to dig out the characteristics of accident causes (Huang & Wu, 2017; Gao et al., 2019). At present, the researches of domestic scholars are generally based on the theory of human safety management (Hao et al., 2013; Wang et al., 2018). These studies mainly focus and discuss the relationship between individual factors, cognitive ability and organizational factors on enterprise safety effectiveness from the qualitative perspective. There are some deficiencies in the above research. On the one hand, the characteristics of power operation itself are not fully reflected, on the other hand, the quantitative analysis of accident causes is not enough.

As a method to describe and study the complex relationship among multiple factors, network research has been applied in many safety management fields. Z. Zhou et al. (2014) constructed the unweighted directional subway construction accident network, and proved that attacking the nodes with high node degree in the network will have a serious impact on the network connectivity. Q. Li et al. (2017) constructed a directional unweighted network model of subway operation security accidents, revealing the network's robustness against random attacks and its vulnerability to intentional attacks. J. Liu et al. (2019) constructed a causative network of railway operation accidents based on ordered pairs, and identified the key factors through network parameter analysis. S. Guo et al. (2020) used a case

P. Li · C. Mi (✉)
College of Economics and Management, Nanjing University of Aeronautics and Astronautics, Nanjing, China
e-mail: cmmi@nuaa.edu.cn

J. Xu
Wuxi Power Supply Branch of State Grid Jiangsu Electric Power Co. Ltd., Wuxi, China

H. Yang et al. (eds.), *AI and Analytics for Public Health*, Springer Proceedings in Business and Economics, https://doi.org/10.1007/978-3-030-75166-1_19

study in the Chinese building construction industry to explore the behavioral risk chains of accidents based on complex network theory, and identified the critical classes and key chains of unsafe behaviors. X. Ma et al. (2013) studied the effect of successive failure processes of each node in the network on the efficiency of the entire network. But network research method has hardly been used in the personal accident of electric power production. Moreover, there is little research on the dynamic propagation of risk or the dynamic process of successive failures of causative nodes in the network.

Considering the complex dynamic relationship between the causes of personal accidents in electric power production, this paper describes the process of accidents in the form of accident chain, and constructs a weighted directional personal accident causation chain network model. Then, based on the capacity load model, the propagation and evolution mechanism of node load is established, and the successive failure formation process of causative node is explored. Through the research, it can get the accident chain which is easy to trigger in the power production, so as to provide targeted guidance and suggestions for the safety management.

2 Accident Data Resource

Based on the data published on the official website of the National Energy Administration, and field investigation of power supply enterprises, and the consultation to books related to electric power safety accidents, 104 cases of personal accidents in power production occurred nationwide from 2001 to 2017 are collected. According to China's "DLT518-2012-Classification and Code of Personal Accidents in Power Production", the accident cases are divided into 9 types as shown in Fig. 1. The number and proportion of different types of accidents can be seen from it, and it shows that he first three types of accidents occur frequently in the work of safety production.

3 Construction of Personal Accident Causative Chain Network

3.1 Model Hypothesis

Hypothesis 1: The nodes and edges of accident causative chain network model

In the accident chain network, nodes represent causative factors or accidents, and the edge represents the causative relationship between the two nodes. There are three forms of causality in the accident chain network, namely, one-cause-one-effect, one-cause-multiple-effect and multiple-cause-one-effect.

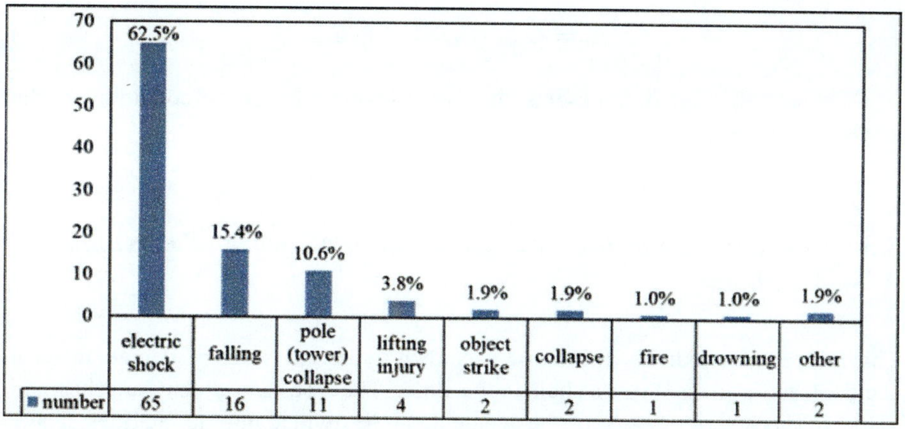

Fig. 1 Classification and statistics of personal accidents in electric power production

Hypothesis 2: The weight of edge

The weight of edge is the possibility that the failure of one node leads to the failure of another node, indicating the causative degree of relationship between nodes. It is also called the propagation probability. The weight calculation method of the edge between any two nodes *i* and *j* is shown in Eq. (1).

$$p_{ij} = \frac{n_{ij}}{n_i} \tag{1}$$

p_{ij} is the probability that causative factor *i* transfers its own dangerous state to causative factor *j*;
n_i is the number of accidents that contains causative factor *i*;
n_{ij} is the number of accidents that contains causative factor *j* under the condition of causative factor *i*.

3.2 Accident Causative Chain Representation

Accident causative chain originates from safety science, and it thinks that the accident is the result of the coupling of many factors and conditions. When several factors or conditions are triggered in a certain situation and connected together to cause an accident, they form an accident chain. The initial factor or condition is called the bottom event of the accident chain, and the accident or catastrophic consequence is called the top event of the accident chain. In a specific accident chain, the relationship between various factors or conditions is logical and the accident chain can be expressed as

$$A_i = C_{i1} \cap C_{i2} \cap \cdots \cap C_{in_i} \tag{2}$$

$C_{ij}(j = 1, 2, \ldots, n_i)$ represents the j-th triggered factor or condition in the accident chain A_i.

3.3 *Construction of the Causative Chain Network of Personal Accidents*

This article first extracts the cause, accident type and severity degree in each accident, and expresses it as a chain in this order. The types of accidents are shown in Table 1. The severity degree of the accident can be divided into no fatalities, heavy casualties (1–2 deaths), and extremely serious casualties (no less than 3 deaths). In this paper, a total of 99 factors are involved, in order to avoid occupying a larger space, this paper will not list them one by one. Each factor corresponds to a node in the network, and the weight of edge is determined according to Eq. (1). Then a directed weighted accident causative chain network as shown in Fig. 2 is constructed, which includes 99 nodes and 287 edges.

Table 1 List of accident causal chains

No.	Accident causal chain	No.	Accident causal chain	No.	Accident causal chain
1	1 → 7 → 3 → 88 → 98	2	3 → 88 → 98	3	4 → 5
4	6 → 21 → 12 → 88 → 98	5	7 → 5	6	8 → 29 → 27 → 52
7	12 → 88 → 98	8	13 → 14 → 43	9	14 → 43
10	15 → 92 → 98	11	18 → 12 → 88 → 98	12	20 → 12 → 88 → 98
13	22 → 28	14	23 → 24 → 18 → 12 → 88 → 98	15	24 → 18 → 12 → 88 → 98
16	25 → 89 → 98	17	27 → 52	18	29 → 27 → 52
19	31 → 32	20	33 → 95	21	36 → 3 → 88 → 98
22	37 → 35	23	40 → 93 → 99	24	41 → 89 → 98
25	42 → 43	26	44 → 43	27	45 → 27 → 52
28	47 → 52	29	48 → 67	30	51 → 52
31	57 → 3 → 88 → 98	32	58 → 50 → 51 → 52	33	61 → 42 → 43
34	62 → 90 → 98	35	64 → 17	36	74 → 86
37	80 → 93 → 99	38	81 → 93 → 99	39	85 → 96 → 98

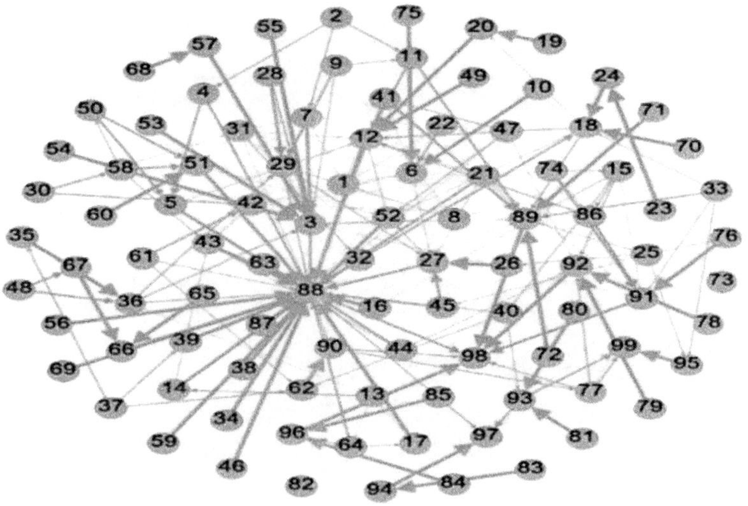

Fig. 2 Causative chain network model of personal accidents in electric power production

4 Research on the Propagation Mechanism of Node Load

4.1 Node Status, Initial Load and Capacity

This paper divides the state of the node into normal state and failure state. The initial load of the node can be understood as the initial risk degree of the node when the entire system is in a relatively stable state. In this paper, the node betweenness in the network is regarded as the initial load of the node. The capacity is the safety threshold of node load, which can be understood as the maximum risk that the node can carry. As the risk of the node continues to shift, when the node load exceeds the safety threshold, the node will change from the normal state to the fault state. Among them, nodes in a failed state are divided into failed nodes on the accident chain and failed nodes outside the accident chain according to whether they can transmit load or risk to other nodes. The path selection mechanism of risk propagation determines whether the failed node is in the accident chain. The specific node state evolution process is shown in Fig. 3.

According to the definition of the load-capacity model, the capacity is related to the initial load of the node and is greater than the initial load, so the capacity is proportional to the initial load and is expressed by Eq. (3).

$$C_i = (1 + \alpha)\ L_i(0) \tag{3}$$

$L_i(0)$ represents the initial load of node i; C_i represents the capacity of node i; α is a tolerance parameter. The larger the α, the larger the node capacity, and the less likely the node will become a failure state.

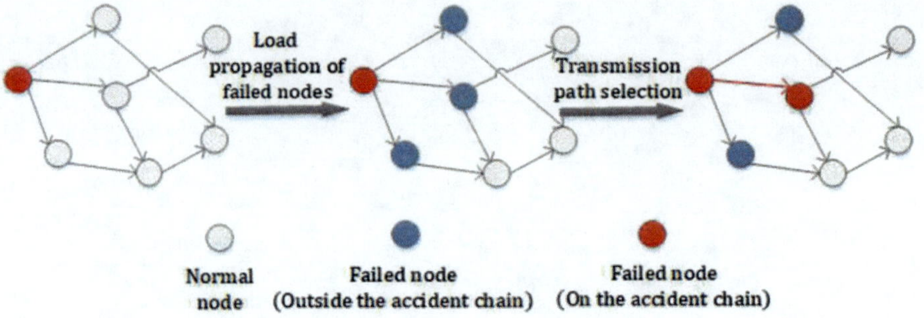

Fig. 3 The specific evolution process diagram of the node state

4.2 The Distribution Mechanism of Node Load

The node load distribution mechanism means that when a node fails, all the load carried by the node must be distributed to the downstream nodes associated with it according to a certain strategy. The paper believes that the spread of node load is not only related to the propagation probability between nodes, but also related to the degree of downstream nodes. The specific load allocated by node i to node j is determined by the propagation intensity between nodes and is proportional to the propagation intensity. In the k-th step, the propagation intensity $I_{ij}(k)$ between node i and node j is as shown in Eq. (4):

$$I_{ij}(k) = \omega_p P_{ij} + \omega_d \frac{d_j(k)}{\sum_{j \in v_k} d_j(k)} \tag{4}$$

ω_p represents the weight of the propagation probability; ω_d represents the weight of the downlink node degree; v_k represents all the downstream nodes of node i at step k; $d_j(k)$ represents the degree of downstream node j at step k. After node i spreads its entire load to the downstream nodes, node i will be separated from the network and no longer have load spreading capability. The load of other downstream nodes is the sum of the original load and the additional load that it accepts from node i.

4.3 The Propagation Evolution Mechanism of Node Load

The formation process of the accident causal chain starts when the state of a certain node in the network changes, that is, from a safe state to a failed state, until the node load distribution and redistribution process is terminated. In this process, the failed node forms an accident causal chain. The following is a detailed description of the initial state, formation and termination conditions and path selection mechanism of the accident causal chain (Dou & Zhang, 2011):

1. Initial state

Select a certain node in the network to be triggered to change from the normal state to the failed state, and record this node as the first node in the accident cause chain. In this paper, the initial failure node is determined by manual selection.

2. Formation and termination conditions

The formation process of the accident causal chain is gradually advanced in units of time step k. Each time step k increases by one unit, a failure node is added to the chain, that is, the node that will be triggered in the next step. The termination condition of the accident causal chain is that the time step k cannot continue to extend, which means that the node has no downstream nodes or the load of the downstream nodes is within the safety threshold.

3. Path selection mechanism

Step 1: Time step $k = 0$. All causative nodes in the network are in a safe state, and the states of the nodes are marked as $\gamma = 0$. The node load matrix L, the node safety threshold matrix C and the propagation probability matrix P are constructed.
Step 2: Artificially select a node in the network as the initial failure node, which is also the first factor in the accident causal chain. Then update the time step k to $k + 1$, and mark the state of this designated node as $\gamma = 1$.
Step 3: Traverse all downlink nodes associated with the failed node. If there is no downlink node, the accident causal chain is terminated. Otherwise, the load of the failed node is distributed to the downlink node according to the distribution mechanism, and update the node load matrix L.
Step 4: For downstream nodes, compare the values in the updated node load matrix L and the values in the node safety threshold matrix C. If there is a situation of $L \geq C$, the corresponding node becomes a failed node, and the state of the node is marked as $\gamma = 1$, and the status of the rest nodes remains unchanged. Otherwise, no node fails, and the chain terminates.
Step 5: If there is a downstream failure node with $\gamma = 1$, all edges between the node and its upstream nodes will fail. And then the corresponding probability in the propagation probability matrix P will be updated to 0.
Step 6: If there are multiple downstream nodes failing, compare the propagation intensity of each edge and select the edge with the highest propagation intensity to extend. The node connected to this edge becomes the next factor in the accident causal chain. And other failed nodes are failed nodes outside the accident chain, and the outgoing edges corresponding to these nodes fail.
Step 7: Return to step 2 until the formation process of the accident causal chain is terminated.

5 Analysis of the Successive Failure Process of Causal Nodes

5.1 Parameter Setting

1. Limiting parameter α.

In order to dig out as many causal chains as possible, α generally does not exceed 0.5. The smaller the value of α, the closer the safety threshold of node load is to the initial load. In the process of propagation, the node load is more likely to exceed the safety threshold, which will promote the node from a safe state to a failed state, and at the same time, the more it can promote the formation of the accident cause chain. Therefore, the value of α is set at 0.2 in this study.

2. The weight of the propagation probability ω_p and the weight of the downlink node degree ω_d.

The propagation probability indicates the extent to which the current failed node affects the downstream node. The greater the propagation probability, the greater the intensity of the impact. However, the downlink node may have multiple other uplink nodes in addition to the one uplink node of the current failed node. Therefore, the degree of the downlink node is not enough to explain the degree of influence of the current failed node on it. Therefore, this paper believes that the propagation probability between nodes is more important than the network structure of the downlink node degree, so the weight of the propagation probability and the weight of the downlink node degree in the propagation intensity are set as $(\omega_p, \omega_d) = (0.7, 0.3)$.

5.2 Demonstration of the Formation Process of the Accident Causal Chain

On the basis of setting the relevant parameters, this paper selects the node with the largest node degree as the initial failure node, that is, node "1" that means lack of safety awareness, and conducts a detailed exploration of the formation process of the accident causal chain.

1. K = 0. Through the statistics of accident causal data, this paper obtains the node propagation probability matrix P, the initial load matrix L and the node capacity matrix C.
2. K = 1. Select node 1 "Poor safety awareness" as the initial failure node, set its state $\gamma = 0$, and mark it as the first factor in the accident causal chain, and then all its own load spreads down the possible path. The downstream nodes of node 1 have 33 nodes such as nodes 3, 4, 6, 7, 8, 9, and 10. The load of node 1 is distributed to all the downstream nodes according to the propagation intensity. After redistribution, 27 of the nodes whose load exceeded the safety threshold

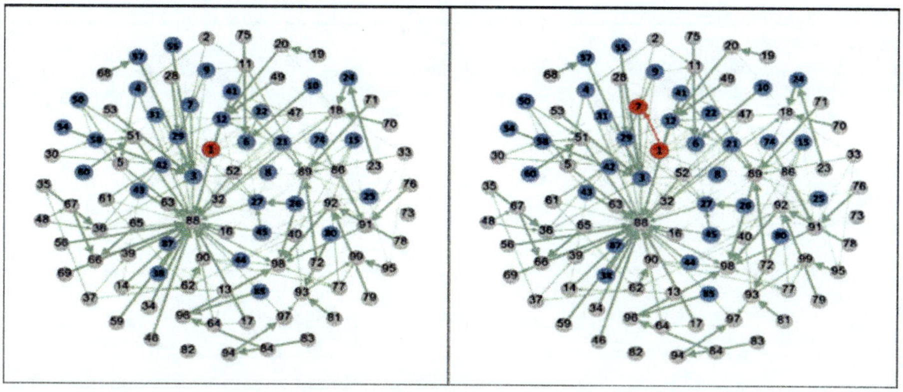

Fig. 4 (K = 1) Load propagation process of node 1

became failed nodes and had the ability to continue to spread the load. The node 7 " Insufficient risk point analysis " with the highest transmission intensity, is selected as the next factor in the accident causal chain.

The two sub-phases of the load propagation process of node 1 at this stage are shown in Fig. 4. The gray node indicates that the node is in a safe state, the blue node indicates the out-of-chain failure node, and the red node indicates the failure node on the cause chain. The red arrow indicates the propagation path of the cause. Same below.

3. K = 2. The causal factor Node 7 "Insufficient risk point analysis" as a failure node in the accident chain continues to spread the load downward. Node 7 has 7 downstream nodes, namely nodes 3, 5, 11, 12, 51, 52 and 88. At this time, the load of node 7 is allocated to 7 downstream nodes according to the propagation intensity. After reallocation, 4 of the nodes whose load exceeds the safety threshold become failed nodes. Among them, the node 3 "Poor implementation of safety technical measures" with the highest transmission intensity is selected as the next factor in the accident causal chain (Fig. 5).

4. K = 3. The causal factor Node 3 " Poor implementation of safety technical measures" as a failed node on the accident chain continues to spread the load downward, and node 3 has 10 downstream nodes. The load of node 3 is distributed to 10 downstream nodes according to the propagation intensity. After redistribution, the load of 8 nodes exceeds the safety threshold and becomes failed nodes, and the node 88 with the highest propagation intensity is selected as the next factor in the accident-causing chain (Fig. 6).

5. K = 4. The causal factor node 88 "Electric shock" as a failed node in the accident chain continues to spread the load downward. There are 3 downstream nodes in node 88, namely nodes 97, 98 and 99. The load of node 88 is distributed to three

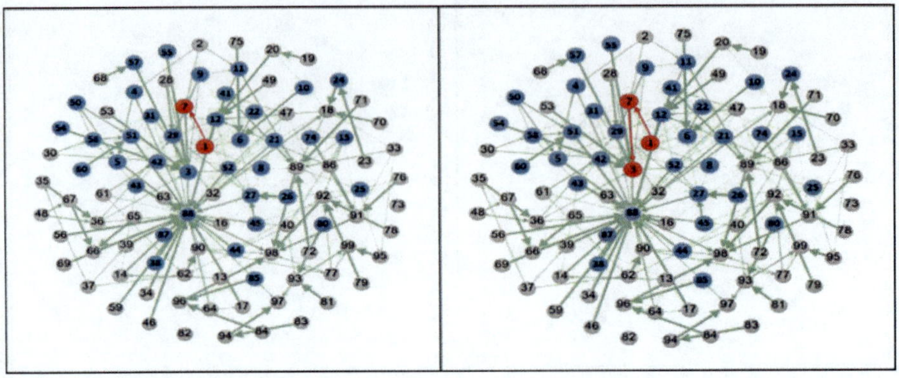

Fig. 5 (K = 2) Load propagation process of node 7

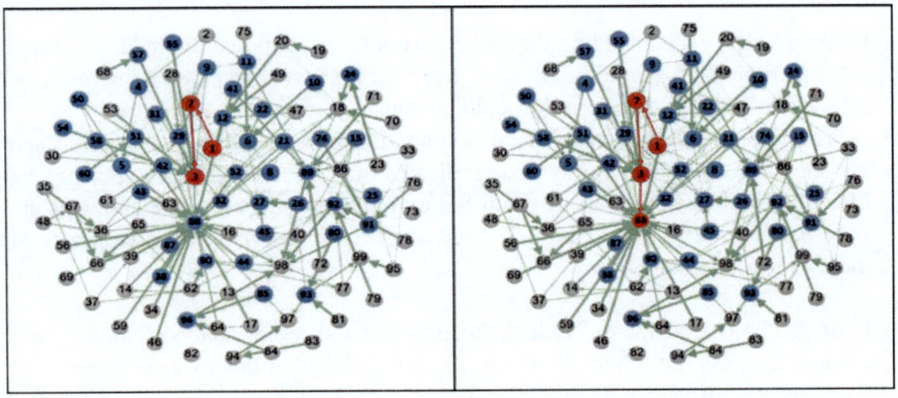

Fig. 6 (K = 3) Load propagation process of node 3

downstream nodes according to the propagation intensity. After redistribution, the loads of the three nodes all exceed the safety threshold and become failed nodes, from which the node with the largest propagation intensity 98 " Heavy casualties (1-2 deaths)" is selected as the next factor in the accident causal chain (Fig. 7).

6. K = 5. There is no downstream node for node 98 " Heavy casualties (1–2 deaths)", so node 98 becomes the last factor in the accident chain, and the accident cause chain formation process is terminated.

According to the above-mentioned load spreading process taking node 1 "poor safety awareness" as an example, an accident causal chain consisting of 5 nodes "1 → 7 → 3 → 88 → 98" is formed, that is, "Poor safety awareness → Insufficient risk point analysis → Poor implementation of safety technical measures → Electric shock → Heavy casualties (1–2 deaths)".

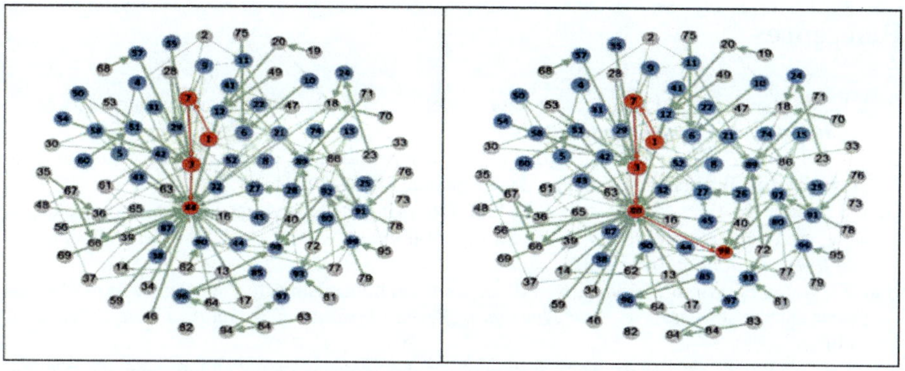

Fig. 7 (K = 4) Load propagation process of node 88

According to the above process, the remaining nodes in the power production personal accident cause chain network are analyzed, and a total of 39 accident cause chains with different lengths are obtained, as shown in Table 1:

6 Conclusions

1. From the perspective of accident types, the number of accident chains involving electric shock is the largest, followed by object strikes. From the point of view of the consequences of accident, it is mainly heavy casualties (1–2 deaths), and extremely serious casualties (no less than 3 deaths). Extremely serious casualties are all object strike accidents. Therefore, special attention should be paid to electric shock accidents and object strike accidents.
2. The length of the cause chain varies. The longest cause chain contains 6 nodes, and the shortest cause chain contains 2 nodes. The length of the cause chain of accidents related to electric shock is generally greater than that of other accidents. It shows that more factors are affected in the process of electric shock.
3. There are some sub-chains frequently appearing in multiple different accident-causing chains, such as $3 \rightarrow 88 \rightarrow 98$ and $18 \rightarrow 12 \rightarrow 88 \rightarrow 98$, indicating that this type of situation is very easy to occur in the actual power production work, and special attention should be paid causes that appear in this type of accident chain. By focusing on prevention of one or several accident causes, the purpose of interrupting the accident cause chain and preventing accidents can be achieved.

Acknowledgements This work was funded in part by Wuxi Power Supply Branch of State Grid Jiangsu Electric Power Co., LTD. Project "Research on security management improvement service based on digital drive" (SGJSWX00AZWT1901493), and it was also funded by The National Social Science Fund of China (Grant No. 17BGL055).

References

Dou, B., & Zhang, S. (2011). Load-capacity model for cascading failures of complex networks. *Journal of System Simulation, 23*(7), 1459–1459.

Gao, Y., Fan, Y., & Wang, J. (2019). Evaluation of governmental safety regulatory functions in preventing major accidents in China. *Safety Science, 120*, 299–311.

Guo, S., Zhou, X., Tang, B., & Gong, P. (2020). Exploring the behavioral risk chains of accidents using complex network theory in the construction industry. *Physica A: Statistical Mechanics and Its Applications, 560*, 125012.

Hao, Z., Wang, S., Wu, S., & Song, S. (2013). Study on humanism safety management of electric power enterprises based on SEM. *Journal of Beijing Institute of Technology (Social Sciences Edition), 15*(2), 77–81.

Huang, L., & Wu, C. (2017). Study on system of accident-causing model and its general modeling methods and development trend. *Journal of Safety Science and Technology, 13*(2), 10–16.

Li, Q., Song, L., List, G. F., Deng, Y., Zhou, Z., & Liu, P. (2017). A new approach to understand metro operation safety by exploring metro operation hazard network (MOHN). *Safety Science, 93*, 50–61.

Liu, J., Schmid, F., Zheng, W., & Zhu, J. (2019). Understanding railway operational accidents using network theory. *Reliability Engineering & System Safety, 189*, 218–231.

Ma, X., Li, K., Luo, Z., & Zhou, J. (2013). Analyzing the causation of a railway accident based on a complex network. *Chinese Physics B, 23*(2), 028904.

Wang, Y., Luo, Y., Pei, J., & Huo, L. (2018). Research on risk management and control mode of violation behavior in electric power enterprise. *Journal of Safety Science and Technology, 14*(4), 173–180.

Zhou, Z., Irizarry, J., & Li, Q. (2014). Using network theory to explore the complexity of subway construction accident network (SCAN) for promoting safety management. *Safety Science, 64*, 127–136.

A Simulation Optimization Approach for Precision Medicine

Jianzhong Du, Siyang Gao, and Chun-Hung Chen

1 Introduction

Precision medicine (PM, also called "personalized medicine" or "stratified medicine") prescribes the best treatment tailored to the patients with the specific characteristics (Collins & Varmus, 2015). This is an emerging direction for improving the medical-care service, compared to the traditional practice of recommending an overall good medication for all patients with the same disease. For example, penicillin, one of the most widely used antibiotics that save millions of lives in the twentieth century, could cause allergy in 10% of the total population. It is obviously not the best choice for allergic patients. For allergic patients with mild reaction due to bacterial infections, medications to reduce symptoms are more favourable than antibiotics.

Allergic response is common among patients, and family members of a patient are prone to share the same risk factors. But the cause of allergy varies a lot among humans. Behind this changing phenotype is the difference in the functioning of metabolism (e.g., the activity of enzyme) in our physical bodies. And medications for the same disease use different mechanisms to take effect. In fact, not only

J. Du (✉)
Department of Advanced Design and Systems Engineering, City University of Hong Kong, Hong Kong SAR, China
e-mail: jianzhodu2-c@my.cityu.edu.hk

S. Gao
Department of Advanced Design and Systems Engineering, City University of Hong Kong, Hong Kong SAR, China

School of Data Science, City University of Hong Kong, Hong Kong SAR, China

C.-H. Chen
Department of Systems Engineering & Operations Research, George Mason University, Fairfax, VA, USA

H. Yang et al. (eds.), *AI and Analytics for Public Health*, Springer Proceedings in Business and Economics, https://doi.org/10.1007/978-3-030-75166-1_20

penicillin, almost every treatment has its limitation and cannot be the best treatment for every patient. PM aims to solve this problem by selecting the most cost-effective alternative for each patient based on the patients' biometric information (e.g., gender, age, body mass index (BMI), results of lab test, etc.). In this work, we call patients' characteristics the covariates.

Treatment evaluation is the key for this personalized decision-making problem. A common and direct method for treatment evaluation is clinical trials (Hopp et al., 2018). In the trials, patients are voluntarily summoned and allocated to the baseline group and the treatment group. The hosts need to watch the development of patients' physical condition closely and make the cost-effectiveness conclusion about the treatment by the experimental statistics. A convincing trial usually takes thousands of patients and several years or even decades, which can be a heavy burden economically. At the same time, the required number of patients in clinical trials for PM could increase linearly with the number of covariate values because we need to have enough samples within each subgroup of patients to reach a confident conclusion. However, the number of covariate values with the same illness could be thousands. Therefore, the amount of time and financial resources to manage the patients needed in clinical trials for PM can be prohibitive (Mok, 2011).

To avoid this difficulty, stochastic models are often used as an alternative of clinical trials for treatment evaluation. Like many other applications of management science (e.g., the design of supply chain networks, the staff allocation method in hospitals), we can observe the principle and dynamics of disease evolvements by the massive data from healthcare practice, and then employ stochastic models to capture them in high fidelity. For a lot of diseases, their progression can be decomposed into various stages, from slight to fatal, with transition rates between stages affected by the treatment methods and biometric characteristics of the patients. The interactions between stages interweave with each other making such stochastic models too complex to have closed-form mathematical solutions. Simulation becomes the powerful tool for this personalized problem.

Many research papers spring up in recent years for solving PM problems of various diseases by simulation (Ramos et al., 2020). Equal allocation, which distributes an equal amount of simulation resources among treatments and covariate values, is a popular method in literature. However, simulation is known to be time-consuming, and it usually takes months to make a high-quality simulation-based decision (Chen & Lee, 2011). Meanwhile, the hardness to find the best medicine under each covariate differs a lot. For example, yearly vaccination is distinctively superior to stockpiling some antiviral flu drugs in case of infection for the flu prevention of elderly people. At the same time, the difference between vaccine and stockpiling can be minor for the middle-aged and should be studied with more efforts. Therefore, equal allocation could waste much efficiency especially for PM problems with complex structures and a huge number of covariate values. For a high-quality decision for PM, we need efficient mechanisms to balance our simulation resources between tougher and easier pairs of treatment and covariate value in terms of the required effort to distinguish.

Under the assumption of a finite number of covariate values and treatments, the optimization of the simulation process falls into the Ranking and Selection with covariates (R&S-C) framework (Shen et al., 2019), which generalizes the traditional R&S problem that selects the best alternatives under the unique covariate value. The procedures in R&S can be classified into two categories (Hunter & Nelson, 2017): the fixed-confidence setting where the performance measure should be guaranteed and the required number of simulation samples is random (e.g., indifference zone (IZ) procedure) and the fixed-budget setting where the total simulation budget is given and the selection quality measure should be optimized (e.g., optimal computing budget allocation (OCBA) and expected value of information). In Shen et al. (2019), they propose procedures that generalize the IZ procedure in the fixed-confidence setting of R&S and can guarantee the average probability of correctly selecting the best treatment for each covariate value is above a pre-specified level (e.g., 99%) when solving a general PM problem. This work differs from Shen et al. (2019) from the very beginning of the setting. Following the style of the OCBA in the fixed-budget setting of R&S (Chen & Lee, 2011; Gao et al., 2017), we solve PM problems under R&S-C framework by optimizing the efficiency of the simulation-based personalized decision process.

Our contributions are two-fold. First, we introduce the quality measure for the simulation process of PM, derive a solution that asymptotically optimizes the performance measure, and give the algorithm that can achieve the optimal solution asymptotically. Second, we apply our new algorithm to a PM problem established in Hoogendoorn et al. (2019) and show the good performance of this new algorithm.

2 Problem Formulation

Suppose we have a different treatments whose performance depends on the value of covariate $\mathbf{c}$. Although some biometric characteristics of patients are continuous in nature, they are treated as discrete by medical staff. For example, weight is recorded as x_1 pounds, and height is recorded as x_2 feet x_3 inches. Therefore, we assume that covariate $\mathbf{c}$ has a finite number of b possible values $\mathbf{c}_1, \mathbf{c}_2, \ldots, \mathbf{c}_b$. Let $S_{ti}(\mathbf{c}_s)$ be the i th simulation sample of treatment t under covariate value $\mathbf{c}_s$, and $\mu_t(\mathbf{c}_s)$ and $\sigma_t^2(\mathbf{c}_s)$ be the expectation and variance of $S_{ti}(\mathbf{c}_s)$, $t = 1, \ldots, a$, $s = 1, \ldots, b$. We need to find $\mathrm{t}^*(\mathbf{c}_s) = \arg\max_t \mu_t(\mathbf{c}_s)$ for each covariate value $\mathbf{c}_s$. For the sake of simplicity, we assume the best treatment under covariate value $\mathbf{c}_s$, $t^*(\mathbf{c}_s)$, is unique for all $s = 1, \ldots, b$. $S_{ti}(\mathbf{c}_s)$ has normal distribution. $S_{ti}(\mathbf{c}_s)$ and $S_{t'i'}(\mathbf{c}_{s'})$ are independent if $t \neq t'$, $i \neq i'$, or $s \neq s'$.

Suppose the total number of simulation samples is n, and $N_{t,s}$ is the number of samples that we allocate to treatment t under covariate value $\mathbf{c}_s$. $\alpha_{t,s} = N_{t,s}/n$ and $\boldsymbol{\alpha} = (\alpha_{t,s}, t = 1, .. a, \; s = 1, .. b)$. Let $\hat{\mu}_t(\mathbf{c}_s)$ and $\hat{\sigma}_t^2(\mathbf{c}_s)$ be the sample mean and variance based on $N_{t,s}$ simulation samples. When the simulation budget runs out, we select treatment $\hat{t}^*(\mathbf{c}_s) = \arg\max_t \hat{\mu}_t(\mathbf{c}_s)$ for each $\mathbf{c}_s$, $s = 1, \ldots, b$. For a given covariate value $\mathbf{c}_s$, due to the samples' randomness, $\hat{t}^*(\mathbf{c}_s) \neq \mathrm{t}^*(\mathbf{c}_s)$ happens with a probability that decreases as the sample means $\hat{\mu}_t(\mathbf{c}_s)$, $t = 1, \ldots a$, are more accurate. By the Law of Large Number, the more samples we give to treatment t under $\mathbf{c}_s$, the more accurate $\hat{\mu}_t(\mathbf{c}_s)$ will be. But a larger $N_{t,s}$ also means the total number of samples for other treatment-covariate value pairs (i.e., $n - N_{t,s}$) will be less and their estimates will be more inaccurate. We need to balance between $N_{t,s}$'s, or equivalently $\alpha_{t,s}$'s, to minimize our probability of false selection (i.e., $\hat{t}^*(\mathbf{c}_s) \neq t^*(\mathbf{c}_s)$) for each $\mathbf{c}_s$.

We introduce the risk-neutral measure $\text{PFS}_E = \sum_{s=1}^{b} p_s \text{PFS}(\mathbf{c}_s)$ to evaluate the performance of allocation $\boldsymbol{\alpha}$, where $\text{PFS}(\mathbf{c}_s) = \mathbb{P}\left(\hat{t}^*(\mathbf{c}_s) \neq t^*(\mathbf{c}_s)\right)$ is the probability that the selected best treatment is <u>*not*</u> the true best one under $\mathbf{c}_s$. PFS_E tells the expected probability that we recommend a non-best treatment based on simulation results to a patient whose covariate value $\boldsymbol{c}$ is random and equals to $\boldsymbol{c}_s$ with probability p_s. In practice, the participants may use a risk-averse measure (e.g., $\text{PFS}_M = \min_{s=1,\ldots,b} \text{PFS}(\boldsymbol{c}_s)$, i.e., the minimum probability that we recommend a non-best personalized treatment to a patient based on simulation results) rather than the risk-neutral measure PFS_E. In Gao et al. (2019), we can show that two common risk-averse measures, including PFS_M, are asymptotically equivalent to PFS_E, and the same optimal $\boldsymbol{\alpha}^*$ can optimize PFS_E and the risk-averse measures simultaneously.

3 The Optimal Budget Allocation and an Asymptotic Optimal Algorithm for Precision Medicine

Note that $\text{PFS}(\mathbf{c}_s) = \mathbb{P}\left(\cup_{t \neq t^*(\mathbf{c}_s)} \hat{\mu}_{t^*(\mathbf{c}_s)}(\mathbf{c}_s) < \hat{\mu}_t(\mathbf{c}_s)\right)$, i.e., the probability that a non-best treatment outperforms the true best under $\mathbf{c}_s$. Since $\hat{\mu}_{t^*(\mathbf{c}_s)}(\mathbf{c}_s) - \hat{\mu}_t(\mathbf{c}_s)$ has normal distribution $\mathcal{N}(\mu_{t^*(\mathbf{c}_s)}(\mathbf{c}_s) - \mu_t(\mathbf{c}_s),\ \sigma^2_{t^*(\mathbf{c}_s)}(\mathbf{c}_s)/N_{t^*(\mathbf{c}_s),s}$ $.+\sigma_t^2(\mathbf{c}_s)/N_{t,s})$, we have $\mathbb{P}\left(\hat{\mu}_{t^*(\mathbf{c}_s)}(\mathbf{c}_s) - \hat{\mu}_t(\mathbf{c}_s) < 0\right)$

$$= \quad \Phi\left(-\left(\mu_{t^*(\mathbf{c}_s)}(\mathbf{c}_s) - \mu_t(\mathbf{c}_s)\right) / \left(\sigma^2_{t^*(\mathbf{c}_s)}(\mathbf{c}_s)/N_{t^*(\mathbf{c}_s),s} + \sigma_t^2(\mathbf{c}_s)/N_{t,s}\right)^{1/2}\right),$$

which decreases at the rate of $\mathcal{R}_{t,s} = \left(\mu_{t^*(\mathbf{c}_s)}(\mathbf{c}_s) - \mu_t(\mathbf{c}_s)\right)^2 / 2(\sigma^2_{t^*(\mathbf{c}_s)}(\mathbf{c}_s)/\alpha_{t^*(\mathbf{c}_s),s}$ $+\sigma_t^2(\mathbf{c}_s)/\alpha_{t,s})$. Here, $\mathcal{R}_{t,s} = \lim_{n\to\infty} -\log \mathbb{P}\left(\hat{\mu}_{t^*(\mathbf{c}_s)}(\mathbf{c}_s) - \hat{\mu}_t(\mathbf{c}_s) < 0\right)/n$ and $\Phi(\bullet)$ is the c.d.f. of standard normal distribution.

It can be shown that $\lim_{n\to\infty} -\log \text{PFS}(\mathbf{c}_s)/n = \min_t \mathcal{R}_{t,s}$ (Gao et al., 2017) and thus $\lim_{n\to\infty} -\log \text{PFS}_E/n = \min_s \min_t \mathcal{R}_{t,s}$. By some mathematical steps (Du et al., 2020), we can show the optimal $\boldsymbol{\alpha}^*$ maximizing $\lim_{n\to\infty} -\log \text{PFS}_E/n$ as follows.

Theorem 1 The optimal budget allocation $\boldsymbol{\alpha}^*$ asymptotically maximizing the measure $\text{PCS}_E = 1 - \text{PFS}_E$ satisfies

$\mathcal{Q}_s^b = \mathcal{Q}_s^n$, $s = 1, 2, \ldots, b$; $\quad \mathcal{R}_{t,s} = \mathcal{R}_{t',s}$, $s = 1, 2, \ldots, b$, $t, t' = 1, 2, \ldots, a$ and $t \neq t' \neq t^*(\mathbf{c}_s)$;

$\mathcal{R}_{t,s} = \mathcal{R}_{t',s'},\ s, s' = 1, 2, \ldots, b,\ t, t' = 1, 2, \ldots, a, t \neq t^*(\mathbf{c}_s)$ and $t' \neq t^*(\mathbf{c}_{s'})$;
where $\mathcal{Q}_s^b = \sigma^2_{t^*(\mathbf{c}_s)}(\mathbf{c}_s)/\alpha^2_{t^*(\mathbf{c}_s),s}$ *and* $\mathcal{Q}_s^n = \sum_{t \neq t^*(\mathbf{c}_s)} \sigma_t^2(\mathbf{c}_s)/\alpha^2_{t,s}$.

The $\mathcal{Q}_s^b = \mathcal{Q}_s^n$ in Theorem 1 balances the simulation efforts between the true best treatment and non-best ones; the $\mathcal{R}_{t,s} = \mathcal{R}_{t',s}$ and $\mathcal{R}_{t,s} = \mathcal{R}_{t',s'}$ balance the probability of falsely selecting a non-best one within and across covariate values. The relationship for the optimal $\boldsymbol{\alpha}^*$ requires the value of true mean $\mu_t(\mathbf{c}_s)$ and variance $\sigma_t^2(\mathbf{c}_s)$, which are unknown during simulation. We can use the sample mean and variance to approximate $\mu_t(\mathbf{c}_s)$, $\sigma_t^2(\mathbf{c}_s)$, and thus $\boldsymbol{\alpha}^*$. The accuracy of approximation will improve with the increase of total simulation samples n. However, the optimal relationship is highly nonlinear, meaning the heavy computing force required and the waste of computing resources in approximating $\boldsymbol{\alpha}^*$ every time we update the sample mean and variance. Next, we introduce an iterative algorithm which is shown to converge to the optimal allocation $\boldsymbol{\alpha}^*$ and avoids solving the nonlinear system.

Let $\hat{\mathcal{R}}_{t,s}$, $\hat{\mathcal{Q}}_{s^r}^b$, and $\hat{\mathcal{Q}}_{s^r}^n$ denote the estimates of $\mathcal{R}_{t,s}$, $\mathcal{Q}_s^b$, and $\mathcal{Q}_s^n$ by plugging-in the sample mean, sample variance and $\hat{\alpha}_{s,t}$. Our proposed algorithm and its convergence analysis are as follows.

1. Initialization. Perform n_0 simulation samples for each treatment t under each covariate value $\mathbf{c}_s$, $t = 1, \ldots, a$, $s = 1, \ldots, b$. Calculate the sample mean and variance $\hat{\mu}_t(\mathbf{c}_s)$ and $\hat{\sigma}_t^2(\mathbf{c}_s)$. Set $r = 0$, $\hat{t}^*(\mathbf{c}_s) = \arg\max_t \hat{\mu}_t(\mathbf{c}_s)$, $N_{t,s} = n_0$, $n^{(r)} = \sum_{s=1}^{b}\sum_{t=1}^{a} N_{t,s}$, and $\hat{\alpha}_{t,s} = N_{t,s}/n^{(r)}$.
2. Iterate until $n^{(r)} = n$.

   (a) Find $(t_*, s_*) \in \arg\min_{s=1,\ldots,b;\ t=1,\ldots,a;\ t \neq \hat{t}^*(\mathbf{c}_s)} \hat{\mathcal{R}}_{t,s}$. Assign $s^r = s_*$.
   (b) If $\hat{\mathcal{Q}}_{s^r}^b < \hat{\mathcal{Q}}_{s^r}^n$, $t^r = \hat{t}^*(\mathbf{c}_{s^r})$; otherwise $t^r = t_*$. Simulate treatment t^r under covariate value $\mathbf{c}_{s^r}$ for one sample. Update $\hat{\mu}_{t^r}(\mathbf{c}_{s^r})$, $\hat{\sigma}_{t^r}^2(\mathbf{c}_{s^r})$, $\hat{t}^*(\mathbf{c}_{s^r})$, N_{t^r,s^r}, $n^{(r+1)}$ and $\hat{\alpha}_{s,t}$. $r \leftarrow r + 1$.

Theorem 2 *For* $\hat{\alpha}_{s,t}$ *generated by the new algorithm,* $t = 1, 2, \ldots a$ *and* $s = 1, 2, \ldots, b$, *we have*

$\hat{\mathcal{Q}}_{\mathrm{s}}^{\mathrm{b}} = \hat{\mathcal{Q}}_{\mathrm{s}}^{\mathrm{n}},\ \mathrm{s} = 1, 2, \ldots, \mathrm{b};\quad \hat{\mathcal{R}}_{\mathrm{t,s}} = \hat{\mathcal{R}}_{\mathrm{t',s}},\ \mathrm{s} = 1, 2, \ldots, \mathrm{b},\ \mathrm{t}, \mathrm{t'} = 1, 2, \ldots, \mathrm{a}$ and $\mathrm{t} \neq \mathrm{t'} \neq \mathrm{t}^*(\mathbf{c}_{\mathrm{s}})$;
$\hat{\mathcal{R}}_{\mathrm{t,s}} = \hat{\mathcal{R}}_{\mathrm{t',s'}},\ \mathrm{s}, \mathrm{s'} = 1, 2, \ldots, \mathrm{b},\ \mathrm{t}, \mathrm{t'} = 1, 2, \ldots, \mathrm{a}, \mathrm{t} \neq \mathrm{t}^*(\mathbf{c}_{\mathrm{s}})$, and $\mathrm{t'} \neq \mathrm{t}^*(\mathbf{c}_{\mathrm{s'}})$.

We skip the proof of Theorem 2 due to the limitation of space. Readers can refer to Du et al. (2020) for more details. Theorem 2 proves our new algorithm, which avoids solving the nonlinear system repeatedly, can realize the optimal $\boldsymbol{\alpha}^*$ of Theorem 1 asymptotically. Thus, the rationale of this new algorithm is rigorously justified. Meanwhile, based on the proof, we have $\lim_{r \to \infty} N_{t,s} = \infty$ for all $t = 1, \ldots a$ and $s = 1, \ldots, b$, which guarantees $\hat{t}^*(\mathbf{c}_s) = t^*(\mathbf{c}_s)$ for all $s = 1, \ldots, b$ asymptotically and indicates we can find the best treatment for every patient when total simulation budget goes to infinity. In Gao et al. (2019), a numerical comparison between this new algorithm and some benchmark procedures is presented.

4 Case Study: Personalized Management of Chronic Obstructive Pulmonary Disease

In this section, we apply our new algorithm to one of the latest established PM problems from real-world data in the literature. The PM problem is about the chronic obstructive pulmonary disease (COPD) that affects more than 2% of the total population worldwide and leads to more than three million mortalities yearly. Symptoms of COPD are long-term breathlessness, cough, and sputum production. The cause of the high morbidity rate of this disease includes smoking and air pollution and is expected to exist and threaten public health for many years. Until now, this disease has no cure, making proper health management especially important. The personalized treatment has shown substantial potential for improving patients' quality of life compared to the traditional approach (Ramos et al., 2020). We need efficient and effective methods from precision medicine to decide the best treatment tailored to each patient.

The discrete event simulation model for COPD management comes from Hoogendoorn et al. (2019). In this Markovian model shown in Fig. 1 (left part), three negative events serve as benchmarks in the health state transition of a COPD patient: exacerbation, pneumonia, and death. Which random event of the three happens first only depends on the current health state. If exacerbation or pneumonia happens, the patient has an extra risk of mortality. If the patient successfully survives from the negative events, he/she is still likely to face the recurrence of the same event. So, the occurrence of events divides a patient's life into irregular time intervals whose lengths are random variables with different Weibull distributions. The scale and shape parameters of the Weibull distribution are regression equations of patients'

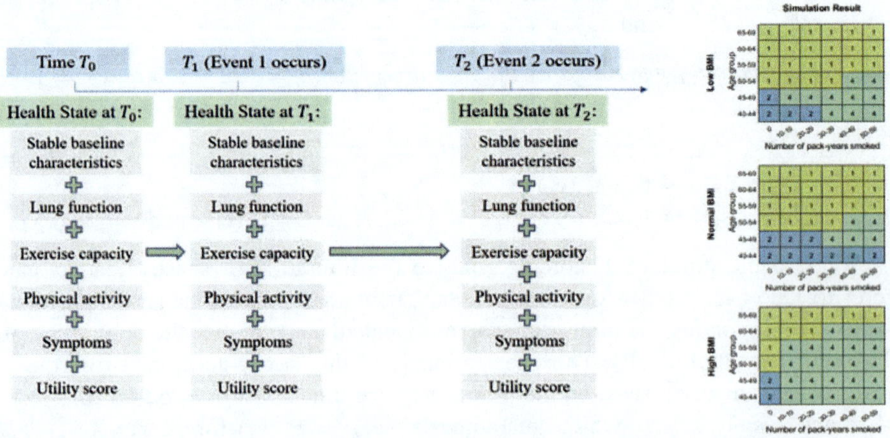

Fig. 1 Discrete event simulation model for COPD (left) and simulation result (right): the number on each cell of the heatmap tells the best treatment based on simulation results for the patient covariate value that the cell represents

health state. Depending on which event happens, whether the event is severe (i.e., the exacerbation is severe, pneumonia needs hospitalization), and whether the patient develops the symptoms (e.g., breathlessness) are Bernoulli random variables whose success probabilities are also regression equations. All these regression equations are estimated by the real-world data in Ramos et al. (2020), Hoogendoorn et al. (2019).

When an event happens, we update the patient's state including lung function, exercise capacity, physical activity, symptoms, and utility score based on the stable baseline characteristics and the realization of the above-mentioned random variables. For illustration purposes, we adopt age, BMI, and the number of pack-years smoked from the stable characteristics (Hoogendoorn et al., 2019) as covariates. The value of age can be 40–44, 45–49, 50–54, 55–59, 60–64, 65–69; BMI can be low, normal, and high; the value of the number of pack-years smoked can be 0 (indicating a non-smoker), 10–19, 20–29, 30–39, 40–49, and 50–59. In total, this problem has 108 covariate values. The value of other patients' physical health factors under the same covariate value are randomly generated based on the statistics in Appendix III of Hoogendoorn et al. (2019) during the simulation. For example, the initial value of the presence of heart failure is a random variable with Bernoulli distribution whose success probability is 0.05.

Like Hoogendoorn et al. (2019), we consider four hypothetic treatments. The first reduces the decline rate in lung function by 12%, the second increases the time to exacerbation by 55%, the third improves the physical activity level by 2 points, and the last treatment reduces the probability of having cough/sputum by 40%. The improvements in treatments are achievable by adjusting the dosage. The effectiveness of treatments is measured by the quality-adjusted life-year (QALY), a weighted sum of the utility score.

In this case study, we set $n_0 = 1.0 \times 10^5$ and $n = 6.4 \times 10^9$ and use the new algorithm in Sect. 3 to decide which pair of treatment and covariate values to sample at each iteration. When simulation ends, the number of samples each pair of treatment and covariate value receives varies from 1×10^5 to 2.4×10^9. The most difficult pair-wise comparison (with the closest sample means) happens between treatments 1 and 4 under the covariate value of age group = 50–54, the number of pack-years smoked = 30–39, and low BMI. The 95% confidence intervals of the two treatments are [3.9260, 3.9262] and [3.9257, 3.9259] respectively, which are sufficiently far away. Consequently, we can reasonably treat the estimated best treatments by simulation as the real best treatments.

Equal allocation (EA), which distributes an equal number of samples to each pair of treatments and covariate values (i.e., $N_{t,s}$ are all equal, $t = 1, \ldots a$ and $s = 1, \ldots, b$), is a popular method in the literature about solving PM by simulation. However, EA has no intelligent mechanism for selecting the best treatment. Note that in the new algorithm, we have given 2.4×10^9 samples to treatment 4 of the most difficult pair-wise comparison. Less samples will lead to the overlap of the two confidence intervals or even a false selection. If we use equal allocation to reach the same confidence as in the most difficult pair-wise comparison, each pair of treatments and covariate values should receive 2.4×10^9 samples as shown in

Table 1 Comparison on the number of simulation samples needed

Name of method	Maximum $N_{t,s}$	Minimum $N_{t,s}$	$\sum_{s=1}^{b}\sum_{t=1}^{a} N_{t,s}$
The new algorithm	2.4×10^9	1.0×10^5	6.4×10^9
Equal allocation	2.4×10^9	2.4×10^9	$2.4 \times 10^9 \times 108 \times 4 = 1036.8 \times 10^9$

Table 1. This means equal allocation will require 16,100% more simulation samples ($1036.8 \times 10^9/(6.4 \times 10^9) - 1$) than the new algorithm.

The final simulation results are shown in Fig. 1 (right part), where the yellow cell with number "1", the blue cell with number "2" and the green cell with number "4" represents the first, second and fourth treatment. In each subfigure of Fig. 1 (right part), the horizontal axis and the vertical axis represent the number of pack-years smoked and the age group. Subfigures on the first, second, and third row tell the simulation results for covariate values that the BMI is low, normal, and high. Here is an example to read the results. We can see a blue cell with number "2" on the left-bottom corner of the subfigure on the third row of Fig. 1 (right part). Since the blue cell with number "2" represents the second treatment, we can tell the best treatment is treatment 2 for patients with age group = 40–44, the number of pack-years smoked = 0, and high BMI. From these results, we can see the first treatment, which reduces the decline rate of lung function, can bring the best effectiveness to elderly patients (e.g., at age 65–70); the second treatment, which increases the time to exacerbation, is sometimes more favorable to non-smokers; the fourth treatment, which reduces the probability of having symptoms, can improve the QALY of young patients with high BMI. The third treatment is not the best choice for any patient considered because its performance is always dominated.

5 Summary

This paper solves the precision medicine (PM) problem by a simulation optimization method. We first introduce the optimal allocation for the simulation-based decision process of PM and then propose a new algorithm that can be shown to achieve the optimal solution asymptotically. The proposed new algorithm is applied to a latest PM case study from the literature. The numerical result indicates when compared with the popular method in literature, the new algorithm achieves much higher simulation efficiency for PM problems.

Acknowledgements The authors thank the editor and two referees for helpful comments that improved this paper.

References

Chen, C. H., & Lee, L. H. (2011). *Stochastic simulation optimization: An optimal computing budget allocation*. World Scientific Publishing.

Collins, F. S., & Varmus, H. (2015). A new initiative on precision medicine. *New England Journal of Medicine, 372*(9), 793–795.

Du, J., Gao, S., & Chen, C. H. (2020). A contextual ranking and selection method for personalized medicine (under review).

Gao, S., Chen, W., & Shi, L. (2017). A new budget allocation framework for the expected opportunity cost. *Operations Research, 65*(3), 787–803.

Gao, S., Du, J., & Chen, C. H. (2019). Selecting the optimal system design under covariates. In *Proceedings of 2019 IEEE 15th International Conference on Automation Science and Engineering (CASE)* (pp. 547–552). Vancouver, Canada.

Hoogendoorn, M., Ramos, I. C., Baldwin, M., Guix, N. G., & Mölken, M. (2019). Broadening the perspective of cost-effectiveness modeling in chronic obstructive pulmonary disease a new patient-level simulation model suitable to evaluate stratified medicine. *Value in Health, 22*(3), 313–321.

Hopp, W., Li, J., & Wang, G. (2018). Big data and the precision medicine revolution. *Production and Operations Management, 27*(9), 1647–1664.

Hunter, S. R., & Nelson, B. L. (2017). Parallel ranking and selection. In *Advances in modeling and simulation* (pp. 249–275). Springer.

Mok, T. S. K. (2011). Personalized medicine in lung cancer: What we need to know. *Nature Reviews Clinical Oncology, 8*, 661.

Ramos, I. C., Hoogendoorn, M., & Mölken, M. (2020). How to address uncertainty in health economic discrete-event simulation models an illustration for chronic obstructive pulmonary disease. *Medical Decision Making, 40*(5), 619–632.

Shen, H., Hong, L. J., & Zhang, X. (2021). Ranking and selection with covariates for personalized decision making. INFORMS Journal on Computing, forthcoming.

Research on Patent Information Extraction Based on Deep Learning

Xiaolei Cui and Lingfei Qian

1 Introduction

Patent text is one of the important technical resources in the industry. It contains the latest technological information and can reflect the technological development of a field. For enterprises, patent text is not only an important competitive intelligence, but also the basis and support of internal product or technological innovation, and its core part mainly exists in unstructured form (Madani & Weber, 2016). Under the background of the integration of information and industrialization, extracting information from patent text can help enterprises integrate technical resources and build a structured knowledge graph, knowledge base and other knowledge network models.

With the promotion of innovation awareness, the number of patent documents has grown rapidly. How to extract effective information from the massive patent texts is one of the problems facing enterprises. Information extraction is one of the important techniques in natural language processing technology, which mainly includes entity extraction, relationship extraction and event extraction. In the field of patent text information extraction, researchers mostly carry out entity extraction and relationship extraction. At the same time, the extraction of patent information mainly depends on manual construction of domain dictionaries and adding artificial features, which requires a lot of manpower and time. Therefore, an effective method of automatic patent text information extraction is proposed, which can improve the efficiency of information extraction, reduce the dependence on labor, and save resources and time costs.

X. Cui (✉) · L. Qian
College of Economics and Management, Nanjing University of Aeronautics and Astronautics, Nanjing, China

H. Yang et al. (eds.), *AI and Analytics for Public Health*, Springer Proceedings in Business and Economics, https://doi.org/10.1007/978-3-030-75166-1_21

Recently, the deep learning model has gradually become the mainstream in the field of information extraction. This paper proposes a patent text information extraction framework based on deep learning. The experimental process is mainly divided into three parts: data pre-processing, triples and feature extraction, and data post-processing. First, the acquired patent documents are cleaned and randomly selected, and then sentence segmentation processing and manual data annotation are performed to complete the data pre-processing process. Then the pre-processed data is input into the model in the form of character embedding to improve the model effect. In the stage of triples and feature extraction, we introduce semantic role tag when extracting information and use sequence labeling model to extract entity words and relation words simultaneously to improve the efficiency of triples extraction. Finally, because the entity words or relational words in the sentence are not unique, the identified triple has noise. In this paper, a data post-processing method is proposed, in which the triple is regarded as a short text to match the corresponding patent text. This method can reduce the noise of triplet data and improve the precision of triplet extraction.

The innovations of this paper are as follows:

1. We introduce a sequence labeling model to extract entity words, relation words and patent features. This model enables us to achieve automatic extraction of patent information without relying on artificial dictionaries or adding artificial features.
2. This paper can identify the situation where the relation word is on the left or right side of two entity words, and fully consider the relationship between the entities of the cross sentence. In the data post-processing experiment, we treat triples as a short sentence, introduce the concept of semantic matching, and improve the performance of the model by improving data input.

2 Related Work

2.1 Open Information Extraction

The concept of open information extraction was first put forward by Banko and Etzioni (2018), which refers to extracting all the semantic relations that can be found from the text without defining the relation type in advance. Most of the traditional researches on entity relationship extraction are based on limited relations. In this method, the category of entity relations is constructed manually, and the problem of entity relationship extraction is transformed into a multi-classification problem. With the emergence of heterogeneous data on the Internet, this method is faced with some problems, such as incomplete category coverage, rough classification, lack of necessary semantic information and so on. Therefore, many researchers begin to pay attention to the field of open information extraction.

With the development of information technology, many researchers have applied statistical machine learning algorithms to open information extraction tasks. Zeng et al. (2014) introduced the convolutional neural network into the relationship extraction task to extract the word level features and sentence level features in the text.Wu and Wu (2017) proposed a semi-supervised Chinese open entity relation extraction method C-COERE based on CRFs. Stanovsky et al. (2018) transformed the open information extraction problem into a sequence marking problem, extended the deep semantic role labeling model based on BiLSTM algorithm. Yan et al. (2019) improved the model proposed by Shu et al. (2015) and used GRU instead of LSTM to make the model more suitable for small data sets.

The application of research methods based on machine learning in the field of information extraction is expanding. Researchers are constantly improving the neural network model by adding artificial features to improve the effectiveness of the model. Some researchers have proposed sequential tagging model for text extraction. Although this method has achieved some results, there is still a lot of room for improvement. For example, complex sentences with nested structure and coordinate structure in Chinese text should be considered in information extraction. To sum up, the research of Chinese open information extraction is still developing, and there is a lot of room for development, so we can learn from its research methods when extracting information from patent texts.

2.2 SAO Structure Extraction of Patent

In the research of patent text information extraction, the application of subject-behavior-object (SAO) structure theory has been more mature. This theory is derived from the invention problem solving theory. S refers to the subject, O refers to the object, and A refers to the relationship between the subject and the object. S and O usually are composed of nouns or nominal phrases, which is an entity or concept in a sentence. An is usually a verb that forms a triple structure with S and O. Since SAO structure only extracts relative words as verb structures, SAO structure extraction can be regarded as a sub-task of open information extraction in patent text domain. The purpose of SAO structure extraction is to identify the relevant information of unstructured text in the organizational relation triple and to identify various relational phrases and their relational objects in any sentence. The development of SAO structure extraction in English patent field is better than that in Chinese patent field, and the extraction tools that can be used are also more mature. Niklaus et al. (2018) compared the effects of several open information extraction systems and found that the portability of the currently developed Open IE system is poor. Due to the differences in grammatical structures between Chinese and English, these tools cannot be applied to the field of SAO structure extraction of Chinese patent texts.

In the task of SAO structure extraction of Chinese patent text, researchers usually divide it into two parts: entity word extraction and relation extraction. Firstly, extract the substantive words from the patent text. Secondly, the candidate triple is obtained by matching the relation words between two entity pairs. Finally, traditional features are introduced to judge whether there is a corresponding relationship between relational words and entity pairs. Based on this research idea, Rao et al. (2015) uses support vector machine algorithm to extract SAO structure, and verifies the effectiveness of introducing SPT phrase syntactic tree features. He et al. (2017) added semantic role tagging features based on Rao's research, which improved the accuracy of the algorithm. Zhang et al. (2019) used He′s method to extract candidate triples and introduced syntactic semantic features into XG-Boost algorithm to verify the validity of syntactic semantic features. The above research mainly applies the traditional machine learning algorithm, and the recognition of entity words is mainly based on artificial dictionary or existing ontology mapping. However, the construction and application of artificial dictionaries is not convenient, and it will cost a lot of manpower.

In the absence of applicable domain ontology and terminology dictionary, Lv (2019) transformed the recognition of entity words into the problem of patent term extraction, and inputted multiple features of words into the BiLSTM-CRF model for entity extraction. Wu et al. (2020) used words as input units in term extraction, which proved that the performance of character embedding was better than that of word embedding in the framework of deep learning (Meng et al., 2019). Luo (2019) uses BiLSTM-CNN model to identify entities in wetland literature, and applies BiLSTM depth model based on semantic role tagging to wetland literature data relation extraction. The sequence tagging algorithm solves the problem that it is difficult to obtain entity words manually and the workload is heavy, so this research will also use the sequence tagging algorithm to extract information from patent texts.

3 Method

3.1 Pre-trained Model

BERT is a pre-trained model based on a bidirectional encoder from Transformer that proposed by the Google team in 2018 (Vaswani et al., 2017). It can learn the grammar, the semantics and inter-sentence relationship of texts through two pre-trained tasks of Masking Language Model (MLM) and Next Sentence Prediction (NSP). After the pre-trained model, the text is mapped into a vector that can be recognized by the computer. Compared with the traditional text mapping method, the vector obtained by the pre-trained model is dynamic. Dynamic word vectors can represent different meanings of words in different contexts, and solve the problem that a word has multiple meanings.

The structure of BERT is shown in Fig. 1.

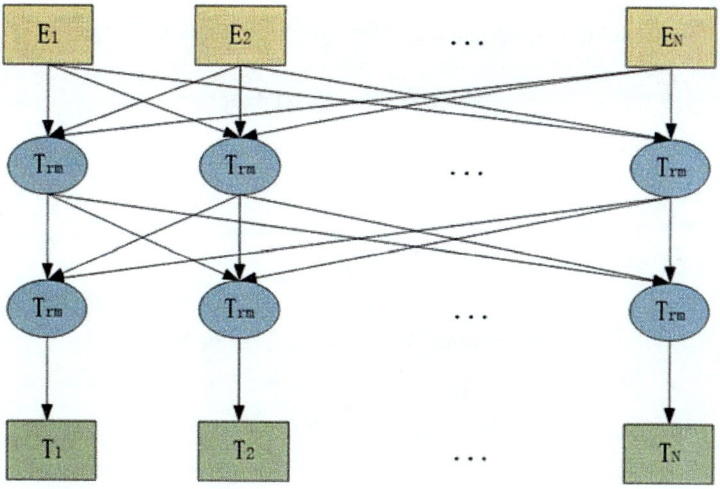

Fig. 1 BERT Model Structure

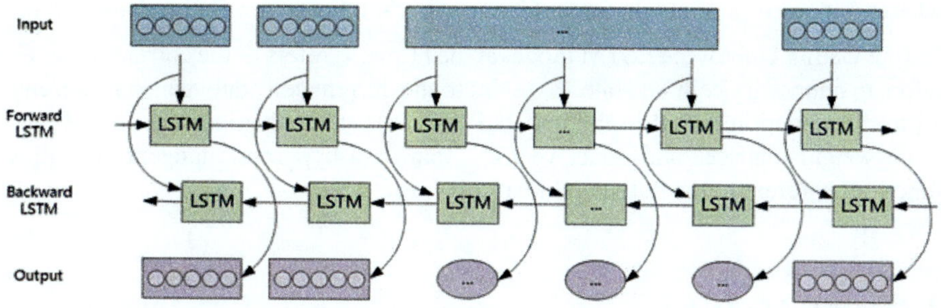

Fig. 2 BiLSTM model structure

3.2 *BiLSTM Model*

BiLSTM is Bi-directional Long Short-Term Memory. LSTM is a model based on memory cells and a gate mechanism. In the Bidirectional Long Short Memory Neural Network (BiLSTM), it mainly contains two reverse LSTM models: the forward LSTM model can store the text information from the previous part, and the backward LSTM model can store the text information behind. The BiLSTM model can use context information to extract features from text data. By analyzing the data set, the semantic information of the context is not only important for the labeling of entity words, relation words, and patent features, but also used for the relationship matching between entities. BiLSTM information makes the model more accurate prediction of the current position and improves the accuracy of the information extraction model. The structure of BiLSTM is shown in Fig. 2.

When the model unit is updated at *t* time, the formula is as follows:

$$i_t = \sigma\left(W_i \cdot [h_{t-1}, x_t] + b_i\right) \tag{1}$$

$$f_t = \sigma\left(W_f \cdot [h_{t-1}, x_t] + b_f\right) \tag{2}$$

$$o_t = \sigma\left(W_o \cdot [h_{t-1}, x_t] + b_o\right) \tag{3}$$

$$\tilde{C}_t = \tanh\left(W_C \cdot [h_{t-1}, x_t] + b_C\right) \tag{4}$$

$$C_t = f_t * C_{t-1} + i_t * \tilde{C}_t \tag{5}$$

$$h_t = o_t * \tanh\left(C_t\right) \tag{6}$$

where, i_t, f_t, o_t represents the Input Gating Unit, the Forgotten Gating Unit and the Output Gating Unit of the LSTM model at the *t* time. x_i refers to the current input. $\tilde{C}_t$ refers to temporary cellular state. C_t refers to the current cell state, which represents all the information stored in the updated cell. σ is the activation function. W and b are weight matrices and offset vectors. Tanh is a hyperbolic tangent activation function. h_t refers to the output value of this layer.

3.3 CRF Model

The conditional random field (CRF) can output a set of probability values, and the model can judge the label of the character according to the size of the probability value. The CRF layer can make up for the shortcomings that the BiLSTM layer cannot handle the relationship between adjacent sequence values. This paper uses the CRF layer after BiLSTM to optimize the sequence labeling results and output the tag value with the highest probability value corresponding to the word sequence.

The specific description is:

$$s(X, Y) = \sum_{i=0}^{n} A_{y_i, y_{i+1}} + \sum_{i=1}^{n} P_{i, y_i} \tag{7}$$

$$p(Y|X) = \frac{e^{s(X,Y)}}{\sum_{\tilde{Y} \in Y_X} s\left(X, \tilde{Y}\right)} \tag{8}$$

$$\log(p(Y|X)) = s(X, Y) - \log\left(\sum_{\tilde{Y} \in Y_X} s\left(X, \tilde{Y}\right)\right) \tag{9}$$

$$Y^{*} = \arg\max_{\widetilde{Y} \in Y_X} s\left(X, \widetilde{Y}\right) \tag{10}$$

where, $P_{i,j}$ is the probability that the word i is mapped to the label j. $A_{i,j}$ represents the probability that tag i is transferred to tag j. $\widetilde{Y}$ represents the real sequence of tags. Y_X represents all possible tag sequences for the input sequence X.

4 Experimental Setup and Results

4.1 Data Set

This article takes the field of new energy vehicle battery technology as the research object, and uses the search method of keyword to obtain 3029 experimental data from CNKI China Patent Database. We mainly extract information from patent abstracts. In order to reduce the noise data, the experimental text is segmented. The specific processing method is as follows: First, using semicolon or period as text segmentation point. Then use the comma as the segmentation point to split the previously segmented sentence. Finally, we judge whether there is a noun structure before the verb in the segmented text. If there is no noun structure before the verb in the text, it is considered that the sentence shares the subject word with the previous sentence and is merged with the previous sentence.

This paper randomly selects 175 of the 2560 patent texts that remaining after cleaning as experimental data, and obtains a total of 6829 experimental corpus after sentence segmentation processing. It is divided into training set and test set at a ratio of 8:2. The statistics of the dataset are shown in Table 1.

4.2 Triad and Feature Extraction Experiment

4.2.1 Data Annotation

This paper uses the BIO labeling method to label the subject, object, relational words, and features. "B" represents the beginning of a word, "I" represents the middle or end of a word, and "O" represents a part that does not belong to any

Table 1 Dataset statistics table

Dataset	Number of texts	Number of characters
Training set	5367	37,276
Test set	1462	10,037

Table 2 Experimental parameters

Parameter name	Value
Number of hidden layer nodes	100
Max sequence length	128
Learning rate	0.001
Optimization algorithm	Adam
Batch size	8
Drop out	0.5
Epoch	150
L2 regularization coefficient	0.006

label. In this article, we have nine types of tags: "B-SUBJ", "I-SUBJ", "B-OBJ", "I-OBJ", "B-PRE", "I-PRE", "B-CHAR", "I-CHAR", "O".

4.2.2 Parameter Settings

This experiment based on the TensorFlow framework, the parameter settings in the training process of this experiment are shown in Table 2.

4.2.3 Comparison Experiment Settings

Based on the previous theoretical research, this part sets up the following comparative experiments for triples and feature extraction:

1. BiLSTM-CRF: In recent years, the combination of BiLSTM and CRF has gradually become the mainstream model in sequence annotation algorithms. We choose its results as a baseline for patent information extraction.
2. BERT-BiLSTM-CRF: We added BERT as a pre-trained model on the basis of BiLSTM-CRF model to analyze the impact of the pre-trained model on the experimental results.

4.2.4 Experimental Results and Analysis

This experiment takes precision rate (P), recall rate (R) and F1 as evaluation indicators, and takes the average of five experiment results as the result.

The experimental results are shown in Table 3.

Experimental results show that the performance of the BERT-BiLSTM-CRF model is better than that of the BiLSTM-CRF model, which prove that the research method proposed in this paper is effective. F1 value increased by an average of 3.85%. At the same time, because the pre-trained model can capture more text features and semantic information, the recognition rate of patent features has increased the most significantly.

Table 3 Results of information extraction model

Entity Type	Experiment	P (%)	R (%)	F1 (%)
Subject	BiLSTM-CRF	79.30	77.60	78.42
	BERT-BiLSTM-CRF	**84.81**	**80.00**	**82.04**
Relationship	BiLSTM-CRF	75.37	71.22	73.23
	BERT-BiLSTM-CRF	**77.52**	**73.17**	**75.11**
Object	BiLSTM-CRF	85.63	79.74	82.58
	BERT-BiLSTM-CRF	**90.12**	**83.22**	**86.52**
Feature	BiLSTM-CRF	67.03	79.67	73.29
	BERT-BiLSTM-CRF	**76.74**	**82.01**	**79.26**

Table 4 Dataset statistics table

Dataset	Number of triples	Number of correct triples	Number of wrong triples
Training set	3610	2063	1547
Test set	902	500	402

4.3 Data Post-processing Experiment

4.3.1 Data Pre-processing

The purpose of this experiment is to obtain the correct triples from candidate triples. It is necessary to preprocess the data obtained in the previous process. We put the extracted subject words, relation words, and object words into sets named S, P, and O respectively. At this time, we set E as the set of candidate triples. The elements in E are formed by permutation and combination of elements in the set S, P, and O. For example, $E_1 = (S_1, P_1, O_1)$. After data preprocessing, this paper obtains a total of 4512 candidate triples, which are divided into training set and test set at a ratio of 8:2. The statistical information is shown in Table 4.

4.3.2 Experimental Settings

This experiment has been improved from two aspects of model and data input format.

The model comparison experiment settings are as follows:

1. SVM (Support Vector Machine): SVM algorithm is a classic machine learning algorithm and one of the commonly used algorithms in relation extraction tasks. We set it as a comparative experiment to compare the performance of machine learning algorithms and deep learning algorithms.
2. BiLSTM: This model is a commonly used classification model in RNN models, which can learn the vector context semantic information of long sequences of text.

3. BERT-BiLSTM: We combine the BERT pre-trained model with BiLSTM to enhance the model's representation of semantic features and improve model performance.

We have verified the effects of the following three data labeling methods on the experimental results:

1. Annotation 1: We treat the triplet as a short text to enhance the semantic representation of the triplet. This method inserts a special symbol between the triple and the patent text, and inputs it into the model as a whole.
2. Annotation 2: This method inserts special symbols between entity words, relation words and patent texts, and at the same time replaces the entity words and relation words with other special symbols at corresponding positions in the patent text.
3. Annotation 3: Combining the above two methods, this article proposes a third annotation method, which can not only enhance the semantic representation of triples, but also annotate location information in text.

4.3.3 Experimental Results and Analysis

Since the positive and negative samples are relatively balanced, this experiment uses the accuracy rate (Acc) as the evaluation standard, which represents the proportion of the number of samples that are correctly predicted to the total number of samples. The experimental results are shown in Table 5.

Based on the experimental results, this paper draws the following two conclusions: First, from the perspective of data labeling, this paper treats triples as a short sentence that can enhance semantic representation, and the proposed data labeling method is effective and can improve the model performance. Second, from the perspective of the model, the effect of the deep learning algorithm is significantly better than that of the traditional machine learning algorithm. After adding the pre-trained model, the model performance has improved, combined with the third After data labeling, the model accuracy rate can reach the optimal value of 93.75%.

Table 5 Results of data post-processing

Model	Annotation 1 (%)	Annotation 2 (%)	Annotation 3 (%)
SVM	77.00	75.00	76.00
BiLSTM	74.72	89.14	90.80
BERT-BiLSTM	**88.21**	**92.37**	**93.75**

5 Conclusion and Future Work

This paper selects patent texts in the field of new energy vehicle battery technology as the research object, proposes an information extraction experimental process based on sequence labeling algorithm and semantic matching, and verifies the feasibility and effectiveness of the experimental process. The experimental process integrates the deep learning algorithm BiLSTM, CRF, and pre-trained model. In the absence of an available domain knowledge base and artificial dictionary, it can automatically identify patent entity relationship triples and feature information. This research improves the efficiency of patent information extraction, saves the time and cost of information extraction, and provides a foundation and technical guarantee for building a knowledge model in the field of new energy vehicle battery technology.

However, this article still has room for improvement. Firstly, the BERT model has a huge number of parameters, which requires a lot of training resources and training time. In the future, we will further explore ways to improve the efficiency of model training. Secondly, future research needs to further increase the amount of data and improve the accuracy of the model. Finally, this article only conducts experimental and applied research on patent texts in the field of new energy battery technology. The proposed experimental process needs to be further tested in terms of universality and expand the field of adaptive research.

References

Banko, M., & Etzioni, O. (2018). The tradeoffs between open and traditional relation extraction. In *Proceedings of ACL-08: HLT* (pp. 28–36).

He, Y., Lv, X., & Liu, X. (2017). Extract non-taxonomic relations between ontological concepts from Chinese patent documents. *Computer Engineering and Design, 38*(01), 97–102.

Luo, Y. (2019). *Research on wetland entity identification and open relationship extraction*. Beijing Jiaotong University.

Lv, X. (2019). *Research on the construction of Chinese patent knowledge graph*. Beijing University of Science and Technology Information.

Madani, F., & Weber, C. (2016). The evolution of patent mining: Applying bibliometrics analysis and keyword network analysis. *World Patent Information, 46*, 32–48.

Meng, Y., Li, X., Sun, X., et al. (2019). Is word segmentation necessary for deep learning of chinese representations. *arXiv preprint arXiv*, 1905.05526.

Niklaus, C., Cetto, M., Freitas, A., et al. (2018). A survey on open information extraction. *arXiv preprint arXiv*, 1806.05599.

Rao, Q., Wang, P., & Zhang, G. (2015). Text feature analysis on SAO structure extraction from Chinese patent literatures. *Acta Scientiarum Naturalium Universitatis Pekinensis, 51*(02), 349–356.

Shu, Z., Dequan, Z., Xinchen, H., et al. (2015). Bidirectional long short-term memory networks for relation classification. In *Proceedings of the 29th Pacific Asia conference on language, information and computation* (pp. 73–78).

Stanovsky, G., Michael, J., & Zettlemoyer, L. (2018). Supervised open information extraction. In *Proceedings of the 2018 Conference of the North American Chapter of the Association for Computational Linguistics: Human Language Technologies* (Vol. 1 (Long Papers), pp. 885–895).

Vaswani, A., Shazeer, N., Parmar, N., et al. (2017). Attention is all you need. *Advances in Neural Information Processing Systems*, 5998–6008.

Wu, J., Cheng, R., & Hao, H. (2020). Automatic extraction of Chinese terminology based on BERT embedding and BiLSTM-CRF model. *Journal of the China Society for Scientific and Technical Information, 39*(04), 409–418.

Wu, X., & Wu, B. (2017). The CRFs-based Chinese open entity relation extraction. In *2017 IEEE Second International Conference on Data Science in Cyberspace (DSC)* (pp. 405–411).

Yan, C., Fu, X., Wu, W., et al. (2019). Neural network based relation extraction of enterprises in credit risk management. In *2019 IEEE International Conference on Big Data and Smart Computing (Big Comp)* (pp. 1–6).

Zeng, D., Liu, K., & Lai, S. (2014). Relation classification via convolutional deep neural network. In *Proceedings of COLING 2014, the 25th International Conference on Computational Linguistics: Technical Papers* (pp. 2335–2344).

Zhang, Y., Lv, X., Shen, Y., & Xu, L. (2019). Chinese patent entity relation extraction based on subject action object structure. *Computer Engineering and Design, 40*(03), 706–712.

Electric Power Personal Accident Characteristics Recognition Based on HFACS and Latent Class Analysis

Zhao Chufan, Mi Chuanmin, and Xu Jie

1 Introduction

Modern electric power industry is an industry equipped with modern science and technology. It is a highly centralized and unified socialized large-scale production system as far as the conversion, transmission and distribution of power energy are concerned. It is a dynamic and complex system involving many factors. In the process of electric power production power supply must be continuous, it's network is quite complex and power storage is difficult, which means that once the large power grid accident occurs, it is likely that uncontrollable chain reaction will happen, and the consequences are so serious that other industries can't match. Therefore, we must attach great importance to the improvement of safe production of power enterprises and reduce the number of power accidents.

According to the nature of the accidents, power production accidents can be divided into three categories: personal injury accidents, equipment damage accidents and power grid collapse accidents. And personal injury accidents account for the majority of the total accidents. With the development of science and technology, the level of automation and standardization of power system is constantly improving, and the proportion of equipment failure in accident causes is becoming smaller and smaller. Some scholars have proposed that in all the causes of industrial accidents, the proportion of human factors has increased from 20% to 80% (Hollnagel, 1993). Therefore, it is very important to analyse the occurrence

Z. Chufan · M. Chuanmin (✉)
College of Economics and Management, Nanjing University of Aeronautics and Astronautics, Nanjing, China
e-mail: cmmi@nuaa.edu.cn

X. Jie
Wuxi Power Supply Branch of State Grid, Jiangsu Electric Power Co., Ltd., Wuxi, China

H. Yang et al. (eds.), *AI and Analytics for Public Health*, Springer Proceedings in Business and Economics, https://doi.org/10.1007/978-3-030-75166-1_22

of power personal accidents from the perspective of human factors to improve the capacity of producing safely of power enterprises.

Up to now, many scholars have discussed the occurrence of power accidents from the perspective of human factors. Wu established the influencing factor model of human factor safety in electric power enterprises, verified the model with structural equation model, and obtained the correlation coefficient between each factor and safety stress (Wu, 2013). Li used HFACS model and grey correlation analysis method to identify the hidden hazard sources of a power grid enterprise, and made a scientific and accurate judgment on its risk degree (Li et al., 2013). Zhang discussed the relevance and significance of the factors in the four levels of HFACS framework of power accident by chi square test and odds ratio analysis, and found the concrete lack of management at the organizational level and the causes of the emergence of unsafe behaviour (Zhang, 2016). Although the existing researches on human factor identification in power accidents is relatively sufficient, few researches have been conducted on the characteristic patterns of accidents and the correlation of causative factors of specific types of accidents.

This paper will continue to use the HFACS model which has been widely used in safety accident investigation and make appropriate improvement. Based on the qualitative analysis of human factor identification, the latent class analysis method is used to identify the hidden characteristics in accidents, and to qualitatively measure the prominent causative factors of different types of accidents and the correlation between factors. According to the results, the management countermeasures and suggestions under different circumstances will be put forward.

2 Methodology

2.1 *Modified Human Factors Analysis and Classification System for Electric Power Personal Accidents*

Human factors analysis and classification system (HFACS) stems from Reason's Swiss Cheese Model. It is a systematic tool for analyzing and investigating human errors of accidents. It is proposed by Scott A. shappell, an expert from the Federal Aviation Administration, and Douglas A. Wiegma from the University of Illinois (Wiegmann & Shappell, 2001). HFACS divides human factors into four levels of failures. From top to bottom, the four levels are organizational influences, unsafe supervision, preconditions for unsafe acts and unsafe acts. Among the four levels of failures, unsafe acts are the direct factors leading to accidents, so the level is called explicit error. The other three levels of failures are indirect factors leading to accidents, so they are called implicit errors. The preconditions of unsafe acts refer to the subjective and objective conditions that lead to unsafe acts, and unsafe supervision and organizational influences are the root causes of accidents. The influence factors of each level in the model are subdivided into several specific influence factors. Although the model seems to have causality from top to bottom, there is no strict domino effect in the framework, and the four levels can completely cause accidents respectively (Cao et al., 2013). Moreover, HFACS model makes up for the one sidedness of classical accident-causing theories such as accident causation theory of Heinrich, the theory on unexpected release of energy and trace intersecting theory, and comprehensively analyzes the causes of accidents from four aspects of human, machine, management and environment. From the perspective of human factors, HFACS model does not stay on the analysis of human errors, but goes deep into the organizational factors in the root. HFACS is so flexible and comprehensive that we can construct index system on the base of it before subsequent analysis.

HFACS is an open and systematic tool designed under the background of analyzing aviation safety accidents. In different fields, its manifestation should follow the principles of being close to industry background and easy to understand (Chen, 2014). Therefore, this paper will modify the original model according to the characteristics and operation mode of the electric power industry, the general causes of accidents involved in accident reports and the improvement ideas of Zhang (Fu & Zhou, 2016) and Liu (Zhang & Fu, 2017). The framework of modified HFACS is shown in Fig. 1. The examples of connotation of four levels are described in detail below.

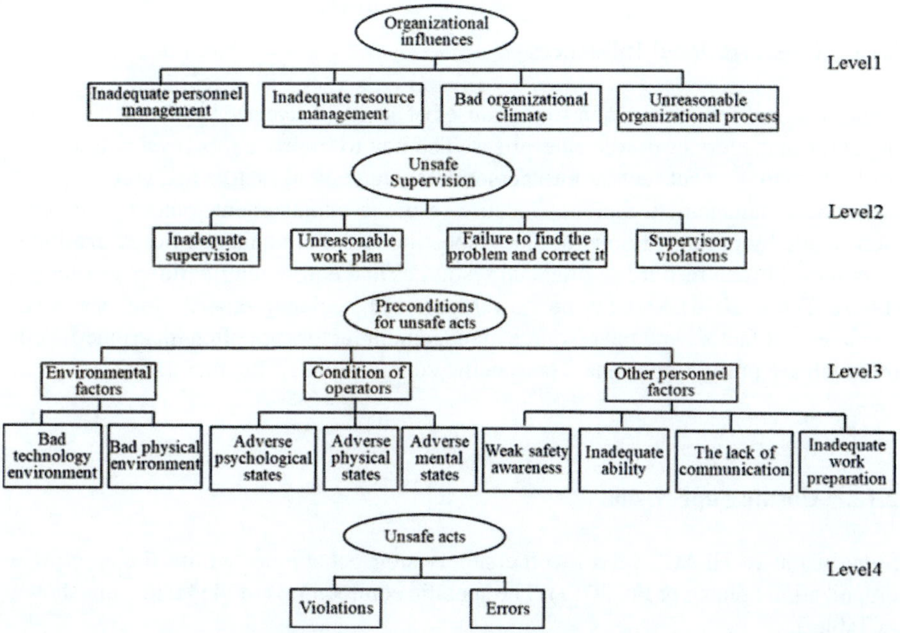

Fig. 1 Framework of modified HFACS

Table 1 Connotations of factors in the level of organizational influences in HFACS

Human factors	Connotation examples
X_1 Inadequate personnel management	Unreasonable employment of people who has inadequate qualifications or professional abilities; Inadequate training and education; Unreasonable management of personnel employment, dismissal, promotion, qualification examination, reward and punishment.
X_2 Inadequate resource management	Inadequate procurement, maintenance, distribution and fault management of equipment and materials; Unreasonable allocation of funds; Inadequate investment; Unreasonable equipment installation (such as device violation, lack of protection device or the device defect was not found).
X_3 Bad organizational climate	The lack of safety awareness; Lax safety management atmosphere; Chaotic site management; Inadequate implementation of safety production responsibility; Inadequate attention to safety in ideology; Random operation on site.
X_4 Unreasonable organizational process	The enterprise system was not perfect; There were defects in the process specification (such as loopholes in the acceptance and approval system); The rules and regulations were not implemented thoroughly; The standardized operation was mere formality.

2.1.1 Organizational Influences

Organizational influences refer to the causes of accidents caused by wrong decisions of senior managers in power enterprises. The key to improve this level is to divide resource management factors into personnel management factors and resource (not personnel) management factors. Because resource management refers to the decisions made by organization about the allocation and maintenance of organizational resources (human resources, financial resources and equipment/facilities resources) (Fu & Zhou, 2016).And in the power personal accident reports, the personnel management factors and the resource management factors are often mentioned, both of which are pretty important. The specific connotations of the factors are shown in Table 1.

2.1.2 Unsafe Supervision

Supervision in HFACS refers to the supervision behaviours within the enterprise organization (Zhang & Fu, 2017). The specific connotations of the factors are shown in Table 2.

Table 2 Connotations of factors in the level of unsafe supervision in HFACS

Human factors	Connotation examples
X_5 Inadequate supervision	The lack of supervision or the lack of whole process supervision; Supervision was mere formality; Supervisors did not urge operators to implement rules and regulations and implement necessary actions; Supervisors lacked sense of responsibility and prestige; Supervisors have not been trained.
X_6 Unreasonable work plan	Unreasonable risk management, personnel allocation and production schedule; Improper work arrangement for the complexity, time, intensity and site change of the operation project.
X_7 Failure to find the problem and correct it	The supervisors did not find improper behavior or unsafe trend; Inadequate inspection of hidden danger; The identification of dangerous points was not comprehensive; The supervisor failed to correct a known problem.
X_8 Supervisory violations	The supervisor was not on duty; The supervisor worked without a certificate; The supervisor violated the rules and commanded against the rules; The supervisor directly participated in the work or left without permission when the task was not completed.

2.1.3 Preconditions for Unsafe Acts

The preconditions for unsafe acts are the direct causes of unsafe acts, including environmental factors, operator status and other personnel factors. In this paper, other human factors in the original model are re-divided into four factors to fit the real causes of power personal accidents. The specific connotations of the factors are shown in Table 3.

2.1.4 Unsafe Acts

This paper remains two factors in the level of unsafe acts: violations and errors. Because most of the violations are out of consciousness and most of the errors are out of unconsciousness in power personal accidents, we no longer subdivide the two factors. The specific connotations of the factors are shown in Table 4.

2.2 *Latent Class Analysis (LCA)*

Latent class analysis (LCA) is a dichotomous attitude measurement method proposed by Lazarsfeld and Henry. A latent class model consists of manifest variables and latent variables. The different levels of manifest variables refer to the categories of subjects in actual measurement. The different levels of latent variables refer to

Table 3 Connotations of factors in the level of preconditions for unsafe acts in HFACS

Human factors		Connotation examples
Environmental factors	X_9 Bad technology environment	Equipment failures; Defects in the workplace.
	X_{10} Bad physical environment	Unfavorable working environment about weather, geology, temperature, lighting, noise, vibration, etc.
Condition of operators	X_{11} Adverse psychological states	The operator was lack, hasty, nervous and flustered or had fluke mind.
	X_{12} Adverse physical states	Bad states such as fatigue, myopia, color blindness and drunkenness.
	X_{13} Adverse mental states	The operator was under great pressure or absent-minded; The lack of thinking ability or operation accuracy.
Other personnel factors	X_{14} Weak safety awareness	The operator was in lack of safety awareness and risk awareness.
	X_{15} Inadequate ability	The operator was in lack of learning ability, professional ability, execution ability and the ability of hazard identification.
	X_{16} The lack of communication	The team was in lack of team cooperation; Information communication was not smooth and clear; There were problems on scheduling or operation.
	X_{17} Inadequate work preparation	There was no site investigation, pre-shift meeting, shift handover or disclosure of technology before work; The operator was not familiar with the operation site and did not take safety measures; Warning signs were not placed; Drawings were not updated.

Table 4 Connotations of factors in the level of unsafe acts in HFACS

Human factors	Connotation examples
X_{18} Violations	The acts against rules and regulations by operators.
X_{19} Errors	The operators' accidental collision, misjudgment, identification error; The operator failed to clear the site and evacuate personnel in time.

the different latent classes obtained after estimation (Qiu, 2008). The basic idea of LCA is that the association between manifest variables of a group can be explained by several mutually exclusive latent classes. Through the estimation process, each sample will be divided into a latent class, and the manifest variables will be locally independent. In this sense, latent class analysis is a clustering method, but its principle is different from the traditional clustering method. It is mainly based on the probabilistic model and uses the estimation and comparison of probability to classify samples. The most important feature or advantage of latent class analysis is that individuals in the same class will be homogeneous, while individuals from different classes will be heterogeneous (Vermunt & Magidson, 2014). Compared with the traditional clustering methods, latent class analysis uses more standardized and scientific standards to judge the number of categories and the validity of the model, and the requirements of the data distribution are more relaxed (Jiao et al., 2010).This study uses the exploratory latent class analysis method, and its main analysis steps include: probabilistic parameterization, parameter estimation and model fitting, classification.

Step1: probabilistic parameterization. The most breakthrough theory of latent class model is to transform the probability of variables into parameters. LCA involves two different parameters, latent class probability and conditional probability. Suppose that there are three manifest variables *A*, *B* and *C* with *I*, *J* and *K* levels respectively, which are not independent of each other. The process of latent class analysis is to finding a latent variable *X* with *T* latent classes which can not only explain the relationship among *A*, *B* and but also maintain the local independence of three manifest variables in each class of *X*. The mathematical model is:

$$\pi_{ijk}^{ABC} = \sum_{t=1}^{T} \pi_t^X \pi_{it}^{\overline{A}X} \pi_{jt}^{\overline{B}X} \pi_{kt}^{\overline{C}X} \tag{1}$$

π_{ijk}^{ABC} represents the joint probability of a latent class model. π_t^X is the latent class probability, which represents the probability of the latent variable *X* at the level of *T* when the manifest variables are locally independent, That is the probability of object randomly selected from the sample belonging to the latent class *T*. The total probability of every latent class is 1

$$\sum_{\mathrm{t}} \pi_{\mathrm{t}}^{\mathrm{X}} = 1.00 \tag{2}$$

$\pi_{it}^{\overline{A}X}$ is the conditional probability, which represents the probability of objects staying in the level of number *i* of manifest variable *A* in latent class number t. Because the levels of each latent variable are independent of each other, the sum of conditional probabilities of each manifest variable is 1.

$$\sum_t \pi_{it}^{\overline{A}X} = \sum_j \pi_{jt}^{\overline{B}X} = \sum_k \pi_{kt}^{\overline{C}X} = 1.00 \ (3)$$

Step2: parameter estimation and model fitting. Usually, the method of maximum likelihood (ML) is used to estimate the parameters, and Expectation-Maximization method (EM) and Newton-Raphson method (NR) are used for model fitting in the iterative process. When examining the degree of model adaptation, the researchers usually determine the number of classes on the basis of Pearson's chi-squared test, likelihood-ratio chi-square test, Akaike Information Criterion (AIC) and Bayesian Information Criterion (BIC). The smaller the values of AIC and BIC, the better the model adapts. When the sample capacity reaches more than thousands of people, we determine the number of classes mainly according to BIC, otherwise AIC.

Step3: classification. The final step of latent class analysis is to classify all the objects into appropriate latent classes. That is to create a new class variable to illustrate the posterior class attribute of every object. The principle of classification is Bayesian theory. The calculation procedure of classification is as follows

$$\hat{\pi}_{tijk}^{\overline{X}ABC} = \frac{\overline{\pi}_{ijkt}^{ABCX}}{\sum_{t=1}^{T} \hat{\pi}_{ijkt}^{ABCX}} \tag{4}$$

3 Model Construction and Implementation

3.1 Procedure of Electric Power Personal Accident Characteristic Pattern Recognition Model

Step1: construct data set based on HFACS. In order to convert the collected accident reports from text to data, we construct a data set to quantify the accidents. The method is analyzing the text of each accident report and comparing the causes of the accident with 19 factors in HFACS model one by one. If the accident involves a certain factor, number "1"will be recorded. If the accident does not involve a certain factor, number"0*//* will be recorded. If the number of accidents collected is *N*, a 0–1 matrix with N rows and 19 columns will be formed. In the matrix, each row represents an accident and each column represents a human factor.

Step2: cluster accidents using LCA. 19 factors in HFACS are taken as manifest variables of LCA. Each manifest variable has two levels that are "0" and "1". LCA generates latent variables by establishing probabilistic model that describes accident samples, which is the latent class hidden in accident samples. The clustering process is as follows: at the very beginning, suppose that the number of accident class is 1, that is to say, all the accidents belong to one class. Then the number of accident classes is gradually increased, and at the same time, the parameters are estimated. Then determine the number of classes when the model fitting is optimal according to AIC, BIC and other indicators. After determining the number of the model classes, each accident t sample will be classified into different classes according to the posterior probability. This step can be completed by LatentGDLD.

Step3: name each class and discuss their characteristics. After determining the number of accident classes, focus on the clustering results of each class, especially the latent class probabilities and conditional probabilities. Latent class probabilities reflect the number of accidents in each class, and conditional probabilities reflect the significant level of each manifest variable (19 factors) in each latent variable (accident classes). The higher the conditional probability, the more likely the factor will occur in the class. We name the classes according to their characteristics and put forward the corresponding management countermeasures.

3.2 Empirical Analysis

3.2.1 Data Sources

In this study, 173 personal injury and death accidents from the "Compilation of National Electric Power Accidents and Electric Power Safety Incidents" compiled by the power safety supervision department of the national energy administration are selected. They all happened from 2015 to 2018.By focusing on the accident description part and accident causes part of each accident report and comparing the text information with 19 influencing factors, a 0–1 matrix with 173 rows and 19 columns will be formed.

3.2.2 Analysis of Model Fitting and Determination of Class Number

Use LatentGOLD4.5 to analyze the latent class of 173 accidents. The fitting results of the model are shown in Table 5. The smaller the likelihood-ratio chi-squared statistic (G^2), AIC and BIC, the better the model fits. When the number of sample is less than 1000, AIC index should be selected as the decision index (Qiu, 2008). Therefore, we decide to cluster the accident samples into five classes.

3.2.3 Name the Classes and Discuss Their Characteristics

According to the latent class probabilities, the numbers of accidents in five classes are 84, 31, 25, 19 and 14 respectively. According to the conditional probabilities, the characteristics of influencing factors of each class can be observed. The higher the conditional probability, the more likely the factor will occur in the class. Take 0.5 as the cut-off point of conditional probability, the factors whose conditional probabilities are higher than 0.5 are regarded as the main human factors of this class of accidents. Figure 2 shows the conditional probabilities of individual factors in the five classes.

From the perspective of whole accidents, we can get the following conclusions:

Table 5 Adaptation indexes of various models

Number of class	AIC	BIC	G^2
1	3245.7782	3305.6907	1446.9200
2	3195.2850	3318.2634	1356.4268
3	3169.6916	3355.7358	1290.8334
4	3165.3383	3414.4484	1246.4802
5	3162.9609	3475.1364	1204.1024
6	3171.1056	3546.3473	1172.2474.
7	3177.0891	3615.3966	1138.2309

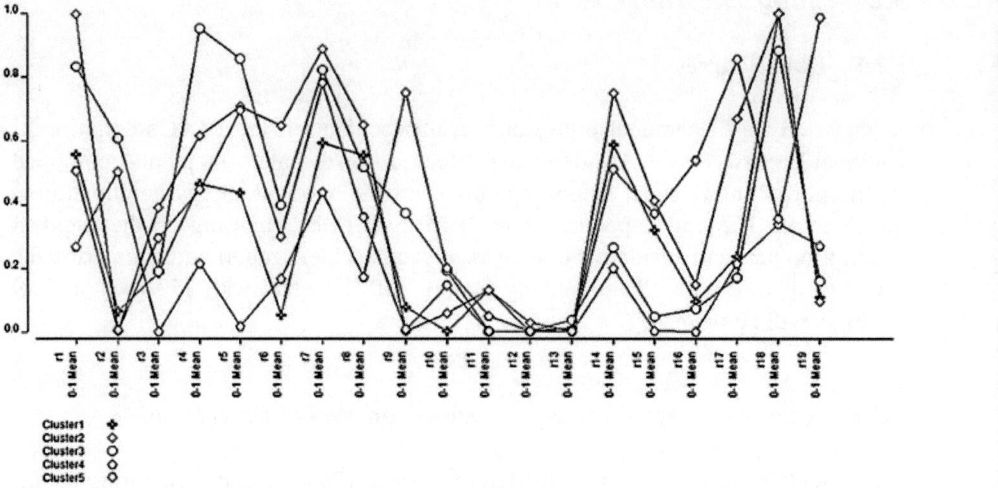

Fig. 2 Conditional probabilities of every factor in five classes

1. The conditional probabilities of three operators' states in the third level are very low. All are lower than 0.2. So the three factors make little contributions to clustering accidents. This shows that the operator's psychological, physical and mental states seldom occur in the accident reports. However, the states of the operators are important causes of accidents according to common sense. Maybe the reality is that when the accident has already occurred, it is difficult to track the psychological, physical and mental states of the operators before the accident happening. Therefore, the conditional probabilities of the three factors can't really reflect the degree they causing an accident.
2. In the first three classes of accidents, the conditional probabilities of "violations" are very high, which are higher than 0.8. And the accidents in first three classes account for 80% of the sample. It is concluded that the causation "violations" is the most important factor of power personal accidents.

From the point of view that each class of accidents has characteristics different from other classes of accidents, we name every class and discuss their characteristics:

Class1: violation accidents in which each factor has an average effect on them. Most of the conditional probabilities of the factors are in intermediate level among the five classes of accidents while the conditional probability of "violations" is the highest reaching 0.9994. There is no prominent cause in the causes of such accidents, but this is the unique characteristics of them. This class of accidents involves 46.12% of the accidents, almost half of the sample, which shows that this class of accidents is prevention emphasis of electric power enterprises.

Class2: accidents caused by operator's weak safety awareness. The characteristics of the second class of accidents are similar to those of the third class. The conditional probabilities of factors in the first and second levels and "violations" are very high. But the obvious factor different from other class of accidents is "inadequate personnel management". Its conditional probability is as high as 0.9951. Additionally the conditional probability of "weak safety awareness" is significantly higher than that of other classes of accidents. From the perspective of organizational influences, the reason for such accidents is the lack of safety education and training for employees. And from the perspective of unsafe supervision, the reason why this kind of accidents happens is that the supervisors fail to find out and correct the unsafe acts of employees in time. Both of the two reasons lead to weak safety awareness of employees, and eventually lead to violations.

Class3: accidents caused by unreasonable organizational process. Although the conditional probabilities of several factors in the first two levels are high, the most prominent factor is "unreasonable organizational process ", and its conditional probability is as high as 0.9524. The probabilities "inadequate personnel management" and "inadequate resource management" occurring are also high, which indicates that there are huge loopholes at the root of organizational influences level. This situation will eventually lead to the operators' violations in constructions.

Class4: accidents directly caused by bad technology environment. The most prominent factor in this class is bad technology environment, and conditional probabilities of other factors are low, especially violations and errors in unsafe acts level. It should be pointed out that the conditional probability of inadequate resource management is 0.5039. Although it is not prominent in the horizontal comparison of five classes of accidents, it is an important cause factor in the vertical comparison of factors about this kind of accidents. This information just explains how "bad technology environment" occurs in this kind of accidents.

Class5: accidents caused by errors. The outstanding feature of this kind of accidents is that the conditional probability of errors is very high, reaching 98.5%. Observing other factors, it is found that the conditional probability of "inadequate supervision", "fail to find the problem and correct it" and "inadequate work preparation" are also high. These situations just explain the causing process to operator's errors. First of all, because the supervisor did not implement adequate supervisions and fail to discover improper acts of relevant personnel and correct them in time, the hidden dangers of inadequate work preparation were left over. Then these factors lead to errors in the construction of the operators.

3.2.4 Management Countermeasures and Suggestions

1. In view of the situation that accident reports can't reflect the states of operators, it is suggested that the power company should strengthen humanistic care for employees, care about their physical and mental health, pay attention to their family difficulties, and create a good working and living environment. On the other hand, when the accident has occurred, the accident investigation should be

as deep as possible, Do not stay on the superficial and direct causes, or blindly attribute the causes to the organization or supervisor.

In view of the frequent occurrence of operators' violations, it is suggested that power enterprises strengthen the standardized management of on-site operations, intensify the crackdown on violations, strengthen the control of high-frequency violations, and effectively eliminate the randomness of field operations.

2. In view of the fact that nearly half of the total accidents are "violation accidents in which each factor has an average effect on them", this paper reminds the enterprises to realize that violations of operators are the most direct and prominent factors leading to the accidents. Whether from the perspective of organizational influences or supervision, enterprises should take the elimination of violations as the core task of improving security capability.

In view of the weak safety awareness of operators can easily be caused by inadequate personnel management, this paper reminds the organization of electric power enterprises to strengthen the safety education and improve the working skills and safety awareness of employees, and heighten staff's awareness of abiding by rules and regulations.

In view of the serious problems in the level of organizational influences, it is reminded that the organizational influencing factors are the root causes of electric power accidents, and the level of safety capability of an electric power enterprise largely depends on the management level of the organization. It is suggested that electric power enterprises should pay attention to the implementation of the main responsibility of safety production and constantly improve the safety production system and measures.

In view of the situation that some accidents may be directly caused by bad technology environment, it is suggested that the power enterprises should strengthen the management of equipment failure and quality, and improve the inspection methods to equipment and tools in dangerous areas.

In view of the fact that some accidents are caused by the operator's errors, it is suggested that the supervisor should refine the division of supervision responsibilities of each operation area and strengthen the on-site daily supervision and investigation on hidden dangers. Supervisors should focus on work preparations such as the disclosure of technology and the notification of hazard sources.

4 Conclusion

1. The classic HFACS model has been modified according to the characteristics of the power industry and the general causes of power personal accidents to construct an index system. We have described the specific connotations of each influencing factor in the model.

2. We have identified the causes of 173 power personal accidents using modified HFACS, used the LCA method to cluster accidents and identified the characteristics of them. In addition to the overall characteristics of accidents, the characteristics of five classes of accidents have been recognized and the internal relationships among the causes have been revealed. According to the overall characteristics and different characteristics of each class of accidents, we have put forward corresponding management countermeasures and suggestions to power enterprises.

Acknowledgements This work was funded in part by Wuxi Power Supply Branch of State Grid Jiangsu Electric Power Co., LTD. Project "Research on security management improvement service based on digital drive" (SGJSWX00AZWT1901493), and it was also funded by The National Social Science Fund of China (Grant No. 17BGL055).

References

Cao, H. F., Lv, R. L., & Huo, Z. Q. (2013). Study on civil aviation accident cause quantized analysis. *Journal of Civil Aviation University of China, 27*(4), 11–13.

Chen, Z. B. (2014). Consistency research on human factors analysis of coal mining accidents. *China Safety Science Journal, 24*(2), 145–150.

Fu, G., & Zhou, L. (2016). Study on correspondence between organizational influences in HFACS and organization behavior in 24Model. *China Safety Science Journal, 26*(11), 25–30.

Hollnagel, E. (1993). Human reliability analysis: Context and control. *Academic Press, 53*(2), 99–101.

Jiao, C., Zhang, J. T., & Guan, D. D. (2010). Latent Class Analysis of basic and comprehensive examination of postgraduates in the major of psychology from 2007 to 2009. *Journal of Chinese examinations, 4*, 145–150.

Li, Y., Jing, N., & Hong, M. L. (2013). Identification and evaluation of hidden hazard sources of human accidents in power grid enterprises based on HFACS and grey correlative method. *Journal of Safety Science and Technology, 9*(2), 157–161.

Qiu, Z. H. (2008). *Latent class modeling principles and techniques*. Beijing Education Press.

Vermunt, J. K., & Magidson, J. (2014). Latent class models for classification. *Computational Statistics & Data Analysis, 41*, 531–537.

Wiegmann, D. A., & Shappell, S. A. (2001). Human error analysis of commercial aviation accidents: Application of the Human Factors Analysis and Classification System (HFACS). *Aviation, Space, and Environmental Medicine, 72*(11), 1006–1016.

Wu, S. (2013). *Research and application of human factor safety in electric power enterprises*. Beijing Jiaotong University.

Zhang, H., & Fu, G. (2017). Comparative analysis between unsafe supervision in HFACS and the unsafe behavior in 24 Model. *Journal of Safety and Environment, 17*(2), 582–586.

Zhang, J. W. (2016). *Research on human error of power accidents based on HFACS*. University.

Sentiment Analysis Based on Bert and Transformer

Tang Yue and Ma Jing

1 Introduction

With the rapid development of the mobile Internet, users are getting closer and closer to the Internet. Netizens can quickly express their opinions and opinions on the network platform. As the main means of network interaction, text messages continue to emerge from the network platform. In the face of a large amount of text information, how to effectively filter and classify valuable text information becomes more and more important.

Mining sentiment information and opinion attitudes in short texts has important practical significance to realize the sentiment classification of texts. Text sentiment classification includes four processes: word segmentation, text representation, feature extraction, and classification.

Sentiment analysis is an application of short text classification. Short texts are short in length and have a certain degree of difficulty in extracting effective feature words; in addition, short texts are updated quickly and are massive. These characteristics of short texts cause short text classification technology to face the following difficulties:

1. Feature sparsity: the use of traditional vector space representation will cause vector space sparsity;
2. Semantic sparsity: it is difficult to effectively extract short texts using traditional methods
3. Word standardization: make the word segmentation dictionary unable to identify unregistered words, resulting in insufficiently accurate representation of the text vector;

T. Yue (✉) · M. Jing
College of Economics and Management, Nanjing University of Aeronautics and Astronautics, Nanjing, Jiangning, China

H. Yang et al. (eds.), *AI and Analytics for Public Health*, Springer Proceedings in Business and Economics, https://doi.org/10.1007/978-3-030-75166-1_23

in view of the above problems, this article has two innovations to improves on the following two aspects:

1. Combining the BERT word vector to construct text vector space, word2vec and other static representation models, their word vectors are static word vectors, which cannot solve problems such as one-time ambiguity. However, BERT word vectors can not only obtain different word vector representations for the same word according to context information, but also it can obtain richer semantic information and syntax information, effectively solving the problem of feature sparseness and semantic sparseness.
2. Improve self-attention. The initial Transformer model only weights the information of the character granularity, but ignores the words, phrases and syntax information in the text. Therefore, this paper tries to use the N-gram model to extract multi-granularity information and improve the first layer encoder of the Transformer model.

2 Related Research

The field of natural language processing involves various research directions such as semantic analysis (Hofmann, 2013), topic tagging (Socher et al., 2013), sentiment analysis (Maas et al., 2011) and machine translation (Wu et al., 2016; Hao et al., 2019). Emotion analysis is one of the most important tasks in natural language processing. Many scholars at home and abroad focus on the following two aspects: short text feature representation and extraction, optimization and improvement of classification model.

Text feature representation is the key problem of emotion analysis task. How to effectively extract the emotional features of text is of great significance to improve the effect of sentiment classification. Hinton proposed a word embedding (Distributed representation) method. The core idea of the method is to map each word into a fixed-length vector through training, and the vectors of all words will be combined to form a word vector space. Mikolov et al. (1986) integrated the word vector representation model Word2vec based on the idea of Word embedding, which can represent words with similar meanings with similar word vectors Pennington et al. (2014). Propose global vector for word representation, The Glove model essentially integrates the latest global matrix factorization and local context window methods at the time, so that both global statistical information and local window information are considered; the results also show, Glove has achieved good performance in word-related tasks. Ma et al. (2018) proposed a weighted Word2vec text classification method. First, the text is trained with word vectors, and the text keywords are divided into overlapping and non-overlapping parts by setting the word similarity threshold, and then calculating the weighted similarity of the two parts. Then use parameterized linear weighting method to calculate text similarity, which improves the classification effect.

At present, sentiment analysis methods are mainly divided into two categories, methods based on traditional machine learning and methods based on deep learning. Shah et al. (2020) designed a BBC news text classification system. The author selects and compares logistic regression, random forest and k-nearest neighbor as classification algorithms. Then test, analyze and compare these classifiers, and finally draw conclusions. The experimental conclusions show that the BBC news text classification model has obtained satisfactory results.

With the development of deep learning, neural network models are widely used in text classification tasks, which are mainly divided into three categories: Convolutional Neural Networks (CNN), Recurrent Neural Networks (RNN), and Attention-based models.

The CNN model first achieved great success in the field of computer vision. Collobert and Weston (2008) first applied CNN to natural language processing tasks such as part-of-speech tagging, named entity tagging, and semantic similarity. Kim (2014) used the CNN model to classify sentence text, and obtained good results, forming the classic model Text-CNN.

RNN is suitable for processing sequence information input. The most commonly used RNN models are LSTM and GRU. Nowak et al. (2017)) used LSTM network to classify text and proved the effectiveness of LSTM on text classification tasks. Wang et al. (2019) integrates word embedding features, word sentiment features, word weight features, etc., and builds a GRU neural network text sentiment classification model.

Attention model can distinguish the importance of different information in the text, and assign different attention weights. Yang et al. (2016)) applied the Attention model to the task of document classification, so that it can pay attention to important content and secondary content separately when constructing text expressions. Experiments show that the Attention mechanism is superior to previous text classification methods. In 2017, Vaswani et al. (2017) proposed a model completely based on the Attention structure, breaking the structure of the traditional Attention model, without using any RNN and CNN structures, and achieved great improvements in the field of machine translation. Letarte et al. (2018) applied the self-Attention mechanism to emotion classification tasks and achieved good results. Subsequently, in 2018, the Google team Devlin et al. (2018) and others proposed the BERT model based on the two-way Transformer structure, and achieved major breakthroughs in 11 natural language processing tasks.

3 Research Framework

Use the International System of Units (SI) only. Never combine SI units and CGS or other units. If you must use other units, always state the units for each quantity that you use in an equation or in a figure.

3.1 Self-Attention Model

In recent years, the attention mechanism has been widely used in various tasks of natural language processing based on deep learning. Most of the traditional Encoder-Decoder models use CNN or RNN as the basic core of the model. Although CNN can perform convolution kernel operations in parallel, It cannot capture the semantic information before and after the sentence; RNN can effectively capture the semantic information of the text sequence, but the input of each unit depends on the output of the previous unit, so it cannot be calculated in parallel. In 2017, Vaswani et al. proposed a Transformer model based entirely on the Attention structure. While efficient parallel computing, attention can be weighted to all sequence information, thereby effectively capturing the information of the entire sequence, thus improving performance at the same time, the training speed is also faster.

The essence of the self-attention mechanism is a weight mapping obtained by a series of key-value pairs through a query. Given the query of an element in the Target, by calculating the similarity between the query and the key, the weight coefficient of each key to the value is obtained, and then the value of the value is weighted and summed to obtain the final Attention weight value. The process is shown in the figure. The query passes a series of key-value weighted calculations in the Source, and finally gets the Attention-value. Its structure is shown in Fig. 1:

Calculating self-attention is usually divided into the following three steps. The first step is to calculate the Query, Key, Value matrix, and calculate the similarity between the input query and each key to obtain the corresponding weight coefficient, as shown in formula (1):

$$f(Q, K) = QK^T \tag{1}$$

The second step is to use a normalization function such as softmax to normalize the weight coefficients obtained, as shown in formula (2):

$$a_{\mathrm{i}} = sofmax\,(f\,(Q, K)) \tag{2}$$

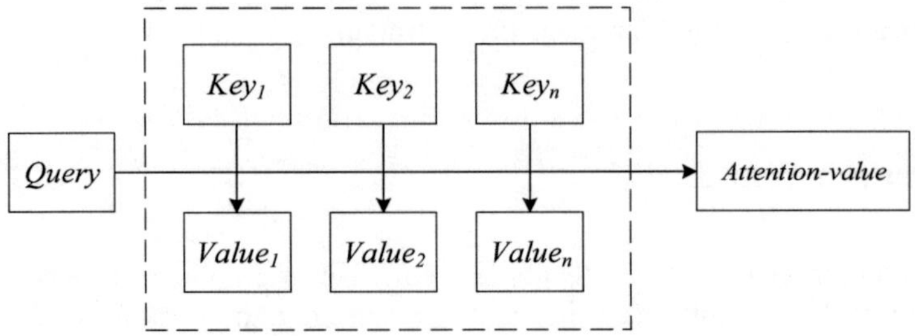

Fig. 1 Attention model

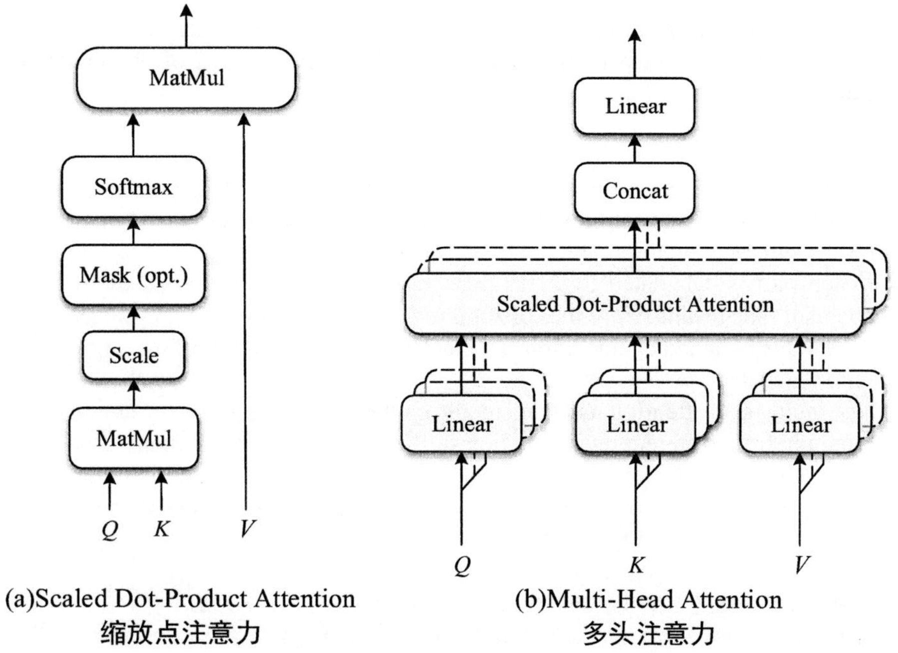

Fig. 2 Scaled Dot-Product attention (**a**) and multi-head attention (**b**)

The third step is to perform a weighted summation of the normalized weight coefficient and value to get the final attention value. The attention calculation formula is shown in formula (3):

$$Attention\,(Q, K, V) = \sum a_i V \tag{3}$$

In the article "Attention is All You Need" published by Google in 2017, a model Transformer based entirely on the attention mechanism was proposed, and it has outstanding performance in the field of neural machine translation. Its core is called the self-attention mechanism, Mainly includes the following three parts: zoom dot product attention, multi-head attention and position coding.

The structure of the scaled dot product attention is shown in Fig. 2a. The input of this structure is composed of Q-dimensional Query, Key and dimensional Value values. By calculating the dot product of Query and all Keys, each dot product value is divided by dk, and finally a softmax function is used for normalization to obtain the weight of these attention values. The calculation formula (4) is as follows:

$$Attention\,(Q, K, V) = softmax\left(\frac{QK^T}{\sqrt{d_k}}\right)V \tag{4}$$

Among them are the three matrices with dimension, and the attention weight is obtained through matrix operation, and finally the weight matrix of dimension is output.

The multi-head attention model is a multi-head mechanism built on the basis of scaling dot product attention. Its structure is shown in Fig. 2b. Query, Key, and Value will first undergo a linear transformation through a Linear layer, each time Q, K, The parameters W of the linear transformation of V are initialized randomly. Before calculating attention, the input sequence is divided into N-head equal parts, and each part is calculated by a separate head. This is the so-called multi-head mechanism. The results of N times of zooming dot product attention are spliced, and then the value obtained by linear transformation is the output result of the multi-head attention model.

The multi-head attention mechanism allows the model to jointly pay attention to information from different representation subspaces at different positions, and learn different attentions, which can effectively prevent the loss of information. While improving the accuracy of the model, it makes the model more robust. it is good. The calculation formula (5) is as follows:

$$\begin{aligned} &MultiHead(Q, K, V) = Concat(head_1, head_2, \ldots, head_n)W^O \\ &Where\ head_i = Attention\left(QW_i{}^Q, KW_i{}^K, VW_i{}^V\right) \end{aligned} \tag{5}$$

It is a projection matrix, and its projection range is. Its function is to project the three matrices of Q, K, and V to different dimensions through linear transformation, and keep the matrix dimension unchanged, the purpose is to allow different heads to learn different attention Strength makes the attention more diverse and improves the accuracy and robustness of the model.

Since the Transformer model does not have the convolution operation of CNN and the cyclic transmission of RNN, it cannot capture the position information of the sequence, that is, the input text will randomly disrupt the order of words, and eventually get the same attention result and model output, so it must be injected. Relative position or absolute position information, encode the position information of the sequence and merge it with the word vector to identify the information of the word at different positions.

Position coding is usually a vector obtained by training for a specific task. The Transformer model uses sine and cosine functions of different frequencies to construct position information, as shown in formula (6):

$$\begin{aligned} PE(pos, 2i) &= \sin\left(pos/10000^{2i/d}\right) \\ PE(pos, 2i+1) &= \cos\left(pos/10000^{2i/d}\right) \end{aligned} \tag{6}$$

Among them, pos represents the position of the word in the text, d represents the dimension of the word vector, and i represents the i-th dimension of the position vector. The formula maps the position pos information to a d position vector, adding sin variables to the even-numbered positions of the word vectors of each word, and cos variables to the odd-numbered positions. The purpose of using cos and sin exchange mapping is to enrich the position information. The value of the i-th element of is the corresponding value in the above formula.

3.2 BERT Model

Most of the current research uses word2vec or Glove to obtain the word vector of the text. It maps a word to a low-dimensional dense semantic space so that similar words can share context information, but the word vectors are all static representations and are ignored. For example, the semantic representation of bank can be bank or river bank. BERT is a new language representation model, which can be widely used in many downstream NLP tasks, such as intelligent question answering, sentiment analysis, automatic summarization, etc. One of the most commonly used is the word vector representation of BERT, which contains rich contextual information, which can effectively improve the quality of text representation, thereby improving the effect of subsequent tasks. BERT essentially uses Transformer to construct a multi-layer bidirectional Encoder network. The training process is shown in Fig. 3.

Where w_i is the vector representation of the text, and T_i is the target word vector obtained through the bidirectional Transformer structure, BERT adopts two preprocessing methods, Masked LM and Next-Sentence Prediction, which can capture the semantic representation of words and sentences respectively. Among them, "two-way" means that when the model is processing a certain word, it can use the two parts of the word before and after the position, and randomly cover

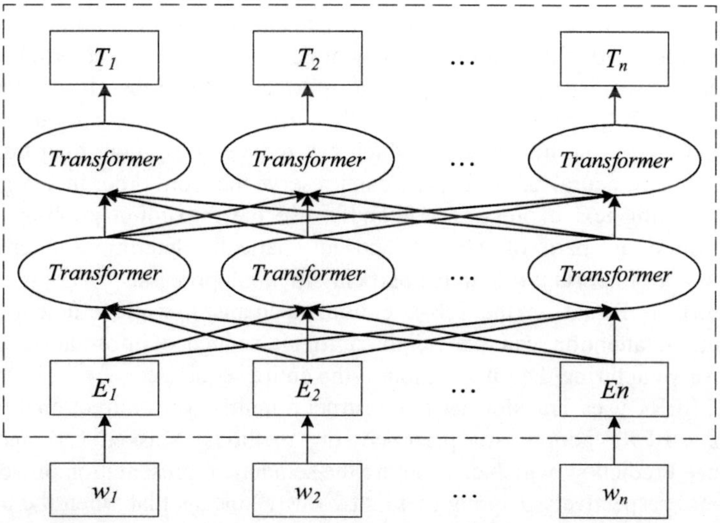

Fig. 3 BERT model

some words, forcing the model to refer to the context information when learning the expression of each word, so as to obtain rich semantic information.

4 Experimental Result and Analysis

4.1 Experimental DATASET

The experimental data are from microblog users' comments, with more than 10ksentiment tagged Weibo, and about 50 k positive and negative comments. The dataset is called Weibo_senti_100k.

In order to verify the effectiveness of the model, this paper adopts the single variable principle and designs the following multiple sets of comparative experiments.

1. TextCNN, Select TextCNN, which has a better effect in the current text classification task, as the first set of experiments. Uses the 300d word vector pre-trained by wiki corpus.
2. TextRNN, Use Bi-LSTM to build a model, word-based text vector representation, combined with bidirectional long-term short-term memory network to capture the emotional semantic information of sentence context.
3. TextRCNN, RNN is good at processing sequence structure and can take into account the context information of the sentence, but RNN belongs to the "biased model". The later words in a sentence are more important. This may affect the final classification result because it has the greatest impact on sentence classification. The word may be anywhere in the sentence. CNN is an unbiased model that can obtain the most important features through maximum pooling. However, the size of the sliding window of CNN is not easy to determine. If it is too small, it will easily cause important information to be lost. If it is too large, it will cause huge parameter space. RCNN use two-way loop structure is used to obtain contextual information, which can reduce noise more than traditional window-based neural networks, and can preserve the word order in a large range when learning text expression. Secondly, use the maximum pooling layer to obtain important parts of the text, and automatically determine which feature plays a more important role in the text classification process.
4. Transformer, Based on the self-attention mechanism, while efficient parallel computing, attention can be weighted to all sequence information, thereby effectively capturing the information of the entire sequence.
5. BERT, BERT uses Transformer to construct a multi-layer bidirectional Encoder network, BERT adopts two preprocessing methods, Masked LM and Next-Sentence Prediction, which can capture the semantic representation of words and sentences respectively. Among them, "two-way" means that when the model is processing a certain word, it can use the two parts of the word before and after the position, and randomly cover some words, forcing the model to refer to the

context information when learning the expression of each word, so as to obtain rich semantic information.
6. BERT + Transformer, BERT's word vector representation, which contains rich contextual information, can effectively improve the quality of text representation, thereby improving the effect of subsequent tasks, Then Transformer use the self-attention model to globally capture the important information of the text and weight the important features.

4.2 Evaluation

Text classification problems usually use accuracy (acc), recall (recall), precision (precision), F1 value and other indicators to measure the effect of the classification algorithm, which is an extension of the confusion matrix. The confusion matrix is shown in Table 1.

The accuracy rate acc is the proportion of the correct result of the classifier to the total observation value. The calculation formula is as follows:

$$acc = \frac{TP + TN}{TP + TN + FP + FN}$$

Precision refers to the proportion of samples that are predicted by the classifier as positive and that the prediction is correct to all samples that are predicted to be positive. The calculation formula is as follows:

$$P = \frac{TP}{TP + FP}$$

The recall rate R refers to the proportion of samples that are predicted by the classifier as positive and that the prediction is correct to all samples that are truly positive. The calculation formula is as follows:

$$P = \frac{TP}{TP + FN}$$

F-Score indicator is a new indicator that combines the two indicators of Precision and Recall. Its value ranges from 0 to 1. The formula is as follows:

Table 1 Confusion matrix

	Positive	Negative
True	TP	TN
False	FP	FN

$$F_\beta = \frac{(1 + \beta^2) * P * R}{\beta^2 * P + R}$$

At this time, $F_\beta$is based on the weighted harmonic average of P and R, which is an index to measure the relative importance of R to P. It is usually taken to obtain the standard F1 value. The formula is specifically:

$$F_1 = \frac{2 * P * R}{P + R}$$

4.3 Experimental Results

Considering that the classification task, all model evaluations in this experiment uniformly use accuracy as the evaluation index. The greater the accuracy, the better the model classification effect. In the evaluation of individual categories, because F1-score takes into account both the accuracy and recall indicators of the classification model, it is more comprehensive, so the F1 value is used as the evaluation indicator. The results of the five groups of comparative experiments are shown in Table 2.

Comparing the experimental results, we can see, There is little difference between CNN, RNN and RCNN, and the error is basically maintained at 0.5%. Transformer based on the self-attention mechanism, while efficient parallel computing, attention can be weighted to all sequence information, and the BERT model contains richer contextual information, which can effectively improve the quality of text representation. Compared to traditional models, the accuracy increased by about 1% and 2%, the combination of the two has increased by about 3%. It shows that the model can fully mine the deep emotional semantic information of the text and improve the effect of sentiment analysis.

Table 2 Experiment results

Model	Accuracy (%)
TextCNN	85.27
TextRNN	85.42
TextRCNN	85.75
Transformer	86.64
BERT	**87.52**
BERT + transformer	**88.27**

5 Summary

In this paper, in order to further improve the emotional feature extraction of text, this paper proposes a sentiment analysis model based on BERT and Transformer. The word vector containing rich semantic information is extracted through the BERT model to enhance the model's semantic understanding of the text; and based on the self-attention model, the important information of the text is captured globally and the important features are weighted. So as to make full use of the hierarchical information of the entire text sequence to deal with the text classification problem more effectively.

The experimental results show that compared with Text-CNN and other models, the model can improve emotion classification by up to %3. It shows that the model can fully mine the deep emotional semantic information of the text and improve the effect of sentiment analysis.

In addition, this method still has a certain space for exploration. For example, it can be combined with the dependency syntax tree structure to obtain the syntax information of the text, which is more in line with human understanding of the text. Therefore, the next research direction lies in how to parse the syntactic structure of the text and integrate semantic features to further improve the quality of text representation.

References

Collobert, R., & Weston, J. (2008). A unified architecture for natural language processing: Deep neural networks with multitask learning. In *Proceedings of the 25th international conference on Machine learning* (pp. 160–167).

Devlin, J., Chang, M. W., Lee, K., et al. (2018). BERT: Pre-training of deep bidirectional transformers for language understanding. arXiv preprint arXiv:1810.04805.

Hao, J., Wang, X., Shi, S., et al. (2019). Multi-granularity self-attention for neural machine translation. arXiv preprint arXiv:1909.02222.

Mikolov, T., Chen, K., Corrado, G., et al. (1986). Efficient Estimation of Word Representations in Vector Space[J]. *Computer Science*, 2013.

Hofmann, T. (2013). Probabilistic latent semantic analysis. arXiv preprint arXiv:1301.6705.

Kim, Y. (2014). Convolutional neural networks for sentence classification. arXiv preprint arXiv:1408.5882.

Letarte, G., Paradis, F., Giguère, P., et al. (2018). Importance of self-attention for sentiment analysis. In *Proceedings of the 2018 EMNLP Workshop BlackboxNLP: Analyzing and Interpreting Neural Networks for NLP* (pp. 267–275).

Maas, A. L., Daly, R. E., Pham, P. T., et al. (2011). Learning word vectors for sentiment analysis. In *Proceedings of the 49th annual meeting of the association for computational linguistics: Human language technologies* (Association for Computational Linguistics) (Vol. 1, pp. 142–150.1631–142–150.1642).

Nowak, J., Taspinar, A., & Scherer, R. (2017). LSTM recurrent neural networks for short text and sentiment classification. In *International Conference on Artificial Intelligence and Soft Computing* (pp. 553–562). Springer.

Pennington, J., Socher, R., & Manning, C. D. (2014). Glove: Global vectors for word representation. In *Proceedings of the 2014 conference on empirical methods in natural language processing (EMNLP)* (pp. 1532–1543).

Shah, K., Patel, H., Sanghvi, D., & Shah, M. (2020). A comparative analysis of logistic regression, random forest and KNN models for the text classification. *Springer Singapore, 5*(4).

Socher, R., Perelygin, A., Wu, J., et al. (2013). Recursive deep models for semantic compositionality over a sentiment treebank. In *Proceedings of the 2013 conference on empirical methods in natural language processing* (pp. 1631–1642).

Vaswani, A., Shazeer, N., Parmar, N., et al. (2017). Attention is all you need. In *Advances in neural information processing systems* (pp. 5998–6008).

Wang, G., Xuejian, H., & Min, L. (2019). Multi-feature fusion GRU neural network text sentiment classification model. *Small Microcomputer System, 40*(10), 2130–2138.

Wang, J., Luo, L., & Wang, D. (2018). Research on Chinese short text classification based on Word2Vec. *Computer System Applications, 27*(05), 209–215.

Wu, Y., Schuster. M., Chen, Z., et al. (2016). Google's neural machine translation system: Bridging the gap between human and machine translation. arXiv preprint arXiv:1609.08144.

Yang, Z., Yang, D., Dyer, C., et al. (2016). Hierarchical attention networks for document classification. In *Proceedings of the 2016 conference of the North American chapter of the association for computational linguistics: Human language technologies* (pp. 1480–1489).

Collection and Analysis of Electricity Consumption Data: The Case of POSTECH Campus

Do-Hyeon Ryu, Young Myoung Ko, Young-Jin Kim, Minseok Song, and Kwang-Jae Kim

1 Introduction

Advanced metering infrastructure (AMI) is to an integrated system of smart meters, communication networks, and data management systems that measure, collect, and analyze various types of data, such as amounts of electricity, gas, heat, and water consumption (Shen et al., 2014). It enables time-based information and frequent collection and transmits such information to various parties in real time (Niyato & Wang, 2012). Therefore, the information provided by the AMI can be used by managers and consumers of electric utilities to make intelligent decisions for the efficient operation of power plants and reduction of energy consumption and cost (Littman et al., 2012).

The AMI has been applied to factories, commercial buildings, and university campuses (Iwayemi et al., 2011). University campuses are major consumers of energy and need uninterrupted power supply for critical infrastructures, such as

D.-H. Ryu
Department of Industrial and Management Engineering, Pohang University of Science and Technology, Pohang, Gyeongbuk, South Korea

Y. M. Ko · M. Song · K.-J. Kim (✉)
Department of Industrial and Management Engineering, Pohang University of Science and Technology, Pohang, Gyeongbuk, South Korea

Open Innovation Big Data Center, Pohang University of Science and Technology, Pohang, Gyeongbuk, South Korea
e-mail: kjk@postech.ac.kr

Y.-J. Kim
Department of Electrical Engineering, Pohang University of Science and Technology, Pohang, Gyeongbuk, South Korea

Open Innovation Big Data Center, Pohang University of Science and Technology, Pohang, Gyeongbuk, South Korea

H. Yang et al. (eds.), *AI and Analytics for Public Health*, Springer Proceedings in Business and Economics, https://doi.org/10.1007/978-3-030-75166-1_24

laboratories and cafeteria (Alrashed, 2020; Chung & Rhee, 2014). Many universities have recently developed AMI in their campuses and utilized it as a testbed to study new research or business ideas and technologies (Talei et al., 2012). For example, the University of California San Diego, the Wesleyan University in Connecticut, and the Princeton University have developed their own AMI and conducted various studies using the data collected. The universities have applied the collected data to understand their energy consumption and devise an effective energy conservation strategy (Farhangi, 2016).

The Pohang University of Science Technology (POSTECH) in Korea has also developed an AMI. POSTECH has many facilities, such as classrooms, offices, laboratories, factories, and graduate student apartments, where rich energy data can be collected. The AMI aims to collect, store, and manage electricity data and make POSTECH a global hub of energy big data, in which internal and external faculty and researchers use the collected data in education, research, and new business development. POSTECH has also developed a platform called Open Innovation Big Data Center (OIBC) to systematically manage, analyze, and share the collected data from the AMI.

In this work, we first describe the POSTECH AMI and the OIBC platform in detail. Then, we introduce analysis of the data collected from campus buildings. The remainder of this paper is structured as follows. Section 2 explains the sensors and an architecture installed in the POSTECH AMI and the systems in the OIBC platform. Section 3 presents the analysis results of electricity consumption data from campus buildings. Section 4 discusses issues in operating the AMI and the OIBC platform and in utilizing the collected data. Section 5 concludes the paper.

2 Infrastructure for Collection and Analysis of Electricity Consumption Data

2.1 AMI Architecture at POSTECH Campus

POSTECH can be divided into four major areas, namely, educational, research, living, and production areas (Fig. 1). In the production area, machines and equipment that consume huge amounts of electricity are used to make materials or products for education and research. Facilities in the educational area include classrooms, meeting rooms, and offices. In the research area, laboratories and offices for research exist. Finally, cafeterias, dormitories, and apartments exist in the living area. Each area holds different buildings for various objectives. Smart meters were installed in the following seven buildings selected from the four areas to collect rich data: School of Environmental Science and Engineering (a), Information Research Laboratorie (b), Science Building IV (c), Tae-Joon Park Digital Library (d), Biotech Center (e), C5 (f), and Graduate Student Apartment (g). C5 is a complex building that comprise a café, classrooms, halls, offices, and laboratories.

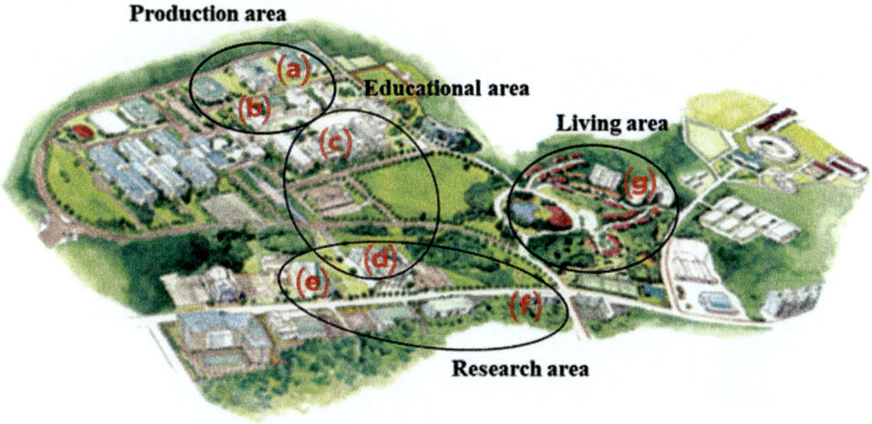

Fig. 1 Location of seven representative buildings in four major areas at POSTECH

We installed four types of smart meters made by Encored Technologies and RETIGRID, which are companies that develop and provide smart meters and data management and monitoring services. An EnerTalk Biz (3ch) sensor can cover large areas, such as an entire or floors of buildings. An EnerTalk (24ch) sensor collects electricity consumption data of specific zones or rooms of buildings. A plugin (1ch) sensor measures the data of an individual appliance. Finally, a high-sampling sensor can be installed in machines or equipment that require a large amount of electricity. We installed 266 sensors in the seven buildings (Table 1). Data consistency was validated by comparing the electricity consumption values measured by the sensors and the actual values measured by a power quality meter. All sensors satisfied 95% in data consistency.

The POSTECH AMI architecture is shown in Fig. 2 in detail. Internal data mean the collected data by smart meters from the seven buildings. The internal data are collected and transmitted to an IoT-based data collection system every second. External data provided by government agencies and private companies, such as weather and electricity prices, are also obtained through the application of programming interfaces (APIs). Such data are collected and transmitted to the collection system every hour. The collection system delivers all data to an IoT-based data monitoring system. In this case, the electricity data are delivered every 15 min for safe delivery. Data consistency was validated by comparing the electricity consumption values measured by the sensors and the values displayed in the IoT-based data collection and monitoring systems. All sensors satisfied 95% in data consistency. As open platforms, IoT-based data collection and monitoring systems can be equipped with different types of sensors and APIs. We attempted to add a meteorological sensor and occupancy sensors to measure environmental conditions, such as temperature, humidity, fine dust, and number of occupants entering and leaving a building. The collection system also delivers all collected data to an

Table 1 Number of sensors installed in seven buildings at POSTECH

Sensors Buildings	EnerTalk Biz (3ch) 	EnerTalk (24ch) 	Plugin (1ch) 	High sampling 	No. of sensors for each building
School of Envi. Sci. and Eng. (a)	26	–	–	4	30
Information Research Laboratorie (b)	–	15	–	4	19
Sci. Building IV (c)	27	–	18	2	47
Tae-Joon Park Digital Library (d)	54	–	–	23	77
Biotech Center (e)	19	–	–	15	34
C5 (f)	43	–	–	12	55
Graduate Student Apart. (g)	–	4	–	–	4
Total	169	19	18	60	266

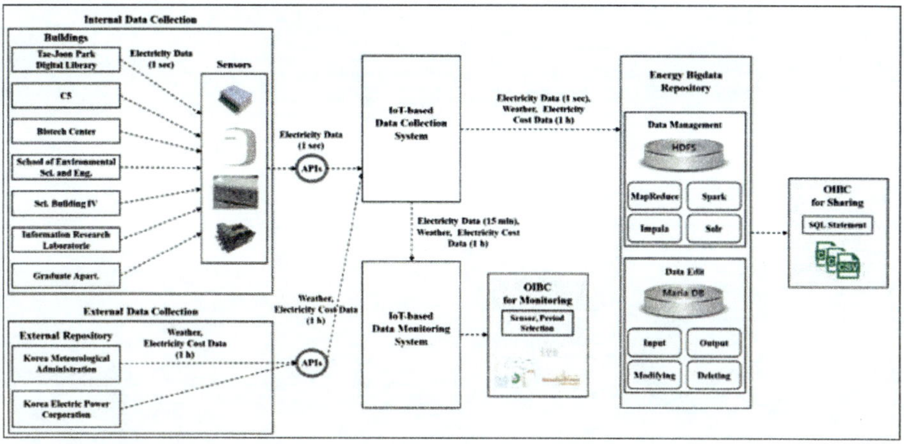

Fig. 2 Schematic of POSTECH AMI architecture

energy big data repository. The repository consists of Hadoop Distributed File System (HDFS) of Hadoop and Maria DB. HDFS is a file system that stores large files in distributed servers and works well with low-specification servers. Maria DB is a relational database management system that deals with the required Structured Query Language (SQL) statements and extracts and delivers data from the repository. We plan to expand the AMI by installing more sensors in other campus buildings.

2.2 *Data Monitoring and Sharing System of the OIBC Platform*

Data collection began from 2017, and the total amount to date is over 2TiB. The OIBC platform monitors and shares the collected data by the POSTECH AMI. People who want to utilize the data can gain authority to access the platform from its website (oibc.postech.ac.kr). OIBC has two types: monitoring platform and sharing platform. The monitoring platform has various functions, such as selecting buildings and sensors and changing periods, to allow users to find the data they want. The platform also provides a visualization function that allows users to view charts through a website (Fig. 3). The selected data are visualized with various charts, such as pie, line, and column charts. Given the user-friendly interface of the platform, users can easily zoom in and out the charts by dragging and clicking; they can also add and remove data from different sensors by clicking. Such visualization functions and interface help users understand various data deeply.

The sharing platform aims to share the collected data. Users who want to utilize the data can visit this platform to call and download the data. All commands are

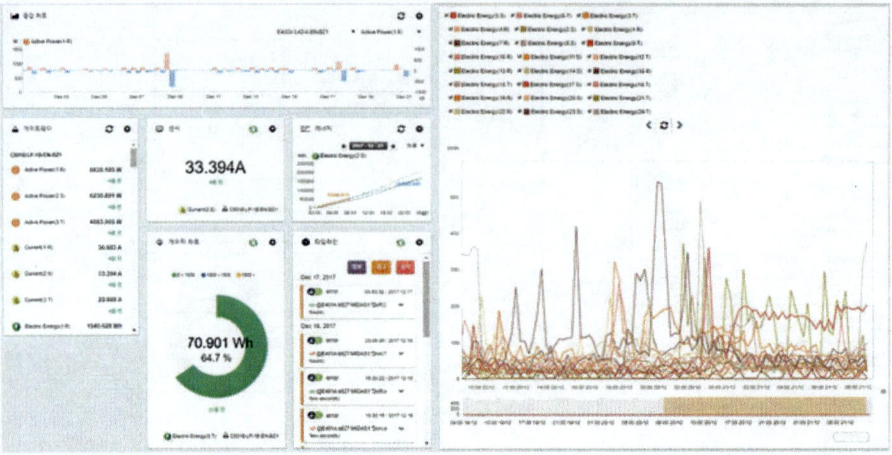

Fig. 3 Monitoring system of the OIBC paltform

entered and worked by SQL statements. Similar to the monitoring platform, the sharing platform allows users to easily find the data as building, floor, or zone levels and change periods of the data by customizing SQL statements properly. The sharing platform provides statement samples that show basic structure, unique codes of each building, and commands, allowing users who are unfamiliar with SQL statements to find and download data, including sensor name, time, voltage, current, active power, and reactive power. These data are delivered as a csv file.

The platforms have problem awareness and altering functions, and the condition of sensors is inspected regularly for the collection, management, and sharing of high-quality data. In addition, all systems are continuously upgraded to make them more user-friendly and stable.

3 Analysis of Electricity Consumption Patterns in POSTECH Campus

Energy efficiency is a critical issue for campus buildings because it is associated with the comfort and satisfaction of students (Jomoah et al., 2013). POSTECH has approximately 3500 students, 650 researchers, and 280 professors. The amount of electricity consumption has been continuously increasing, and POSTECH alerts campus members about the increase through bulletin boards. Hence, it is useful that POSTECH applies the collected data in the OIBC platform to understand the electricity consumption patterns and promote electricity conservation in the campus. To achieve this goal, we analyze the collected data to show and characterize the electricity consumption patterns of the seven buildings.

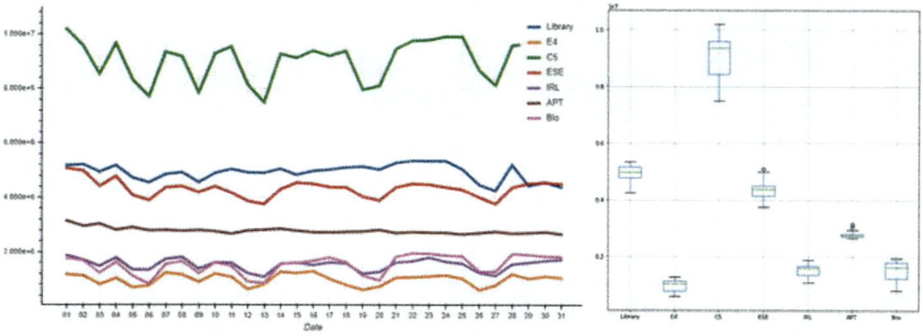

(a) (b)

Fig. 4 (**a**) Line chart and (**b**) box plots showing the amounts of electricity consumption of the seven buildings in October 2019

We utilized the data collected in October 2019, the month during a semester and before the COVID-19 outbreak. Thus, it would be a proper timeframe to extract the typical pattern of electricity consumption in the campus. Figure 4a shows the daily total amount of electricity consumption of the seven buildings. The amount of electricity consumption varies depending on the buildings. Given its large size and various facilities, C5 building (C5) consumes the largest amount of electricity. C5 has eight floors, and its floor space is twice that of other buildings. Moreover, C5 has a café, many halls where many people visit for various events, and start-ups that uses various machines and equipment. Tae-Joon Park Digital Library (Library), School of Environmental Science and Engineering (ESE), and Graduate Student Apartment (APT) consume moderate amounts of electricity. Biotech Center (Bio), Information Research Laboratorie (IRL), and Science Building IV (E4), are relatively smaller than other buildings and hence consume low amounts of electricity. This result implies that the amounts of electricity consumption of buildings vary depending on building size.

Figure 4a indicates that the patterns similarly changed depending on the date. The amounts of electricity consumption decreased every weekends (5th, 6th, 12th, 13th, 19th, 20th, 26th, and 27th). In addition, the amounts decreased on the third and the ninth because the 2 days are national holidays. Table 2 shows the statistical results of comparisons of the total amounts of electricity consumption between weekends and holidays and weekdays of the seven buildings. The normality of the utilized data was analyzed. Results showed that the data of E4, C5, Bio, and IRL showed normal distribution, whereas those of the other buildings were not normally distributed. Therefore, we conducted t-test and Wilcoxon signed-ranked test for the comparison. Table 2 shows that all buildings, except for APT, consume significantly different amounts between weekdays and weekends and holidays at $\alpha = 0.05$. This statistical result explains the flat pattern of electricity consumption of APT during the month in Fig. 4a.

Table 2 Statistical comparison of electricity consumption amounts of seven buildings at POSTECH between working days and days off

(a) Comparison results using t-test					(b) Comparison results using Wilcoxon signed-rank test.			
t-test	E4[a]	C5[a]	Bio[a]	IRL[a]	Wilcoxon signed-rank test	ESE[a]	Library[a]	APT
t	9.1687	12.4017	11.1617	9.7214	*v*	0.0000	5.0000	20.0000
p-value	0.0000	0.0000	0.0000	0.0000	*p-value*	0.0020	0.0195	0.4922

[a]Significant difference at $\alpha = 0.05$

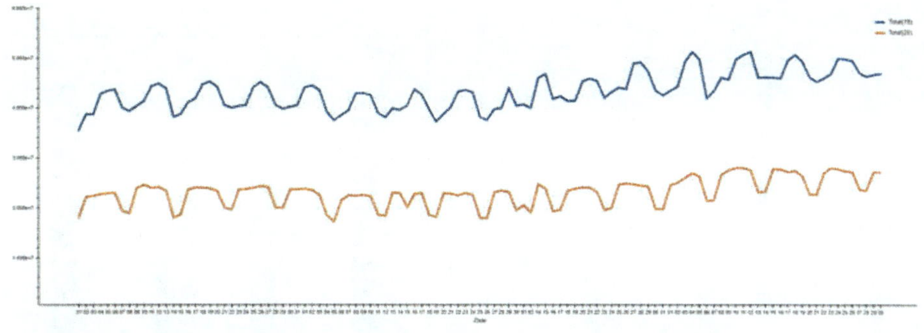

Fig. 5 Daily total amounts of electricity consumption of the seven buildings during the 4 months in 2019 (blue) and 2020 (orange)

Library shows an unusual pattern from the 14th to the 25th in Fig. 4a, which is attributed to long opening hours during the midterm examination season. The official midterm season was from the 21st to the 25th when the amount was the highest, but most students prepared the exams a week before. After the season, most students leave the campus and take a rest, and these students' behaviors were reflected to the amount of electricity consumption of the library. The amount was the lowest during the month and then recovered again. Unlike other buildings, APT showed no variation in the amount of electricity consumption regardless of weekdays or weekends. This result suggests that the amount of electricity consumed by graduate students and researchers who live in the apartment is unaffected by academic schedule. Figure 4b shows box plots that present the median and first and third quartiles for the seven buildings. As shown in Fig. 4a, the amount of electricity consumption of C5 is variable, whereas that of APT shows a slight change. This result explains that the amounts of electricity consumption of the buildings differ depending on occupant type and their behaviors in buildings.

The amounts of electricity consumption before and after the COVID-19 outbreak were also compared. The COVID-19 pandemic has forced universities to transition their classes to non-face-to-face delivery. As a result, many students did not return to the campus after the winter vacation, which became a major factor affecting the amount of electricity consumption. We utilized the data from March to June in 2019 and 2020 to compare the amounts of electricity consumption. The 4 months correspond to the spring semester of POSTECH. Figure 5 shows the total amounts of electricity consumption of the seven buildings. Data from the 3rd to the 13th of May do not exist because of a sensor problem. Blue and orange lines indicate the amounts during the 4 months in 2019 and 2020, respectively. The trends look similar, but a large big gap exists between the two lines. The amount of electricity consumption almost halved after the COVID-19 outbreak. This result is attributed mainly to the POSTECH policy that all members work from home at this time.

Figure 6a shows that most buildings, except E4, C5, and Bio, consumed greater electricity amounts before the COVID-19 outbreak. Table 3 displays the statistical

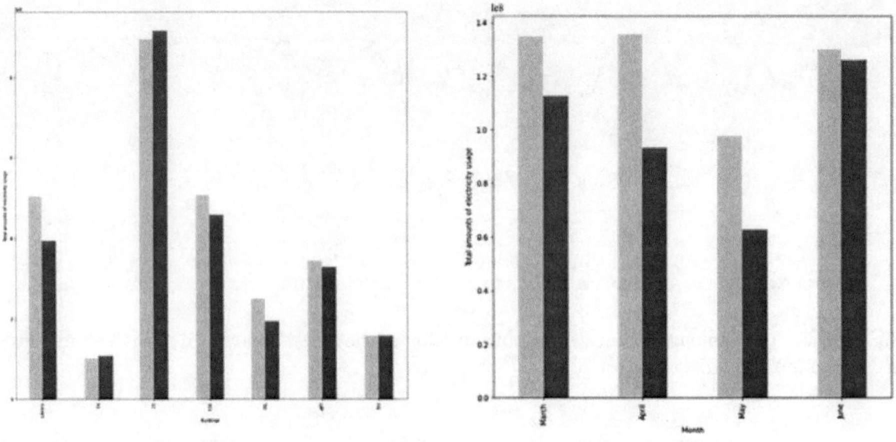

(a) (b)

Fig. 6 Column charts showing the total amounts of electricity usage of (**a**) the seven buildings and (**b**) Library before (light gray) and after (dark gray) the COVID-19 outbreak

comparison between the total amounts of electricity consumption of the seven buildings before and after the COVID-19 outbreak. The data of E4, C5, IRL, APT, and Bio showed normal distribution, but those of the other buildings were not normally distributed. Therefore, we conducted t-test and Wilcoxon signed-ranked test for the normal and non-normal data groups, respectively. Table 3 shows that Library, ESE, IRL, and APT consumed significantly different amounts before and after the COVID-19 outbreak at $\alpha = 0.05$. Buildings where undergraduate students use many facilities are Library, E4, and C5. However, E4 and C5 did not significantly differ in the amounts of electricity consumption. The increased usage of meeting rooms equipped with various IT systems, such as beam projectors and video conference systems, might have raised the electricity consumption. In fact, many meetings were switched from face-to-face to virtual meetings. Library showed the largest gap before and after the COVID-19 outbreak possibly because of the changed operation policy of Library (Fig. 6b). Given the lack of undergraduate students staying at POSTECH at that time, the university decided to open only several important zones and reduce their operation time. For example, the group study rooms, which account for about 40% of the fifth floor, was not open although they had been operated 24 hours a day. This result implies that the changed operation policy due to COVID-19 considerably affected the amount of electricity consumption at the campus.

Table 3 Statistical comparison of the electricity amounts of the seven buildings before and after the COVID-19 outbreak

(a) Comparison results using t-test						(b) Comparison results using Wilcoxon signed-rank test		
t-test	E4	C5	Bio	APT[a]	IRL[a]	Wilcoxon signed-rank test	ESE[a]	Library[a]
t	−2.0474	−1.2095	−0.0620	9.2907	10.9975	*v*	1041.0	429.0
p-value	0.0518	0.2278	0.9506	0.0000	0.0000	*p-value*	0.0000	0.0000

[a]Significant difference at $\alpha = 0.05$

4 Discussion

POSTECH developed an AMI and an OIBC platform that represent the addition to a campus of new technologies and devices. Operating them requires the integration of new skills and proficiencies from various domains. Therefore, cooperation with internal and external expertise is important. In this respect, POSTECH can use the developed AMI and the OIBC platform as a testbed to attract external partners. The partners can implement their new technologies, products, and services in the open platform and then collect data from the POSTECH campus. Additionally, the partners can utilize rich data to test and upgrade their technologies, products, and services. Campus members also must be encouraged to participate in collecting and utilizing data to activate the AMI and the OIBC platform. For the participation, POSTECH can establish various strategies. For example, POSTECH can also adopt such a reward system to encourage participation of members. If members participate surveys or develop ideas to improve the AMI and the OIBC platform, POSTECH can provide prizes or coupons as a reward. Blockchain-based virtual money, which passes as cash in POSTECH, is also considered as an alternative for reward.

Although this study conducted descriptive analyses to uncover the electricity consumption patterns of seven buildings at POSTECH, a systematic analysis to deeply understand the patterns has not been conducted yet. As a first step, determinants of the electricity consumption patterns should be identified. There may be a number of campus-related determinants such as class schedules, academic calendar, and seasonality. As explained in Sect. 3, some buildings showed no significant reduction in electricity consumption, although the number of occupants in the buildings decreased because of the COVID-19 pandemic. This implies the existence of other major factors affecting the electricity consumption of the university. Existing studies have found various determinants which have a high correlation with electricity consumption. It would be interesting to check which determinants are critical to the electricity consumption patterns in the POSTECH campus. Such a study would help to utilize the pattern information in developing strategies and policies for a more efficient usage of electricity in campus.

The identified patterns have a high potential to be utilized for the campus. First, the information can be employed for the efficient electricity consumption. The electricity consumption patterns of buildings represent occupants' behaviors. Normally, more occupants and their activities lead to more electricity consumption. Therefore, operators can figure out when and where campus members gather based on the electricity consumption patterns. With this information, operators can reduce a waste of electricity by stopping facilities (e.g., turning off the lights and heating) of buildings where a few campus members exist. Additionally, operators can consume more electricity to operate facilities of buildings where a lot of campus members do activities. This strategy enables efficient operations of facilities with saving electricity as well as improving the quality of campus life of students and faculty. Second, the electricity consumption patterns can be utilized to support the management of important equipment and devices. Campus buildings have numerous expensive and sensitive equipment and devices for research. The non-intrusive load monitoring methods enable analysts to disaggregate the total electricity consumption into those by each equipment and device. The information of electricity consumption of each equipment and device can be utilized to monitor and diagnose their conditions. When the conditions become worse, the electricity consumption patterns would change over time. With this information, operators can detect significant faults and prevent failures through proactive maintenance.

5 Conclusion

This paper describes the POSTECH AMI and the OIBC platform. In detail, we explained the installed sensors, the schematics of AMI, and the functions of the OIBC platform. The AMI and the platform would be a useful testbed or a living laboratory for various users. Researchers from various industries can simulate and validate their technologies, products, services, and studies by embedding them in the AMI and the OIBC platform. As the AMI and the OIBC platform are expanded, POSTECH is expected to play a pivotal role in providing a global hub of energy big data and related research.

We also introduced some applications that show the electricity consumption patterns of seven buildings at POSTECH. The results show that the patterns of electricity consumption vary depending on building size, occupant type and their behaviors, building uses, and the COVID-19 outbreak. In this study, we utilized only electricity consumption data. However, additional information on facility usage features, such as the number of occupants and operating time of each facility, need to be considered for detailed analysis. Various types of data could help POSTECH devise strategies and make decisions to optimize electricity consumption of their buildings.

References

Alrashed, S. (2020). Key performance indicators for smart campus and microgrid. In *Sustainable cities and society* (Vol. 60).

Chung, M. H., & Rhee, E. K. (2014). Potential opportunities for energy conservation in existing buildings on university campus: A field survey in Korea. *Energy and Buildings, 78*, 176–182.

Farhangi, H. (2016). *Smart microgrids: Lessons from campus microgrid design and implementation*. CRC Press.

Iwayemi, A., Wan, W., & Zhou, C. (2011). Energy management for intelligent buildings. *Energy Management System, 22*, 123–144.

Jomoah, I. M., Al-Abdulaziz, A. U., & Kumar, R. S. (2013). Energy management in the buildings of a university campus in Saudi Arabia—A case study. In *4th International Conference on Power Engineering, Energy and Electrical Drives* (pp. 659–663).

Littman, A., Lyon, G., Shah, A., & Vogler, J. (2012). Exploring advanced metering infrastructure deployments for commercial and industrial sites. In *Energy sustainability (American Society of Mechanical Engineers)* (Vol. 44816, pp. 979–989).

Niyato, D., & Wang, P. (2012). Cooperative transmission for meter data collection in smart grid. *IEEE Communications Magazine, 50*(4), 90–97.

Shen, B., Ghatikar, G., Lei, Z., Li, J., Wikler, G., & Martin, P. (2014). The role of regulatory reforms, market changes, and technology development to make demand response a viable resource in meeting energy challenges. *Applied Energy, 130*, 814–823.

Talei, H., Zizi, B., Abid, M. R., Essaaidi, M., Benhaddou, D., & Khalil, N. (2012). Smart campus microgrid: Advantages and the main architectural components. In *3rd International Renewable and Sustainable Energy Conference* (pp. 1–7).

Balance Between Pricing and Service Level in a Fresh Agricultural Products Supply Chain Considering Partial Integration

Peihan Wen and Jiaqi He

1 Introduction

Fresh agricultural products (FAP) are perishable and can suffer losses, including a quality loss and a quantity loss, during long-distance transportation (Cai et al., 2013; Yang et al., 2015; Zhou et al., 2015). Adopting the cold chain service provided by the third party logistics (3PL) enterprise can effectively reduce the transport loss of agricultural products. However, the low level of cold chain service and the difficulty in making profit is the key problem in fresh agricultural products supply chain (FAPSC).

Relevant literatures are still limited, and supply chain structures adopted are basically no integration (NI) and complete integration (CI) (Song et al., 2019; Song & He, 2019; Yu et al., 2019; Yang & Tang, 2019; Zhang et al., 2018; Ma et al., 2019).There is few literature studying pricing and cold chain service level decision-making in FAPSC under partial integration (PI) even PI is common in reality, which means that a member of a supply chain owns a part of the shares of another member to obtain the corresponding profit of the other party (Li et al., 2020; Chen et al., 2017). PI can increase the total profit of the supply chain and alleviate the double marginal effect. Sometimes it is even better than NI or CI (Zhang et al., 2018; Serbera, 2019; Fiocco, 2016). However, the motivation of PI is more complex than that of CI, that is, the supply chain after PI needs to meet certain conditions to obtain

P. Wen (✉)
School of Management Science and Real Estate, Chongqing University, Chongqing, P. R. China
e-mail: wenph@cqu.edu.cn

J. He
College of Mechanical Engineering, Chongqing University, Chongqing, P. R. China

H. Yang et al. (eds.), *AI and Analytics for Public Health*, Springer Proceedings in Business and Economics, https://doi.org/10.1007/978-3-030-75166-1_25

greater profit (Levy et al., 2018). On the other hand, most of existing researches on pricing and decision-making of cold chain service level in FAPSC consider the integration between suppliers and retailers, and few studies were conducted from the perspective of 3PL enterprises. Hence, we studied the influence of PI on pricing and cold chain service level decision in FAPSC from the perspective of 3PL enterprise.

Contributions for literature are in the following two aspects. Firstly, we considered NI, PI and CI with participation of the 3PL enterprise and discussed the influence of three supply chain structures on decisions of cold chain service and profit of each player in FAPSC. Secondly, we identified conditions regarding service sensitivity, price sensitivity, basic deterioration rate and share proportion, under which PI can be better than NI in terms of cold chain service level and profit of each player.

2 Supply Chain Profit Model Considering Integration

2.1 Supply Chain Structures

Given notations in Table 1, considering a FAPSC composed of a supplier, a 3PL enterprise and a fresh agricultural products e-commerce (FAPE) enterprise, the supplier produces FAP, which are transported by the 3PL enterprise, and then sold to the market through the FAPE enterprise. We considered both the quality loss and the quantity loss of FAP, which were expressed by freshness level and quantity loss rate respectively. Following (Cai et al., 2010; Wu et al., 2015) where the freshness level is a strictly increasing differentiable function of the cold chain service level, we denote the freshness level as $\theta(e) = \kappa e$, taking values within [0,1], which $e \in [0, 1]$ is the cold chain service level. And there is a basic quantity loss rate μ ($\mu \in [0, 1]$) in the transportation without the cold chain service. Higher values of μ represents more perishable agricultural products. Similar to (Song et al., 2019; Zhang et al., 2015), the quantity loss rate with the cold chain service can be translated into $(1 - e)\mu$. For a demand function of a product, the linear demand function was widely used in the literature (Zhou et al., 2015; Song et al., 2019) (Zhang et al., 2018). And the cold chain service cost function of 3PL enterprise is $C(e) = \lambda e^2 / 2$ (Song et al., 2019), which λ is the cold chain service cost coefficient.

Figure 1 presents three types of structures of the agricultural products supply chain as follows: (1) There is no integration between supply chain players, so they make decision independently to maximize their individual profits; (2) A α percentage of the supplier's shares holds by the 3PL enterprise, but they still make decision independently to maximize their individual profits; (3) A supplier and a 3PL enterprise form an alliance named S-T alliance, and jointly determines the wholesale price charged to the FAPE enterprise and the cold chain service level to maximize the profit of the S-T alliance.

Table 1 Notations used in model

Parameter	Description
C_R	The unit operation cost of FAPE enterprise
C_S	The unit production cost of supplier
P_S	The unit spot price of product in the spot market
D	The market demand of product
a	Market potential
τ	Price sensitivity of product $\tau > 0$
κ	The ability of cold-chain service to keep agricultural products fresh, $\kappa > 0$
δ	Consumers' sensitivity to fresh agricultural product's freshness, $\delta > 0$
h	The cold chain service sensitivity, $h > 0$
μ	The basic quantity loss rate, $\mu \epsilon [0, 1]$
λ	The cold chain service cost coefficient, $\lambda > 0$
α	The percentage of supplier's shares holds by 3PL enterprise, $\alpha \leq 0.5$
$\theta(e)$	The freshness level function of product, $\theta(e) \epsilon [0, 1]$.
$C(e)$	The cold chain service cost function
Decision variable	**Description**
e^{j*}	The cold chain service level of 3PL enterprise, j = NI, PI, CI, $e^{j*} \in [0, 1]$
P_R^{j*}	The unit retail price of fresh produce e-commerce enterprise, j = NI, PI, CI
P_W^{j*}	The unit wholesale price of supplier, j = NI, PI, CI
P_T^{j*}	The unit cold chain service price of 3PL enterprise, j = NI, PI, CI
Profit function	**Description**
Π_T^j	The 3PL enterprise's profit, j = NI, PI, CI
Π_R^j	The supplier's profit, j = NI, PI, CI
Π_S^j	The FAPE enterprise's profit, j = NI, PI, CI

2.2 *Assumptions*

Assumptions from literature are as follows: (1) The market demand is influenced by retail price and freshness (Cai et al., 2013; Song et al., 2019; Yang & Tang, 2019; Fiocco, 2016; Levy et al., 2018); (2) PI means that one company holds no more than 50 percent of the other's shares and gets the corresponding amount of profit, and they make independent decisions (Ma et al., 2019); (3) To simplify the analysis but still capture the essence of PI, the supplier's production cost is normalized to zero (Song et al., 2019); (4) To satisfy the FAPE enterprise's order quantity, assuming that there is a spot market with ample capacity, where the supplier can purchase sufficient agricultural products with the exogenous unit spot price to offset the deteriorated agricultural products, and the unit cold chain service price charged to the supplier is less than the unit spot price (Song et al., 2019). Besides, assumption (5) is presented: the quantity loss rate and freshness of agricultural products after

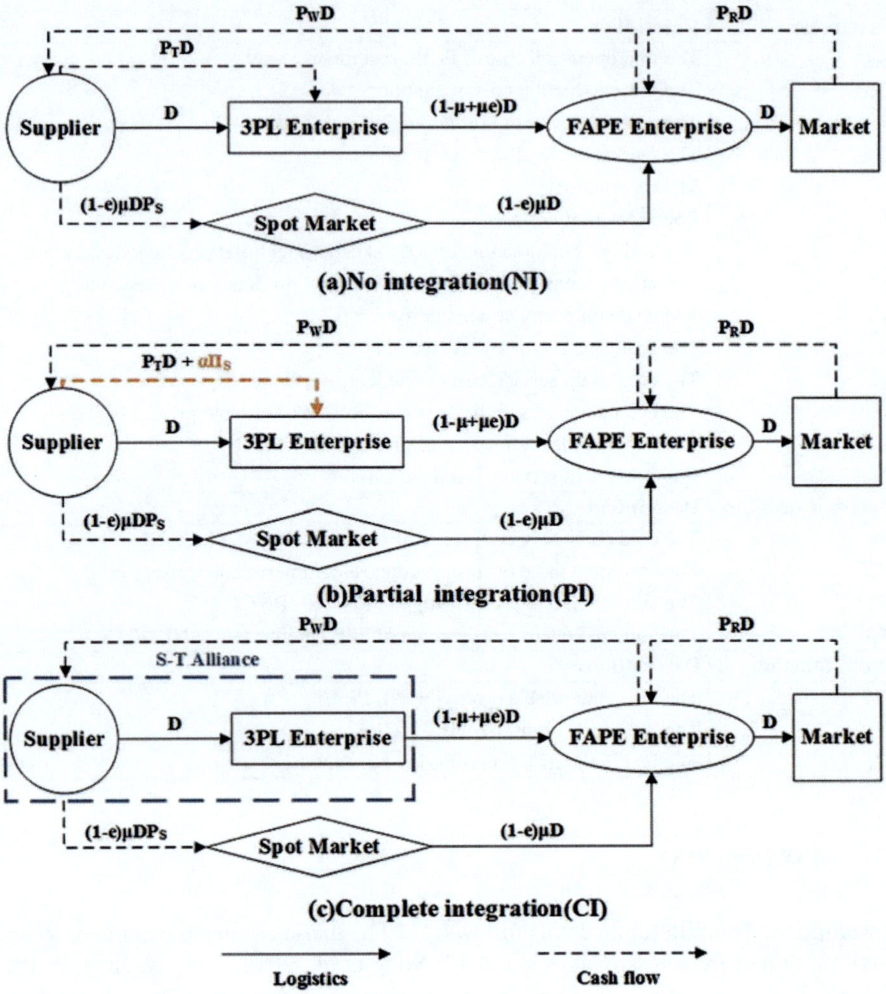

Fig. 1 The supply chain structures

arrival are affected by the cold chain service level. According to the assumption (1), the market demand of product is

$$D = a + \delta\theta(e) - \tau \mathrm{P_R} = a + \delta\kappa e - \tau P_R = a + he - \tau P_R,$$

where $\mathrm{h} = \delta\kappa$ is the service sensitivity.

2.3 *Profit Model Under NI, PI and CI*

The Stackelberg game between the supplier, the 3PL enterprise and the FAPE enterprise works as follows. Firstly, the 3PL enterprise decides the unit cold chain service price and the cold chain service level; and then the supplier chooses the wholesale price; finally, the FAPE enterprise determines the retail price.

Under NI, the profit functions for the FAPE enterprise, the supplier, and the 3PL enterprise are shown as Formula (1) to (3), respectively.

$$\Pi_R^{\mathrm{NI}}(P_W, P_R, e) = (P_R - P_W - C_R)\,\mathrm{D} \tag{1}$$

$$\Pi_S^{\mathrm{NI}}(P_W, P_T, P_R, e) = (P_W - P_T)\,\mathrm{D} - (1-e)\,\mu \mathrm{DP_S} \tag{2}$$

$$\Pi_T^{NI}(e, P_T, P_R) = P_T D - \frac{\lambda e^2}{2} \tag{3}$$

Under PI, the profit functions for the FAPE enterprise, the supplier, and the 3PL enterprise are shown as Formula (4) to (6), respectively.

$$\Pi_{\mathrm{R}}^{\mathrm{PI}}(P_W, P_R, e) = (P_R - P_W - C_R)\,D \tag{4}$$

$$\Pi_{\mathrm{S}}^{\mathrm{PI}}(P_W, P_T, P_R, e) = (1-\alpha)\,[(P_W - P_T)\,D - (1-e)\,\mu \mathrm{DP_S}] \tag{5}$$

$$\Pi_T^{\mathrm{PI}}(P_W, P_T, P_R, e) = P_T D - \frac{\lambda e^2}{2} + \alpha\,[(P_W - P_T)\,D - (1-e)\,\mu \mathrm{DP_S}] \tag{6}$$

Under CI, the profit functions for the FAPE enterprise and the S-T alliance are shown as Formula (7) and (8), respectively.

$$\Pi_{\mathrm{R}}^{\mathrm{CI}}(P_W, P_R, e) = (P_R - P_W - C_R)\,D \tag{7}$$

$$\Pi_{\mathrm{S-T}}^{\mathrm{CI}}(P_W, P_R, e) = P_W \mathrm{D} - (1-e)\,\mu \mathrm{DP_S} - \frac{\lambda e^2}{2} \tag{8}$$

3 Comparison Analysis

Based on the profit functions, the equilibrium results under NI, PI and CI are presented in Table 2 with a backward induction method (See Appendix for the detailed solution).

Table 2 Equilibrium results under NI, PI, and CI

Structure	NI	CI
P_T^*	$\frac{4\lambda(a-C_R\tau-\mu P_S\tau)}{8\lambda\tau-(h+\mu P_S\tau)^2}$	\
e^*	$\frac{(h+\mu P_S\tau)(a-C_R\tau-\mu P_S\tau)}{8\lambda\tau-(h+\mu P_S\tau)^2}$	$\frac{2\mu P_S\tau(a-h-C_R\tau)+(h-\mu P_S\tau)(a-C_R\tau+\mu P_S\tau)}{4\tau(\lambda-\mu P_S h)-(h-\mu P_S\tau)^2}$
P_W^*	$\frac{(a-C_R\tau+\mu P_S\tau)\left[8\lambda\tau-(h+\mu P_S\tau)^2\right]+(a-C_R\tau-\mu P_S\tau)\left(h^2-\mu^2P_S^2\tau^2+4\lambda\tau\right)}{16\lambda\tau^2-2\tau(h+\mu P_S\tau)^2}$	$\frac{2(a-C_R\tau+\mu P_S\tau)(\lambda-\mu P_S h)+\mu P_S(h-\mu P_S\tau)(a-h-C_R\tau)}{4\tau(\lambda-\mu P_S h)-(h-\mu P_S\tau)^2}$
P_R^*	$\frac{(3a+C_R\tau+\mu P_S\tau)\left[8\lambda\tau-(h+\mu P_S\tau)^2\right]+\left(3h^2+2h\mu P_S\tau-\mu^2P_S^2\tau^2+4\lambda\tau\right)(a-C_R\tau-\mu P_S\tau)}{32\lambda\tau^2-4\tau(h+\mu P_S\tau)^2}$	$\frac{a+he^{CI*}+P_W^{CI*}\tau+C_R\tau}{2\tau}$
Structure	**PI**	
P_T^*	$\frac{(1-\alpha)\left[(a-C_R\tau-\mu P_S\tau)\left(4\lambda\tau-\alpha h^2-2\alpha h\mu P_S\tau-\alpha\mu^2P_S^2\tau^2\right)+(h+\mu P_S\tau)\bullet\left(-\alpha ah+\alpha hC_R\tau-3\alpha h\mu P_S\tau+\alpha a\mu P_S\tau-\alpha\mu^2P_S^2\tau^2-\alpha\mu P_SC_R\tau^2\right)\right]}{(2\tau-\alpha\tau)\left(4\lambda\tau-\alpha h^2-2\alpha h\mu P_S\tau-\alpha\mu^2P_S^2\tau^2\right)-\left(h\tau+\mu P_S\tau^2-h\alpha\tau-\alpha\mu P_S\tau^2\right)(1-\alpha)(h+\mu P_S\tau)}$	
e^*	$\frac{(a-C_R\tau-\mu P_S\tau)(h+\mu P_S\tau)-2\alpha h(2-\alpha)(a+\mu P_S\tau-C_R\tau)}{8\lambda\tau-4\alpha\lambda\tau-(h+\mu P_S\tau)^2}$	
P_W^*	$\frac{a+he^{PI*}-C_R\tau+P_T^{PI*}\tau+\mu P_S\tau-e^{PI*}\mu P_S\tau}{2\tau}$	
P_R^*	$\frac{3a+3he^{PI*}+C_R\tau+P_T^{PI*}\tau+\mu P_S\tau-e^{PI*}\mu P_S\tau}{4\tau}$	

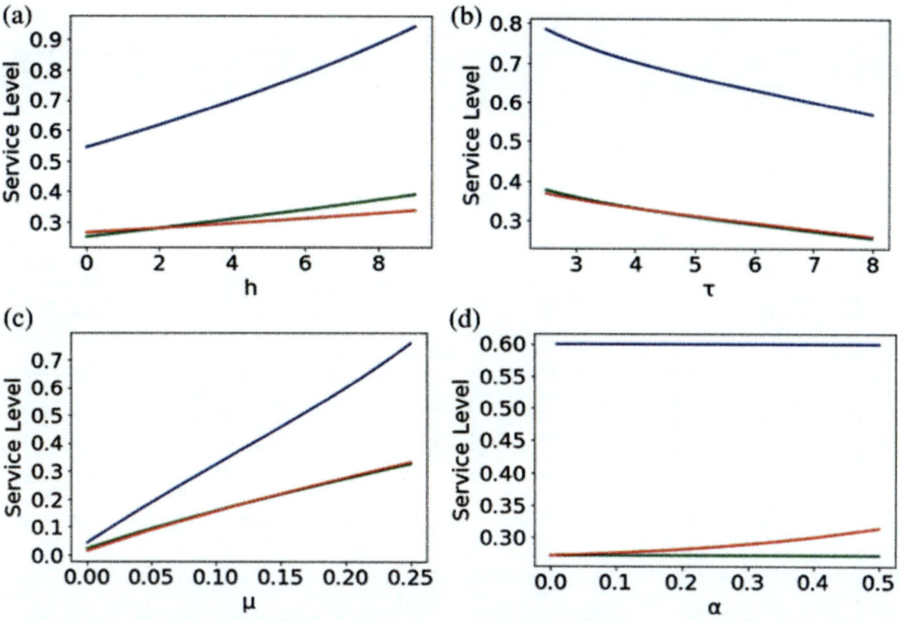

Fig. 2 Impacts on service level. (**a**) The impacts of service sensitivity h on service level e. (**b**) The impacts of price sensitivity τ on service level e. (**c**) The impacts of basic deterioration rate μ on e (**d**) The impacts of share proportion α on service level e

Following (Yu et al., 2019; Chen et al., 2017),because of the complexity of analyzing equilibrium results directly, numerical experiments were used to analyze the influence of service sensitivity, price sensitivity, basic deterioration rate and the share proportion on the cold chain service level of 3PL enterprise and the profit of each player, respectively. Parameters were initialized as follows: $a = 100$, $\lambda = 110$, $P_S = 100$, $C_R = 1.5$, $h = 1.5$, $\tau = 7$, $\alpha = 0.1$, $\mu = 0.2$, which guarantee the service level between [0,1] and a positive profit for each player. Results are shown in Figs. 2 and 3, and two observations can be derived from Figs. 2 and 3, respectively.

Observation 1: From Fig. 2, (1) The service level in CI is higher than PI or NI; (2) The service level in PI is higher than NI when service sensitivity h is low ($h \leq 1.5$), price sensitivity τ is high ($\tau \geq 7$) and basic deterioration rate μ is high ($\mu \geq 0.14$), or service sensitivity h is low ($h \leq 1.5$), price sensitivity τ is high ($\tau \geq 7$), but basic deterioration rate μ is low ($0.1 < \mu < 0.14$) and the share proportion α is high ($\alpha \geq 0.3$). If not, the service level in NI is higher than PI.

Observation 1 can be explained as follows: (1) Because profit and cost sharing with the supplier, the 3PL in CI is more motivated to provide higher service level than NI or PI; (2) When h is low, τ is high, or μ is high but h is low, the 3PL in NI or PI chooses to lower the service level to cut down the cold chain cost, but the 3PL in PI is able to provide higher service level than that in NI because of the share

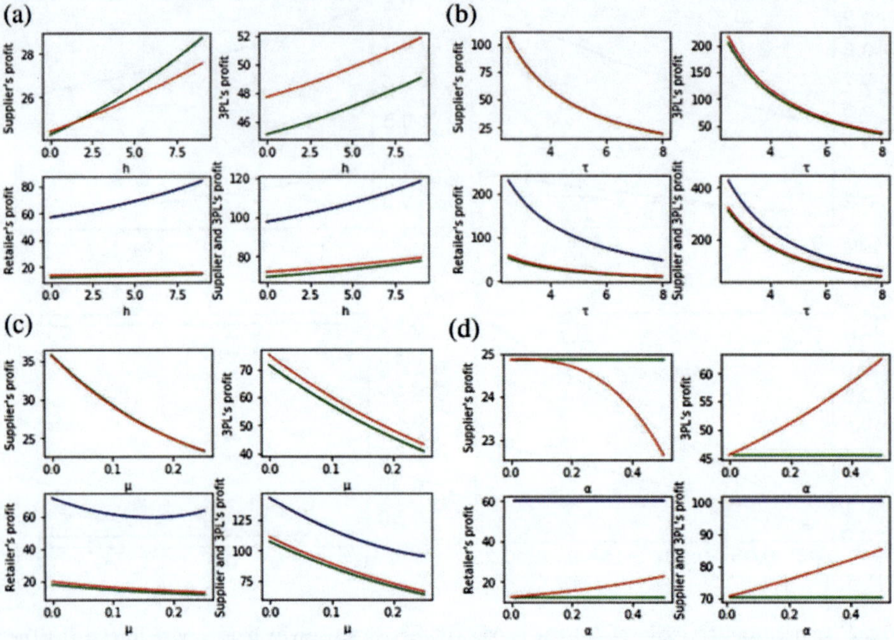

Fig. 3 Impacts on profit. (**a**) The impacts of service sensitivity h on profit. (**b**) The impacts of price sensitivity τ on profit. (**c**) The impacts of basic deterioration rate μ on profit. (**d**) The impacts of share proportion α on profit

profit from supplier. And when h is low ($h = 1.5$), τ is high ($\tau = 7$), and μ is high ($\mu = 0.2$), the service level in PI is higher than NI no matter how α changes. But when μ is low, only if α is high, the 3PL is able and more willing to provide higher service level than that in NI on account of more share profit that the 3PL receives from supplier.

Observation 2: From Fig. 3, (1) Each player is better off with CI than PI or NI; (2) If service sensitivity h and share proportion α is low ($h \leq 1.5$ and $\alpha \leq 0.15$), and price sensitivity τ and basic deterioration rate μ is high ($\tau \geq 7$ and $\mu \geq 0.25$), each player is better off PI than NI. If not, PI is only better for 3PL and FAPE enterprise, while supplier gains more profit under NI.

Observation 2 can be explained as follows: (1) CI eliminates the dual marginal effect between the supplier and the 3PL, so the retail price is lower than NI or PI. With cost and profit sharing, 3PL has more incentive to improve service level than NI or PI. The lower retail price and the higher service level leads to better sales results and the higher profit of each player than NI or PI; (2) PI mitigates the dual marginal effect between supplier and 3PL, so the wholesale price is lower than NI, which is in retailer's favor. For 3PL, it can get share profit from supplier under PI;

(3) When h is low, τ and μ is high, the 3PL in PI is more willing to provide higher service level than that in NI. Because the 3PL in NI chooses to reduce service level and service price to lower the service cost and increase market demand for products, and the 3PL in PI chooses to provide an appropriate service level to reduce the service cost and deterioration cost borne by the supplier to maintain own profit and supplier's profit. And when α is low, the profit due to PI is bigger than the share profit given to 3PL, supplier can benefit from PI.

4 Conclusions

A FAPSC composed of a supplier, a 3PL enterprise, and a FAPE enterprise was studied considering three conditions of NI, PI, and CI. Related profit models were constructed and solved with a backward induction method. Through the numerical analysis of equilibrium results, we found that CI could benefit all players, and all players could be better off PI than NI only if h and α were low and μ and τ were high, which was different from our intuition. Likewise, the level of cold chain service under CI was the highest, and the service level under PI was higher than that under NI when h was low, τ was high, and μ was high, or when h was low, τ was high, μ was low and α was high. The above findings can be used for FAPSC management. Further research will consider studying real life data or scenarios to explore more management advice.

Appendix

Proof of Table 2: Under NI, from (1), we acquire $\frac{\partial^2 \Pi_R^{NI}}{\partial P_R^2} = -2\tau < 0$; that is, Π_R^{NI} has maximum value. Solving the first-order conditions $\frac{\partial \Pi_R^{NI}}{\partial P_R} = 0$, we derive

$$\mathrm{P}_R^{NI}(P_W, e) = \frac{a + he + P_W\tau + C_R\tau}{2\tau} \tag{9}$$

Substituting (9) into (2), we can acquire $\Pi_S^{NI}(P_W, P_T, e)$, and derive $\frac{\partial^2 \Pi_S^{NI}}{\partial P_W^2} = -\tau < 0$; that is, Π_S^{NI}

has maximum value. Solving the first-order conditions $\frac{\partial \Pi_S^{NI}}{\partial P_W} = 0$, we obtain

$$\mathrm{P}_W^{NI}(P_T, e) = \frac{a + he - C_R\tau + P_T\tau + (1 - e)\,\mu P_S\tau}{2\tau} \tag{10}$$

Substituting (9) and (10) into (3), we can acquire $\Pi_T^{NI}(P_T, e)$. Further, the Hessian matrix of $\Pi_T^{NI}(P_T, e)$ over (P_T, e) is as follows:
$\mathbf{H} = \begin{bmatrix} -\frac{\tau}{2} & \frac{h+\mu P_S\tau}{4} \\ \frac{h+\mu P_S\tau}{4} & -\lambda \end{bmatrix}$, $|\mathbf{H_1}| = -\frac{\tau}{2} < 0$, $|\mathbf{H_2}| = \frac{\tau\lambda}{2} - \left(\frac{h+\mu P_S\tau}{4}\right)^2 > 0$ when $\lambda > \frac{(h+\mu P_S\tau)^2}{8\tau}$.

H is negatively definite. Solving the first-order conditions $\frac{\partial \Pi_T^{NI}(P_T,e)}{\partial P_T} = 0$, $\frac{\partial \Pi_T^{NI}(P_T,e)}{\partial e} = 0$, we derive

$$P_T^{NI*} = \frac{4\lambda (a - C_R\tau - \mu P_S\tau)}{8\lambda\tau - (h + \mu P_S\tau)^2} \tag{11}$$

$$e^{NI*} = \frac{(h + \mu P_S\tau)(a - C_R\tau - \mu P_S\tau)}{8\lambda\tau - (h + \mu P_S\tau)^2} \tag{12}$$

Substituting (11) and (12) into (9) and (10), we derive (P_W^{NI*}, P_R^{NI*}). Similarly, we can derive equilibrium results under PI and CI in Table 2.

References

Cai, X. Q., Chen, J., Xiao, Y. B., et al. (2010). Optimization and coordination of fresh product supply chains with freshness-keeping effort. *Production and Operations Management, 19*(3), 261–278.

Cai, X. Q., Chen, J., Xiao, Y. B., et al. (2013). Fresh-product supply chain management with logistics outsourcing. *Omega, 41*(4), 752–765.

Chen, J. G., Hu, Q. Y., & Song, J. S. (2017). Effect of partial cross ownership on supply chain performance. *European Journal of Operational Research, 258*(2), 525–536.

Fiocco, R. (2016). The strategic value of partial vertical integration. *European Economic Review, 89*, 284–302.

Levy, N., Spiegel, Y., & Gilo, D. (2018). Partial vertical integration, ownership structure, and foreclosure. *American Economic Journal: Microeconomics, 10*(1), 132–180.

Li, J., Yang, S. L., Shi, V., et al. (2020). Partial vertical centralization in competing supply chains. *International Journal of Production Economics, 224*, 107565.

Ma, X. L., Wang, S. Y., Islam, S. M. N., et al. (2019). Coordinating a three-echelon fresh agricultural products supply chain considering freshness-keeping effort with asymmetric information. *Applied Mathematical Modelling, 67*, 337–356.

Serbera, J. P. (2019). Separation versus affiliation with partial vertical ownership in network industries. *International Journal of the Economics of Business, 26*(3), 383–397.

Song, Z. L., & He, S. W. (2019). Contract coordination of new fresh produce three-layer supply chain. *Industrial Management & Data Systems, 119*(1), 148–169.

Song, Z. L., He, S. W., & Xu, G. S. (2019). Decision and coordination of fresh produce three-layer e-commerce supply chain: A new framework. *IEEE Access, 7*, 30465–30486.

Wu, Q., Mu, Y. P., & Feng, Y. (2015). Coordinating contracts for fresh product outsourcing logistics channels with power structures. *International Journal of Production Economics, 160*, 94–105.

Yang, C. T., Dye, C. Y., & Ding, J. F. (2015). Optimal dynamic trade credit and preservation technology allocation for a deteriorating inventory model. *Computers & Industrial Engineering, 87*, 356–369.

Yang, L., & Tang, R. H. (2019). Comparisons of sales modes for a fresh product supply chain with freshness-keeping effort. *Transportation Research Part E: Logistics and Transportation Review, 125*, 425–448.

Yu, Y. L., Xiao, T. J., & Feng, Z. W. (2019). Price and cold-chain service decisions versus integration in a fresh agri-product supply chain with competing retailers. *Annals of Operations Research, 287*(1), 465–493.

Zhang, J. X., Liu, G. W., Zhang, Q., et al. (2015). Coordinating a supply chain for deteriorating items with a revenue sharing and cooperative investment contract. *Omega, 56*, 37–49.

Zhang, Y. J., Rong, F., & Wang, Z. (2018). Research on cold chain logistic service pricing—Based on tripartite Stackelberg game. *Neural Computing and Applications, 32*(1), 213–222.

Zhou, Y. W., Cap, Z. H., & Zhong, Y. G. (2015). Pricing and alliance selection for a dominant retailer with an upstream entry. *European Journal of Operational Research, 243*(1), 211–223.

A Stacking-Based Classification Approach: Case Study in Volatility Prediction of HIV-1

Mohammad Fili, Guiping Hu, Changze Han, Alexa Kort, and Hillel Haim

1 Introduction

In this study, we examine the relationship between the in-host sequence variance (volatility) at specific Env positions and volatility at their closest neighbors. To measure the volatility, one can take a blood sample from a patient and observe if a particular position has the same amino acid or not in all viruses isolated from the sample. If there is no variability, we assign the value of 0 and call it non-volatile; otherwise, it is volatile, and we assign the value 1 for that position. Therefore, we have a binary classification problem.

The importance of this study is the ability to apply this tool to predict the future volatility at specific positions of Env targeted by therapeutics. Therefore, as part of a bigger research idea, this study tries to assess this neighborhood relationship. For this purpose, a stacking-based classification method is proposed to predict the current state of a position based on the neighbors' volatility.

The rest of this paper is organized as the following: Section 2 is dedicated to the research background. In this section, we show similar studies in both the HIV-1 part and the classification method. Then, in the materials and methods section, first, we describe the datasets used in this study. The ensemble network method will be explained in detail, as well. In Sect. 4, the numeric results are presented, and the performance of the proposed algorithm is compared with that of the base learners.

M. Fili · G. Hu (✉)
Department of Industrial and Manufacturing Systems Engineering, Iowa State University, Ames, IA, USA
e-mail: gphu@iastate.edu

C. Han · A. Kort · H. Haim
Department of Microbiology and Immunology, Carver College of Medicine, University of Iowa, Iowa City, IA, USA
e-mail: hillel-haim@uiowa.edu

H. Yang et al. (eds.), *AI and Analytics for Public Health*, Springer Proceedings in Business and Economics, https://doi.org/10.1007/978-3-030-75166-1_26

The paper concludes with Sect. 5 on the summary of the findings and discussion of future research directions.

2 Research Background

HIV-1 is the primary cause of acquired immune deficiency syndrome (AIDS). The genetic diversity of HIV-1 presents considerable challenges to the development of an efficient vaccine. Antiviral medications have been developed and used to suppress the level of the virus, which can prevent the development of the disease. There are nearly one million Americans diagnosed with HIV, and most of those individuals are treated by antiretroviral therapeutics (Centers for Disease Control and Prevention, 2020). A fraction of HIV-infected individuals generate antibodies that can effectively suppress virus levels in a broad range of viruses (Klein et al., 2013; Hraber et al., 2014; West et al., 2014; Mikell et al., 2011; Burton & Mascola, 2015; Haynes et al., 2016; Moore et al., 2015). Accordingly, such broadly-acting antibodies have also been tested as therapeutics to suppress virus levels in patients (Caskey et al., 2017). However, a concern with long-term antiviral therapy is the high mutation rate of the virus. In some patients, such mutations create resistance to medications. A major concern is the spreading of these resistant mutants in the population.

The envelope glycoproteins (Envs) within HIV-1 show the greatest degree of inter-host and intra-host diversity. Since the Env is contained on the surface of the virus particle, it is the primary target in AIDS vaccine design (Snoeck et al., 2011). It has been shown that for position of the Env protein, the level of variance in amino acid sequence between viruses that co-circulate in the host (designated volatility) is highly conserved among different patients (DeLeon et al., 2017). Volatility is a metric measures the intra-host variance for any position. Volatility at any position can vary between patients at any time point; however, the average volatility in any two groups of patients infected by the same HIV-1 subtype (clade) is similar. That volatility at any position changes continuously in each patient serves as the major motivation for this study. We try to predict the current volatility of an Env position based on the volatility at the closest neighbors on the three-dimensional structure of Env.

In this work, we proposed a stacking-based classification method. This method can help to decrease the generalization error rate and deducting the biases of the learners (Wolpert, 1992). This technique works in two steps, where, in the beginning, a group of classifiers predicts for each observation, and then these predicted values are given to a new learner called meta-learner to predict the class labels.

There have been existing studies that incorporate clustering as part of their learning stage for prediction. In (Agrawal et al., 2019), an ensemble classification approach is utilized after the clustering phase. In (Yang et al., 2019), a clustering algorithm divides the data into subsets in which experts are trained, which differs from the proposed method, where the base learners will be applied throughout the dataset to capture the pattern as much as possible. In (Yuxian et al., 2009; Kan et al., 2018), a clustering algorithm groups data into similar clusters, and an artificial neural network (ANN) is used as the learner afterward, but no stacking ensemble has been used. In (Sadrawi et al., 2018; Ibrahim & Far, 2016; Bernas & Płaczek, 2015; Kim & Seo, 2015; Ressom et al., 2006), an ensemble of ANNs is created after the data is clustered. Also, PCA is used in (Adhikari & Saha, 2014) to determine the underlying clusters along with an ensemble of k-nearest neighbors (KNN) and multi-layer perceptron (MLP) for classification. In (Wang et al., 2020), clustering has been incorporated, and then support vector machine (SVM), ANN, and gradient-boosting decision tree (GBDT) were stacked for the prediction. In (Amini et al., 2014), fuzzy clustering is performed on the dataset and having radial bias functions (RBFs) constructed as base classifiers, and the MLP is then used to aggregate the predictions.

In this paper, the proposed method utilizes a clustering algorithm for the purpose of creating similar regions alongside a diverse set of base classifiers that are then aggregated by an MLP, which is the meta learner in this stacking organization. A unique contribution of this algorithm is the feature creation procedure that helps the meta learner captures the relation between the base learners and the data points in the space based on the classifiers' performance in different regions.

In this study, we used the k-means technique to divide the space into separate regions. Training the base classifiers over the training set, we can evaluate the performance of each classifier within a region and use this information as a new feature and give it to the meta learner. For the meta learner, MLP, as a class of feedforward artificial neural network has been used. Besides the region evaluation, a neighborhood evaluation procedure is also applied, which creates another set of new features for the MLP by evaluating each classifier within a neighborhood of a data point.

The novelty of the proposed method is the feature creation procedure from the regional and neighborhood sections in order to have better predictions. By adding the evaluation information, the meta learner is able to find any meaningful relationship between groups of observation and the performance of each learner; therefore, it can emphasize more on the best learners in each region of the problem predict the class label of the new instances. In our case, the MLP can do it by assigning higher weights to those learners.

3 Materials and Methods

The appearance of mutations in the HIV-1 genome can render the virus resistant to medications. The existence of such resistant viruses can increase morbidity and the likelihood of transmission to other individuals. Understanding the relationship between sequence variance (volatility) at a position of interest on Env and sequence volatility at neighboring positions can contribute to the prediction of future states (e.g., the loss of sequence conservation at a position of Env critical for activity of a therapeutic).

In this study, we developed a stacking-based classification method to predict the current volatility state of a position based on that of physical neighbors. We used amino acid sequence data of HIV-1 Env for this purpose.

3.1 Datasets

In this study, sequences from the Los Alamos National Lab (LANL) database (https://www.hiv.lanl.gov) and National Center for Biotechnology Information (NCBI) database (https://www.ncbi.nlm.nih.gov) were used to attain the nucleotide sequences of the HIV-1 Env gene.

For this analysis, 191 patients are selected, and for each patient there are at least five Env sequences from the same blood sample. The sequences are attained by cloning the Env gene using the single genome amplification method (Salazar-Gonzalez et al., 2008). Non-functional Envs sequences, as well as the sequences with nucleotide ambiguities or large deletions in conserved regions, are removed (DeLeon et al., 2017; Han et al., 2020). A hidden Markov model is used to align the nucleotide sequences using HMMER3 software (Gaschen et al., 2001) and is translated into the amino acid sequence, which has been used for this study.

The Env positions described in this study follow the standard HXBc2 numbering of the Env protein (Korber et al., 1998). For each patient, the sequences are compared at every position to determine whether variance in amino acid sequence exists or not. For this purpose, a value 1 is assigned to positions with variance in the amino acid sequence and a value 0 for positions with the same amino acid sequence at that specific position. Therefore, a vector containing 856 features describing the variance at 856 positions is created.

Because of the folded structure of the Env protein, positions from different domains can be adjacent to each other in its three-dimensional structure. To identify the neighbors of a position, we used the data from the cryo-electron microscopy image of HIV-1 Env strain JRFL (Lee et al., 2016). To calculate the distance between any two positions, we used the coordinates of the nearest two amino acid atoms. Using these data, we found the 10 closest neighbors to a specific position.

In this study, we tested four different positions – 146, 144, 188, and 148 – to predict the current level of volatility for each position based on the 10 nearest neighbors on the three-dimensional structure of the protein as the features.

3.2 The Ensemble Network Method

The Ensemble network algorithm is a stacking-based classification method which is consisted of two stages. In the first stage, a pool of base classifiers will be selected.

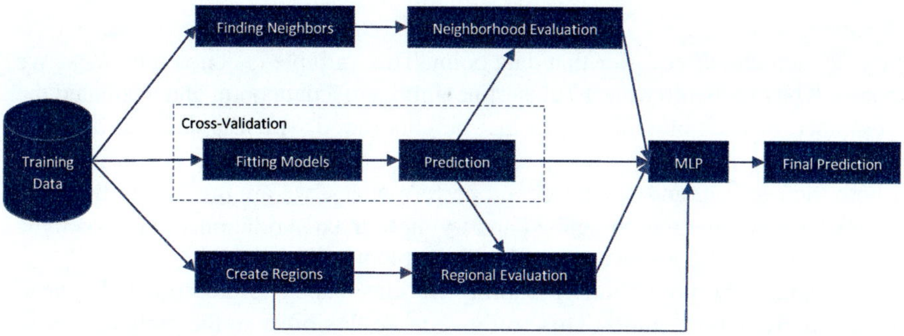

Fig. 1 Flowchart of the ensemble network algorithm

At the same time, a clustering algorithm will be used to partition the data points into similar regions (clusters). Then based on the result of selected classifiers and their performance in each region, as well as their performance in the neighborhood of each observation, a new dataset will be constructed. This new dataset will be fed into a meta learner – here, MLP – to classify new instances. In this algorithm, the meta learner tries to learn the relation between the data points in the space and the classifiers' performance to be able to decide how to choose the correct class labels. Figure 1 shows the flowchart for the ensemble network algorithm.

3.2.1 Stage One: Training Base Classifiers

In the first step, a pool of classifiers will be defined. The selected classifiers should be accurate and diverse for overall model performance. For this study, five classifiers are selected: Extra-tree classifier (ET), support vector machines (SVM), Adaboost, naïve Bayes, linear discriminant analysis (LDA). The selection of these classifiers is due to the fact that they showed good results for the HIV-1 datasets in previous experiments.

In the second step, we divide the data into train and test sets. A 75–25% split is applied in this work. We denote the features for the training and test sets by X_{train}, X_{test} respectively. Given X_{train} to the k-means algorithm, we clustered the observation into similar regions. Using the elbow method, the best number of clusters is obtained, and a region number is assigned to each observation $X_j \in X_{train}$ $(j = 1, \ldots, N)$.

We used cross-validation to predict the probability of being volatile for each observation. For this purpose, the dataset is divided into five folds, and the classifier L_i is trained on four folds while using the remaining one-fold for the prediction. In the process of training each classifier, principal component analysis (PCA) has been used, not for the purpose of dimension reduction, but to make the explanatory variables mutually orthogonal (uncorrelated). We also tuned the hyper-parameters of each classifier to achieve the best results possible.

Knowing which region, the data point X_j belongs to, we measured the performance of each classifier L_i for that data point. This variable is denoted by RP_{ij}. We also used KNN to identify the 10 closest neighbors to a data point and evaluated the performance of each classifier L_i in that neighborhood which is denoted by NP_{ij}. For the purpose of assessing the base learners' performance, accuracy is selected. For both the regional and neighborhood performances, we are measuring the local accuracy based on the region defined and trying to make a relationship between the classifiers' results and a specific area in the problem space.

In this study, the prediction probability for class labels is incorporated instead of the crisp prediction results. This will give more flexibility to the meta learner to understand the relations between the predicted results and the true class labels.

Finally, we gather all features generated so far together to make a new dataset X_{train}^{new}. This new dataset consists of the predicted probabilities for each classifier, the region in which each data point belongs to, regional performances, and the neighborhood performances. For a general model with M base learners, we will have $3M + 1$ features.

3.2.2 Stage Two: Training Meta-Learner

After generating X_{train}^{new}, the MLP can be trained on the new dataset. We used hyper-parameter tuning for this phase to obtain the best results possible. The meta-learner tries to find the relationship between the generated results and the classifiers' performance with the true class labels.

In order to classify a new instance, we need to generate X_{test}^{new}. For this purpose, we need to find the region in which the new observation is more likely to be contained. Therefore, we can measure the distance of a new observation from the centers of the clusters and assign the group number for which the distance is the minimum.

Since the models are available from the first stage, predicting the class probability will not be a problem. Furthermore, because the regions are known, we can assign the expected performance from the previous part to each new instance. However, for the neighborhood performance evaluation, we need to find the closest neighbors for each data point. Knowing the neighbors, one can measure the local accuracy in the neighborhood of $X_q \in X_{test}$ from the available points.

4 Results and Discussions

As mentioned earlier, we compared the performance of Ensemble Network with that of the single base learners used in training the model. For the evaluation metrics, we used accuracy, precision, recall or sensitivity, and area under the receiver operating characteristic (ROC) curve (AUC).

Table 1 Comparison between the ensemble network with the base learners for position 188 (%)

Model	Accuracy	Precision	Recall	F1 Score	AUC
Extra-trees classifier	54	58	90	71	47
Support vector machine	58	60	97	74	53
Adaboost	54	58	90	71	44
Naïve Bayes	38	43	10	16	47
Linear discriminant analysis	52	57	86	69	36
Ensemble network	**60**	**60**	**100**	**75**	**60**

Table 2 Comparison between the ensemble network with the base learners for position 148 (%)

Model	Accuracy	Precision	Recall	F1 Score	AUC
Extra-trees classifier	62	65	90	75	52
Support vector machine	65	65	97	78	50
Adaboost	62	65	9	75	52
Naïve Bayes	33	0	0	–	49
Linear discriminant analysis	56	62	84	71	45
Ensemble network	**65**	**65**	**100**	**79**	**56**

Table 3 Comparison between the ensemble network with the base learners for position 146 (%)

Model	Accuracy	Precision	Recall	F1 Score	AUC
Extra-trees classifier	69	68	65	66	73
Support vector machine	71	70	70	70	70
Adaboost	73	73	70	71	74
Naïve Bayes	67	65	65	65	71
Linear discriminant analysis	71	70	70	70	84
Ensemble network	**75**	69	**87**	**77**	75

The comparison for position 188 is summarized in Table 1. Maintaining the same precision, it improved all other metrics. As we can see, the accuracy improved by 2%, compared to the best available that belongs to SVM. However, the margin is more considerable when it is compared to the other methods. For precision, we can observe the same performance as SVM, but it is superior to the rest of the classifiers regarding the other metrics. It also reached the maximum performance for recall and improved the F1 score as well. In addition, it depicts a significant improvement in AUC.

Table 2 includes the result for position 148. Maintaining the same accuracy and precision as the best one, SVM, ensemble network model improved the recall, which led to a better F1 score. It also increased the AUC by 6% and 4% compared to SVM and Adaboost, respectively.

In Table 3, a comparison has been made for position 146. Although the precision dropped by 4%, we can observe an increase in accuracy and a significant improvement in recall, which resulted in a better F1 score in the end.

Table 4 Comparison between the ensemble network with the base learners for position 144 (%)

Model	Accuracy	Precision	Recall	F1 Score	AUC
Extra-trees classifier	60	63	94	41	75
Support vector machine	60	63	94	52	75
Adaboost	60	63	94	40	75
Naïve Bayes	40	62	16	55	25
Linear discriminant analysis	44	83	16	54	27
Ensemble network	**65**	65	**100**	**62**	**79**

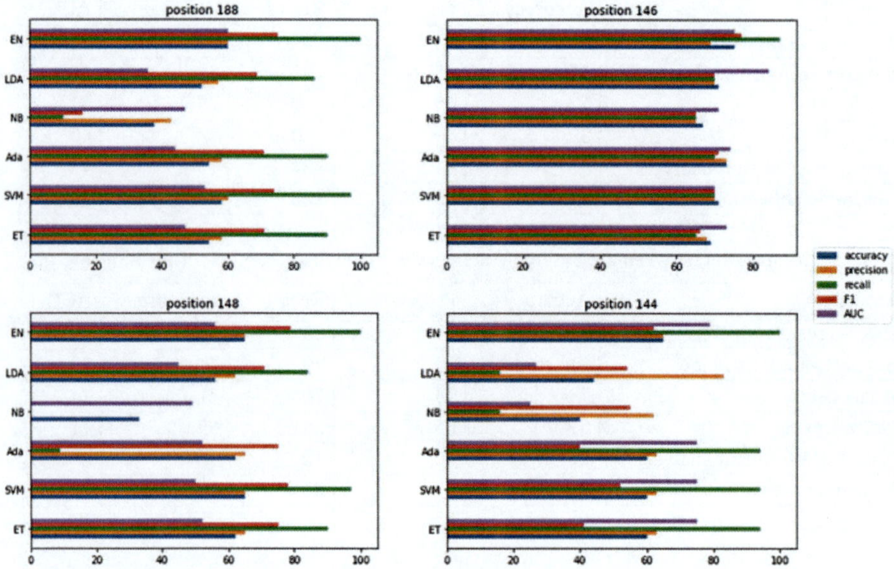

Fig. 2 Comparing the ensemble network with the base classifiers used in the study. EN: ensemble network, LDA: linear discriminant analysis, NB: naive Bayes, Ada: AdaBoost, SVM: support vector machine, ET: extra-trees classifier

For position 144, as Table 4 depicts, improvement can be observed for accuracy, recall, and F1-score. The AUC metric is also better than the base classifiers. These analyses show that the ensemble network improved the results for all four positions explored. It can be seen that the recall improved in all cases. For accuracy, one showed equal performance, and the rest improved. F-1 score, which is a harmonic mean of precision and recall, also increased when using the ensemble network for all four positions, which shows even if the precision decreased in some cases, the improvement in recall was such significant that F1-score for the proposed method surpassed the rest.

The results for these four positions are plotted in Fig. 2.

5 Conclusions

In this study, we proposed a new stacking-based classification technique to predict the in-host sequence variance (volatility) state of different positions of HIV-1 Env. We used five different base learners and incorporated MLP as the meta-learner. We tested four different datasets, each corresponding to a specific position in the amino acid sequence of Env. The data represented the absence or presence of volatility at each position.

We compared the results of the proposed method with the individual performance of each base learner. Accuracy, precision, recall, F1-score, and AUC have been used as the evaluation metrics. Improvements have been observed in all four cases, with recall improved in all datasets when the ensemble network is used. Although the precision decreased in some cases, the increase in recall was so significant that the F1-score, which is the harmonic mean of precision and recall, improved for all cases. In addition, recall is a very important metric since it shows the percentage of times when a truly volatile position is classified correctly. As shown in this study, the recall was improved in all four different positions.

Looking at the AUC, we can see that this metric is low for some positions, but considering this fact that due to the complex structure of the problem, any result better than a random guess is valuable. Also, as mentioned earlier, this work is a part of a bigger research idea in which the future volatility will be explored; therefore, the models built upon the information from this research will work better since they will use more valuable features besides the neighborhood information. In this study, we tried to improve the performance of the learner. Looking at the models that the ensemble network is compared with, we can see that some of them are powerful and flexible enough to be used alone; however, we showed that the proposed method could push them even further.

The findings in this study, first, support that that variability in the amino acid sequence of the protein is grouped on the three-dimensional structure of the molecule since we utilized the physical neighbors of a position as the primary features for the initial stage of stacking. This finding is intuitive due to the fact that in each patient, selective pressures applied on Env likely focus on a specific part of the molecule and lead to variation (Caskey et al., 2017; DeLeon et al., 2017; Han et al., 2020). Second, the proposed method – ensemble network – improved classification metrics comparing to the base learners that have been used in this algorithm, supporting that it is a good option for predicting the volatility state of Env positions.

In this study, we used k-means to create the regions and evaluated the performances in those regions. The choice of the technique to create the regions is of high importance. As a future research direction, one can investigate the effect of creating the regions on the overall performance of this stacking algorithm. Optimizing the performance by carefully defining the boundaries of regions is another topic to explore. This can link the performance of classifiers with the choice of districts and can help to achieve the best results possible.

Acknowledgments This work was supported by Magnet Grant 110028-67-RGRL from The American Foundation for AIDS Research (amfAR).

References

Adhikari, S., & Saha, S. (2014). *Multiple classifier combination technique for sensor drift compensation using ANN & KNN*. https://doi.org/10.1109/IAdCC.2014.6779495

Agrawal, U., et al. (2019). Combining clustering and classification ensembles: A novel pipeline to identify breast cancer profiles. *Artificial Intelligence in Medicine*. https://doi.org/10.1016/j.artmed.2019.05.002

Amini, M., Rezaeenour, J., & Hadavandi, E. (2014). Effective intrusion detection with a neural network ensemble using fuzzy clustering and stacking combination method. *Journal of Computer Security, 1*(4), 293–305.

Bernas, M., & Płaczek, B. (2015). Fully connected neural networks ensemble with signal strength clustering for indoor localization in wireless sensor networks. *International Journal of Distributed Sensor Networks*. https://doi.org/10.1155/2015/403242

Burton, D. R., & Mascola, J. R. (2015). Antibody responses to envelope glycoproteins in HIV-1 infection. *Nature Immunology*. https://doi.org/10.1038/ni.3158

Caskey, M., et al. (2017). Antibody 10-1074 suppresses viremia in HIV-1-infected individuals. *Nature Medicin*. https://doi.org/10.1038/nm.4268

Centers for Disease Control and Prevention. (2020). *Estimated HIV Incidence and Prevalence in the United States 2014–2018*. [Online]. Available: http://www.cdc.gov/hiv/library/reports/hiv-surveillance.htm

DeLeon, O., et al. (2017). Accurate predictions of population-level changes in sequence and structural properties of HIV-1 ENV using a volatility-controlled diffusion model. *PLoS Biology*. https://doi.org/10.1371/journal.pbio.2001549

Gaschen, B., Kuiken, C., Korber, B., & Foley, B. (2001). Retrieval and on-the-fly alignment of sequence fragments from the HIV database. *Bioinformatics*. https://doi.org/10.1093/bioinformatics/17.5.415

Han, C., et al. (2020). Key positions of HIV-1 Env and signatures of vaccine efficacy show gradual reduction of population founder effects at the clade and regional levels. *mBio*. https://doi.org/10.1128/MBIO.00126-20

Haynes, B. F., et al. (2016). HIV-host interactions: Implications for vaccine design. *Cell Host and Microbe*. https://doi.org/10.1016/j.chom.2016.02.002

Hraber, P., Seaman, M. S., Bailer, R. T., Mascola, J. R., Montefiori, D. C., & Korber, B. T. (2014). Prevalence of broadly neutralizing antibody responses during chronic HIV-1 infection. *AIDS*. https://doi.org/10.1097/QAD.0000000000000106

Ibrahim, H., & Far, B. H. (2016). Clustering and artificial neural network ensembles based effort estimation. *SEKE*, 301–308.

Kan, G., et al. (2018). A novel hybrid data-driven model for multi-input single-output system simulation. *Neural Computing and Applications, 29*(7), 577–593.

Kim, S. E., & Seo, I. W. (2015). Artificial Neural Network ensemble modeling with conjunctive data clustering for water quality prediction in rivers. *Journal of Hydro-environment Research*. https://doi.org/10.1016/j.jher.2014.09.006

Klein, F., Mouquet, H., Dosenovic, P., Scheid, J. F., Scharf, L., & Nussenzweig, M. C. (2013). Antibodies in HIV-1 vaccine development and therapy. *Science*. https://doi.org/10.1126/science.1241144

Korber, B. T., Foley, B. T., Kuiken, C. L., Pillai, S. K., & Sodroski, J. G. (1998). Numbering positions in HIV relative to HXB2CG. *AIDS Research and Human Retroviruses*.

Lee, J. H., Ozorowski, G., & Ward, A. B. (2016). Cryo-EM structure of a native, fully glycosylated, cleaved HIV-1 envelope trimer. *Science*, (80). https://doi.org/10.1126/science.aad2450

Mikell, I., Sather, D. N., Kalams, S. A., Altfeld, M., Alter, G., & Stamatatos, L. (2011). Characteristics of the earliest cross-neutralizing antibody response to HIV-1. *PLoS Pathogens*. https://doi.org/10.1371/journal.ppat.1001251

Moore, P. L., Williamson, C., & Morris, L. (2015). Virological features associated with the development of broadly neutralizing antibodies to HIV-1. *Trends in Microbiology*. https://doi.org/10.1016/j.tim.2014.12.007

Ressom, H. W., Turner, K., & Musavi, M. T. (2006). Estimation of ocean water chlorophyll-a concentration using computational intelligence. In *OCEANS 2006* (pp. 1–6).

Sadrawi, M., Sun, W. Z., Ma, M. H. M., Yeh, Y. T., Abbod, M. F., & Shieh, J. S. (2018). Ensemble genetic fuzzy neuro model applied for the emergency medical service via unbalanced data evaluation. *Symmetry (Basel)*. https://doi.org/10.3390/SYM10030071

Salazar-Gonzalez, J. F., et al. (2008). Deciphering human immunodeficiency virus type 1 transmission and early envelope diversification by single-genome amplification and sequencing. *Journal of Virology*. https://doi.org/10.1128/jvi.02660-07

Snoeck, J., Fellay, J., Bartha, I., Douek, D. C., & Telenti, A. (2011). Mapping of positive selection sites in the HIV-1 genome in the context of RNA and protein structural constraints. *Retrovirology*. https://doi.org/10.1186/1742-4690-8-87

Wang, Y., Feng, L., Li, S., Ren, F., & Du, Q. (2020). A hybrid model considering spatial heterogeneity for landslide susceptibility mapping in Zhejiang Province, China. *Catena*. https://doi.org/10.1016/j.catena.2019.104425

West, A. P., Scharf, L., Scheid, J. F., Klein, F., Bjorkman, P. J., & Nussenzweig, M. C. (2014). Structural insights on the role of antibodies in HIV-1 vaccine and therapy. *Cell*. https://doi.org/10.1016/j.cell.2014.01.052

Wolpert, D. H. (1992). Stacked generalization. *Neural Networks*. https://doi.org/10.1016/S0893-6080(05)80023-1

Yang, Y., Zheng, K., Wu, C., Niu, X., & Yang, Y. (2019). Building an effective intrusion detection system using the modified density peak clustering algorithm and deep belief networks. *Applied Sciences*. https://doi.org/10.3390/app9020238

Yuxian, Z., Min, L., Jianhui, W., Dan, W., & Yunfei, M. (2009). A hybrid modeling using clustering algorithm for textile slashing process. In *2009 Chinese Control and Decision Conference* (pp. 5751–5754).

Social Relations Under the Covid-19 Epidemic: Government Policies, Media Statements and Public Moods

Wangzhe, Zhongxiao Zhang, Qianru Tao, Nan Ye, and Runjie Xu

1 Introduction

Hitherto, no social studies have assessed the relationship between government policy, media opinion and public sentiment during the outbreak from the perspective of emotional preference. The early research on the new epidemic situation mainly analyzed the transmission mode and economic impact of the virus, which was reflected in the fields of health environment, government policy, financial risk and so on (Goodell, 2020; Zhang et al., 2020; Falato et al., 2020; Lin et al., 2020; Fang et al., 2020). Early predictions and simulations have shown that, in addition to its serious health consequences, the epidemic will shrink global economic growth and increase global poverty and unemployment (Ashraf, 2020).

In fact, major public health events have also had a certain impact on social stability. For example, problems such as rush purchase of living goods, rising prices and rumor spread in the early stage of the epidemic are all aspects that need to be focused on by the government and relevant departments in addition to economic and health factors (Cinelli et al., 2020).

As the leader of controlling public emergencies, the government is responsible for analyzing the current situation and formulating reasonable policies (Cheng et

Wangzhe · Q. Tao · N. Ye
College of Economics and Management, Nanjing University of Aeronautics and Astronautics, Nanjing, China

Z. Zhang
Dalian University of Foreign Languages, Business School, Dalian, China

R. Xu (✉)
College of Economics and Management, Nanjing University of Aeronautics and Astronautics, Nanjing, China

Nanjing Boya Blockchain Research Institute, Nanjing, China

H. Yang et al. (eds.), *AI and Analytics for Public Health*, Springer Proceedings in Business and Economics, https://doi.org/10.1007/978-3-030-75166-1_27

al., 2020). Under the epidemic situation, a large number of related policies mainly focus on restrictions, including sacrificing economic growth to control the epidemic. These policies often affect the public mood, and the public is also very likely to speculate on the current epidemic trend through the relevant policies of the government (Anderson et al., 2020). Compared with the policy documents, the opinions released by the media play a more important role in the public health events and social and economic development. On the one hand, the public is more likely to access and understand the opinions released by the media, on the other hand, the emotions of the masses interfere with the implementation progress of government policies to a certain extent (Jacob et al., 2020; Bonaccorsi et al., 2020; Depoux et al., 2020).

China's novel coronavirus pneumonia is the most important part of the study. From the perspective of sentiment analysis, this paper collected monitoring data from government official website, newspapers and social media, and tracked and evaluated the social mood in the time before the outbreak of the new crown pneumonia in China. (the data set used in this study was from February 1st to April 6th. It has been extended to December 1 to June 1).

By using mf-dcca model, we explore the relationship between government and official media in major public health emergencies in China, the United States and the United Kingdom. Combined with the actual situation of the three countries' epidemic development, we analyze the optimal relationship between the government and the official media, which provides a further theoretical basis for the relationship between the government and the public, and puts forward another media under the emergency This paper discusses the relevant methods of media hedging government policies in social emotion.

We explain how government policies affect people's behavior consciousness through the media of news; what roles social media and news media play in risk communication, rumors and false information; and how the public's behavior response affects the government's bias in policy choices. The main conclusions are as follows

1. During the epidemic period, there is a significant cross correlation between the national government policies and the emotions reflected by the media. At this time, the media, as an intermediary role, conveys the emotional preference for the future expectation of the epidemic, and thus affects the public mood and mobilizes the enthusiasm of the whole people to fight the epidemic.
2. When the mood index fluctuates little, the mood of the media and the government is consistent, and the public sentiment is less affected at this time.
3. When the emotion index fluctuates greatly, obvious emotional words will appear in government policies and media speeches. The media will give emotional hedging language explanations to these restrictive policies of the government, trying to resolve the negative emotions of the people.

2 Analysis Process and Method

2.1 Data Acquisition and Preprocessing

1. Government policy data (December 1, 2019 – June 1, 2020)

This paper selects the text data of government and official media in three countries from February 1, 2020 to April 8, 2020 for empirical research. China was the first country in the world to suffer a large-scale epidemic. However, the timely prevention and control of the epidemic by the Chinese government quickly alleviated the epidemic.Wuhan, the most severely affected city in China, lifted the lockdown measures on April 8, 2020, indicating the gradual end of the epidemic in China.the United States is one of the world's economic and cultural centers, the outbreak was not serious in early February, but the optimism maintained by the United States government in the early stage of the epidemic caused the epidemic to continue to this day, and it will continue to spread. Three countries in this period of prevention and control measures have a more distinct representative, so this paper selected China, the United States, the United Kingdom as the experimental object.

2. News media data (December 1, 2019 – June 1, 2020)

In terms of the selection of textual data, this study selected the policies on the epidemic issued daily by the central government of each country and some key regions, with relatively comprehensive data. In the text data of official media, this paper selects the traditional and old commercial and critical media whose opinions are regulated by relevant national authorities to publish relevant opinions about the epidemic. Chinese media chose the FT, FT is the financial times, the only Chinese business critical financial information web site, the site the thorough analysis of the influential events in the Chinese economy, and because the site is mainly targeted at China, compared with general English newspapers and magazines, FT on some sensitive issues more cautious, more can reflect the actual situation of China. The media in the UK choose Guardian. Guardian is a national comprehensive daily newspaper in the UK. Together with The Times and The Daily Mail, guardian is jointly named as the three major newspapers in the UK. American media chose Bloomberg, the world's leading provider of information services, news and media.

3. Sentiment data (December 1, 2019 – June 1, 2020)

In this study, we explored people's emotions from three perspectives: search engine, social media and online shopping. Specifically, people obtain the relevant information about the epidemic situation and protection through search engines, spread ideological and even false information about the epidemic through social media, and purchase scarce resources around the community (such as masks, hand sanitizers, disinfectants, etc.) through online shopping.

Search engine: search engine we chose "Baidu", which is equivalent to China's Google, with more than 1 billion Chinese users. We have statistically analyzed novel coronavirus pneumonia, COVID-19 and Novel Coronavirus Pneumonia from the search engine. In addition, the Chinese Center for Disease Control and prevention has issued a series of guidelines to recommend residents to take personal protective measures, mainly including: respiratory protection, hand hygiene, Because people often get and understand this kind of popular science content through search engines, we include "Mask", "hand sanitizer", "disinfectant" and "temperature" into the search keywords.

Online shopping: we chose Alibaba for online shopping. Alibaba platform, including Taobao and tmall, is an e-commerce platform similar to Amazon in China, with more than 1 billion users. Novel coronavirus pneumonia, hand washing solution, disinfectant and thermometer were used to evaluate the public's behavioral responses to the new crown pneumonia epidemic.

2.2 *Emotional Analysis*

In this study, the affective analysis dictionary based on BosonNLP Chinese semantic open platform is used to analyze text data. BosonNLP is an automatic emotional polarity dictionary based on a large number of data sources such as news, forum, Weibo and so on. BosonNLP Chinese semantic Open platform provides industry-leading emotional analysis. Its machine learning model is based on hundreds of thousands of news balanced corpora and millions of social network balanced corpora. At the same time, new things can be learned as part of the corpus according to the content of the analysis. The accuracy of BosonNLP analysis can reach 85–90% (Ying et al., 2017). In addition, compared with the general emotion dictionary, the BosonNLP has a better performance in dynamic text analysis (Hu, 2017).

Since the text data selected in the study come from the government policy and the official media strictly supervised by the state, the corpus in the BosonNLP can basically cover the extracted text. Therefore, this paper uses Boson emotion analysis as a tool for emotional analysis of this study text. In addition, if the index is greater than 0.5, the text is positive. If the index is less than 0.5, the text appears negative.

2.3 *Cross Correlation Analysis Method*

In order to study the relationship between the government and the official media in China, the United States and the United Kingdom. In this paper, we first test the cross correlation of time series from the qualitative point of view, so we use the test time series cross correlation statistic $Q_{cc}(m)$ proposed by podobnik (Podobnik et al., 2009). Suppose that two time series $x(t)$and $y(t)$ of the same length, $t = 1, 2 \cdots, N$, where N is the length of the time series, and the cross correlation statistics are as follows:

$$Q_{cc}(m) = N^2 \sum_{i=1}^{m} \frac{C_i^2}{N - i}$$

Especially,

$$C_i = \frac{\sum_{k=i+1}^{N} x_k y_{k-i}}{\sqrt{\sum_{k=1}^{N} x_k^2 \sum_{k=1}^{N} y_k^2}}$$

m is the degree of freedom because C is subject to the mean value of 0, and the variance is $\frac{1}{N}(N - i)$, so the cross correlation statistic $Q_{cc}(m)$ is approximately a chi square distribution with m degrees of freedom. If there is no cross correlation between the two time series, the cross correlation test is consistent with chi square distribution. If the cross correlation test statistic exceeds the chi square distribution $\chi(m)$ at a certain confidence level, the cross correlation between the two time series is significant.

2.4 MF-DCCA Model

Podobnik and Stanley (2008) first proposed the de trend correlation analysis method (DCCA), and then Zhou (2008) and others mixed the detrended fluctuation analysis model (DFA) model with DCCA model to form a multifractal de trend cross correlation analysis model (MF-DCCA). The model is a quantitative method to study the internal cross correlation and multifractal characteristics of two non-stationary time series. The basic principle of mf-dcca is as follows:

Suppose there are two time series $x(t)$and $y(t)$, $t = 1, 2\cdots, N$, where N is the length of the time series. In order to avoid artificial instability of time series, two new time series are constructed by calculating the cumulative dispersion of time series:

$$X(t) = \sum_{k=1}^{t} (x(k) - \overline{x}), \quad t = 1, 2, \cdots, N$$

$$Y(t) = \sum_{k=1}^{t} (y(k) - \overline{y}), \quad t = 1, 2, \cdots, N$$

Here,$\overline{x} = \frac{1}{N}\sum_{i=1}^{N} x_i$,$\overline{y} = \frac{1}{N}\sum_{i=1}^{N} y_i$.

Two new time series X (t) and Y (t) are divided into uncorrelated equal length subsequences. The length of each sequence is $N_s = \mathrm{int}\left(\frac{N}{s}\right)$ subsequence. Because the time series is not necessarily exactly divided into n_ In order to ensure that the information of the time series can be fully utilized, the time series is inverted and processed in the same way, so the final $2N_s$ subsequence. In order to ensure the stability of the final results, it is generally assumed that $10 < s < \frac{N}{4}$.

The $2N_s$ subsequences were fitted by polynomial regression using the least square method. Finally, the local trend fitting values $X_v(i)$ and$Y_v(i)$ of each subsequence were obtained, and each subsequence was de trended to obtain the local covariance function.

The local covariance function of positive cut sequence, i.e. $V = 1, 2, \cdots, N_s$, is as follows:

$$F^2(s, v) = \frac{1}{s}\sum_{i=1}^{s}\left|X((v-1)s+i) - X_v(i))\right| \times \left|Y((v-1)s+i) - Y_v(i))\right|$$

The local covariance function of $V = N_s + 1, 2, \cdots, 2N_s$ is as follows:

$$F^2(s, v) = \frac{1}{s}\sum_{i=1}^{s}\left|X(N-(v-N_s)s+i) - X_v(i))\right| \times \left|Y(N-(v-N_s)s+i) - Y_v(i))\right|$$

The q-order wave function is obtained by averaging the local covariance of all subsequences:

$$F_q(s) = \left(\frac{1}{2N_s}\sum_{v=1}^{2N_s}\left(F^2(s, v)\right)^{q/2}\right)^{1/q}$$

When $q = 0$, the wave function of the time series is obtained by the law of l'urbida:

$$F_0(s) = \exp\left(\frac{1}{4N_s}\sum_{v=1}^{2N_s} LnF^2(s, v)\right)$$

If there is a long-range correlation between time series, the relationship between wave function $F_q(s)$ and time scale s should be as follows:

$$F_q(s) \sim s^{h_{12}(q)}$$

So,

$$LnF_q(s) = h_{12}(q)\,Ln(s) + Ln(C)$$

Then $h_{12}(q)$ is the slope of the straight line $LnF_q(s)$ and $Ln(s)$ obtained by the least square method for each q value, also known as the generalized Hurst index. If $h_{12}(q)$ is not a constant, but a value varying with q, then the cross correlation between the two time series has multifractal characteristics. When $q = 2$, MF-DCCA model becomes a single fractal DCCA model to study the cross correlation of two time series, namely $H_{12} = h_{12}(2)$. If $1 > H_{12} > 0.5$, the two time series have long-range positive correlation; if $0.5 > H_{12} > 0$, the two column time series have long-range inverse correlation; if $H_{12} = 0.5$, there is no long-range correlation.

Shadkhoo proposed the multifractal index $\tau(q)$ to further test whether there is multiple cross correlation between the two time series:

$$\tau_{12}(q) = qH_{12}(q) - 1$$

If $\tau(q)$ is a linear function that changes with the change of q, then the cross correlation of two time series has multifractal characteristics, otherwise, the cross correlation of two time series is single fractal.

The singular value intensity function α_{12} and multifractal spectrum function $f_{12}(\alpha)$ are obtained by Legendre transformation:

$$\alpha_{12} = \tau'_{12}(q) = h_{12}(q) + qh'_{12}(q)$$

$$f_{12}(\alpha) = q(\alpha_{12} - h_{12}(q)) + 1$$

If α_{12} is not a constant, the cross correlation between the two time series has multifractal characteristics, otherwise, it is a single fractal feature. In addition, the multifractal intensity can be reflected by the width of multifractal spectrum

$$\Delta\alpha_{12} = (\alpha_{12})_{max} - (\alpha_{12})_{min}$$

The wider the multifractal spectrum is, the stronger the multifractal intensity of the two time series is.

3 Empirical Analysis and Discussion

3.1 Emotional Analysis Results

In order to study the relationship between the government and the media in major public emergencies, according to the boson sentiment analysis tool, this paper obtains the emotional state and degree reflected by the government policies and official media statements during the epidemic period in China, the United States and the United Kingdom, as shown in Fig. 1.

On the whole, the sentiment reflected by Chinese government policies fluctuates little, and the proportion of negative emotions is relatively small. Only the sentiment index of the government policies on February 16 and March 15 is in a negative state. In contrast, the mood reflected by the US and UK governments fluctuates greatly, with a higher proportion of negative emotions, but still dominated by positive

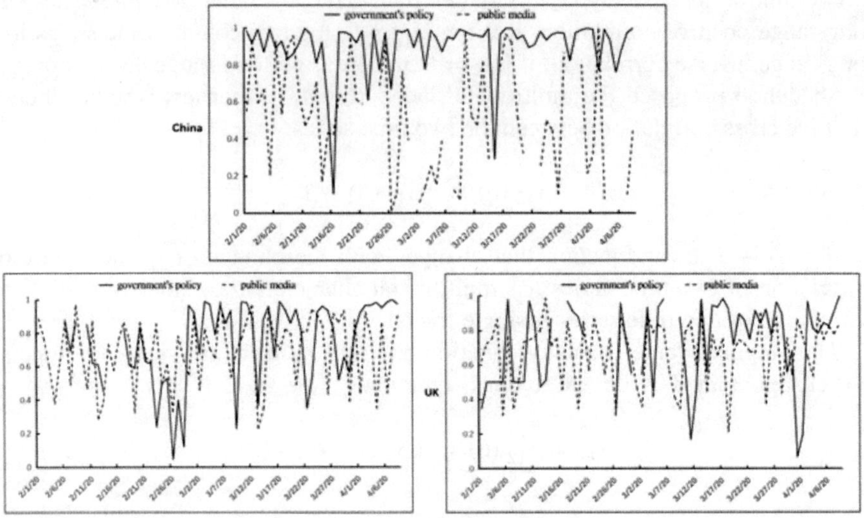

Fig. 1 The emotional analysis results of policies and media

emotions, which indicates that the governments of China, the United States and the United Kingdom are in a negative state During the epidemic period, the government of the United States and the United Kingdom basically adopted a more positive policy towards the public, but the negative policy of the United States and the United Kingdom accounted for a larger proportion than that of China. In contrast to the more positive emotions of the Chinese government policies, the negative comments account for a large proportion of the emotions reflected by the Chinese government's statements, which further proves that there is a negative correlation between government policies and official media opinions in order to alleviate the public's emotions in reality. However, the mood reflected by the official media in the United States and the United Kingdom is generally less volatile than that in China.

From Fig. 1, we can see that the policy index and media index of China, the United States and the United Kingdom overlap at some time points, and the trend basically coincides in some periods. Therefore, this paper believes that there is a certain correlation between the policy and the official media. In order to prove the correlation between policies and media in the three countries and study the characteristics of the correlation, this paper first uses $Q_{cc}(m)$ test to verify the cross correlation between policy index and media index, and then uses MF-DCCA model to verify and analyze the multifractal characteristics of cross correlation.

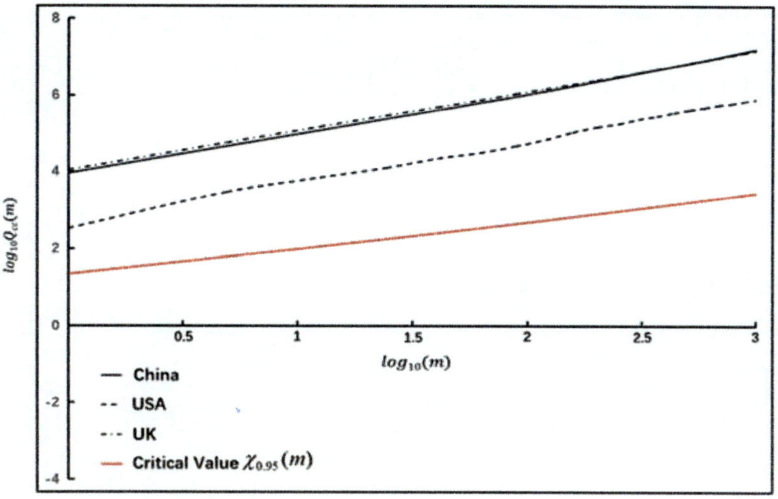

Fig. 2 Cross correlation statistics

3.2 *Proof of Cross Correlation*

In order to prove the cross correlation between the policy index and the media index, this section uses the $Q_{cc}(m)$ test method to test the cross correlation of the three countries. If the $Q_{cc}(m)$ statistic is higher than the chi square distribution $\chi_\alpha(m)$ with confidence level α and degree of freedom m, then there is cross correlation between the government and the media; otherwise, there is no cross correlation between them.

Referring to the experimental process in other studies (Ruan et al., 2018), the chi square distribution $\chi_{0.95}(m)$ with m degree of freedom at 5% significance level was added as the critical value. As can be seen from Fig. 2, the cross correlation statistics $Q_{cc}(m)$ of policy index and media index of China, the United States and the United Kingdom are all above the critical value, which verifies the cross correlation between the policy index and media index of the three countries from a qualitative perspective, and further proves that there is a certain correlation between government policies and official media statements during the epidemic period.

3.3 *Multifractal Analysis*

The above $Q_{cc}(m)$ test method has proved the cross correlation between government policies and media opinions in the three countries from a qualitative perspective. In order to quantitatively study the characteristics and connotation of cross correlation, MF-DCCA model is used to analyze the multifractal characteristics of correlation.

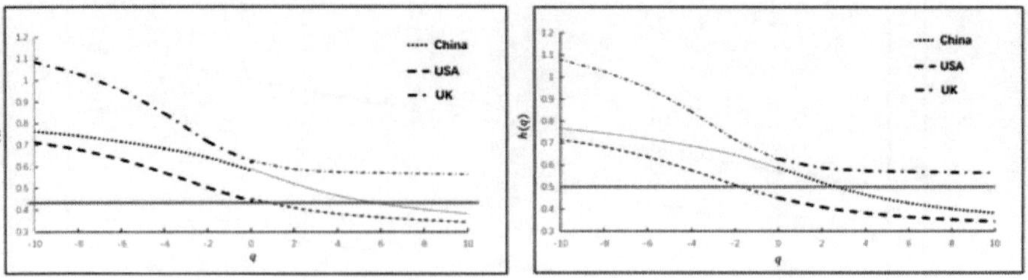

Fig. 3 Generalized Hurst exponent graph

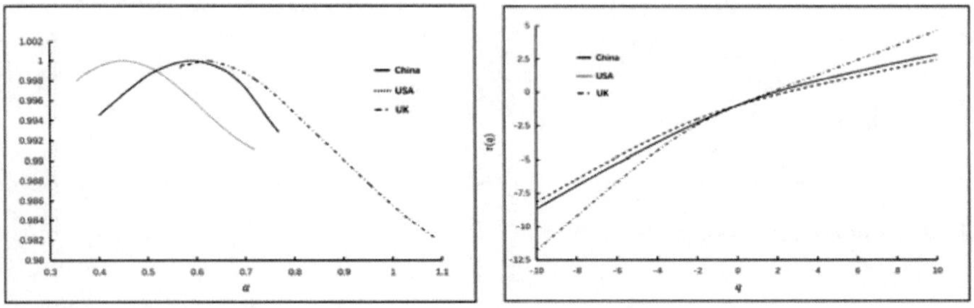

Fig. 4 Indicators of correlation characteristics for the three countries, τ(q) and f(α)

MF-DCCA model includes generalized Hurst index h(q), mass index τ(q) and multifractal spectrum index f(α). Among them, the generalized Hurst index proves that the cross correlation between the two time series has multifractal characteristics, and reflects the positive (negative) correlation characteristics of time series under different fluctuation amplitude, and the quality index further proves the multifractal characteristics of cross correlation. Multifractal spectrum index reflects the intensity of cross correlation between policy index and media index.

Figure 3 shows the generalized Hurst index h(q) of the cross correlation between the policy index and the media index of the United States, China and the United Kingdom. It can be seen from Fig. 3 that when $-10 \leq q \leq 10$, the generalized Hurst index h(q) decreases with the increase of q, and is not a constant. It shows that the cross correlation between the policy index and the media index of the three countries has multifractal characteristics, that is, the correlation between government policy and official media speech has some characteristics.

In order to further prove the multifractal characteristics of the cross correlation between government policies and the emotions reflected by media opinions, this paper adopts the quality index τ(q), where $-10 \leq q \leq 10$. It can be seen from Fig. 4 that τ(q) is non-linear dependent on q rather than a constant, which once again makes clear that the cross correlation between the emotions reflected by government policies and media policies in the United States, China and the United Kingdom is multifractal.

In order to further explore the multifractal characteristics of the cross correlation between media opinions and the emotions reflected by government policies, this

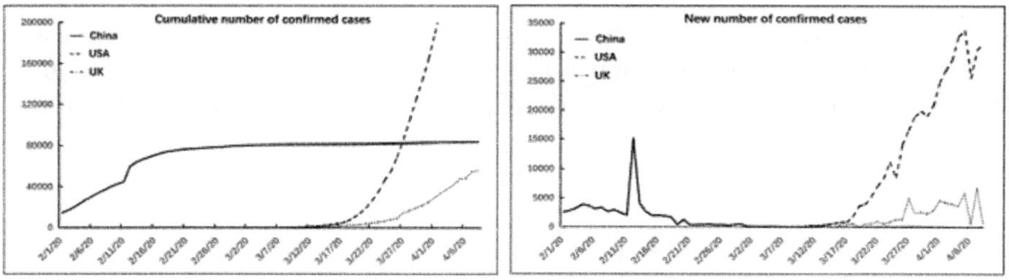

Fig. 5 The trend of epidemic accumulation and new number of confirmed cases

paper uses singular index α and multifractal spectrum f(α) to describe them. It can be seen from Fig. 4 that the multifractal functions of China, the United States and China all show a bell shape, which further illustrates that the cross correlation between the government policies of the three countries and the emotions reflected by media opinions has multifractal characteristics. In addition, the top of the multifractal spectrum curve is relatively flat, the opening is wide, the curve distribution range is large, and the curve of the three countries all shows the phenomenon of right deviation, and the multifractal spectrum shows a hook shape, which indicates that there is a cross correlation between the government policies of China, the United States and the United Kingdom and the emotions reflected by media opinions, but the cross correlation is not large.

The major public health emergencies, the government policy and media speech in the mood fluctuations, keep the same attitude is the best In this way, the masses can realize the seriousness of the problem at the first time, and then take corresponding measures. In the case of small mood fluctuations, the two can appropriately maintain the opposite attitude. When the attitude reflected by the government policy is negative, the more optimistic attitude of the official media can appropriately alleviate the mood of the masses; on the contrary, when the attitude reflected by the government policy is positive, the government can take appropriate measures The negative attitude of the media can timely remind the public of the fact that the epidemic is not over, and promote the development of the epidemic in a better direction (Fig. 5).

4 Conclusion and Discussion

This paper analyzes the cross correlation between the government policies and the official media statements of China, the United States and the United Kingdom during the epidemic period by using the emotional analysis, and makes a detailed analysis of the cross correlation between the government policies of China, the United States and the United Kingdom and the opinions of the official media, and draws the following conclusions:

During the epidemic period, there is a significant cross correlation between the national government policies and the emotions reflected by the media. At this time, the media, as an intermediary role, conveys the emotional preference for the future expectation of the epidemic, and thus affects the public mood and mobilizes the enthusiasm of the whole people to fight the epidemic.

When the mood index fluctuates little, the mood of the media and the government is consistent, and the public sentiment is less affected at this time.

When the emotion index fluctuates greatly, obvious emotional words will appear in government policies and media speeches. The media will give emotional hedging language explanations to these restrictive policies of the government, trying to resolve the negative emotions of the people.

In the event of an outbreak, the government should timely grasp the overall situation of the epidemic situation and take corresponding policy measures. At the same time, the official media under the supervision of the state should release information according to the policy situation, so as to achieve the purpose of promoting the alleviation of the epidemic situation. The government's rational state in policy can make the masses have a clear and accurate judgment on the epidemic situation, which is conducive to the alleviation of the epidemic situation. However, the official media should achieve the synergy effect with the government policy, but the coordination means to keep consistent. The official media should determine their attitude of speech on the same day according to the mood fluctuation reflected by the government the day before and on the same day If the government's policy reflects less emotional fluctuation, the official media should release the same opinion as the government's policy attitude on the same day, so as to ensure that the epidemic information can be timely and accurately transmitted to the masses. On the contrary, if the government's policy reflects a large fluctuation of emotions, the official media should release comments opposite to the government's attitude on that day to reduce the public's over tension or optimism So as to ensure that the epidemic emergencies can be solved as soon as possible.

Acknowledgments This work is supported by the National Social Science Foundation of China (17BGL055), Jiangsu Graduate Education and Teaching Reform Project (JGLX19_014), Central University Basic Scientific Research Operating Fund Special Fund (NJ2020044) and Innovation Project Fund of NUAA (2019EC01, 2019EC09, 2020CX009040).

References

Anderson, R. M., Heesterbeek, H., Klinkenberg, D., & Hollingsworth, T. D. (2020). How will country-based mitigation measures influence the course of the COVID-19 epidemic? *The Lancet, 395*(10228), 931–934.

Ashraf, B. N. (2020). Economic impact of government interventions during the COVID-19 pandemic: International evidence from financial markets. *Journal of Behavioral and Experimental Finance, 27*, 100371.

Bonaccorsi, G., Pierri, F., Cinelli, M., Flori, A., Galeazzi, A., Porcelli, F., ... Pammolli, F. (2020). Economic and social consequences of human mobility restrictions under COVID-19. *Proceedings of the National Academy of Sciences, 117*(27), 15530–15535.

Cheng, C., Barceló, J., Hartnett, A. S., Kubinec, R., & Messerschmidt, L. (2020). COVID-19 government response event dataset (CoronaNet v. 1.0). *Nature Human Behaviour, 4*(7), 756–768.

Cinelli, M., Quattrociocchi, W., Galeazzi, A., Valensise, C. M., Brugnoli, E., Schmidt, A. L., et al. (2020). The covid-19 social media infodemic. arXiv preprint arXiv:2003.05004.

Depoux, A., Martin, S., Karafillakis, E., Preet, R., Wilder-Smith, A., & Larson, H. (2020). The pandemic of social media panic travels faster than the COVID-19 outbreak. *J Travel Med, 27*(3), taaa031.

Fang, Y., Nie, Y., & Penny, M. (2020). Transmission dynamics of the COVID-19 outbreak and effectiveness of government interventions: A data-driven analysis. *Journal of Medical Virology, 92*(6), 645–659.

Falato, A., Goldstein, I., & Hortaçsu, A. (2020). *Financial fragility in the COVID-19 crisis: The case of investment funds in corporate bond markets (No. w27559).* National Bureau of Economic Research.

Goodell, J. W. (2020). COVID-19 and finance: Agendas for future research. *Finance Research Letters, 101512.*

Hu, X. (2017). *Research on Named Entity Recognition and Visualization Methods for Social Aspect Dynamic Analysis [D].* National University of Defense Technology. (in Chinese).

Jacob, L., Smith, L., Butler, L., Barnett, Y., Grabovac, I., McDermott, D., ... Tully, M. A. (2020). COVID-19 social distancing and sexual activity in a sample of the British Public. *The Journal of Sexual Medicine.*

Lin, Q., Zhao, S., Gao, D., Lou, Y., Yang, S., Musa, S. S., ... He, D. (2020). A conceptual model for the outbreak of Coronavirus disease 2019 (COVID-19) in Wuhan, China with individual reaction and governmental action. *International Journal of Infectious Diseases.*

Podobnik, B., Grosse, I., Horvatić, D., et al. (2009). Quantifying cross-correlations using local and global detrending approaches. *The European Physical Journal B, 71*(2), 243.

Podobnik, B., & Stanley, H. E. (2008). Detrended cross-correlation analysis: A new method for analyzing two nonstationary time series. *Physical Review Letters, 100*(8), 084102.

Ruan, Q., Yang, H., Lv, D., et al. (2018). Cross-correlations between individual investor sentiment and Chinese stock market return: New perspective based on MF-DCCA. *Physica A: Statistical Mechanics and its Applications, 503*, 243–256.

Ying, K., Jingchang, P., & Minglei, W. (2017). Research on sentiment analysis of micro-blog's topic based on TextRank's abstract. In *Proceedings of the 2017 International Conference on Information Technology* (pp. 86–90).

Zhang, D., Hu, M., & Ji, Q. (2020). Financial markets under the global pandemic of COVID-19. *Finance Research Letters, 101528.*

Zhou, W. X. (2008). Multifractal detrended cross-correlation analysis for two nonstationary signals. *Physical Review E, 77*(6), 066211.

A Machine Learning Approach to Understanding the Progression of Alzheimer's Disease

Vineeta Peddinti and Robin Qiu

1 Introduction

A healthy brain has about 100 billion neurons. All the neurons communicate with each other by passing information via their extensions known as synapses. Due to Alzheimer's (Korolev, 2014), accumulation of proteins called amyloid plaques and tau/tangles takes place inside and outside the neurons in the patients' brains (Alzheimer's Association, 2020), which contribute to the damage of the neurons by interfering with the communication between the neurons. One such strategy which the researchers are concentrating on is measuring the levels of these proteins to predict the loss of brain cells. Few autopsy results also show that the people with Alzheimer's also suffer from other dementia disorders as well.

There are more than 5.8 million Americans aged 65 and older suffering from Alzheimer's Disease (AD) as of 2020 (Alzheimer's Association, 2020). It is the sixth leading cause of death in United States, which is worse than the breast and the prostate cancer combined. When it comes to the expenses, it costs the nation a total of $305 billion which could reach $1.1 trillion by 2050. People around the age 70 have 61% chance that they die due to dementia compared to the people who do not have dementia. Also, the AD progression rate is uncertain, varying from patient to patient. Some may live for an average of 4–8 years, yet some may live for 20 years after AD is diagnosed. As the elderly population increase, so are new cases. Note

V. Peddinti (✉)
Big Data Lab, Division of Engineering and Information Science, The Pennsylvania State University, Malvern, PA, USA
e-mail: vkp5111@psu.edu

R. Qiu (✉)
Division of Engineering & Info Sci, Pennsylvania State University, Malvern, PA, USA
e-mail: robinqiu@psu.edu

H. Yang et al. (eds.), *AI and Analytics for Public Health*, Springer Proceedings in Business and Economics, https://doi.org/10.1007/978-3-030-75166-1_28

that although it is the elderly people who are mostly affected by AD, statistics show that people of all ages can develop AD (Alzheimer's Association, 2020).

The most common cause of dementia is the Alzheimer's disease accounting for the majority of the cases. As already mentioned, it is a degenerate process meaning it slowly progresses. Also, it is only known to show symptoms only after 20 years it has started. By then the part of the neurons that are responsible for memory, learning and thinking capabilities are damaged. It slowly kills the other nerve cells responsible for eating and other functional activities. AD patients at the severe stage require extensive care and service.

Although there has been a lot of research (Alzheimer's Association, 2020) going on this neurological disorder, the cause is still unknown. AD patients suffering from dementia often ask doctors about the progression rate and the doctors have very less information to suggest a figure. Prediction of the number of months for a stage change in AD is extremely valuable for the doctors to design and provide the treatment strategies and guidance to slow down the progression in the stipulated time. Giving treatments to AD patients and working on their lifestyle changes through early diagnosis and intervention prove effective in slowing down their disease progressions (Qiu & Qiu, 2018; Solomon et al., 2014). Recently, machine learning techniques have been widely applied in AD modeling to better understand AD progression for improved treatments. Therefore, by leveraging the advances of machine learning techniques, this study focuses on predicting the actual time an AD patient will take to progress to the next stage and understanding the factors contributing to AD progressions at different stages.

The remaining paper is organized as follows. Section 2 briefly provides the background for this study. Section 3 presents the different stages that a patient will undergo with AD. Section 4 presents the dataset and methods used for data pre-processing. Section 5 talks about the regression model used to predict the number of months and the graph visualizations of the predicted factors. Section 6 features the results and the study of feature importance deduced from the trained regression model. Section 7 ends with the conclusion.

2 Literature Review

Machine learning techniques have recently gained considerable attention to predict the progression of AD (Wang et al., 2018). Unsupervised approaches have been used to find the progression stage after a known period. Fisher et al. (2019) has used an unsupervised machine learning model called Conditional Restricted Boltzmann Machine to forecast the disease progression. Satone et al. (2019) proposed machine learning techniques to cluster participants in distinct AD progression groups using unsupervised clustering Gaussian Mixture Modelling. Satone et al. (2019) have extended their research using the supervised algorithms like Naïve Bayes, Logistic Regression, and Random Forest. Albright and Initiative (2019) utilized various machine learning classifiers like logistic regression, SVMs, and neural networks and revealed that MLP (Multi-Layer Perceptron) and RNN performed the best for forecasting the progression of AD. Oriol et al. (2019) contributed to the research for late on-set AD (LOAD) prediction using machine learning models. The data for this research is taken from the Alzheimer's Disease neuroimaging Initiative (ADNI). Several classification models like Random Forest, KNN, SVM, LASSO was used. They showed that LASSO combined with ensemble techniques yielded nearly 72% of area under the ROC curve. They further identified the factors contributing to LOAD progression.

Lee et al. (2019) used multi modal deep learning approach on the ADNI data which includes MRI (Magnetic Resonance Imaging), PET (Position Emission Tomography) and other biological markers. The combined dataset was fed into an (Recurrent Neural Network) RNN model to assess the progression of mild cognitive impairment and early AD after a specific time interval. AD biomarkers are modeled using SVM by Franzmeier et al. (2020) to predict the decline in cognitive levels. The results show that the patients suffering from sporadic AD are at greater risk for cognitive decline. El-Sappagh et al. (2020) proposed an ensemble deep learning technique, a stacked convolutional neural network (CNN) and bi-directional long short-term memory (Bi-LSTM), to predict progression values of multiple cognitive scores within a fixed time period. CNN was used to extract the local features whereas Bi-LSTM was used to extract the temporal features. Several modalities were considered like the MRI, demographics, cognitive scores, FAQs for predicting the future AD stage values. They concluded that single modality cannot well determine the prediction of Alzheimer's progression. Instead of predicting the progression stages, researchers have also developed a machine learning model that could predict the cognition decline rate in the memory loss of a patient suffering from AD for up to 2 years in the future (Cohut, 2019).

The literature work mentioned above was focused on the qualitative assessment of the progression stage or the cognition decline after known time. More specifically, the works mentioned above focused on predicting the cognitive decline after certain period or calculating what could be the future stage given the current stage of the patient. Our study mainly investigates understanding the most important factors that might impact the progression. By predicting the number of months that a patient might take to progress to the next stage, this paper identifies the factors contributing to the progression. We believe if we know the time it might take for an AD patient to progress to further stage and contributing factors, doctors can effectively plan the treatment procedure and lifestyle changes for the patient, aimed at helping downtrend his/her progression in the predicted time interval.

3 Alzheimers Disease Stages

The CDR (Clinical Dementia Rating) is a numerical value representing the dementia stage (O'Bryant et al., 2008). It is a 5-point scale calculated by testing six different cognitive and behavioral domains namely memory, orientation, judgement and problem solving, community affairs, home and hobbies performance, and personal care. To calculate all these scores, the information is collected from the patient and the co-participant (e.g., Family-Member). Once all these scores are obtained, an overall CDR score (Global CDR / CDRGLOB) can be calculated by an algorithm designed by National Alzheimer's Coordinating Center (NACC). The Global CDR is standardized as a clinical Dementia Metric applicable to Alzheimer's Disease and other dementias.

AD Stages describes the progression of the Alzheimer's in a patient right from the phase when the symptoms are not yet visible to the phase where the memory cells start getting damaged and disturbs the functional motions in the body. The stages involved in the Global CDR are divided into five categories (O'Bryant et al., 2008), which a patient can be in. The time that a patient stays in any stage varies, largely depending on the age, health, and lifestyle of the patient. The following list shows the five values from the least to the most severe AD stages.

- Stage 1: 0 (No Impairment)
- Stage 2: 0.5 (Questionable Impairment)
- Stage 3: 1 (Mild Impairment)
- Stage 4: 2 (Moderate Impairment)
- Stage 5: 3 (Severe Impairment)

The dataset taken from NACC has the CDRGLOB feature which will be considered for calculating the number of months taken for progressing from a current stage to its next stage. Other clinical dementia metrics can also be used if needed.

4 Materials and Methods

4.1 Data Description and Analysis

The data utilized in the study is provided by NACC, comprising information from patients' demographics, co-patient demographics, subject medications, subject health history, physical data, Global Staging- Clinical Dementia Rating, Clinical Diagnosis, Functional Activities Questionnaire (FAQ) and other dementia markers. The original dataset consisted of a total of 147,565 medical records belonging to 42,022 patients with irregular visit times. Since this study focused on AD patients, we considered only the patients who are marked with probable AD/possible AD (**PROBAD/POSSAD**). As a result, 12,733 PROBAD patients during the period 1st September, 2005 to 12th May, 2015 have been extracted. We removed the patient records who did not have any progression or who had less than two visits since our research required the patient made a progression for the next visit. The final dataset, after applying the above-mentioned conditions, had 16,097 records with 4228 patients from 1st September, 2005 to 13th March, 2015.

There are 643 features in the original dataset. But since our focus was on AD, we identified 79 features that could be important for answering our research questions.

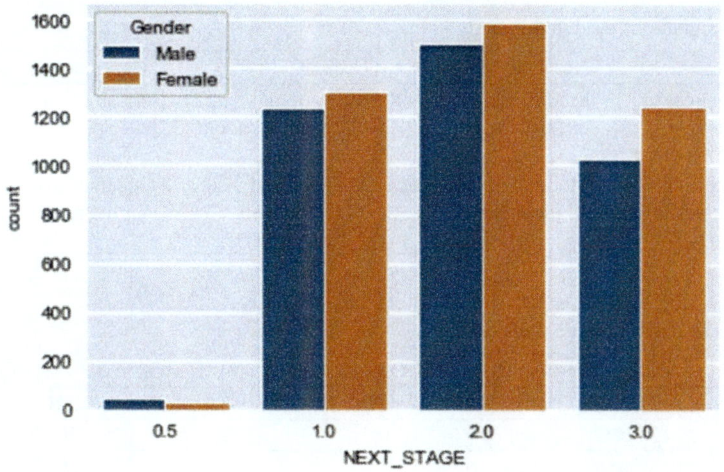

Fig. 1 AD patients' stage distribution

Different missing data schemes have been used for all these features depending on their types which will be discussed in the pre-processing section. In addition to the existing features, we have introduced a few of user defined features such as age that was calculated using the birth year details, number of months that was calculated using the successive visit details.

Figure 1 shows the AD stage distribution of all the patients based their genders. There were more female patients when compared to the male patients who have progressed to the next stage in their successive visits. The distribution shows that the dataset has more patients who have progressed to moderate impairment stage. The dataset clearly shows that there is an imbalance in the future stages' distribution. In particular, the scarcity data in the early stage reflects that people with fewer symptoms unlikely go to see doctors. Self-diagnosis tools available to the public can surely promote and support patient-centered approaches in the field of AD (Qiu & Qiu, 2018).

4.2 Data Preprocessing

Since there are different datatypes, different imputation procedures were employed. There are features which are specific to a patient and such features were imputed by aggregated values belonging to all the visits of that patient. Table 1 summarizes the imputations applied in this study.

Feature selection was performed based on the types of input and output. For a categorical input and a numerical output, ANOVA feature selection was performed by selecting the top 100 features. For a numerical input and a numerical output,

Table 1 Imputations used in this study

Feature Type	Nominal	Ordinal	Continuous
Feature changes at every visit and is specific to a patient	Mode of all visits of the specific patient	Mode of all visits of the specific patient	Mean of all visits of the specific patient
Feature does not change with different visits	Mode of first visit of all the patients	Mode of first visit of all the patients	Mean of first visit of all the patients

Table 2 The dataset before transformation

ID	Data (148 columns)	CDRGLOB	VISITMO	VISITYR
12,345	11xxx	1	8	2011
12,345	12xxx	1	12	2012
12,345	13xxx	1	5	2014
12,345	21yyy	2	6	2015
12,345	31zzz	3	7	2016

Table 3 The dataset after transformation

ID	Data1 (148)	Data2 (148)	CDRGLOB1	CDRGLOB2	**No. of months**
12,345	11xxx	21yyy	1	2	**34**
12,345	12xxx	21yyy	1	2	**18**
12,345	13xxx	21yyy	1	2	**11**
12,345	21xxx	31zzz	2	3	**13**

Pearson feature selection was performed. Six (6) features out of eleven (11) continuous features were identified. The continuous features are checked for any correlation. We found that the correlation between the features were less than 0.5. As a result, all the features are left as is.

There are certain continuous features which had values that had huge variances in their scales. We have normalized such features by applying min-max scaling. All the categorical variables irrespective of their types have been preprocessed through one-hot encoding. To study AD progressions, we must transform the dataset of every visits into a dataset that reveals AD progressions over time. Hence, a dataset that shows AD stages sequentially degenerated between medical visits must be created. In other words, a transformed dataset must show the time it took and the corresponding medical records from the "present" stage to the "future" stage. Tables 2 and 3 provide an exemplary case of the transformation employed in this study. Columns ending with one (1) represent the "present" visits, and columns ending with two (2) are the "future" or next visits when stages are degenerated to the next level.

5 Regression Model to Analyze the Features Contributing to AD Progressions

5.1 Regresion Modeling of AD Porgressions

Random forest modeling is an ensemble technique which combines the predictions from multiple decision trees to make improved predictions, which performs usually better when compared to the prediction of an individual model. It belongs to the family of supervised learning algorithm that can be used for regression and classification. We have used the Random Forest regressor implemented in the scikit-learn library to build our regression model that predicts the time for an AD patient to progress to the next stage.

The dataset is split into training and testing data in the ratio of 80:20. The training data have a total of 6370 records with 298 features and the testing data have a total of 1593 records. The dependent variable will be the "Number of Months". The transformed dataset including 4228 subjects is further categorized into four groups based on the CDRGLOB2 column i.e., the next stage of the patient namely 0.5, 1.0, 2.0, 3.0. The four groups are named as Cluster 0.5, Cluster 1.0, Cluster 2.0, Cluster 3.0 corresponding to the next stage. The data modelling is done both on the entire dataset and the four groups.

The first group includes 59 subjects diagnosed with questionable impairment in their next stage. The second group includes 1467 subjects with mild impairment in their next stage. The third group includes 1630 subjects progressed to moderate impairment stage. The fourth group includes 899 subjects progressed to severe impairment stage. Each random forest model we build was specific to the number of records in the dataset as the number of trees must be set based on the dataset size. The cluster for next_stage 0.5 had a limited number of records. Therefore, we did not take that cluster into consideration.

RandomForestRegressor is a class provided by scikit library to solve regression related problems. Randomized Search CV was used to search for the best hyperparameters using five-fold cross validation (Bergstra & Bengio, 2012). Following is one of the random forest models build on the entire dataset. We used the following configuration to build the random forest model given by the random search:

```
{'n_estimators': 1000,
 'min_samples_split': 2,
 'min_samples_leaf': 1,
 'max_features': 'sqrt',
 'max_depth': 25}
```

Figures 2, 3, and 4 are the prediction results for all the clusters:

Although the error is low, suggesting the model should be great in prediction, the R2 was on an average 30%, for all the clusters which was marginally acceptable.

Fig. 2 Cluster 1.0 metrics

```
MAE: 0.11242051271931479
MSE: 0.022400455128324982
RMSE: 0.14966781594025144
```

Fig. 3 Cluster 2.0 metrics

```
MAE: 0.11803286610172718
MSE: 0.024106568531668617
RMSE: 0.15526290133727572
```

Fig. 4 Cluster 3.0 metrics

```
MAE: 0.12735821228989327
MSE: 0.0275353614999924
RMSE: 0.16593782419928374
```

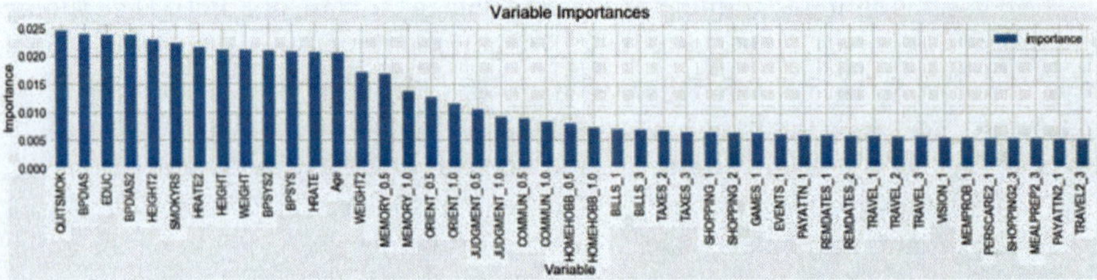

Fig. 5 Variable importance in Cluster 1.0

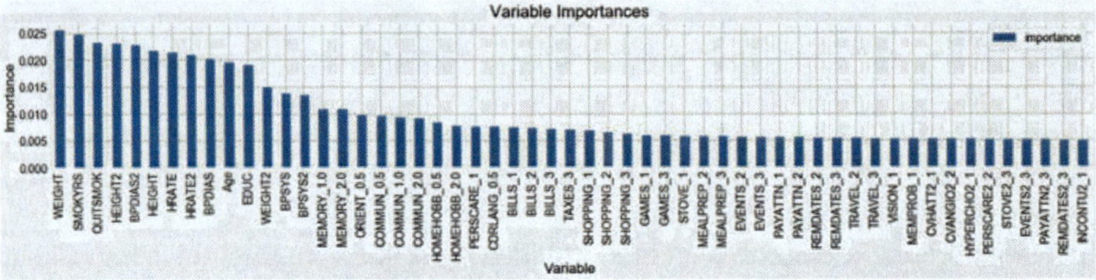

Fig. 6 Variable importance in Cluster 2.0

This suggests that additional features such as biomarkers for measuring AD stages should be included if available, which could help improve further the performance of the model.

5.2 *Feature Importance Analysis*

Figures 5, 6, and 7 are the features which have most significant impact on AD progressions in Cluster 1.0, Cluster 2.0, and Cluster 3.0, respectively. The variables with suffices are the encoded representations of the categorical features in the transformed dataset.

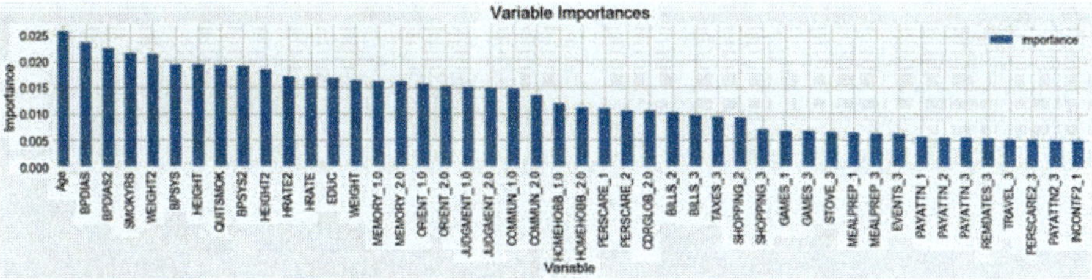

Fig. 7 Variable importance in Cluster 3.0

As shown in Fig. 5, QUITSMOK, BPDIAS, EDUC, HEIGHT, SMOKYRS, WEIGHT in Cluster 1.0 are the significant factors contributing to the progression of the disease from stage 0.5 to stage 1.0. From Fig. 6, WEIGHT, SMOKYRS, QUITSMOK, HEIGHT, HRATE, BPDIAS can be found to be the significant factor contributing to the progression of the disease from stage 1.0 to stage 2.0. For cluster 3.0, Age, BPDIAS, SMOKYRS, WEIGHT, BPSYS, HEIGHT, QUITSMOK, HRATE, EDUC, MEMORY are the significant factors contributing to the progression of the disease from stage 2.0 to stage 3.0 as shown in Fig. 7.

Table 4 is the summary of the list of all the significant features contributing to the progression of the disease in all the three clusters. The highlight features in bold are the most important predictors derived from our developed regression models.

Figure 8 shows the feature importance analysis derived from the whole dataset. **SMOKYRS, Age, BPSYS, BPDIAS, HRATE, HEIGHT, QUITSMOKE, EDUC, MEMORY, ORIENT** and **JUDGMENT** are identified as the top contributing features for AD progressing to the next stage. The patient must be continuously examined for the mentioned features.

6 Conclusions

This study proposed two-step modeling approach to predicting the time that an AD patient might take to progress to the next stage. By leveraging the clinical data on known AD stages, AD patients were clustered based on their "future" (i.e., next) stages. The results derived from cluster-based machine learning models seem practically acceptable in terms of predicting the time for an AD patient to degenerate from one stage to another. Feature analyses were conducted. Specifically, features that significantly contribute to AD progressions were explored either by cluster or as a whole. The uncovered insights can help patients know how to make lifestyle changes to help improve their ongoing treatments.

In the future study, we will apply deep learning techniques to further improve our modeling performance. When more AD patients' medical data become available, we would also like to explore approaches to study AD onset and progression at

Table 4 Summary of the significant features contributing to the AD progressions in Clusters

Cluster 1.0	Cluster 2.0	Cluster 3.0
QUITSMOK	**WEIGHT**	**Age**
BPDIAS	**SMOKYRS**	**BPDIAS**
EDUC	**QUITSMOK**	**SMOKEYRS**
HEIGHT	**HEIGHT**	**WEIGHT**
SMOKYRS	**BPDIAS**	**BPSYS**
HRATE	**HRATE**	**HEIGHT**
WEIGHT	**BPDIAS**	**HRATE**
BPSYS	**Age**	**EDUC**
Age	**EDUC**	**WEIGHT**
MEMORY	**BPSYS**	**MEMORY**
ORIENT	**MEMORY**	**ORIENT**
JUDGMENT	**ORIENT**	**JUDGMENT**
COMMUN	COMMUN	COMMUN
HOMEHOBB	HOMEHOBB	HOMEHOBB
BILLS	PERSCARE	PERSCARE
TAXES	CDRLANG	CDRGLOB
SHOPPING	BILLS	BILLS
GAMES	TAXES	TAXES
EVENTS	SHOPPING	SHOPPING
PAYATTN	GAMES	GAMES
REMDATES	STOVE	STOVE
TRAVEL	MEALPREP	MEALPREP
VISION	EVENTS	EVENTS
MEMPROB	PAYATTN	PAYATTN
PERSCARE	REMDATES	REMDATES
	TRAVEL	TRAVEL
	VISION	PERSCARE
	MEMPROB	PAYATTN
	CVHATT	INCONTU
	CVANGIO	
	HYPERCHO	
	PERSCARE	
	STOVE	
	INCONTU	

the early stage, which would be more beneficial to patients, caregivers, and health professionals given that early AD treatments prove more effective (Maliszewska-Cyna et al., 2017).

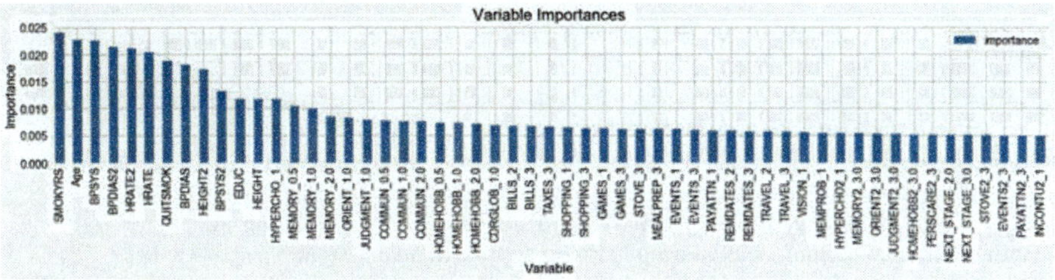

Fig. 8 Variable importance for whole dataset

Acknowledgments The dataset was support by NACC (Proposal ID #776). The NACC database is funded by NIA/NIH Grant U01 AG016976. NACC data are contributed by the NIAfunded ADCs: P30 AG019610 (PI Eric Reiman, MD), P30 AG013846 (PI Neil Kowall, MD), P50 AG008702 (PI Scott Small, MD), P50 AG025688 (PI Allan Levey, MD, PhD), P50 AG047266 (PI Todd Golde, MD, PhD), P30 AG010133 (PI Andrew Saykin, PsyD), P50 AG005146 (PI Marilyn Albert, PhD), P50 AG005134 (PI Bradley Hyman, MD, PhD), P50 AG016574 (PI Ronald Petersen, MD, PhD), P50 AG005138 (PI Mary Sano, PhD), P30 AG008051 (PI Steven Ferris, PhD), P30 AG013854 (PI M. Marsel Mesulam, MD), P30 AG008017 (PI Jeffrey Kaye, MD), P30 AG010161 (PI David Bennett, MD), P50 AG047366 (PI Victor Henderson, MD, MS), P30 AG010129 (PI Charles DeCarli, MD), P50 AG016573 (PI Frank LaFerla, PhD), P50 AG016570 (PI MarieFrancoise Chesselet, MD, PhD), P50 AG005131 (PI Douglas Galasko, MD), P50 AG023501 (PI Bruce Miller, MD), P30 AG035982 (PI Russell Swerdlow, MD), P30 AG028383 (PI Linda Van Eldik, PhD), P30 AG010124 (PI John Trojanowski, MD, PhD), P50 AG005133 (PI Oscar Lopez, MD), P50 AG005142 (PI Helena Chui, MD), P30 AG012300 (PI Roger Rosenberg, MD), P50 AG005136 (PI Thomas Montine, MD, PhD), P50 AG033514 (PI Sanjay Asthana, MD, FRCP), P50 AG005681 (PI John Morris, MD), and P50 AG047270 (PI Stephen Strittmatter, MD, PhD).

References

Albright, J., & Initiative, A.'s. D. N. (2019). Forecasting the progression of Alzheimer's disease using neural networks and a novel preprocessing algorithm. *Alzheimer's & Dementia: Translational Research & Clinical Interventions, 5*, 483–491.

Alzheimer's Association. (2020). *Facts and figures*. Available: https://www.alz.org/alzheimers-dementia/facts-figures

Bergstra, J., & Bengio, Y. (2012). Random search for hyper-parameter optimization. *The Journal of Machine Learning Research, 13*(1), 281–305.

Cohut, M. (2019). *Alzheimer's: Researchers create model to predict decline*. Available: https://www.medicalnewstoday.com/articles/325955

El-Sappagh, S., Abuhmed, T., Islam, S. R., & Kwak, K. S. (2020). Multimodal multitask deep learning model for Alzheimer's disease progression detection based on time series data. *Neurocomputing, 412*, 197–215.

Fisher, C. K., Smith, A. M., & Walsh, J. R. (2019). Machine learning for comprehensive forecasting of Alzheimer's disease progression. *Scientific Reports, 9*(1), 1–14.

Franzmeier, N., Koutsouleris, N., Benzinger, T., Goate, A., Karch, C. M., Fagan, A. M., McDade, E., Duering, M., Dichgans, M., Levin, J., & Gordon, B. A. (2020). Predicting sporadic Alzheimer's disease progression via inherited Alzheimer's disease-informed machine-learning. *Alzheimer's & Dementia, 16*(3), 501–511.

Korolev, I. O. (2014). Alzheimer's disease: A clinical and basic science review. *Medical Student Research Journal, 4*(1), 24–33.

Lee, G., Nho, K., Kang, B., Sohn, K. A., & Kim, D. (2019). Predicting Alzheimer's disease progression using multi-modal deep learning approach. *Scientific Reports, 9*(1), 1–12.

Maliszewska-Cyna, E., Lynch, M., Jordan Oore, J., Michael Nagy, P., & Aubert, I. (2017). The benefits of exercise and metabolic interventions for the prevention and early treatment of Alzheimer's disease. *Current Alzheimer Research, 14*(1), 47–60.

O'Bryant, S. E., Waring, S. C., Cullum, C. M., Hall, J., Lacritz, L., Massman, P. J., Lupo, P. J., Reisch, J. S., & Doody, R. (2008). Staging dementia using Clinical Dementia Rating Scale Sum of Boxes scores: A Texas Alzheimer's research consortium study. *Archives of Neurology, 65*(8), 1091–1095.

Oriol, J. D. V., Vallejo, E. E., Estrada, K., Peña, J. G. T., & Alzheimer's Disease Neuroimaging Initiative. (2019). Benchmarking machine learning models for late-onset alzheimer's disease prediction from genomic data. *BMC Bioinformatics, 20*(1), 1–17.

Qiu, R. G., & Qiu, J. L. (2018). Patient-centered deep learning model and diagnosis service for persons with Alzheimer's disease. In *Proceedings of the International Conference on Industrial Engineering and Operations Management* (Vol. 2018, pp. 1841–1847).

Satone, V. K., Kaur, R., Leonard, H., Iwaki, H., Sargent, L., Scholz, S. W., Nalls, M. A., Singleton, A. B., Faghri, F., Campbell, R. H., & Alzheimer's Disease Neuroimaging Initiative. (2019). Predicting Alzheimer's disease progression trajectory and clinical subtypes using machine learning. *bioRxiv*, 792432.

Solomon, A., Mangialasche, F., Richard, E., Andrieu, S., Bennett, D. A., Breteler, M., Fratiglioni, L., Hooshmand, B., Khachaturian, A. S., Schneider, L. S., & Skoog, I. (2014). Advances in the prevention of Alzheimer's disease and dementia. *Journal of Internal Medicine, 275*(3), 229–250.

Wang, T., Qiu, R. G., & Yu, M. (2018). Predictive modeling of the progression of Alzheimer's disease with recurrent neural networks. *Scientific Reports, 8*(1), 1–12.

Modelling the COVID-19 Epidemic Process of Shenzhen and the Effect of Social Intervention Based on SEIR Model

Wenjie Zhang and Wai Kin (Victor) Chan

1 Introduction

At the beginning of 2020, the outbreak of COVID-19 affected quite a few countries all over the world (Wuhan Municipal Health Commission Infection data, 2020). By October 31, there are over 45,428,731 cumulative COVID-19 confirmed cases over the world, and over 1,185,721 deaths cases (https://www.who.int/emergencies/diseases/novel-coronavirus-2019). Governments are searching for powerful policies to contain the epidemic and reduce loss. China government has adopted emergency policies, including large-scale social intervention, restriction of travelling, etc. But the impacts of the policies are still unclear, and essential for statisticians to research on.

Alerted from Central Committee on 20th Jan, Shenzhen government immediately implemented health intervention including social distancing, neighbourhood lockdown, wearing mask and cutting down on offline meeting. Those implements, namely social interventions, will lead to less face-to-face contracts among citizens. Simulating the epidemic process of Shenzhen and measure the effect of social intervention are significant for health epidemic prevention of densely populated cities all over the world.

To study the epidemic diseases spreading process, mathematical modelling has gained more attention (Anderson, 1999; Levin et al., 1997). Mathematical compartment model like SIR, SEIR have been done to study the COVID-19 epidemic process (Ng et al., 2003; Iannelli et al., 2005; Kermack & McKendrick, 1991; Cooper et al., 2020; Mwalili et al., 2020), figure out the characteristics of

W. Zhang · W. K. (Victor) Chan (✉)
Shenzhen Environmental Science and New Energy Technology Engineering Laboratory, Tsinghua-Berkeley Shenzhen Institute, Tsinghua Shenzhen International Graduate School, Tsinghua University, Shenzhen, Guangdong, People's Republic of China
e-mail: chanw@sz.tsinghua.edu.cn

H. Yang et al. (eds.), *AI and Analytics for Public Health*, Springer Proceedings in Business and Economics, https://doi.org/10.1007/978-3-030-75166-1_29

2019-nCOV (Scheiner et al., 2020),make prediction on future days (Shengli et al., 2020; Scheiner et al., 2020; Cardoso et al., 2020), measure the effect of health intervention (Yang et al., 2020; Sharov, 2020) and the uncovered asymptomatic number of Wuhan (Arcede et al., 2020). In this paper, we will use SEIR model to simulate the COVID-19 epidemic process in Shenzhen and measure the effect of social interventions. We first perform the estimation of parameters in the model Under different situations relative to social intervention, the parameters are adapted. Such like the recovery rate, we treat as a function of time. From this perspective, we add another variable into the SEIR differential equation. Based on daily cases data of Shenzhen, we build up the model to simulate the real process of COVID-19 in Shenzhen, and provide the results of the impact of social intervention.

This study is a retrospective research which focus on the epidemic numerical simulation of Shenzhen, one of Chinese first-tier city. The main contribution of our work is that we simulate the real epidemic process of Shenzhen effectively, compared it with the epidemic process under no-social-intervention and measure the effect of social intervention. This work provides certain reference significance to densely populated urban.

2 Data and Method

COVID-19 first presents at Wuhan without any perception needless to say the government intervention. So the first month confirmed data of COVID-19 in Wuhan is valuable for analyzing the characteristic of COVID-19 without any health intervention. Once the government ordered implement the health intervention, Shenzhen immediately adopted anti-epidemic. That means Shenzhen begins to social intervention initially fast to response to COVID-19. From this perspective, Shenzhen gives a good case of COVID-19 epidemic under the social intervention condition. We develop our model base on simple SIR model (Zhong et al., 2020) implemented to predict the outbreak of Mainland Chain.

2.1 Data Description

We use Shenzhen daily cases released from Shenzhen Health Commission (http://wjw.sz.gov.cn/yqxx/), which includes daily cumulative confirmed cases, daily cumulative cured cases and daily cumulative dead cases from JAN 20th to March 9th in Shenzhen.

2.2 Modified SEIR Model

We modified SEIR model based on the follow characteristics of COVID-19:

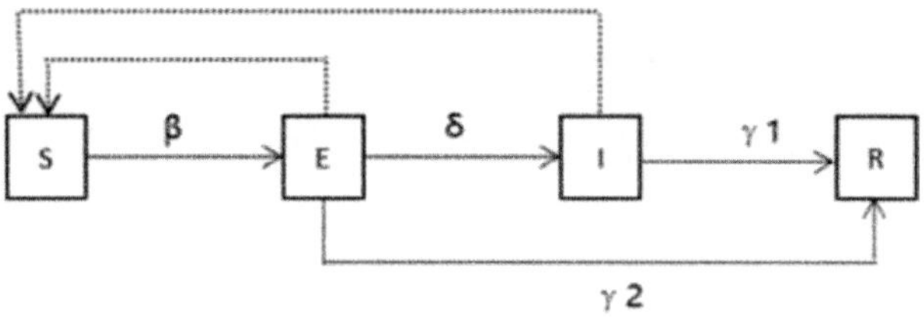

Fig. 1 Modified SEIR Model for COVID-19 Epidemic

1. The virus can transfer to others when the host doesn't get sick. COVID-19 carrier in the exposure phase has certain infectivity. Even didn't show any symptoms, they still have the ability to transmit 2019nCoV (He et al., 2020).
2. There exist asymptomatic carrier. They have infectivity but still not get sick. So the asymptomatic are treated as people in the exposure phase.

The main difference of our model from the original SEIR model is that the exposure possess infectivity and represents two part, the latent phase of the symptomatic and the asymptomatic. Part of them(the asymptomatic) can directly recovery without showing any symptoms.

The modified SEIR model is conceptually shown as (Fig. 1):

We use β representing the transmitting process from the susceptible to the exposure. Specifically, β can be subdivided into two form, namely β_1 and β_2. The higher dotted line denotes the susceptible contact with the infective which possess strong infectivity and then become the exposure. The contact transmission rate is denoted by β_1. The lower dotted line denotes the susceptible contact with the exposure which is proved can trans the virus and then become the exposure. The contact transmission rate is denoted by β_2. The solid lines denote the transfer directions between different compartments.

The equations of our model is as follow:

$$\frac{\mathrm{dS}\,(\mathrm{t})}{\mathrm{dt}} = \frac{-\beta_1 S(t)I(t)}{N} + \frac{-\beta_2 \mathrm{S}(t)\mathrm{E}(t)}{N} = \frac{-\mathrm{k}_1\mathrm{b}S(t)I(t)}{N} + \frac{-\mathrm{k}_2\mathrm{bS}(t)\mathrm{E}(t)}{N} \tag{1}$$

$$\frac{\mathrm{dE}\,(\mathrm{t})}{\mathrm{dt}} = \frac{\beta_1 S(t)I(t)}{N} + \frac{\beta_2 \mathrm{S}(t)\mathrm{E}(t)}{N} - \delta\mathrm{E}(t) - \gamma_2(t)E(t) \tag{2}$$

$$\frac{\mathrm{dI}\,(\mathrm{t})}{\mathrm{dt}} = \delta\mathrm{E}(t) - \gamma_1(t)\mathrm{I}(t) \tag{3}$$

$$\frac{\mathrm{dR}\,(\mathrm{t})}{\mathrm{dt}} = \gamma_1(t)\mathrm{I}(t) + \gamma_2(t)\mathrm{E}(t) \tag{4}$$

Table 1 Parameter definition of modified SEIR model

Notations	Meanings
S(t)	The number of susceptible people
E(t)	The number of exposed people
I(t)	The number of infective people
R(t)	The number of recovered or dead people
N	The total number of citizens
β_1	The rate of transmission for the susceptible to exposed by contact with the infections
β_2	The rate of transmission for the susceptible to exposed by contact with the exposed
k_1	The contact people of the infections per day
k_2	The contact people of the exposed per day
b	The probability of the susceptible trans into the exposed
δ	The incubation rate
$\gamma_1(t)$	The recovery (include death) rate from infective individual
$\gamma_2(t)$	The recovery (include death) rate from exposed individual

Here, the susceptible population can be infected by the exposed individual and the infective individual. The transmission rate β_1 denotes the susceptible who contact with the infected catch the virus 2019-nCOV, then they become the exposure. The transmission rate β_2 denotes the susceptible who contact with the exposure who also are contagious catch the virus 2019-nCOV, then they become the exposure. The incubation rate δ is described as the rate by which the exposed individual develops symptoms. The recovery rate γ_1 denotes some exposed recovery without any symptoms. Those people can be viewed as the asymptomatic who can transfer directly from the exposed phase to the recovery phase. And recovery rate γ_2 denotes infective(symptomatic) individual recovery. Nucleic acid testing changing from positive to negative indicates the virus carriers, both the asymptomatic and the symptomatic, recovered from COVID-19.

Our modified SEIR model is given by (Table 1).

3 Parameter Estimation

3.1 Transmission Rate

Referring to patient information released from the Shenzhen Health Commission, all the patients come from Hubei province or have close contact with the people who come from Hubei province. The national virus alter started on Jan 20th 2020, which is exactly the date that the first COVID-19 patient is recorded. With the alarm from central government, the social intervention is implemented immediately in Shenzhen. So the COVID-19 data of Shenzhen is content for

Table 2
The epidemiological data of COVID-19 in Hubei province

Date	Number of cumulative infections	β(t)
20-Dec-19	5	0.3218
21-Dec-19	7	0.2779
22-Dec-19	8	0.2599
23-Dec-19	13	0.1973
24-Dec-19	14	0.1885
25-Dec-19	16	0.1732
26-Dec-19	17	0.1666
27-Dec-19	19	0.1549
16-Jan-20	44	0.1574
17-Jan-20	62	0.1379
18-Jan-20	121	0.1110
19-Jan-20	198	0.0981
20-Jan-20	270	0.0921
21-Jan-20	375	0.0872
22-Jan-20	444	0.0851
23-Jan-20	549	0.0829
24-Jan-20	729	0.0804

studying COVID-19 spreading under the social intervention condition, but not good enough for calculating the transmission rate without any health intervention.

We use the data of cumulative infections of Wuhan from Jan 16th 2020 to Jan 25th 2020 to calculate the natural transmission rate. At the early stage of COVID-19 spreading, the number of infections (I) and the exposed (E) are at a very small level, which means the susceptible(S) is close to the number of total population (N). The average recovery time is 14 days which is statistical data released from WHO (WHO, 2020). We treat γ_1 as 1/14. The latent asymptomatic cases are not recorded at that time. And without widely spreading may the asymptomatic not exist or very very few. So we treat the γ_2 as 0. The skewed SEIR model can be solved approximately as:

$$\frac{\mathrm{dI}}{\mathrm{dt}} = \beta\frac{\mathrm{IS}}{\mathrm{N}} - \gamma\mathrm{I} \approx (\beta - \gamma_1)\, I \tag{5}$$

$$\mathrm{I}(t) \approx e^{(\beta-\gamma_1)I} \tag{6}$$

$$\beta = \frac{\ln I(t)}{I(t)} + \gamma_1 \tag{7}$$

Referring (Yang et al., 2020; Huang et al., 2020), we get the following Table 2:

Transmission rate is the product of the number of people exposed to each day (k) and the probability of transmission (b) when exposed. The effect of social intervention will be revealed by the the descent degree of k.

$$\beta_{\mathrm{i}} = \mathrm{k_i b},\ \mathrm{i} = 1, 2 \tag{8}$$

3.2 Recovery Rate

As time goes by, the public paid more and more attention on this easier-spreading virus which will result in grasping more and more comprehensive knowledge about it. The government put a lot of effort and the medical workers spend time and work hard to save life from death of COVID-19. From this perspective, the recovery rate will increase along time and being stable after a while. So we assume the recovery rate will be stable after 50 days. The actual recovery rate of Shenzhen calculated by the formula is shown below:

$$\gamma(t) = \frac{R(t + \Delta t) - R(t)}{I(t)\Delta t} \tag{9}$$

Here we consider the asymptomatic and the symptomatic possess same recovery rate. By assuming the recovery rate change along time will not affected by region and timing, we calculate the recovery rate of Hubei Province from 2020 Jan 21th to Mar 10th. The recovery rate line goes as follow (Fig. 2):

Power regression, logarithmic regression, linear regression, exponential regression and polynomial regression are applied to fit the real recovery rate (Table 3).

Where γ denotes the recovery rate of tth day, t denotes the sequence day after the outbreak, R^2 is the goodness of fit. Considering over-fitting, we choose exponential regression as the best-fit. Considering the recovery rate of the asymptomatic is lower than the symptomatic, we give the recovery rate of the first 50 days as follow(after 50 days it will remain same value),where α is a hyperparameter manifesting the differentiation of recovery rate of the asymptomatic:

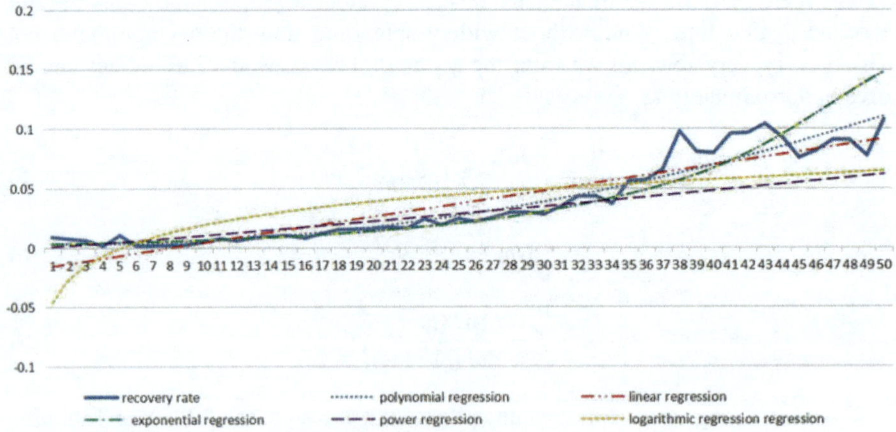

Fig. 2 Modified SEIR Model for COVID-19 Epidemic

Table 3 Fitting results of recovery rate

Fitting method	Fitting formula	R^2
Power regression	$\gamma = 0.0007\, t^{1.1331}$	0.6624
Logarithmic regression	$\gamma = 0.0284\ln(t) - 0.465$	0.5315
Linear regression	$\gamma = 0.0022\, t - 0.0179$	0.8446
Exponential regression	$\gamma = 0.0028e^{0.0793t}$	**0.8824**
Polynomial regression(quadratic)	$\gamma = 5*10^{-5}\, t^2 - 0.0003\, t + 0.0035$	0.9137

$$\gamma_1 = 0.0028e^{0.0793t},\ t \leq 50 \tag{10}$$

$$\gamma_2 = \alpha \cdot 0.0028e^{0.0793t},\ t \leq 50 \tag{11}$$

3.3 *Incubation Rate*

The exposed are not all transfer to the infective. That is the tricky side of COVID-19. A large proportion of the exposed shows symptoms after incubation time. A very small part of the exposed doesn't show any symptoms from catching 2019nCoV to recovery. As the statistics released from Shenzhen Health Commission, we got the number of the asymptomatic and the infective.

Incubation rate of the symptomatic and asymptomatic:

$$\delta_{symptom} = \frac{N_{symptom}}{N_{total}} * \frac{1}{D} \tag{12}$$

$$\delta_{asymptomatic} = \frac{N_{asymptomatic}}{N_{total}} * \frac{1}{D} \tag{13}$$

Where, D is the average incubation time of COVID-19. It is 7 day according to the CDC.

4 Results and Discussions

Shenzhen has a population of 13,438,800. Under normal conditions, the cumulative number of close contacts was stabilized at about eight times as much as infected individuals in Beijing (Suo-yi et al., 2020). Shenzhen is first-tier city as well as Beijing. Therefore, we let the average contact rate of citizen in Shenzhen be 8 on account of that Shenzhen is populated closely to Beijing. The epidemic start from Jan 20th 2020 for it is the day first COVID-19 patient shown in Shenzhen.

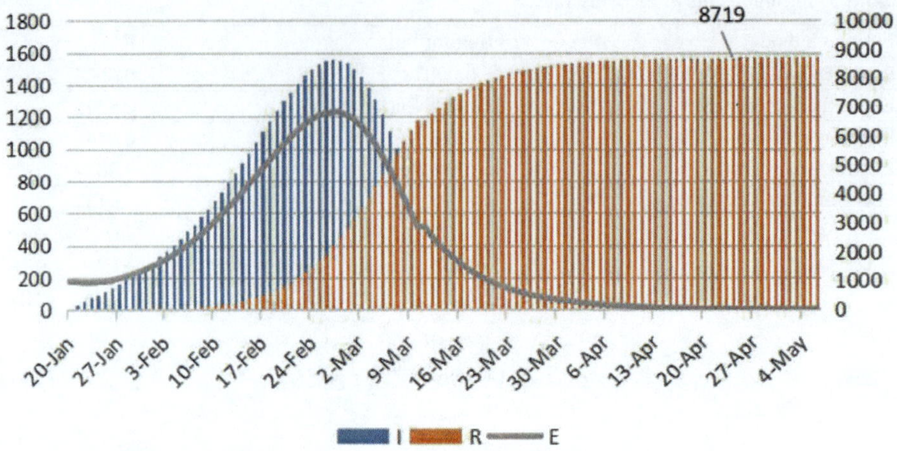

Fig. 3 Modelling the Epidemic in Shenzhen without any Health Intervention

The estimated parameters are calculated under circumstance of no-health-intervention implementation. We use those estimated parameters to simulate the epidemic process in Shenzhen without social intervention.

The number goes as follow (Fig. 3):

Where, k1 = 8, k2 = 8, b = 0.0126, $\delta = 0.128$.

Without any protection measure, there will be added up to 8719 citizens being recovery from this epidemic or death. The epidemic would have been last about 100 days and went to the peak in late February. The daily confirmed cases would reach up to 1589 at Feb 28th.

Social intervention aims to reduce the directly contact among people. Therefor its implementation will reduce the close contact rate of each citizens. In order to measure the effect of social intervention, we change the transmission rate β_1 and β_2 by changing the close contact rate to simulate the real epidemic process in Shenzhen based on our modified SEIR model.

At the early stage of Shenzhen epidemic, there were people from Hubei Province arriving constantly for visiting family and friends or avoiding COVID-19 or seeking medical advises. Those people contributed large part of 2019-nCOV carriers. On Jan 20th 2020, COVID-19 already spread for a while in Shenzhen. That means there are some people get contact with the infective and become the exposed. We assume there are 20 exposed people at Jan 20th, 120 exposed people at Jan 26th, 350 at Jan 30th and 0 at Feb 20th. We choose Mar 09th 2020 as the end day of our simulating epidemic.

The simulating results shown as Fig. 4.

where, k1 = 2, k2 = 4, b = 0.0126, $\delta = 0.128$, $\alpha = 0.96$.

The simulation result shows the simulating daily hospitalized COVID-19 cases goes along the real daily in-care cases. And there are sum up to 440 cases at Mar 09th 2020, which is very close to the actual total infected number 419. The

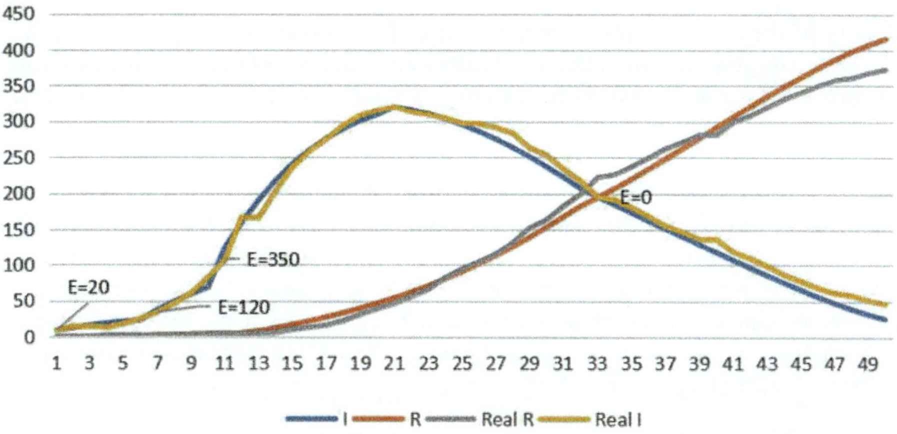

Fig. 4 The Real Epidemic Process Simulation in Shenzhen

simulating recovery curve fit the real recovery line approximately. By the end of the epidemic, simulation gives 364 recovery cases while the real total number is 373 by May 9th. That indicates our simulation is effective.

The above results indicate that the contact rate between the susceptible and the infections is reduced to 1/4 of its original value and the contact rate between the susceptible and the exposed is reduced to 1/2 of its original value after health intervention. That shows the effect of social intervention which lead to less contact will reduce the infections from 8719 to no more than 500 in Shenzhen. It dramatically reduces the number of the infective cases. The whole pandemic lasts about 50 days which largely shorten the duration of COVID-19 epidemic under no protection. The peak value of infective curve is about 320 which is a fifth of the true peak value of Fig. 4.

However, the more strictly social intervention measure taken, the more economical cost will be paid. We change the contact rate between the susceptible and the infections to 1/10 of its original value and the contact rate between the susceptible and the exposed to 1/6 of its original value, the total number of infections merely reduces from 440 to 233, but that will consume multifold resources and human inputs.

5 Conclusion

This paper develops a mechanistic modified SEIR model that captures the dynamics of covid-19 to analyze the spread of COVID-19 and the impact of social intervention. And then optimize the model parameters so that the theory and observation are consistent. The simulation results shows our model can simulate the real epidemic process in Shenzhen appropriately. The social intervention in Shenzhen reduces the activity of the infective to its general level's 1/4 and decreases the activity of the exposed to its general level's 1/2. The health interventions implemented in Shenzhen has protected about 22,000 citizens from this disease. It shows social intervention has great impact on control the epidemic process. It will shorten the duration of this pandemic, lower the peak value and dramatically lower the total infective cases.

The factors considered in this article are limited, including differences in medical standards in various places. Another shortcoming is that the SEIR model uniformly mixes infective individuals and exposed individuals and susceptible individuals. The real world is not in this case. People stay in their social network. They are not random walking points which have the chance to come up with the rest of the people in Shenzhen. We will do more research about 2019-nCOV transmission process at individual level of a certain social network.

Acknowledgements This research was funded by the National Natural Science Foundation of China (Grant No. 71971127) and the Hylink Digital Solutions Co., Ltd. (120500002).

R.B.G. thanks Ann.

References

Anderson, R. M. (1999). The pandemic of antibiotic resistance. *Nature Medicine, 5*, 147–149.

Arcede, J. P., Caga-Anan, R. L., Mentuda, C. Q., et al. (2020). Accounting for Symptomatic and Asymptomatic in a SEIR-type model of COVID-19. *Mathematical Modelling of Natural Phenomena*, 15.

Cardoso, E. H. S., Silva, M. S. D., Júnior, F. E. D. A. F., et al. (2020). Characterizing the impact of social inequality on COVID-19 propagation in developing countries. *IEEE Access*.

Cooper, I., Mondal, A., & Antonopoulos, C. G. (2020). A SIR model assumption for the spread of COVID-19 in different communities. *Chaos Solitons & Fractals*, 139.

He, X., Lau, E. H. Y., Wu, P., Deng, X., et al. (2020). Temporal dynamics in viral shedding and transmissibility of COVID-19. *Nature Medicine, 26*(5), 672–675.

http://wjw.sz.gov.cn/yqxx/

https://www.who.int/emergencies/diseases/novel-coronavirus-2019

Huang, C., et al. (2020). Clinical features of patients infected with 2019 novel coronavirus in Wuhan, China. *The Lancet, 395*, 10223.

Iannelli, M., Martcheva, M., & Li, X. Z. (2005). Strain replacement in an epidemic model with super-infection and perfect vaccination. *Mathematical Biosciences, 195*, 23–46.

Kermack, W., & McKendrick. (1991). Contributions to the mathematical theory of epidemics – II. The problem of endemicity. *Bulletin of Mathematical Biology, 53*(1–2), 57–87.

Levin, S. A., Grenfell, B., Hastings, A., & Perelson, A. S. (1997). Mathematical and computational challenges in population biology and ecosystems science. *Science, 275*, 334–343.

Mwalili, S., Kimathi, M., Ojiambo, V., et al. (2020). SEIR model for COVID-19 dynamics incorporating the environment and social distancing. *Nature Public Health Emergency Collection*, 13.

Ng, T. W., Turinici, G., & Danchin, A. (2003). A double epidemic model for the SARS propagation. *BMC Infectious Diseases, 3*, 19.

Scheiner, S., Ukaj, N., & Hellmich, C. (2020). Mathematical modeling of COVID-19 fatality trends: Death kinetics law versus infection-to-death delay rule. *Elsevier Public Health Emergency Collection*, 136.

Sharov, K. S. (2020). Creating and applying SIR modified compartmental model for calculation of COVID-19 lockdown efficiency. *Chaos, Solitons & Fractals*, 141.

Shengli, C., Peihua, F., & Shi, P. (2020). Application of modified SEIR epidemic dynamic model in prediction and evaluation of COVID-19 in Hubei". Zhejiang da xue xue bao. Yi xue ban = Journal of Zhejiang University. *Medical Sciences, 49*(2), 178–184.

Suo-yi, T. A. N., Zi-qiang, C. A. O., Shuo, Q. I. N., et al. (2020). Inferring the Trend of COVID-19 Epidemic with Close Contacts Counting. *Journal of University of Electronic Science and Technology of China, 49*(5), 788–793.

WHO. (2020). *Report of the WHO-China Joint Mission on Coronavirus Disease 2019 (COVID-19)* (p. 14).

Wuhan Municipal Health Commission Infection data [Online]. (2020). Available: http://wjw.wuhan.gov.cn/front/web/list2nd/no/710

Yang, Z., Zeng, Z., Wang, K., Wong, S., Liang, W., Zanin, M., et al. (2020). Modified SEIR and AI prediction of the epidemics trend of COVID-19 in China under public health interventions. *Journal of Thoracic Disease, 12*(3), 165–174.

Zhong, L., Mu, L., Li, J., Wang, J., Yin, Z., & Liu, D. (2020). Early prediction of the 2019 novel coronavirus outbreak in the mainland China based on simple mathematical model. *IEEE Access, 8*, 51761–51769.

Artificial Intelligence – Extending the Automation Spectrum

Stephen K. Kwan and Maria Cristina Pietronudo

1 Introduction

Automation, defined as a technique or method or system of operation and control of business processes by mechanical or electronic means that replaces human labour (Nof, 1999), is becoming increasingly widespread. Thanks to the improvement of the techniques with which it is implemented (i.e. artificial intelligence, machine learning, robotics) it has assumed characteristics of efficiency and versatility, which have made it applicable in sectors from industry to services and in activities from operational to strategic ones.[1] At present, the risk of replacing "men with machines" no longer concerns those routine and repetitive activities typical of assembly line operators, but more sophisticated and complex activities, which (Gorry & Morton, 1971) defined as unstructured or semi-structured. With advances in AI, the recent history of automation has been completely intertwined with the management of business processes and has provided support in several phases of business activity, entering the core of decision-making processes. Currently a series of debates and discussions have arisen concerning both the activities that could be automated and

[1]For example, Deep Knowledge Venture (Hong Kong) has appointed a robot, Vital, among its board members, which on the basis of a sophisticated system of algorithms, is able to analyse the risks and investment areas of the company's projects.

S. K. Kwan
Lucas College & Graduate School of Business, San José State University, San José, CA, USA

M. C. Pietronudo (✉)
Department of Management and Quantitative Studies, Parthenope University of Naples, Naples, Italy
e-mail: mariacristina.pietronudo@uniparthenope.it

H. Yang et al. (eds.), *AI and Analytics for Public Health*, Springer Proceedings in Business and Economics, https://doi.org/10.1007/978-3-030-75166-1_30

computerized, and the relevance of the role of some workers in organizations, to rather which roles can a machine replace.

2 Historical Background of Decision-Making

Reference (Gorry & Morton, 1971) provided a framework for decision making in the early stages of automation in the enterprise environment. The authors classified decisions based on how well the decision rules were understood:

- structured if decision rules are known and can be programmed;
- semi-structured if some rules are known, but there are still unknowns, so decision need to be supported by humans;
- unstructured if rules are pretty much unknown, thus the decision-maker must provide judgment and evaluation as well as insights into solving the problem.

The authors also aligned these decisions with control types in the enterprise (from (Anthony, 1965)). These are depicted in Fig. 1 with a stereotypical pyramidal form to illustrate the volume and frequency of each type of decision at various levels of the enterprise. Gorry and Morton (1971) also stipulated that structured and semi-structured decisions could be automated if the decision rules were well understood and could be programmed. As more and more decisions in the enterprise are studied and analysed and understood, the more decisions could become automated. This is depicted as the Line of Automation that moves up in time in Fig. 1. This framework has become one of the fundamental building blocks of the study of Management Information Systems (MIS).

Along with the framework of (Gorry & Morton, 1971) researchers have also studied the decision-making process in order to identify opportunities for refinement, investment and automation. This is illustrated in Fig. 2a (Fig. 2b will be discussed in a later section). Data, raw facts usually in isolation, are processed

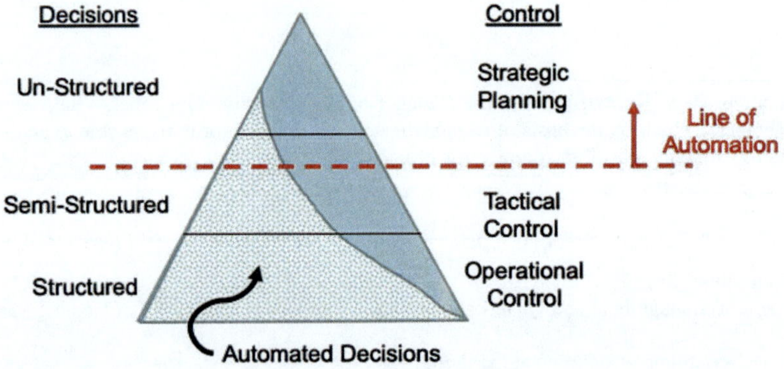

Fig. 1 Summary of enterprise decision making concepts from. (Huang & Rust, 2018)

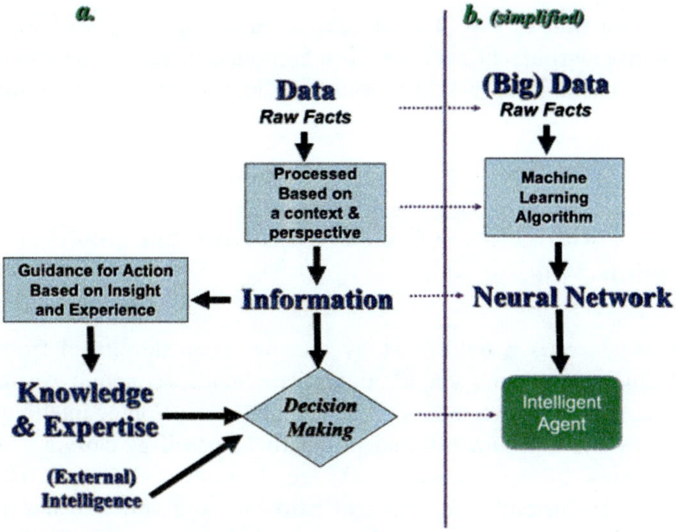

Fig. 2 Traditional Model of Decision Making vs. Machine Learning. (**a**) Traditional concept of decision Making; (**b**) Machine Learning Example

based on a decision context or perspective into information. Information, as the term implies, inform the recipient, i.e. tell him/her something new. Decision making usually involve one or more criteria that the decision maker had to make judgements. Information processed based on the criteria of the decision can help reduce the uncertainty of the parameters facing the decision maker so that he/she could apply some decisions rules to improve outcomes (e.g., evaluating the value of information (Laudon & Laudon, 1999)). Information can also be used as guidance for action based on insight and experience of an expert (could be the decision maker or someone else). This could be stored formally for later use in a knowledge base or informally retained as personal knowledge and expertise. These knowledge and expertise can be fed back to the decision maker to help reduce uncertainty, apply historical perspective and corporate knowledge. The decision maker could also acquire information about adversaries and the market environment in the form of intelligence from outside the organization. Again, these could be fed into the decision-making process to reach conclusions.

Many parts of Fig. 2a have been automated as part of systems for data processing, information processing, decision support, knowledge management, business intelligence, etc. The decision-making process requires the intelligence of the decision maker to evaluate alternatives based on the criteria and rules with information, expertise, insight, and knowledge (Keeney & Raiffa, 1993). Automating the decision-making process basically simulates the human actions by imitating his/her way of thinking, approaches and problem solving, very often improving performance by eliminating subjectivity and human bias. As more and more decisions are studied and understood together with advances in technology, the line of automation in Fig. 1 continues to move up. As the world and the business/societal environment become more complex, there will be an abundance of new decision-making that would become candidates for automation throughout the enterprise.

3 Artificial Intelligence – A Contemporary Perspective of Decision-Making

Artificial Intelligence[2] is a field of study that has been developed from various disciplines including computer science, cognitive science, economics, mathematics, psychology, etc. It included various approaches such as developing intelligent agents that could think and act humanly and rationally (Russell & Norvig, 2015). For an excellent review of the history of AI, see (Russell & Norvig, 2015; Rouse & Spohrer, 2018). Currently, advances in hardware and software had made the application of AI viable and practical (Williams, 2019). summarized AI succinctly into traditional AI where machines are programmed to be smart (e.g., expert systems) and contemporary AI trains machines to be smart (such as machine learning systems).

AI is currently employed in all the components of decision-making depicted in Fig. 1a as well as the selection and implementation of decisions. The decisions that are being automated with AI are increasingly complex including unstructured ones that once seemed only to be performed by humans. The extent of automating decision-making with AI has also gone beyond the enterprise environment to include helping consumers and citizens make decisions that affect their daily lives. (e.g., consumer shopping, health care, driving, collaborative filtering, recommendation systems, etc.) Figure 1b shows an example of how machine learning is used to create an intelligent agent that will be used to automate decision-making. This example is simplified and does not show that other components such as knowledge, expertise and external intelligence could well be employed by the intelligent agent.

4 Extending the Automation Spectrum

In the following, we review some recent contributions to AI to illustrate the extension of automation spectrum for decision-making.

Reference (Huang & Rust, 2018) develop a theory of job replacement as a result of automation with AI. The authors assert that the replacement occurs fundamentally at the task level, rather than the job level, particularly for lower intelligence tasks that

[2]The use of the term Artificial Intelligence often implies reference to Artificial Intelligence System and its capabilities.

are easier for AI to perform. The authors also address the way firms should decide between humans and machines across the four intelligences required for service tasks:

- Mechanical intelligence: performs routine and repeated tasks and has a minimal degree of learning, or adaption. It concerns automation, repetitive tasks for which a large-scale job replacement is expected.
- Analytical intelligence: processes information for problem-solving and learn from it; it learns and adapts systematically based on data. This is a unique capability of human workers, but recently starts to become a capability of more advanced AI.
- Intuitive Intelligence: thinks creatively and adjusts effectively to novel situations; it learns and adapts intuitively based on understanding. AI is developing this intelligence already.
- Empathetic intelligence: it recognizes and understand other peoples' emotions, respond appropriately emotionally, and influence others' emotions; it learns and adapt empathetically based on experience. Empathetic AI is currently being developed,[3] consequently this intelligence is still human's prerogative.

Therefore, some of that intelligence are greater or lesser adapted to carry out some tasks. Considering AI ability in the near future, for tasks that needs of analytical, intuitive and empathetic intelligence, firms should decide between humans and machines on the basis of the intelligence required. It means that a firm needs to think of the task portfolio of a job and optimize the division of labour or integration between human workers and AI.

Reference (Rouse & Spohrer, 2018) focus their study on the integration of human and machine intelligence clarifying the concept of intelligence augmentation (Davenport & Kirby, 2015). The authors address the ways in which human intelligence in such can be augmented rather than replaced, since not all tasks or job may be replaced. For example, deep learning works well when trained with large numbers of examples, but this is not feasible for many tasks such as reasoning about and solving novel problems. Thus, human intervention is required in applying technology, and in the meantime, AI intervention is required to support humans in improving their decisions. Particularly, (Davenport & Kirby, 2015) distinguish the automation with the augmentation continuum; the first refers to digital systems, the latter is a mix of two different types of cognitive systems – biological and digital. Therefore, it is no longer a question of what should humans do and what should computers do. The question now concerns creating the best cognitive team or cognitive organization to address the problems at hand. This portends some creatively different solutions from what has been developed in the past. Furthermore, authors show a realistic view of advantages and limitation of what the AI can do, describing new capabilities now possible, and design an architecture for augmenting intelligence for humans.

[3]The Journal of Service Research had just issued a call for papers for a special issue on AI Service and Emotion (10/01/20).

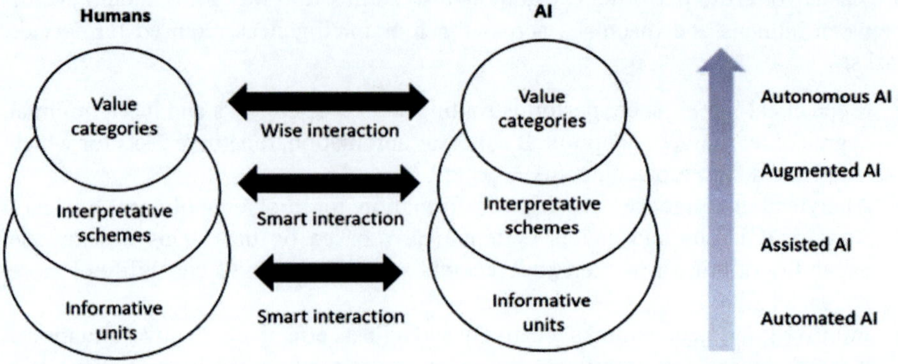

Fig. 3 Variety Information Model applied to AI. (Adapted from Refs. (Barile et al., 2019; Bassano et al., 2020))

The concept of augmented intelligence as to how AI support human in carry out decisions is addressed in (Barile et al., 2019). In analysing the exchange between human and machines concerning what the two entities exchange at different level of interaction, the authors define the intelligence augmentation (IA) as an intelligence equipped with cognition, common sense reasoning, context comprehension and knowledge based. The authors use the Viable System Approach to investigate the interaction between machine and human, stating interaction depends on the cognitive distance between interactive parts (Barile & Saviano, 2012). To assess cognitive distance, interaction dynamics and knowledge exchange, authors use the Variety Information Model (VIM) (Barile, 2009; Barile et al., 2012) composed by (i) informative units that represent data and what can be perceived and elaborated; (ii) interpretative schemes that represent how generic information is transformed into specific (Calabrese et al., 2011); (iii) value categories that represent values and strong beliefs of the viable systems, synthesize the knowledge. In other words, to operate in an effective and efficient way human and machines, in certain circumstances, need to share value rather than data, information or knowledge. Hence, the model considers that not all AI technology are able to share knowledge and value. Many of AI technology operates at the level of informative units. The authors also assert the following as illustrated in Fig. 3 (Barile et al., 2019; Bassano et al., 2020; Rao & Verweij, 2017):

- automated and assisted intelligent are mechanical or analytical intelligence, they operate at the first two levels of information variety being data and information-intensive, but they do not understand the environment and cannot adapt automatically to it;
- augmented or autonomous intelligence, potentially, can acts on value categories striving in understanding beliefs and value of the system in which they operate, but need humans.

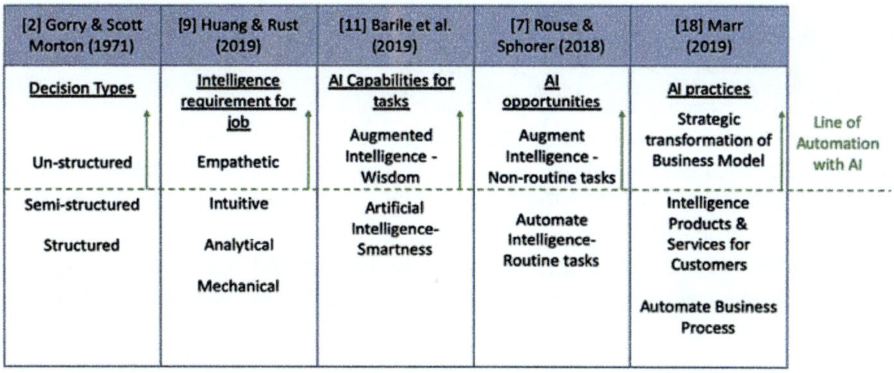

Fig. 4 Extending the Automation Spectrum

From the studies reviewed in the previous section, it is evident that the interpretative approaches of AI and automation have changed over time (Fig. 4).

The adaptability of automation has led academics to move the focus outside the activities of organization and towards the concept of intelligence, intelligence required for tasks. Beyond routine and no routine job, automation concern actions that do not requires an empathetic or intuitive intelligence (Huang & Rust, 2018); intelligences more related to cognitive thinking aspect of human intelligence. This perspective puts the spotlights to what humans do well themselves, in addition to what machines do themselves and paves the way for the new concept of intelligence augmentation. Reference (Rouse & Spohrer, 2018) juxtapose automation with augmentation introducing the automation-augmentation continuum that is from a mix of two different types of cognitive systems – biological and digital, each of them playing different but complementary roles. Pursuing this view, the line of the augmentation would move upward transforming AI from assistant to collaborator, from coach to mediator. However, responsibility for both automation and augmentation, remains with humans, as well as the wiser outcomes remains the higher purpose of humans. Reference (Barile et al., 2019) in fact looked at AI in terms of smartness and wise. They claim the concept of augmentation as a need in doing things that actually humans and machine alone can't do and in doing things in a wiser way.

Reference (Marr, 2019) provides many examples of how businesses take advantage of AI technology to (i) change the way they understand and interact with customers; (ii) offer more intelligent products and services; (iii) improve and automate business processes; (iv) strategically transform their business with new business models (ibid., pg. 6–7). These examples show that the businesses are extending their automation spectrum with AI towards strategic use (cf. (Gorry & Morton, 1971)) as well as reaching out to their customers. That is, the technology is being employed for backstage (from process automation to strategic decision making) along with frontstage (customer interaction) operations (Teboul, 2005).

5 Conclusion

The incredible capacity and ability of machines had grown tremendously in the recent past leading to concerns of academics, policy makers and managers about the possibility of machines replacing much of the work done by humans. This is prompting scholars to shift their focus towards ways to enhance human capabilities through integration with artificial intelligence to achieve augmented intelligence. Studies presented show how augmentation of human and machine intelligence is extending the automation spectrum and machines with artificial intelligence capabilities can now:

- exhibit empathy
- support unstructured and strategic decisions
- support front stage intelligent services with customers
- make decisions autonomously
- demonstrate wise behaviour beyond smartness
- understand human value systems and apply in decision making
- augment instead of just support human decision making
- learn to be smart

This point of view is encouraging stakeholders to have a more positive sense towards AI: AI could help us to work faster and smarter, boosting productivity and creating as many – if not more – jobs than it displaces in the coming decades (Thomas, 2019). Fear of automation of jobs is not the real problem: the issue is how to augment human with machine capability integrating them for a better future.

This research has laid the ground work showing how AI technology had extended the automation spectrum and also touched on the issue related to automation of jobs. The next steps will include:

1. categorize the many examples of AI from (Marr, 2019; WSJ, 2020) according to their capabilities listed above to help us understand the extent of these practices in businesses. A sample of these examples are shown in Appendix A.
2. explore more about the tension between automation of jobs and the seemingly unstoppable advances in AI capabilities. In particular, how businesses are dealing with this before and during the current pandemic with reflection on the future of automation with AI (e.g., see examples from Amazon (Wingfield, 2020) and BMW (Unerti, 2020)).

A.1 Appendix A – Sample AI Applications

Item	Source	Business	Product/service – AI capabilities	Back stage	Front stage	Automated decision	Augmentation	Unstructured decision	Learn to be smart
1	(WSJ, 2020) 8/11/20	OpenAI LP	GBT-3: Q & A, automated text generation		x	x		x	x
2	(WSJ, 2020) 8/11/20	Entrupy	Fake luxury product detection		x	x			
3	(WSJ, 2020) 8/11/20	Royal Bank of Canada	Fraud and money laundering detection	x			x	x	
4			Market, risk analysis	x			x	x	
5			Credit limit setting		x	x			
6			NOMI: Personalized advice		x		x	x	
7	(Marr, 2019)	Alibaba	Personalized shopping pages		x	x			x
8			Chatbot		x	x			
9			Content generation	x		x			
10	(Marr, 2019)	Alphabet/Google	Smarter search		x	x			

(continued)

(continued)

Item	Source	Business	Product/service – AI capabilities	Back stage	Front stage	Automated decision	Augmentation	Unstructured decision	Learn to be smart
11			Image labeling and search		x	x			
12			Language translation		x	x			
13			Ad servicing	x		x			
14			Spam detection	x	x	x			
15			Google Assistant		x		x	x	
16			Self driving cars		x	x			
17			Video Captioning		x	x			
18			Diagnosing Disease		x		x	x	
19			Google Brain – machine and deep learning	x	x				x
20			Deep Mind – human brain simulation	x	x				x

21	(Marr, 2019)	Amazon	Recommendation engine		x	x			
22			Alexa		x	x			x
23	(Marr, 2019)	IBM	Chatbots		x	x			
24			Clinical treatment		x		x	x	
25	(Marr, 2019)	American Express	Fraud detection	x			x	x	
26	(Marr, 2019)	Uber Technologies	Improve wait time, join time, routes	x		x			
27	(Marr, 2019)	John Deere	Reduction in use of pesticide		x	x			
28			Precision agriculture		x	x			
29	(Marr, 2019)	Apple	Face ID		x	x			x
30			Siri and NLP		x	x			x

(continued)

(continued)

Item	Source	Business	Product/service – AI capabilities	Back stage	Front stage	Automated decision	Augmentation	Unstructured decision	Learn to be smart
31	(Marr, 2019)	Baidu	Self-driving cars		x	x			
32			Face ID		x	x			x
33			Real time translation on mobile devices		x	x		x	
34	(Marr, 2019)	Facebook	Monitoring contents	x		x			x
35			Facial recognition	x		x			x
36			Understanding text	x		x			x
37			Suicide Prevention		x		x	x	
38	(Marr, 2019)	JD.com	Facial recognition	x		x			x
39			Automated deliveries		x	x			
40	(Marr, 2019)	Tencent	Miying – medical imaging & diagnosis		x		x		x

References

Anthony, R. N. (1965). *Planning and control systems: A framework for analysis [by]*. Division of Research, Graduate School of Business Administration, Harvard University.

Barile, S. (2009). *Management sistemico vitale* (Vol. 1). Giappichelli.

Barile, S., Bassano, C., Spohrer, J. C., Piciocchi, P., Pietronudo, M. C., & Saviano, M. (2019, July 8th). *AI & Value co-creation: An integrated VSA and SS perspective*. Presented at AIRIS 2019, Universidad Zaragoza.

Barile, S., Pels, J., Polese, F., & Saviano, M. (2012). An introduction to the viable systems approach and its contribution to marketing. *Journal of Business Market Management, 5*(2), 54–78.

Barile, S., & Saviano, M. (2012). Oltre la partnership: un cambiamento di prospettiva. In S. Esposito De Falco & C. Gatti (Eds.), *La consonanza nel governo dell'impresa. Profili teorici e applicazioni* (pp. 56–78). Franco Angeli.

Bassano, C., Barile, S., Saviano, M., Pietronudo, M. C., & Cosimato, S. (2020). AI technologies & value co-creation in luxury context. In *Proceedings of the 53rd Hawaii International Conference on System Sciences*.

Calabrese, M., Iandolo, F., & Bilotta, A. (2011). From requisite variety to information variety through the information theory the management of viable systems. In *Service Dominant logic, Network & Systems Theory and Service Science, Giannini, Napoli*.

Davenport, T. H., & Kirby, J. (2015). Beyond automation. *Harvard Business Review, 93*(6), 58–65.

Gorry, G. A., & Morton, M. S. (1971). A framework for management information systems. *Sloan Management Review, 13*(1), 55–70.

Huang, M., & Rust, R. T. (2018). Artificial intelligence in service. *Journal of Service Research, 21*(2), 155–172.

Keeney, R. L., & Raiffa, H. (1993). *Decisions with multiple objectives: Preferences and value tradeoffs*. Cambridge University Press.

Laudon, K. C., & Laudon, J. P. (1999). *Management information systems*. Prentice Hall PTR.

Marr, B. (2019). *Artificial Intelligence in Practice – How 50 successful companies used AI and Machine learning to solve problems*. Wiley.

Nof, S. Y. (Ed.). (1999). *Springer handbook of automation*. Springer Science & Business Media.

Rao, A. S., & Verweij, G. (2017). *Sizing the prize, What's the real value of AI for your business and how can you capitalise?* PwC Publication, PwC. URL: https://www.pwc.com/gx/en/issues/analytics/assets/pwcaianalysis-sizing-the-prize-report.pdf. Accessed 07 2018.

Rouse, W. B., & Spohrer, J. C. (2018). Automating versus augmenting intelligence. *Journal of Enterprise Transformation*, 1–21.

Russell, S. J., & Norvig, P. (2015). *Artificial intelligence, A modern approach* (3rd ed.). Pearson.

Teboul, J. (2005). *Service is front stage. We are all in services ... more or less!* INSEAD, Fontainebleau.

Thomas, D. (2019, May 13). *Automation is not the future, human augmentation is*. RACONTEUR. URL: https://www.raconteur.net/technology/ai-human-augmentation

Unerti, D. (2020). How BMW used pandemic plant stoppages to boost artificial intelligence. *The Wall Street Journal*.

Williams, J. (2019). *ML/AI models that have the potential to transform our health*. Presented at Women in Data Science Conference, American University, Beirut.

Wingfield, N. (2020). As Amazon pushes forward with robots, workers find new roles. *The New York Times*.

WSJ. (2020). Artificial intelligence daily and artificial intelligence weekly. *The Wall Street Journal*. May 1st, 2019 to date.

Robust Portfolio Optimization Models When Stock Returns Are a Mixture of Normals

Polen Arabacı and Burak Kocuk

1 Introduction

Financial services provides a key service in finance industry, which has a significant contribution to the world economy. Nowadays, due the the emergence of the big data concept and the ever-increasing nature of uncertainty, there is high demand to use sophisticated analytical tools to improve financial services. One of these tools is optimization, which promises fast and reliable decisions in financial services, provided that it accurately captures the uncertain nature of the market.

Portfolio selection problem is one of the most significant problems in financial decisions which seeks to determine the best investment to be made from a number of risky assets given a certain amount of fund. Due to the uncertain nature of asset returns, investors also need to consider the risk associated with their decisions. Therefore, portfolio optimization has become one of the most popular methods used in financial portfolio decisions. In the early years of 1950s, the theory of optimal portfolio selection was developed in Markowitz (1952). According to the theory, the optimal portfolio problem aims to construct a portfolio which achieves maximum expected return with a minimum risk. However, the existence of these two conflicting objectives has become one of the most challenging aspects of the optimal portfolio problem. Thus, risk adjusted models are considered to combine risk and return to present a trade-off.

Although the Markowitz model has been used as a framework to find the optimal portfolios for decades, it suffers from a number of shortcomings. As one of the shortcomings, variance is considered as not an adequate risk measure. Therefore, models with different measures of risks such as Value-at-Risk and Conditional Value-at-Risk (CVaR) are considered in literature (Ghaoui et al., 2003; Chen

P. Arabaci (✉) · B. Kocuk
Industrial Engineering Program, Sabancı University, Istanbul, Turkey

H. Yang et al. (eds.), *AI and Analytics for Public Health*, Springer Proceedings in Business and Economics, https://doi.org/10.1007/978-3-030-75166-1_31

& Yu, 2013; Krokhmal et al., 2002; Rockafellar & Uryasev, 2002; Kocuk & Cornuéjols, 2020). Moreover, despite the importance of the Markowitz model in theory, portfolios determined by this model are sensitive to the estimations of the parameters and perform poorly in an out-of-sample test. In order to overcome this sensitivity problem, some of the studies focused on robust portfolio construction (Ceria & Stubbs, 2006; Ghaoui et al., 2003; DeMiguel & Nogales, 2009; Ceria & Sivaramakrishnan, 2013).

In this paper, we use a robust optimization approach to address these issues. We provide the single-stage equivalents of the standard portfolio optimization models as conic programs with different conic representable uncertainty sets and two risk measures: a distribution independent measure variance, and a distribution dependent measure CVaR. Since the theoretical efficient frontiers obtained for the variance and CVaR measures are the same when the stock returns are assumed to be normal, we concentrate on the case where the returns are modelled as a mixture of normals (we also refer the reader to Kocuk and Cornuéjols (2020) for further discussion about this probabilistic model). We also carry out a computational study on the Standard & Poor's (S&P) 500 data set and compare the resulting optimal portfolios. Our empirical findings suggest that using robust optimization models may provide portfolios with higher returns for a risk-taker investor. Moreover, using robust models with budgeted uncertainty for the same levels of risk may result in higher returns. Finally, the robust optimization models are not very sensitive to the uncertainty in the covariance matrix.

2 Standard Optimization Models

In this section, we review the standard optimization problems for portfolio construction. We assume throughout this section that we have n risky assets, and their mean vector of returns and their covariance matrix are given as μ and Σ, respectively. In addition, decision variable x denotes a portfolio vector.

In an ideal situation, the aim of an investor is to achieve minimum risk and maximum expected return. However, since these two objectives might be conflicting, a compromise has to be made. We will now review some of the well-known optimization problems proposed for this purpose.

2.1 *Markowitz Model*

The theory of optimal selection of portfolios is developed by Markowitz (1952). Markowitz porfolio optimization problem, also called mean-variance problem, adopts variance as the risk measure. The theory presents a trade-off between risk and return as follows:

$$\min_{x} \; x^T \Sigma x - \tau \mu^T x \tag{1a}$$

$$\text{s.t. } e^T x = 1, \;\; x \geq 0. \tag{1b}$$

The first and the second parts of the objection function (1a) refer to risk, which is measured by the variance of the return, and the expected return of the portfolio, respectively. Since minimizing risk and maximizing expected return at the same time might be conflicting, the expected return is multiplied with a constant factor $\tau > 0$ to combine risk and return into a single objective function. Here, $\frac{1}{\tau}$ is a risk-aversion constant used to quantify the trade-off between the expected return and risk. The constraint (1b) corresponds to the summation of the proportions of the total funds invested in portfolio vector x_i equals to 1. In order to prevent short sales, we also introduce the non-negativity constraint in (1b).

2.2 *Conditional Value-At-Risk Model*

Although VaR is a popular risk measure, it is neither coherent nor convex in general. Instead, many practitioners prefer to use Conditional Value-at-Risk (CVaR), which has these two desirable features. Let us give the formal definition of CVaR using the following formula:

$$\text{CVaR}_\alpha(X) = -\,\text{E}[X \mid X \leq -\text{VaR}_\alpha(X)] \;\text{ where }\; \text{VaR}_\alpha(X) := \min\{\gamma : P(X \geq \gamma) \leq 1-\alpha\}. \tag{2}$$

As an example of a portfolio optimization problem involving CVaR, let us assume that the return vector is denoted by r and replace the variance term in the Markowitz model with CVaR. Then, we obtain the following convex program:

$$\min_{x} \; \{\text{CVaR}_\alpha\left(r^T x\right) - \tau \mu^T x : \;\; e^T x = 1, x \geq 0\}. \tag{3}$$

The objective function refers to minimizing the CVaR as a risk measure and combining the risk with the expected return for some $\tau > 0$.

3 Robust Optimization Models

Robust optimization considers uncertainty in problem parameters described through uncertainty sets. Our main motivation to use robust optimization approaches in portfolio optimization is to overcome the sensitivity problems caused by the uncertainty in data since we do not have complete knowledge on parameters of portfolio problem in real life. Therefore, estimating these unknown parameters can result in errors which have negative effects on the *optimal* portfolios obtained through optimization.

The purpose of this section is to present an analysis of robust portfolio optimization problems involving uncertain parameters. We show how to build robust portfolio problems where objective function has robustness and minimizes the risk with a trade-off between risk and return. Even tough we cannot know the exact value of true parameters in reality, we also cannot expect to solve portfolio optimization problems with high accuracy with fully unknown parameters (Lobo & Boyd, 2000). Therefore, we build the two-stage problems where the parameters are partially known with different uncertainty sets and then obtain their single-stage equivalents as conic programs.

Throughout this section, we will denote the sample mean and sample covariance as $\hat{\mu}$ and $\hat{\Sigma}$, respectively.

3.1 Markowitz Model

Let us recall problem (1) we stated as in the Markowitz framework that combines the expected return and variance in the objective function. In order to incorporate robustness into the objective function, we consider the following general form of a two-stage problem:

$$\min_{x} \quad \max_{(\Sigma,\mu)\in\mathcal{S}} \quad \{x^T \Sigma x - \tau \mu^T x : \ e^T x = 1, x \geq 0\}. \tag{4}$$

Here, $\mathcal{S}$ denotes the uncertainty set and τ is a given positive number. In the sequel, we reformulate problem (3.1) as a single-stage conic program for various types of uncertainty sets.

3.1.1 Polyhedral Uncertainty for Mean

In this section, we look at a generic polyhedral uncertainty for mean μ while assuming that Σ is known (or estimated from data) as $\hat{\Sigma}$ and $\bar{\mu} = \hat{\mu}$ (although other choices are allowed). In particular, let us consider the following uncertainty set:

$$\mathcal{S} := \{(\mu, \Sigma) : A\mu \leq b, \ \Sigma = \hat{\Sigma}\},$$

where $A \in \mathbb{R}^{m\times n}$ and $b \in \mathbb{R}^n$ are given. Assuming that the set $\mathcal{S}$ is feasible and bounded, we utilize linear programming duality to obtain a single-stage reformulation of problem (3.1) as the following convex quadratic problem:

$$\min_{x,\lambda} \ \{x^T \Sigma x + \tau \lambda^T b : A^T \lambda = -x, e^T x = 1, \lambda, x \geq 0\}. \tag{5}$$

3.1.2 Uncertainty for Mean and Covariance Matrix

In this section, we consider uncertainty for both covariance matrix Σ and mean vector μ. Our uncertainty set involves ellipsoidal uncertainty for μ and the upper and lower bounds for Σ (in matrix sense). We particularly consider the following uncertainty set:

$$\mathcal{S} := \{(\mu, \Sigma) : (\mu - \hat{\mu})^T \Sigma^{-1}(\mu - \hat{\mu}) \leq \Upsilon^2,\ (1-\beta)\hat{\Sigma} \preceq \Sigma \preceq (1+\beta)\hat{\Sigma}\},$$

where the matrix $\hat{\Sigma}$ and $\hat{\mu}$ are the sample covariance and mean estimated from data, $\beta \in (0, 1]$ and Υ are positive scalars, controlling the robustness level. Note that the set $\mathcal{S}$ is strictly feasible and bounded. Then, we utilize conic programming duality to obtain a single-stage equivalent of problem (3.1) as the following semidefinite program:

$$\min_{x,\Lambda^+,\Lambda^-,\gamma} \quad (1+\beta)\mathrm{Tr}(\hat{\Sigma}\Lambda^+) - (1-\beta)\mathrm{Tr}(\hat{\Sigma}\Lambda^-) + \Upsilon^2\gamma_{22} - 2\hat{\mu}\gamma_{12} \tag{6a}$$

$$\text{s.t.} \quad \begin{bmatrix} \Lambda^+ - \Lambda^- - \gamma_{11} & x \\ x^T & 1 \end{bmatrix} \succeq 0, \quad -2\gamma_{12} = -\tau x,\ \gamma = \begin{bmatrix} \gamma_{11} & \gamma_{12} \\ \gamma_{21} & \gamma_{22} \end{bmatrix} \succeq 0, \quad \Lambda^+, \Lambda^-, \gamma \succeq 0, \quad \text{(1b)}. \tag{6b}$$

3.2 *Conditional Value-At-Risk Model Under Mixture Distribution*

In this section, we focus on the robust optimization version of the CVaR model presented in Sect. 2.2 under the assumption that the return vector is distributed according to a mixture of two multivariate normals. The reasons we consider a mixture distribution with the CVaR optimization model are three-fold: i) The Markowitz model is distribution independent, ii) the frontiers obtained for the Markowitz model and CVaR model under normal distribution are the same theoretically, iii) the merits of mixture distribution with the CVaR optimization model are discussed in a recent paper by Kocuk and Cornuéjols (2020). In this probabilistic model, random returns come from the normal distributions $\mathrm{N}\left(\mu^1, \Sigma^1\right)$ with probability ρ_1 and $\mathrm{N}\left(\mu^2, \Sigma^2\right)$ with probability ρ_2. The motivation for such a model is that although most of the time (with probability ρ_1) stock returns behave as normally distributed as $\mathrm{N}\left(\mu^1, \Sigma^1\right)$, every once in a while (with probability ρ_2) a shock happens and shifts the mean of the normal distribution to the left with a higher variance as $\mathrm{N}\left(\mu^2, \Sigma^2\right)$ (see the discussions in Kocuk and Cornuéjols (2020) for details). We note that if a data set is given, we can compute the parameters of a mixture distribution by using the Expectation-Maximization (EM) Algorithm (Dempster et al., 1977). Since the

CVaR function does not have a closed form expression in this case, we utilize a second-order cone representable approximation proposed in Kocuk and Cornuéjols (2020).

Throughout this section, we will assume that the parameters of problem (2.2), that is μ^1, Σ^1, μ^2 and Σ^2, are uncertain. In order to incorporate robustness into the objective function, we formulate the following general form of a two-stage problem:

$$\min_{x} \Big\{ \sum_{i=1}^{2} \max_{(\mu^i,\Sigma^i)\in\mathcal{S}^i} \big(z_i(\rho_i)\sqrt{x^T\Sigma^i x} - {\mu^i}^T x - \tau\rho_i {\mu^i}^T x \big) \,:\, (1\text{b}) \Big\}. \tag{7}$$

Here, we have $z_i(\rho_i) := \frac{\phi(\Phi^{-1}(\alpha/\rho_i))}{\alpha/\rho_i}$, for $i = 1, 2$ with $\alpha < 0.5$. In the sequel, we reformulate problem (7) as a single-stage conic program for a variety of different uncertainty sets.

3.2.1 Polyhedral Uncertainty for Mean

In this section, we consider a general polyhedral uncertainty for mean vectors μ^i while covariance matrices Σ^i are assumed to be known (or estimated from data) as $\hat{\Sigma}^i$. Let us consider the following uncertainty sets:

$$\mathcal{S}^i := \{(\mu^i, \Sigma^i) : A_i\mu^i \le b_i,\ \Sigma^i = \hat{\Sigma}^i\},$$

where $A_i \in \mathbb{R}^{m\times n}$ and $b^i \in \mathbb{R}^n$ are given for $i = 1, 2$. Assuming that the sets $\mathcal{S}^1$ and $\mathcal{S}^2$ are feasible and bounded, we obtain a single-stage reformulation as the following convex quadratic problem:

$$\min_{x,\lambda_1,\lambda_2} \sum_{i=1}^{2} \{[z_i(\rho_i)\sqrt{x^T\Sigma^i x} + (\tau\rho_i + 1){\lambda_i}^T b^i] : (1\text{b}),\ {A_i}^T\lambda_i = -x,\ \lambda_i \ge 0, i = 1, 2\}. \tag{8}$$

3.2.2 Uncertainty for Mean and Covariance Matrix

In this section, we consider uncertainty for both covariance matrices Σ^i and mean vectors μ^i. Our uncertainty set involves ellipsoidal uncertainty for μ^i and the upper and lower bounds for Σ^i (in matrix sense). In particular, let us consider the following uncertainty sets:

$$\mathcal{S}^i := \{(\mu^i, \Sigma^i) : (\mu^i - \hat{\mu}^i)^T {\Sigma^i}^{-1} (\mu^i - \hat{\mu}^i) \le {\Upsilon_i}^2,\ (1-\beta_i)\hat{\Sigma}^i \preceq \Sigma^i \preceq (1+\beta_i)\hat{\Sigma}^i\},$$

where the matrix $\hat{\Sigma}^i$ is the sample covariance estimated from data, $\beta_i \in (0, 1]$ and Υ_i are positive scalars, controlling the robustness level for $i = 1, 2$. Note that the sets $\mathcal{S}^1$ an $\mathcal{S}^2$ is strictly feasible for positive and bounded. Then, we utilize conic programming duality to obtain a single-stage reformulation of the problem (7) as the following semidefinite program (see Arabacı (2020) for the details of the derivation):

$$\min_{x,\Lambda^+,\Lambda^-,a,b,c} \sum_{i=1}^{2}[(1+\beta_i)\mathrm{Tr}(\hat{\Sigma}^i \Lambda^{i+}) - (1-\beta_i)\mathrm{Tr}(\hat{\Sigma}^i \Lambda^{i-}) + b_i + c_i + \Upsilon_i{}^2 \gamma^i{}_{22} - 2\hat{\mu}^{i^T} \gamma^i{}_{12}] \tag{9a}$$

$$\text{s.t.} \quad \begin{bmatrix} \Lambda_i{}^+ - \Lambda_i{}^- - \gamma^i{}_{11} & x \\ x^T & \frac{4}{z_i^2(\rho_i)}(b+c) \end{bmatrix} \succeq 0, \ a_i = -\frac{z_i(\rho_i)}{2}, -2\gamma^1{}_{12} = -(\tau\rho_i + 1)x, \quad i = 1, 2 \tag{9b}$$

$$\gamma^i = \begin{bmatrix} \gamma^i{}_{11} & \gamma^i{}_{12} \\ \gamma^i{}_{21} & \gamma^i{}_{22} \end{bmatrix} \succeq 0, \begin{pmatrix} a_i \\ b_i \\ c_i \end{pmatrix} \in L^3, \ \Lambda^{i+}, \Lambda^{i-}, \gamma^i \succeq 0, \ \text{(1b)} \quad i = 1, 2. \tag{9c}$$

4 Computational Experiments

In this section, we present the results of our computational experiments, which have been conducted to investigate the effectiveness of robust optimization approaches and assess their impact on optimal portfolios. We use a real data set provided by Kocuk and Cornuéjols (2020) from the S&P 500 index spanning 30 years between January 1987–December 2016 with monthly resolution. This gives us 360 asset return realizations for 11 sectors considered.

We utilize a rolling horizon based evaluation approach with an out-of-sample analysis to evaluate the performance of the robust optimization models on the data set. Let T be the number of data points available, and H and m be positive integers. The main idea of the rolling horizon based evaluation scheme is to use the return vectors of the last H periods, namely, $r^{t-H+1}, \dots, r^t$ to estimate the parameters of a portfolio optimization problem, which we use to determine the portfolio decision $x^{*,t}$ for time period $t+1$. Then, we evaluate the performance of this decision using the return vector r^{t+1} as $\gamma^{t+1} = r^{t+1^T} x^{*,t}$. Finally, we repeat this procedure for $t = T/2, \dots, T$ and for different values of τ and evaluate the overall performance. We fix $H = T/2$ and $\alpha = 0.01$. In the remainder of this section, we provide a comparison of the various robust optimization models constructed in Sect. 3 with respect to different uncertainty sets and robustness levels.

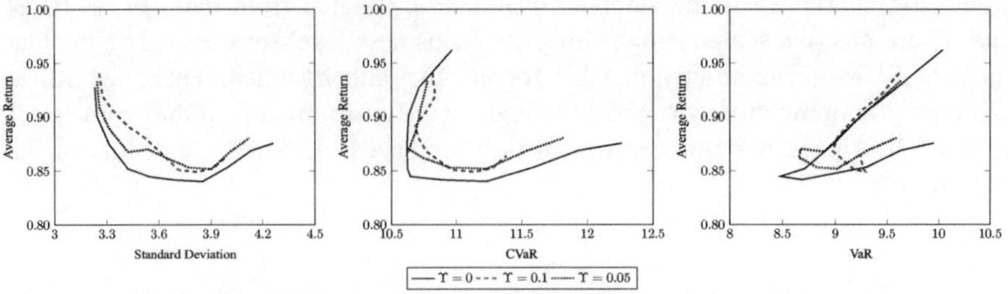

Fig. 1 Performance comparison of the robust Markowitz model with budgeted uncertainty for mean for the S&P 500 data set

4.1 Performance of the Markowitz Models

In this subsection, we will present the performance of the standard Markowitz model (1) and the robust Markowitz models in Sect. 3.1 on the data set. In the following figures, we will provide the average expected return of the optimal portfolios with respect to risk measures; standard deviation, $\text{CVaR}_{0.01}$, and $\text{VaR}_{0.01}$ for different values of risk aversion constant τ.

4.1.1 The Standard Model

As a benchmark, we first solve the standard Markowitz model (1), which can be seen from Fig. 1 with $\Upsilon = 0$. According to the resulting performance, we observe that as the value of τ increases, the average return starts to decrease while the standard deviation starts to increase. Counter-intuitively, the resulting performance shows that $\tau = 0$ (which puts no weight at all to mean return) yields the best mean return. We attribute this to the fact that mean returns are very hard to estimate. In the remainder of this subsection, we solve several robust versions of the Markowitz optimization model and provide the corresponding results in comparison with the results of the standard Markowitz model.

4.1.2 The Robust Model with Budgeted Uncertainty

In this subsection, we will present the performance of the robust Markowitz model with a special case of uncertainty set in Sect. 3.1.1 called *budgeted uncertainty* (Ben-Tal & Nemirovski, 2001), which can be treated as the intersection of the infinity-norm and the 1-norm for mean μ. In particular, we consider the uncertainty set defined as $\mathcal{S} := \left\{(\mu, \Sigma) : \sum_{j=1}^{n} \frac{|\mu_j - \bar{\mu}_j|}{\bar{\mu}_j} \leq \Upsilon,\ \Sigma = \hat{\Sigma}\right\}$, where Υ is a positive scalar controlling the robustness level and $\bar{\mu}$ is assumed to be positive.

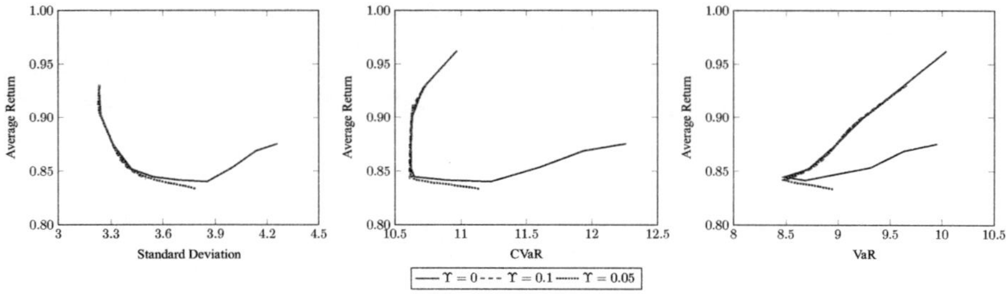

Fig. 2 Performance comparison of the robust Markowitz model with ellipsoidal uncertainty for mean for the S&P 500 data set

In Fig. 1, one can say that as Υ increases, for the same level of standard deviation, the average return of the robust models' portfolios dominates the standard model slightly.

4.1.3 The Robust Model with Ellipsoidal Uncertainty

In this subsection, we present the performance of the robust Markowitz model with ellipsoidal uncertainty for mean discussed in Sect. 3.1.2 when the β, which controls the robustness level, is set to 0.

According to Fig. 2, one can say that even the slightest increase in the robustness level Υ results in a lower average return and risk. Although not reported, we also observe that resulting portfolios obtained from the robust model remain nearly unchanged for the different values of β.

4.2 Performance of the CVaR Models Under Mixture Distribution

In this subsection, we solve the robust versions of the CVaR model assuming the random return vector is distributed as a mixture of normals discussed in Sect. 3.2. In order to estimate the parameters of the two mixtures, we employ the EM Algorithm on S&P 500 data set. We obtain the following relations between the parameters for the S&P 500 data set (see Sect. 3.2 and Kocuk and Cornuéjols (2020) for the significance of these relations):

$$\hat{\rho}_1 > \hat{\rho}_2,\ \hat{\mu}^1 > \hat{\mu}^2,\ \hat{\Sigma}^2 > \hat{\Sigma}^1.$$

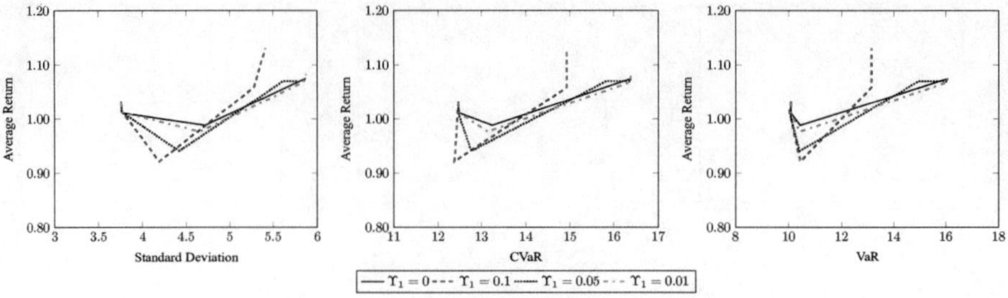

Fig. 3 Performance comparison of the robust CVaR model with budgeted uncertainty for mean for the normal with ρ_1 of the mixture for the S&P 500 data set

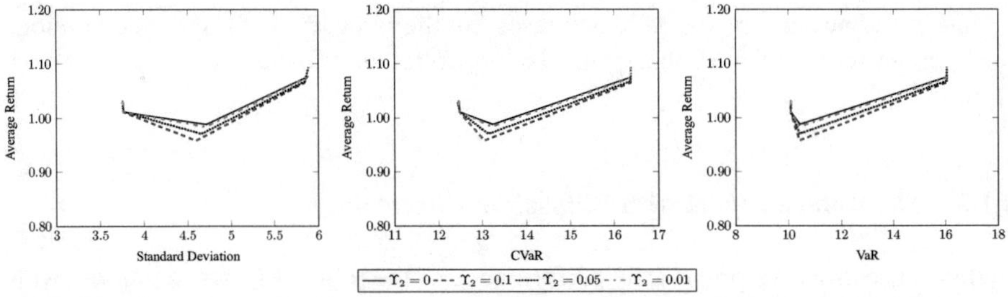

Fig. 4 Performance comparison of the robust CVaR model with budgeted uncertainty for mean for the normal with ρ_2 of the mixture for the S&P 500 data set

4.2.1 The Robust Model with Budgeted Uncertainty

In this subsection, we will present the performance of the robust model with budgeted uncertainty for mean can be described as a special case of Sect. 3.2.1. In particular, we use the uncertainty set defined in Sect. 4.1.2 for each of the mixture distributions with different levels of uncertainty Υ_1 and Υ_2.

As can be seen from Fig. 3, as Υ increases, the average return of the robust model starts to dominate the standard model while risk stays at similar levels for greater values of τ. By contrast, Fig. 4 shows that robust models result in poor portfolios as the robustness level increases.

4.2.2 The Robust Model with Ellipsoidal Uncertainty

The following figures show the performance of the robust model with ellipsoidal uncertainty for mean discussed in Sect. 3.2.2 when the robustness levels β_1 and β_2 are set to 0.

In Fig. 5, one can say that the resulting robust portfolios are inversely correlated with the increment in the robustness level in general. Moreover, for the same level of risk, the robust problem may result with a lower return. On the other hand, Fig. 6 shows that for some same risk levels, the robust model results with higher returns.

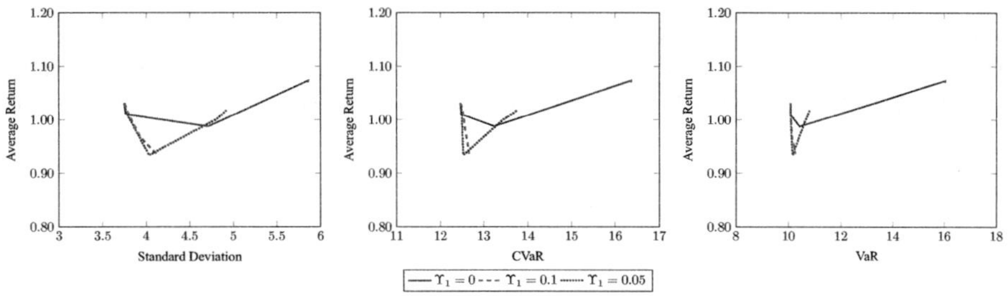

Fig. 5 Performance comparison of the robust CVaR model with ellipsoidal uncertainty for mean for the normal with ρ_1 of the mixture for the S&P 500 data set

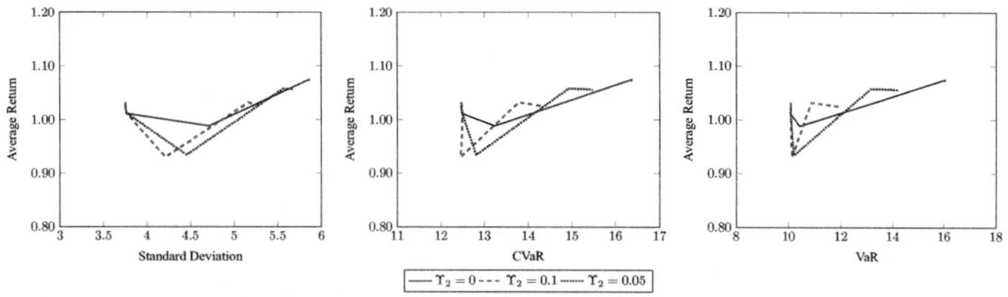

Fig. 6 Performance comparison of the robust CVaR model with ellipsoidal uncertainty for mean for the normal with ρ_2 of the mixture for the S&P 500 data set

This may be due to the fact that the random returns come from the different normals of the mixture. We again note that when the values of β_1 and β_2 change, the resulting portfolios obtained from the robust models remain nearly unchanged.

5 Conclusion

In this paper, we address a number of challenging aspects of portfolio optimization problem including conflicting objectives, inadequate risk measures, and sensitivity issue in parameter estimation. We provide an overview of risk adjusted optimization models with different risk measures including a distribution independent measure variance, and a distribution dependent measure CVaR. We then adapt robust optimization to incorporate estimation errors or perturbations into the standard portfolio optimization problem where the returns are modelled as a mixture of normals. We present an analysis on robust portfolio optimization problems with uncertainty sets involving polytopic, ellipsoidal, or budgeted uncertainty for either mean or covariance or both, and cast these problems as conic programs. Moreover, we provide computational experiments on a real data set and compare the performances of the resulting portfolios. Our computational experiments show that optimal portfolios constructed with the robust optimization models yield higher returns and higher risk than the standard portfolio models. Furthermore, employing Markowitz robust models with budgeted uncertainty for the same levels of risk results in portfolios with higher returns. Finally, we conclude that changes in the robustness levels for covariance matrices have relatively limited effect on the resulting portfolios than the mean return vector.

References

Arabacı, P. (2020). *Comparison of robust optimization models for portfolio optimization*. MS Thesis.

Ben-Tal, A., & Nemirovski, A. (2001). *Lectures on modern convex optimization: analysis, algorithms, and engineering applications*. SIAM.

Ceria, S., & Stubbs, R. A. (2006). Incorporating estimation errors into portfolio selection: Robust portfolio construction. *Asset Management* (pp. 270–294). Springer.

Ceria, S., & Sivaramakrishnan, K. (2013). *Portfolio optimization* (pp. 429–464).

Chen, R., & Yu, L. (2013). A novel nonlinear value-at-risk method for modeling risk of option portfolio with multivariate mixture of normal distributions. *Economic Modelling, 35*, 796–804.

DeMiguel, V., & Nogales, F. J. (2009). Portfolio selection with robust estimation. *Operations Research, 57*(3), 560–577.

Dempster, A. P., Laird, N. M., & Rubin, D. B. (1977). Maximum likelihood from incomplete data via the EM algorithm. *Journal of the Royal Statistical Society: Series B, 39*(1), 1–22.

Ghaoui, L. El., Oks, M., & Oustry, F. (2003). Worst-case value-at-risk and robust portfolio optimization: A conic programming approach. *Operations Research, 51*(4), 543–556.

Kocuk, B., & Cornuéjols, G. (2020). Incorporating Black-Litterman views in portfolio construction when stock returns are a mixture of normals. *Omega, 91*, 102008.

Krokhmal, P., Palmquist, J., & Uryasev, S. (2002). Portfolio optimization with conditional value-at-risk objective and constraints. *Journal of Risk, 4*, 43–68.

Lobo, M. S., & Boyd, S. (2000). The worst-case risk of a portfolio. *Unpublished manuscript*. Available from http://faculty.fuqua.duke.edu/%7Emlobo/bio/researchfiles/rsk-bnd.pdf.

Markowitz, H. (1952). Portfolio selection. *The Journal of Finance, 7*(1), 77–91 (1952).

Rockafellar, R. T., & Uryasev, S. (2002). Conditional value-at-risk for general loss distributions. *Journal of Banking & Finance, 26*(7), 1443–1471.

Two-Stage Chance-Constrained Telemedicine Assignment Model with No-Show Behavior and Uncertain Service Duration

Menglei Ji, Jinlin Li, and Chun Peng

1 Introduction

Telemedicine, largely relying on advanced remote communication technology and medical equipment, offers the necessary healthcare services for the patients in rural areas to access specialty care (i.e. consulting, remote diagnosis, treatment, monitoring and follow-ups and so on) that is often unavailable for them (Myers, 2003). Teleconsultation can effectively reduce patients' waiting time, improve the quality of medical services, and prevent unnecessary hospital visits, as well as save medical system costs, especially in rural areas (e.g. Zanaboni et al., 2009; Ward et al., 2015; Rajan et al., 2018). Moreover, hospital-based applications of telemedicine present a potentially important solution for small and rural hospitals where access to local specialists is rarely available (Jetty et al., 2017), which has been applied in many situations, such as controlling patients with heart diseases, specialist care for neonates, emergency rooms. More importantly, under the current context of the global pandemic of COVID-19, people are usually advised to stay at home in particular during the lockdown and quarantine to avoid the spread of the coronavirus and the medical resources (e.g. doctors, nurses, and other personal staffs) are rare for the hospitals. In this regard, telemedicine can provide consulting services for the unurgent patients (e.g. patients with chronic diseases) and the ones who might be infected by coronavirus or flu (since they have very similar symptoms) while they have light symptoms, which is expected to deliver timely care while

M. Ji · J. Li
Beijing Institute of Technology, Beijing, China
e-mail: jimenglei@bit.edu.cn; jinlinli@bit.edu.cn

C. Peng (✉)
HEC Montréal & GERAD, Montréal, QC, Canada
e-mail: chun.peng@hec.ca

H. Yang et al. (eds.), *AI and Analytics for Public Health*, Springer Proceedings in Business and Economics, https://doi.org/10.1007/978-3-030-75166-1_32

minimizing exposure to protect medical practitioners and patients during these hard times (Jnr, 2020; Latifi & Doarn, 2020; Loeb et al., 2020).

National Telemedicine Center of China (NTCC) has explored various forms of collaborative services with provincial hospitals (called tertiary hospitals) and 120 Regional Cooperative hospitals (called primary hospitals). Compared with the provincial hospitals, regional Cooperative hospitals are relatively small and have less skilled doctors and less advanced medical equipment. Currently, the number of teleconsultations is up to nearly 150 cases a day, especially during the global pandemic of COVID-19. Once both primary doctors and patients apply for the teleconsultation service, NTCC will assign the doctors from tertiary hospitals to patients and schedule a teleconsultation service according to the severity of the disease and the available resources in tertiary hospitals. However, at the same time, most of the doctors in tertiary hospitals still provide traditional outpatient medical services during the weekday, so they might not show up at the teleconsultation in the case of some emergency events. Such a case is regularly occurred in practice, based on our survey in several tertiary hospitals of China. In this regard, this paper incorporates the no-show behavior of the doctors by using a specific uncertainty set in the two-stage chance-constrained telemedicine assignment model, which does not rely on any distributional assumptions of no-show behaviour.

As we know that Charnes and Cooper (1959) firstly introduce chance-constrained programs (CCPs) in the context of stochastic optimization. In recent years, CCPs have been extensively studied in terms of new methodological developments and wide applications in various fields, including transportation, healthcare operations, supply chain, energy, machine learning, and so on. However, due to the non-convex feasible region of CCPs, it is generally hard to solve CCPs. A common way is to derive the conservative approximations of chance constraint (Ahmed & Shapiro, 2008; Birge & Louveaux, 2011) by using stochastic programming and robust optimization techniques. Note that chance-constrained formulation is also widely proposed in healthcare, especially for operating rooms (ORs) planning and surgery allocation problems that make the decisions on the open of ORs and assignment of surgeries (or doctors), which is close to our topic. In terms of dealing with chance constraints by robust optimization/distributionally robust optimization, the existing literature mainly considers the uncertainty of surgery duration that can be captured by an uncertainty set or an ambiguity set (e.g. Wang et al., 2017; Zhang et al., 2018; Deng et al., 2019; Wang et al., 2019; Zhang et al., 2020). Our work is highly related to the stochastic programming chance-constrained models, especially for the two-stage setting. Jebali and Diabat (2017) present a novel two-stage chance-constrained stochastic programming model for operating room planning by considering random surgery duration, random patient length of stay in the ICU, and random resource capacity reserved for emergency cases. Wang et al. (2021) study a chance-constrained multiple bin packing problem to operating room planning that minimizes the number of open operating rooms. Kamran et al. (2018) propose different formulations of two-stage stochastic programming and two-stage chance-constrained stochastic programming with multiple objectives, including minimization of the patients waiting time, tardiness, cancellation, block overtime, and the number of surgery days of each surgeon within the planning horizon. Noorizadegan and Seifi (2018) based on a decomposition of set-partitioning formulation of an integrated surgery planning and scheduling problem with chance constraints considering uncertain time, propose an efficient solution method. For more details about the healthcare resources planning and scheduling for the traditional service, we refer the interested readers to the recent review papers (e.g. Zhu et al., 2019; Dai and Tayur, 2020; Ahmadi-Javid et al., 2017). However, all the above-aforementioned studies study the ORs planning and surgery (or doctor) allocation problems for the traditional medical service, instead of telemedicine.

From the operations management viewpoint, there are very few existing studies about telemedicine in the literature. Rajan et al. (2019) investigate the effect of telemedicine on chronic care in the service systems with heterogeneous customers by maximizing revenue and welfare of the specialists. Saghafian et al. (2018) develop a novel model of agent knowledge to manage the workload in telemedical physician triage. More recently, Wang et al. (2020) study the price and capacity decisions in a telemedicine service system from the government subsidy policy perspective. Qiao et al. (2020) establish a queuing simulation system to search the most reasonable resource allocation combination of doctors. The most relevant work to us is Erdogan et al. (2018), who present a two-stage stochastic linear program to study the optimal scheduling of telemedicine patients (i.e. the arriving time and waiting time of patients, the overtime of a session) by considering the uncertainties of service duration and patients' no-show behavior. They employ a set of finite scenarios to capture the uncertain service duration and patients' no-show behavior.

Unlike Erdogan et al. (2018) that consider the no-show behaviour of patients for optimal telemedicine scheduling, we address the telemedicine assignment between the doctors and patients by using a two-stage min-max chance-constrained model and employ an uncertainty set to capture the no-show behavior of telemedical doctors, which finally gives rise to a two-stage binary integer program with binary recourse problem. We then propose an enumeration-based column-and-constraint generation (C&CG) solution method to solve the resulting two-stage binary integer program. To the best of our knowledge, this is the first attempt to incorporate the no-show behavior of telemedical doctors and uncertain service duration for the telemedicine assignment problem in the literature. We expect that this could open an avenue in terms of telemedicine services from an operations management viewpoint.

The structure of this paper is presented as follows: Sect. 2 proposes a novel two-stage chance-constrained model and Sect. 3 presents an enumeration-based column-and-constraint generation solution method to solve the resulting two-stage binary integer program with integer decisions in the recourse problem. We conduct a numerical experiment in Sect. 4 and finally conclude the paper by Sect. 5.

2 Two-Stage Chance-Constrained Programming Model

In this section, we explore the optimal assignment decisions of doctors (i.e. $\boldsymbol{y}$) and patients (i.e. $\boldsymbol{x}(z, \boldsymbol{d})$) when scheduling a remote telemedicine service when the realizations of both no-show doctors and the uncertain service duration are observed. During the time period $[0, T]$, a set of patients $\mathcal{I}$ are assigned to a set of telemedical doctors $\mathcal{J}$. We assume the uncertain service duration with distribution $\mathbb{Q}$. The goal is to minimize the total expected cost of assignment cost of telemedical doctors and patients and the penalty cost for the unassigned patients. All the related notations that are used in this paper can be given in Table 1.

In this paper, we incorporate the no-show behavior of doctors for the telemedicine assignment. More specifically, we use uncertainty set $\mathcal{Z} := \left\{ z \in \{0, 1\}^{|\mathcal{J}|}, \ \sum_{j \in \mathcal{J}} z_j = \Gamma \right\}$ to represent the number of doctors that are no-show, where Γ is a integer, $0 \leq \Gamma \leq K - 1$, and $z_j = 1$ if doctor j is no-show and otherwise 0, which captures the level of robustness for no-show telemedical doctors. Note that, if $\Gamma = 0$, it indicates that all the telemedical doctors show up. The larger the value of Γ is, the more conservative the model is.

In this regard, we consider the following two-stage min-max chance-constrained model with binary decisions in the recourse problem and adversary over $z \in \mathcal{Z}$, which specifically can be represented as follows,

Table 1 Notations

Sets	
$\mathcal{I}$	The set of patients, $i \in \mathcal{I}$
$\mathcal{J}$	The set of telemedical doctors, $j \in \mathcal{J}$
Ω	The set of scenarios of random service duration for each scenario $\omega \in \Omega$
Parameters	
T	The length for the telemedicine service block
c_{ij}	Assignment cost, doctor j who is scheduled to service patient i
b_j	Overtime cost for the doctor j
r_i	Penalty cost for patient i who can't be assigned at the time block
h_j	Assignment cost for doctors j
d_i^{ω}	Random service duration for the patient i under scenario ω
K	The number of assigned doctors
Decision variables	
x_{ij}^{ω}	Binary variables, $x_{ij}^{\omega} = 1$ if assignment for patient i to doctor j under scenario ω, and otherwise 0
y_j	Binary variables, $y_j = 1$ if doctor j is assigned for the telemedicine service, and otherwise 0

$$\min_{y} \sum_{j\in\mathcal{J}} h_j y_j + \min_{x(\cdot)} \max_{z\in\mathcal{Z}} \mathbb{E}_{\mathbb{Q}}\left[\sum_{i\in\mathcal{I}}\sum_{j\in\mathcal{J}} c_{ij}x_{ij}(z,\boldsymbol{d}) + \sum_{i\in\mathcal{I}} r_i\left(1-\sum_{j\in\mathcal{J}} x_{ij}(z,\boldsymbol{d})\right)\right] \tag{1a}$$

$$\text{s.t.} \sum_{j\in\mathcal{J}} y_j \le K \tag{1b}$$

$$\sum_{j\in\mathcal{J}} x_{ij}(z,\boldsymbol{d}) \le 1 \qquad \forall i\in\mathcal{I} \tag{1c}$$

$$x_{ij}(z,\boldsymbol{d}) \le y_j(1-z_j) \qquad \forall i\in\mathcal{I}, j\in\mathcal{J} \tag{1d}$$

$$\mathbb{P}_{\mathbb{Q}}\left\{\sum_{i\in\mathcal{I}} d_i x_{ij}(z,\boldsymbol{d}) \le T y_j(1-z_j)\right\} \ge 1-\eta \qquad \forall j\in\mathcal{J} \tag{1e}$$

$$y_j\in\{0,1\}, x_{ij}(z,\boldsymbol{d})\in\{0,1\} \qquad \forall i\in\mathcal{I}, j\in\mathcal{J}. \tag{1f}$$

For model (1), we aim to minimize the total expected worst-case cost, including the assignment cost of patients and telemedical doctors and the penalty cost of patients who are not assigned. Constraint (1b) requires that the total number of assigned telemedical doctors is less than K ($K \le |\mathcal{J}|$). Constraint (1c) indicates that a patient either be served by a doctor or he/she is not served. Constraint (1d) represents that a patient only can be served by a show-up telemedical doctor. Probabilistic constraint (1e) ensures that the probability of overtime for telemedical doctor j who is show-up, is no less than $1-\eta$ ($0 \le \eta < 1$). Constraint (1f) presents all the assignment decisions of patients and doctors to be binary variables.

Note that, the second-stage decisions $x_{ij}(z,\boldsymbol{d})$ are decided, once the uncertainty of telemedical doctors' no-show behavior and the uncertainty of uncertain service duration are realized. For the ease of exposition, we remark that, in the remaining discussion we use the notations of x_{ij} for short.

In the following, we assume that distribution $\mathbb{Q}$ is finitely known with N empirical samples, i.e. $\{\boldsymbol{d}_\omega\}_{\omega=1}^N$, and each with probability p_ω, such that $\sum_{\omega=1}^{N} p_\omega = 1$. Therefore, given the finite scenarios of random service duration $\{\boldsymbol{d}_\omega\}_{\omega=1}^N$, $\omega\in\Omega := \{1,2,\cdots,N\}$, we introduce the binary variable ρ_ω to reformulate constraint (1e), in which $\rho_\omega = 1$ if $\sum_{i\in\mathcal{I}} d_i^\omega x_{ij}^\omega(z,\boldsymbol{d}) \le T y_j(1-z_j)$ holds under distribution $\mathbb{Q}$, and otherwise 0. In doing so, using a big-M method, we can reformulate probabilistic constraint (1e) as the following two constraints,

$$\sum_{i\in\mathcal{I}} d_i^\omega x_{ij}^\omega \le T y_j(1-z_j) + M(1-\rho_j^\omega) \qquad \forall j\in\mathcal{J}, \omega\in\Omega \tag{2}$$

$$\sum_{\omega\in\Omega} p_\omega \rho_j^\omega \ge 1-\eta. \tag{3}$$

Under the assumption of uniform scenario probability, i.e. $p_\omega = 1/N$, this gives rise to the following model,

$$\min_{y} \sum_{j\in\mathcal{J}} h_j y_j + \min_{x,\rho} \max_{z\in\mathcal{Z}} \frac{1}{N} \sum_{\omega=1}^{N} \left(\sum_{i\in\mathcal{I}} \sum_{j\in\mathcal{J}} c_{ij} x_{ij}^{\omega} + \sum_{i\in\mathcal{I}} r_i (1 - \sum_{j\in\mathcal{J}} x_{ij}^{\omega}) \right) \tag{4a}$$

$$\text{s.t.} \sum_{j\in\mathcal{J}} y_j \le K \tag{4b}$$

$$\sum_{j\in\mathcal{J}} x_{ij}^{\omega} \le 1 \qquad \forall i \in \mathcal{I}, \omega \in \Omega \tag{4c}$$

$$x_{ij}^{\omega} \le y_j(1 - z_j) \qquad \forall i \in \mathcal{I}, j \in \mathcal{J}, \omega \in \Omega \tag{4d}$$

$$\sum_{i\in\mathcal{I}} d_i^{\omega} x_{ij}^{\omega} \le T y_j (1 - z_j) + M(1 - \rho_j^{\omega}) \qquad \forall j \in \mathcal{J}, \omega \in \Omega \tag{4e}$$

$$\sum_{\omega\in\Omega} \rho_j^{\omega} \ge N(1 - \eta) \qquad \forall j \in \mathcal{J} \tag{4f}$$

$$y_j, x_{ij}^{\omega}, \rho_j^{\omega} \in \{0, 1\} \qquad \forall i \in \mathcal{I}, j \in \mathcal{J}, \omega \in \Omega. \tag{4g}$$

3 Enumeration-Based C&CG Solution Method

In this section, we propose an enumeration-based C&CG solution method with some specific enhancements. Note that, a two-stage problem with the structure of model (4) can not be directly solved by the existing decomposition-based algorithm (e.g. Benders decomposition and classical C&CG (Zeng & Zhao, 2013)), given that all the decisions are binaries (especially for the recourse decisions) and the subproblem takes the form of max-min (bi-level problem) over $z \in \mathcal{Z}$. Inspired by C&CG method (Zeng & Zhao, 2013), we propose an enumeration-based C&CG method to solve the min-max model with binary decisions in the recourse problem. We first describe how our method works, and then present two simple enhancement strategies.

Let m be the iteration index. Model (4) can be solved iteratively by enumeration-based C&CG algorithm within a master-sub-problem framework. At each iteration m, for the master problem (5), the decisions of telemedical doctors' assignment $\hat{\boldsymbol{y}}$ are decided. We use $\boldsymbol{x}^l$ and $\boldsymbol{\rho}^l$ to represent the variables associated the l-th scenario ($l \in \{1, \cdots, m\}$), and $\boldsymbol{z}^l$ to represent the conditional scenarios (show-up or no-show) of the telemediccal doctors in the l-th scenario. We summarize the specific steps in Algorithm 1.

Algorithm 1 Enumeration-based C&CG Solution Algorithm

Initialize A tolerance $\epsilon \geq 0$ and maximum run time *timelimit*
Initialize $m = 0$, $\text{LB} = -\infty$, $\text{UB} = +\infty$ for all ω.
while ($runtime \leq timelimit$ and $|\frac{\text{UB}-\text{LB}}{\text{UB}}| > \epsilon$) **do**
  Solve the MP (5)
  Record optimal solution ($\boldsymbol{x}^m$, $\boldsymbol{y}^m$, $\boldsymbol{\rho}^m$, η^m) and optimal objective $lobj^m$.
  Update LB := $lobj^m$.
  Fix $\boldsymbol{y} := \boldsymbol{y}^m$, solve SP (6) and obtain the cost information of all the potential scenarios.
  Obtain the worst-case cost θ^m with the selected scenarios.
  Update UB: $=\min\{\text{UB}, \theta^m + \sum_{j \in \mathcal{J}} h_j y_j\}$.
  Create variables $\boldsymbol{x}^{m+1}$, $\boldsymbol{\rho}^{m+1}$ and add the related constraints (5b), (5c), (5d), (5e), (5f) to MP.
  Set $m := m + 1$.
end while
return UB and corresponding optimal solution ($\boldsymbol{x}^*$, $\boldsymbol{y}^*$) for which $obj^* = \text{UB}$.

$$[\textbf{MP}]: \quad \min_{\boldsymbol{y},\boldsymbol{x},\boldsymbol{\rho} \in \{0,1\}; \eta \geq 0} \sum_{j \in \mathcal{J}} h_j y_j + \eta \tag{5a}$$

s.t. (4b)

$$\sum_{j \in \mathcal{J}} x_{ij}^{l\omega} \leq 1 \qquad \forall i \in \mathcal{I}, \omega \in \Omega \tag{5b}$$

$$x_{ij}^{l\omega} \leq y_j(1 - z_j^l) \qquad \forall i \in \mathcal{I}, j \in \mathcal{J}, \omega \in \Omega \tag{5c}$$

$$\sum_{i \in \mathcal{I}} d_i^{\omega} x_{ij}^{l\omega} \leq T y_j (1 - z_j^l) + M(1 - \rho_j^{l\omega}) \qquad \forall j \in \mathcal{J}, \omega \in \Omega \tag{5d}$$

$$\sum_{\omega \in \Omega} \rho_j^{l\omega} \geq N(1 - \eta) \qquad \forall j \in \mathcal{J} \tag{5e}$$

$$\eta \geq 1/N \sum_{\omega=1}^{N} \left(\sum_{i \in \mathcal{I}} \sum_{j \in \mathcal{J}} c_{ij} x_{ij}^{l\omega} + \sum_{i \in \mathcal{I}} r_i (1 - \sum_{j \in \mathcal{J}} x_{ij}^{l\omega}) \right) \tag{5f}$$

For ant given $\hat{\boldsymbol{y}}$ from MP, we could enumerate all the potential worst-case scenarios and solve SP (6) for every scenario to obtain the maximum recourse cost. Since the unassigned patients will be penalized in any scenarios, the recourse problem is always feasible. Next a set of recourse variables and corresponding constraints w.r.t. this scenario will be added to MP. Given that $\hat{\boldsymbol{y}}$ is finite, the algorithm finally converges and terminates in a finite number of iterations. We assume the number of assigned doctors is more than the no-show doctors, i.e. $\Gamma \leq K - 1$, if not, all patients will not be assigned, the problem becomes meaningless.

$$[\textbf{SP}]: \quad \max_{z \in \mathcal{Z}} \min_{\boldsymbol{x},\boldsymbol{\rho} \in \{0,1\}} 1/N \sum_{\omega=1}^{N} \left(\sum_{i \in \mathcal{I}} \sum_{j \in \mathcal{J}} c_{ij} x_{ij}^{\omega} + \sum_{i \in \mathcal{I}} r_i (1 - \sum_{j \in \mathcal{J}} x_{ij}^{\omega}) \right) \tag{6a}$$

$$\text{s.t.}\ (4c), (4f)$$

$$x_{ij}^{\omega} \leq \hat{y}_j(1 - z_j) \qquad \forall i \in \mathcal{I}, j \in \mathcal{J}, \omega \in \Omega \qquad (6b)$$

$$\sum_{i \in \mathcal{I}} d_i^{\omega} x_{ij}^{\omega} \leq T\hat{y}_j(1 - z_j) + M(1 - \rho_j^{\omega}) \qquad \forall j \in \mathcal{J}, \omega \in \Omega \qquad (6c)$$

As we all know that MP is generally hard to solve, so we present some strategies to enhance the efficiency to get the MP's solution and improve the convergence of the enumeration-based C&CG algorithm.

- **Good solutions of MP before convergence.** MP evolves into a large-scale integer program, which is generally considered to be computationally time-consuming. When the relative gap of low bound and upper bound is large, it is unnecessary to get an optimal solution of MP, a good feasible solution could be able to identify significant scenarios. Thus, at the beginning of the algorithm, we set the relatively larger optimality tolerance to MP. when the relative gap becomes smaller, we set the optimality tolerance smaller.
- **Valid inequality.** If $\sum_{j \in \mathcal{J}} y_j \leq \Gamma$, for the worst-case scenarios, all assigned telemedical doctors are no-show, and all patients will be unassigned, which makes problem meaningless. In this regard, we introduce the valid inequality $\sum_{j \in \mathcal{J}} y_j \geq \Gamma + 1$ to avoid this.

4 Numerical Experiment

In this section, we try to explore some basic results to illustrate our modeling framework. We consider a pair of 20 patients and 4 telemedical doctors. For simplicity, we assume that $|\mathcal{J}| = K$. The budget of uncertainty in terms of telemedical doctors takes values by $\Gamma \in \{0, 1, 2, 3\}$. We also consider $N \in \{10, 20, 30, 40\}$ empirical samples for the random service duration. For each scenario ω, the service duration d_i^{ω} is randomly generated by a uniform distribution [5, 10] in minutes. We consider the length for each time block by $T = 50$ minutes. We set the big-M by $M := \max_{\omega} \sum_{i \in \mathcal{I}} d_i^{\omega}$. We assume that the assignment cost c_{ij} is set to 4, working cost of telemedical doctors $h_j = 20$, penalty cost for unserved patients $r_i = 100$.

All the experiments are implemented in MATLAB 2019a and YALMIP toolbox using CPLEX 12.71 as the solver with the default setting. All the computations are executed on a Desktop machine equipped with Intel(R) Xeon(R) 3.30 GHz processor and 128 GB RAM in a Windows 64-bit system. The current iteration will be finished instead of terminating the algorithm immediately if the time limit is achieved within an hour. We report the CPLEX CPU time in seconds. All the performances are averaged over three instances.

Table 2 reports the average performance and total expected cost with $\eta = 0.05$, including the CPLEX CPU time (time) and the number of iterations (# of iter) and the optimal total expected cost (obj) in terms of different N and Γ for the two-stage chance-constrained telemedicine assignment model. As we can clearly see from Table 2, on the one hand, for a certain Γ value, the CPU time is roughly increasing with respect to the increase of N from 10 to 40, but we can solve them optimally within an hour. On the other hand, given the number of scenarios N, it seems that model with $\Gamma \in \{1, 2\}$ (about half hour on average) is harder to solve when compared with $\Gamma \in \{0, 3\}$ (only a few seconds on average). Similar observations can be derived for the number of iterations. In terms of the optimal expected total cost, the budget of uncertainty Γ for behavior seems to have a big impact on the total cost, which can be explained by the fact that the model would be more conservative if Γ becomes large, while the number of empirical samples N might have a slight impact on the total cost, given that N does not change drastically if varying from 10 to 40.

5 Concluding Remarks

Telemedicine is expected to deliver timely care while minimizing exposure to protect medical practitioners and patients, especially in the current context of the global pandemic of COVID-19. In this paper, we study the telemedicine assignment problem for the telemedical doctors and patients when addressing the different sources of uncertainty (i.e. behavior of telemedical doctors and uncertain service duration). To our best knowledge, this is the first attempt to address such a problem. We propose a two-stage chance-constrained model with integer decisions in the recourse problem, which can be reformulated into a two-stage binary integer program. We then develop an enumeration-based C&CG method to solve it, and finally illustrate our framework by a simple numerical experiment. However, we believe that the proposed framework leaves space for many future research directions, for instance, incorporating the no-show behavior of the patients in the optimization model, which is being addressed in our another work. We expect that this paper could open an avenue for this streamline of research by exploring different sources of uncertainty in telemedicine operations management in a data-driven context.

Acknowledgments This study is fully supported by the project of "Scheduling and Optimization of Telemedicine Resource from a Data-Driven Perspective" from Natural Science Foundation of China [grant 71972012]. We thank this grant support very much.

Table 2 Computational performance and optimal expected total cost

Γ	0			1			2			3		
N	Obj	Time	# of iter	Obj	Time	# of iter	Obj	Time	# of iter	Obj	Time	# of iter
10	160.0	0.4	1.0	201.6	39.8	6.3	716.8	12.7	3.3	1324.8	2.7	2.0
20	160.0	0.6	1.0	147.5	1315.7	5.3	651.2	1602.4	3.7	1228.8	2.6	2.3
30	160.0	0.7	1.0	225.1	1346.0	5.7	689.1	2512.8	3.3	1258.7	6.3	3.0
40	160.0	0.9	1.0	158.7	2448.9	2.7	616.3	3047.0	3.7	1244.8	6.4	2.0
Average	160.0	0.6	1.0	183.2	1287.6	5.0	668.3	1793.7	3.5	1264.3	4.5	2.3

References

Ahmadi-Javid, A., Jalali, Z., & Klassen, K. J. (2017). Outpatient appointment systems in healthcare: A review of optimization studies. *European Journal of Operational Research, 258*(1), 3–34.

Ahmed, S., & Shapiro, A. (2008). Solving chance-constrained stochastic programs via sampling and integer programming. In: *INFORMS TutORials in operations research: State-of-the-art decision-making tools in the information-intensive age. Informs* (pp. 261–269).

Birge, J. R., & Louveaux, F. (2011). *Introduction to stochastic programming*. Springer Science & Business Media.

Charnes, A., & Cooper, W. W. (1959). Chance-constrained programming. *Management Science, 6*(1), 73–79.

Dai, T., & Tayur, S. (2020). Om forum—healthcare operations management: A snapshot of emerging research. *Manufacturing & Service Operations Management*, 22.

Deng, Y., Shen, S., & Denton, B. (2019). Chance-constrained surgery planning under conditions of limited and ambiguous data. *INFORMS Journal on Computing, 31*(3), 559–575.

Erdogan, S. A., Krupski, T. L., & Lobo, J. M. (2018). Optimization of telemedicine appointments in rural areas. *Service Science, 10*(3), 261–276.

Jebali, A., & Diabat, A. (2017). A chance-constrained operating room planning with elective and emergency cases under downstream capacity constraints. *Computers & Industrial Engineering, 114*(DEC.), 329–344.

Jetty, A., Moore, M. A., Coffman, M., Petterson, S., & Bazemore, A. (2017). Rural family physicians are twice as likely to use telehealth as urban family physicians. *Telemedicine & E Health, 24*(4), 268–276. https://doi.org/10.1089/tmj.2017.0161.

Jnr, B. A. (2020). Use of telemedicine and virtual care for remote treatment in response to covid-19 pandemic. *Journal of Medical Systems, 44*(7), 1–9.

Kamran, M. A., Karimi, B., & Dellaert, N. (2018). Uncertainty in advance scheduling problem in operating room planning. *Computers & Industrial Engineering, 126*(DEC.), 252–268.

Latifi, R., & Doarn, C. R. (2020). Perspective on covid-19: Finally, telemedicine at center stage. *Telemedicine and e-Health, 26*(9), 1106–1109.

Loeb, A. E., Rao, S. S., Ficke, J. R., Morris, C. D., Riley III, L. H., & Levin, A. S. (2020). Departmental experience and lessons learned with accelerated introduction of telemedicine during the covid-19 crisis. *The Journal of the American Academy of Orthopaedic Surgeons, 28*(11), e469–e476.

Myers, M. R. (2003). Telemedicine: an emerging health care technology. *Health Care Management, 22*(3), 219–223.

Noorizadegan, M., & Seifi, A. (2018). An efficient computational method for large scale surgery scheduling problems with chance constraints. *Computational Optimization & Applications, 69*(2), 535–561.

Qiao, Y., Ran, L., & Li, J. (2020). Optimization of teleconsultation using discrete-event simulation from a data-driven perspective. *Telemedicine & e-Health, 26*(1), 112–123.

Rajan, B., Tezcan, T., & Seidmann, A. (2018). Service systems with heterogeneous customers: Investigating the effect of telemedicine on chronic care. *Management Science, 65*(3), 955–1453.

Rajan, B., Tezcan, T., & Seidmann, A. (2019). Service systems with heterogeneous customers: investigating the effect of telemedicine on chronic care. *Management Science, 65*(3), 1236–1267.

Saghafian, S., Hopp, W. J., Iravani, S. M., Cheng, Y., & Diermeier, D. (2018). Workload management in telemedical physician triage and other knowledge-based service systems. *Management Science, 64*(11), 5180–5197.

Wang, S., Li, J., & Peng, C. (2017). Distributionally robust chance-constrained program surgery planning with downstream resource. In *2017 international conference on service systems and service management* (pp. 1–6). IEEE.

Wang, S., Li, J., & Mehrotra, S. (2019). A solution approach to distributionally robust chance-constrained assignment problems. *INFORMS Journal on Optimization*. http://www.optimization-online.org/DB_FILE/2019/05/7207.pdf.

Wang, X., Zhang, Z., Yang, L., & Zhao, J. (2020). Price and capacity decisions in a telemedicine service system under government subsidy policy. *International Journal of Production Research*, 1–14.

Wang, S., Li, J., & Mehrotra, S. (2021). Chance-constrained bin packing problem with an application to operating room planning. *INFORMS Journal on Computing*. https://doi.org/10.1287/ijoc.2020.1010.

Ward, M. M., Jaana, M., & Natafgi, N. (2015). Systematic review of telemedicine applications in emergency rooms. *International Journal of Medical Informatics, 84*(9), 601–616.

Zanaboni, P., Scalvini, S., Bernocchi, P., Borghi, G., Tridico, C., & Masella, C. (2009). Teleconsultation service to improve healthcare in rural areas: acceptance, organizational impact and appropriateness. *BMC Health Services Research, 9*(1), 238.

Zeng, B., & Zhao, L. (2013). Solving two-stage robust optimization problems using a column-and-constraint generation method. *Operations Research Letters, 41*(5), 457–461.

Zhang, Y., Jiang, R., & Shen, S. (2018). Ambiguous chance-constrained binary programs under mean-covariance information. *SIAM Journal on Optimization, 28*(4), 2922–2944.

Zhang, Z., Denton, B. T., & Xie, X. (2020). Branch and price for chance-constrained bin packing. *INFORMS Journal on Computing, 32*, 547–564.

Zhu, S., Fan, W., Yang, S., Pei, J., & Pardalos, P. M. (2019). Operating room planning and surgical case scheduling: a review of literature. *Journal of Combinatorial Optimization, 37*(3), 757–805.

Exploring Social Media Misinformation in the COVID-19 Pandemic Using a Convolutional Neural Network

Alexander J. Little, Zhijie Sasha Dong, Andrew H. Little, and Guo Qiu

1 Introduction

Misinformation on social media has shifted the way American people receive information and can impact public perception during a disease outbreak. In 2019, 79% of Americans used social media (Social media usage in U.S., n.d.) and 54% consumed news from social media sources (Shearer, 2020). Sometimes look upon as benign, health care related social media misinformation has already affected the geographic spread and temporal increase of the infectious disease outbreaks. Oyeyemi and Gabarron explored misinformation's impact on the Ebola outbreak and found the fake saltwater cures led to the deaths of multiple individuals (Oyeyemi et al., 2014). The ability to address misinformation on social media represents an opportunity for government or public health agencies to increase the volume and impact of accurate and beneficial information. Social Media as a data source for solution methodologies in the public health sector has been a very relevant topic in recent times and has been explored in response to various health care emergencies. Dai and How used Twitter data to explore how electronic cigarette ads were impacting different socio-economic groups (Dai et al., 2017). During the Zika virus outbreak Ghenai and Mejova used a machine learning methodology to successfully identify posts containing misinformation (Ghenai & Mejova, 2017).

A. J. Little · Z. S. Dong (✉)
Ingram School of Engineering, Texas State University, San Marcos, TX, USA
e-mail: sasha.dong@txstate.edu

A. H. Little
Network Surveillance Engineering, Consolidated Communications, Sacramento, CA, USA

G. Qiu
Ingram School of Engineering, Texas State University, San Marcos, TX, USA

College of Engineering and Applied Sciences, Nanjing University, Nanjing, China

H. Yang et al. (eds.), *AI and Analytics for Public Health*, Springer Proceedings in Business and Economics, https://doi.org/10.1007/978-3-030-75166-1_33

Pulido et al. explored the social impact of healthcare related misinformation post from various social media outlets (Pulido et al., 2020).

COVID-19 represents the worst disease outbreak in recent history and has been identified as the first social media pandemic (Murthy, 2020). During the pandemic, World Health Organization posts only received hundreds of thousands of interaction while some misinformation sources received 52 million interactions (AP Fact Check. AP NEWS. [Online]. Available: https://apnews.com/hub/ap-fact-check, n.d.). This illustrates that accidental or purposeful diffusion of misinformation is pervasive in the landscape of the response effort. This consumption of misinformation leads to a misinformed response effort and may have cause an increase of unnecessary COVID-19 cases. Studying the impact of misinformation on a pandemic is important because it can help mitigate cases while developing a more robust response effort for government and healthcare entities. Showing an association between misinformation and COVID-19 cases can help to focus more attention on the negative effects of social media misinformation and provide some insights on how to address misinformation.

Organizations' response efforts hinge on the ability to identify the misinformation. Currently public identification of social media misinformation is difficult and relies on crowd sourced volunteers for manual classification. There are countless groups working to manually address misinformation in a general sense (Chen, 2020; MFCA, 2020; Mitra & Gilbert, 2015; Liu & Wu, 2018; Yu et al., 2017) but there are fewer groups focusing specifically on social media. Due to the volume of data it is unreasonable to assume organizations will be able to police information manually so one avenue of exploration is to develop Artificial Intelligence methodologies trained for this purpose. Specifically, in the case we focus on the development of a Convolutional Neural Network (CNN). This method is a relatively complex methodology that is commonly used for image processing but has seen recent interest in Natural Language Processing. It is a supervised machine learning method that restructures the weights of the pathways and learns based on a labeled data set. Most high performing CNN models are very complex and hard to implement. The focuses when designing our algorithm is on simplicity for application in government, industry, and the non-profit sectors.

The contributions of this paper are summarized as follows:

- Developing a set of pandemic misinformation keywords which can be used for other disease outbreaks.
- Proposing a CNN model for identifying Twitter misinformation automatically with a focus on industry usability and application.
- Exploring how the misinformation diffuses from a geo-temporal perspective and providing managerial insights.

The remainder of the paper is organized as follows. Section 2 describes data collection and processing, and Sect. 3 introduces the Convolutional Neural Network (CNN) model. In Sect. 4, we conduct numerical experiments based on a case study to explore the diffusion of COVID-19 misinformation Tweets over time in the United States. Section 5 summarizes the conclusions and suggests some future scopes based on the proposed model and approach.

2 Data

2.1 Data Collection

The data was rigorously processed using Python and a high-performance Linux cluster. A team at the Information Sciences Institute, University of Southern California has been continuously compiling a COVID-19 specific Twitter data set starting January 28, 2020 (Chen, 2020). The data set was published and updated in the form of Twitter IDs and requires a Tweet Hydration step to collect all the relevant contextual information. Besides texts, there are several other important harvested features associated, and Table 1 summarizes the most relevant ones with a short description.

2.2 Preprocessing

The outline of the cleaning procedure is described in Fig. 1. After the data was hydrated it was then converted from JSONL format to CSV. In this step relevant features were mapped to a simpler more user-friendly form using a python algorithm. This step consists of translating indexed JSON values to CSV values that could be viewed and managed on a graphical user interface. During the converting process, an indexed language value was chosen to only select English Tweets, and

Table 1 Relevant harvested Tweet features

Location	A user specified home location
Date/Time	The specific Date/Time the Tweets was posted
Hash Tags	Contains all hash tags associated with the Tweet
URLs	URLs associated with the Tweet
Followers	Number of the users' followers
Friends	Number of the users' friends
Favorites	Number of Tweets favorited
Statuses	Number of Statuses posted
Retweeted	Amount of times the post was retweeted
User Creation Date	The date the user made the account
Verified	Whether the user is verified or not
Text	All the text included in the Tweet

Fig. 1 Flowchart of Twitter Data Cleaning Process

Table 2 Developed key phrases used to refine the dataset

Developed key phrases
Common cold, flu, bovine, low mortality, 99.7%, 0.1%, 36%, 0.03%, old people, least deadly, the damn flu, holding breath, perfectly healthy, fake crisis, empty hospitals, fake, made up, overblown, lower death rate, there is no, hyped, hype, isn't deadly, no virus, con, unrelated causes, herd immunity, 360,000, 80,000

later these Tweets were filtered for only United State Location Tweets based on the location place object value.

The last step is to further refine the data set and simplify the classification process. We developed a set of misinformation keywords for that, which is shown in Table 2. The development was based on a COVID-19 fact checking database put forth by The Corona Virus Facts Alliance (MFCA, 2020), and we particularly focus on people denying the importance of COVID-19, minimizing the potential threat, and incorrectly comparing to the seasonal flu. The keywords developed are shown in Table 2. Some keywords are straight forward like common cold, overblown, and hyped. These are terms that diminish and undermine the public health response by equating COVID-19 to the common cold or flu. Some of the key words like fake crisis, empty hospitals, and made up reflect the fact that some users were denying the Pandemic in its entirety. The percent terms like 99.7% and 0.1% represented commonly misrepresented and incorrect statistics used misrepresenting the death rates. The numbers 360,000 and 80,000 represent misquoted flue death rates used to make inaccurate comparisons to early COVID-19 death rates. After further cleaning the dataset was refined to 148,901 Tweets spanning from Jan 28, 2020 to May 31, 2020.

2.3 *Labeling*

To train supervised machine learning methods, a labeled data set needs to be utilized. There are several labeled misinformation data sets available, but it is important to make a pandemic specific misinformation data set that has the possibility of being applied to future pandemics. The classification structure used was originally developed by Mitra and Gilbert in their development of a misinformation database (Mitra & Gilbert, 2015). The labeling criteria is shown in Table 3. Values labeled as 1 and 0.5 represent Certainly Accurate and Probably Accurate, respectively. Certainty Accurate is 100% accurate containing no misinformation, and Probably

Table 3 Labeling criteria

Labeled values	Labeling criteria
1	Certainly Accurate: an informative Tweet with useful information for the public
0.5	Mostly Accurate: an informative Tweet that may contain partially correct or difficult to check information but positively impacts the public
0	Uncertain: a Tweet that does not inform or impact the public
−.5	Mostly Inaccurate: a Tweet that contains incorrect information or may contain some correct information but negatively impacts the public
−1	Certainly Inaccurate: a Tweet that contains no correct information and negatively impacts the public

Accurate contains mostly correct information but impacts and informs the public in a positive manner. This is to say that this Tweet encourages people to take proactive and healthy measures in response to COVID-19. Tweets with a 0 value do not have pertinent information about COVID-19. The −.5 value represents mostly inaccurate information. The class may have some factual information, but it is being misrepresented in a way that may be negatively impacting the public. This class is the most difficult to identify. The final value and the easiest to identify is −1 or certainly inaccurate. This is a Tweet that contains 100% incorrect information. We manually combed through the data and individually labeled each Tweet in the training set. The total number of labeled Tweets is 5436.

3 Convolutional Neural Network (CNN) Model

Considering the potentials for natural language processing, we propose a CNN model to verify the success in misinformation classification (Liu & Wu, 2018; Yu et al., 2017; Nguyen et al., 2017), and we plan to utilize the developed key phrase data set to achieve the above standard results. This method reads subsections of a vectorized sentence matrix built from the Tweet. The Word2Vec library was used to convert the text after stop words and punctuation were removed. It vectorizes the text to a certain size and a size of 50 was chosen for the model based on manual testing. In the future, a grid search method will be employed. A sentence matrix of size 3 and the original text values are shown in Fig. 2. The Tweet text values are actual posted words with the stop words being removed. The resultant vectorized sentence matrix uses Word2Vec mapping to convert the text values into a matrix of vectors corresponding to the text. What is worth mentioning is that some of the inherent language structure of text values is contained in the matrix with some transformation of word vectors corresponding to changes in language structures of words.

After building the matrix, it is then processed by the CNN model. The model reads and averagely pools the values with a pool size of 3 for the input layer. The next hidden layer is another average pool layer with a pool size of 2 and a dropout

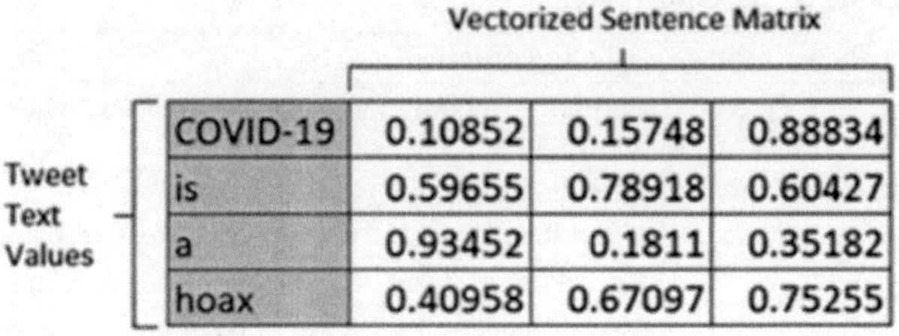

Fig. 2 Example of vectorized sentence matrix of size 3

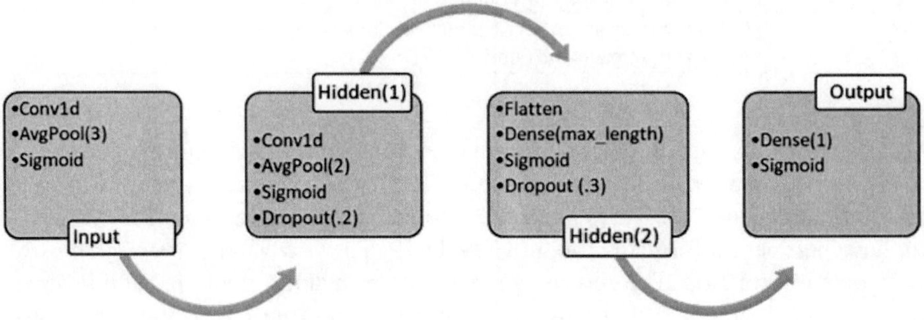

Fig. 3 Process chart of CNN model

Table 4 Metrics for CNN

Metric	Value
Accuracy	77.68%
True positive	64
True negative	785
False positive	51
False negative	193
Recall	24.90%
Precision	55.65%

of 0.2. A dense flatten layer is then used with a dropout of 0.3 and finally a single neuron is used for the output layer. All layers use a sigmoid activation function and to simply the classification process a converted binary target was used. A graphic representation of the model is shown in Fig. 3. The model was then trained with stratified test split of 0.1 on the labeled dataset and the performance indicators are shown in Table 4.

The algorithm is currently in development and the scores reflect that. The algorithm underperforms in precision and recall but these scores are expected to increase drastically with a larger data set, algorithm development, and hyper parameter tuning.

4 Case Study

A case study was conducted to explore the diffusion of COVID-19 misinformation Tweets over time in the United States. This allows us to graphically illustrate the distribution of the Tweets alluding to the natural structure formed in online social networks. This structure is very interesting, and more importantly, the analysis can offer insights on how the misinformation spreads allowing for targeted response efforts and other response methodologies. In this case study, the labeled training data set from Sect. 2 was used, and all the instance of retweets were captured with the Tweets between February and May of 2020. The geoinformation was utilized from these Tweets to map the misinformation and this distribution is shown in Fig. 4.

This figure has metropolitan areas colored in dark grey and Tweet instances colored in maroon. What is notable is that fact the very few Tweets are outside major metropolitan areas. The implication being that there may be no specific geographic focal point for misinformation Tweets besides major metropolitan areas. We expected to see a shift in the geographic region over time representing a generalized geographic spreader for COVID-19 misinformation but in a general sense the misinformation seems to be pervasive through all geographic regions. When examining the mapping it shows an overall decline in misinformation the further the pandemic progresses. To further explore this idea four states were chosen and graphed in Fig. 5. California, Texas, New York, and Florida were chosen because of the high volume of Tweets from those locations. The relative percentages of the total data set are 15.26%, 8.75%, 8.67%, and 6.24%, respectively. This high Tweet percentage may be due to high population density or high social media

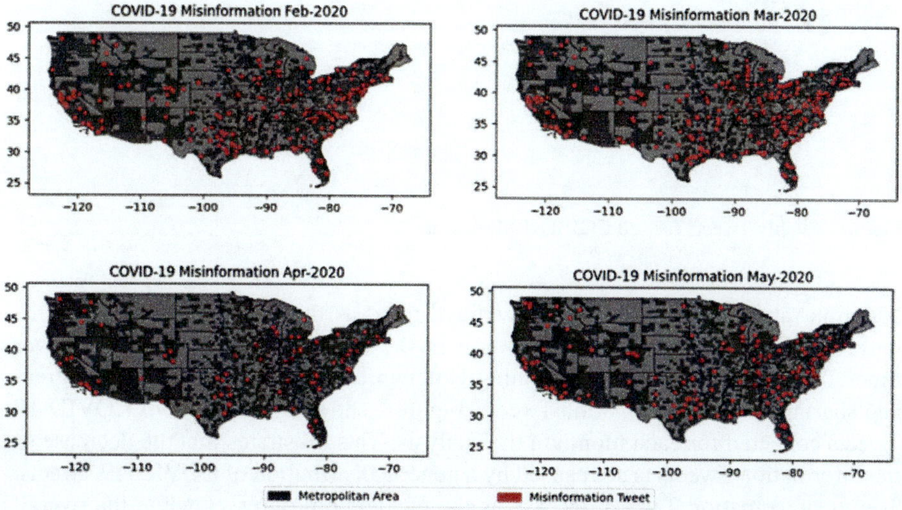

Fig. 4 Monthly distribution of COVID-19 misinformation Tweets in the United States from Feb to May of 2020

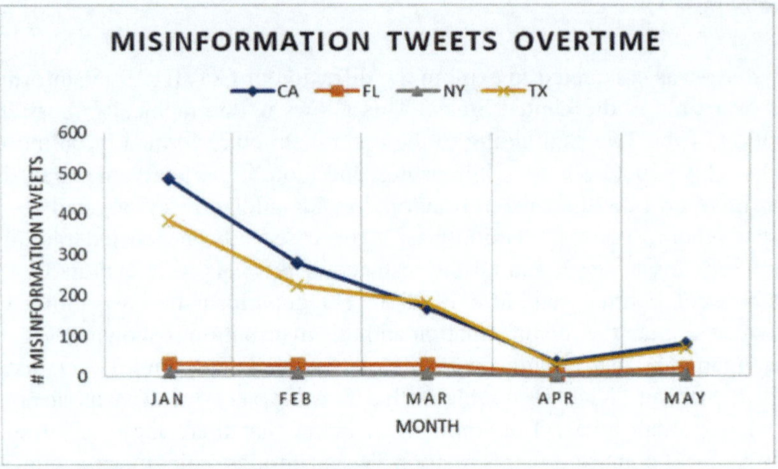

Fig. 5 Monthly Misinformation Tweets for California, Texas, New York, and Florida

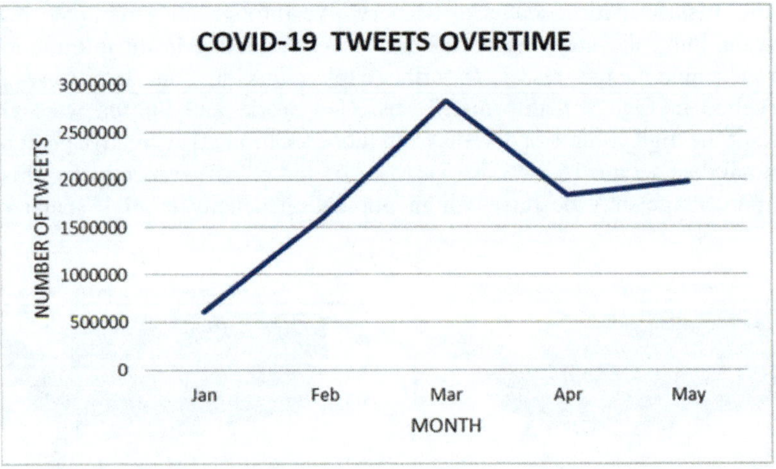

Fig. 6 Monthly Tweets for full COVID-19 Data Set

adoption rate. Figure 5 shows a downward trend as the pandemic progress. This may be attributed to an actual decrease in the quantity of misinformation over time. The public is likely be becoming better informed as the pandemic progress and sharing less misinformation Tweets. Figure 6 shows the quantity of COVID-19 Tweets collected for each month in the analysis. This illustrates that the decrease of misinformation Tweets is not caused by a general distribution of COVID-19 Tweets. The misinformation Tweets decease as the pandemic progresses while the overall number of Tweets increases.

This distribution also shows that there may not be one specific geographic area to target to reduce overall misinformation. Not having a central geographic area speaks to the necessity to map and understand diffusion in the inherent online structure. Where there is no physical geographic location that acts as a super spreader there is almost certainly specific spaces or grouping in the online social construct that misinformation is developed and disseminated. These groups can be purposeful dissemination or natural development of misinformation. This downward trend of misinformation also shows addressing misinformation early in the cycle could most effectively mitigate the invalid posts.

5 Conclusion and Future Work

A larger sample size is needed to verify but as a pandemic progresses misinformation may decline since the public becomes more informed. The beginning of the pandemic was chaotic and lacking a central message and direction. For the government, or healthcare responders, one way to mitigate the initial misinformation spread is to develop a clear and concise message about the event focusing on where and how people could get the information they need. This illustrates that channels need to be developed to address information concerns before events like this happen so individual know where they will be able to receive accurate information. This work represents a classification methodology and an initial exploration of misinformation distribution using the pandemic as a case study.

For future work, we plan to include the developed CNN model in an ensemble method with a Random Forest algorithm and design more contextual features for that method. Some other ways this research could be expanded on include: (1) exploring a less general data set and mapping the path of individual misinformation Tweets to explore the diffusion structures, (2) testing this conclusion with other disaster events set to see if the same downward trend is present in most cases, and (3) conducting analysis to find misinformation hubs in the social network that act as focal points for misinformation spreading.

Acknowledgements Part of the data collection and pre-processing work was under the support of National Science Foundation Research (NSF) Experiences for Undergraduates (REU) program associated with grant CISE-1948159.

References

AP Fact Check. AP NEWS. [Online]. Available: https://apnews.com/hub/ap-fact-check

Chen E. *Echen102/COVID-19-TweetIDs*. https://github.com/echen102/COVID-19-TweetIDs. Accessed 17 Aug 2020.

Dai, H., Deem, M. J., & Hao, J. (2017). Geographic variations in electronic cigarette advertisements on Twitter in the United States. *International Journal of*

Public Health, 62(4), 479–487. [Online]. Available: https://EconPapers.repec.org/RePEc:spr:ijphth:v:62:y:2017:i:4:d:10.1007_s00038-016-0906-9

Ghenai, A., & Mejova, Y. (2017). Catching zika fever: Application of crowdsourcing and machine learning for tracking health misinformation on Twitter. *2017 IEEE International Conference on Healthcare Informatics (ICHI)*, 518–518.

Liu, Y., & Wu, Y. (2018). Early detection of fake news on social media through propagation path classification with recurrent and convolutional networks. In *AAAI*.

MFCA. *IFCN Covid-19 Misinformation*. https://www.poynter.org/ifcn-covid-19-misinformation/. Accessed 14 Aug 2020.

Mitra, T., & Gilbert, E. (2015). *CREDBANK: A large-scale social media corpus with associated credibility annotations* (pp. 258–267). AAAI Press. [Online]. Available: http://libproxy.txstate.edu/login?url=http://search.ebscohost.com/login.aspx?direct=true&db=edselc&AN=edselc.2-52.0-84960983952&site=eds-live&scope=site

Murthy, R. (2020). *First social media pandemic*. (in en-US), *The Statesman*, 2020/04/01/T14:14:31+05:30. [Online]. Available: https://www.thestatesman.com/opinion/first-social-media-pandemic-1502872544.html

Nguyen, T., Li, C., & Niederée, C. (2017). On early-stage debunking rumors on Twitter: Leveraging the wisdom of weak learners. *ArXiv, abs/1709.04402*.

Oyeyemi, S. O., Gabarron, E., & Wynn, R. (2014). Ebola, Twitter, and misinformation: A dangerous combination? *BMJ: British Medical Journal, 349*, g6178. https://doi.org/10.1136/bmj.g6178

Pulido, C. M., Ruiz-Eugenio, L., Redondo-Sama, G., & Villarejo-Carballido, B. (2020). A new application of social impact in social media for overcoming fake news in health. *International Journal of Environmental Research and Public Health, 17*(7), 2430. https://doi.org/10.3390/ijerph17072430

Shearer, E. (2020). *Americans are wary of the role social media sites play in delivering the news*. https://www.journalism.org/2019/10/02/americans-are-wary-of-the-role-social-media-sites-play-in-delivering-the-news/. Accessed 1 Sept 2020.

Social media usage in U.S. (n.d.), (in en), *Statista*. [Online]. Available: https://www.statista.com/statistics/273476/percentage-of-us-population-with-a-social-network-profile/

Yu, F., Liu, Q., Wu, S., Wang, L., & Tan, T. (2017). A convolutional approach for misinformation identification. In *Presented at the Proceedings of the 26th International Joint Conference on Artificial Intelligence, Melbourne, Australia*.

Personalized Predictions for Unplanned Urinary Tract Infection Hospitalizations with Hierarchical Clustering

Lingchao Mao, Kimia Vahdat, Sara Shashaani, and Julie L. Swann

1 Introduction

Clinical predictive analysis is of increasing interest to policy-makers, healthcare providers, and researchers with the potential to reduce healthcare costs and improve care quality (Choi et al., 2016; Hasan et al., 2010; Donzé et al., 2013). Even a small reduction in potentially avoidable hospitalizations (PAH) would result in substantial savings in economic and human costs (Walsh et al., 2012). This study aims to build personalized prediction models for unplanned hospital admissions for Urinary Tract Infection (UTI). UTI is the most frequent and preventable healthcare-associated infection (HAI) in the US, one of the five most common ambulatory care-sensitive conditions (ACSC), and an important cause of morbidity and excess healthcare costs (Walsh et al., 2012; Saint, Meddings, et al., 2009; Saint et al., 2009; Billings et al., 1993; Unroe et al., 2018). UTI not only results in patient discomfort but also increases the risk of PAH and discharge delays (Saint, Meddings, et al., 2009; Unroe et al., 2018).

One of the main challenges of predictive modeling using healthcare data is the large heterogeneity of patient profiles coupled with low disease occurrence rates. This heterogeneity and sparsity pose difficulties for conventional classification algorithms to achieve good classification performance without becoming overly complex. Bertsimas et al. partitioned the study population into five cost buckets to alleviate the heterogeneity of cost patterns; and used classification trees and clustering methods to divide data into more uniform groups, which improved predictions for healthcare costs (Bertsimas et al., 2008). Elbattah et al. used unsupervised learning to find coherent clusters of patients and showed that the

L. Mao (✉) · K. Vahdat · S. Shashaani · J. L. Swann
Edward P. Fitts Department of Industrial and Systems Engineering, North Carolina State University, Raleigh, NC, USA
e-mail: lmao3@ncsu.edu

H. Yang et al. (eds.), *AI and Analytics for Public Health*, Springer Proceedings in Business and Economics, https://doi.org/10.1007/978-3-030-75166-1_34

clustering-aided models achieved higher accuracy in predicting the length of stay of hip fractures (Elbattah & Molloy, 2017). Beyan et al. proposed a hierarchical decomposition method that partitions data into smaller subsets, each with a different feature space, and showed improvement in classification performance with over 20 imbalanced data sets (Beyan & Fisher, 2015). Therefore, we claim that by dividing the population into similar risk groups based on patient demographics, medical history, care quality, and environmental factors we can build more effective prediction models for each group.

Although a number of clustering algorithms have been applied to healthcare data, such as partition clustering, agglomerative clustering, and density clustering, they are purely data-driven and highly dependent on the major patterns of the data (Choi et al., 2016; Hasan et al., 2010; Donzé et al., 2013; Beyan & Fisher, 2015; Nithya et al., 2015; Ogbuabor & Ugwoke, 2018). In addition to traditional data-driven algorithms, we aim to leverage existing knowledge from literature and domain to define representative patient groups. Our proposed framework adopts a hierarchical structure because of its advantage to model dependence relationships between levels of the hierarchy.

The contributions of this paper are two-fold:

- a hierarchical clustering framework that leverages both existing knowledge and data-driven patterns to group patients with similar risk levels with respect to unplanned UTI admissions. This approach can also be applied to non-healthcare problems where data is highly heterogeneous and imbalanced, and domain knowledge can be used to guide focused modeling; and
- monthly probabilistic predictions for Medicare beneficiaries' risk for unplanned UTI admissions and interpretable insights about the most relevant variables, which may facilitate the design of interventions.

To our best knowledge, our study is the first to predict UTI hospitalization as small as monthly intervals. The closest in the literature is a study by Carter, which focuses on the nursing home population and provides quarterly predictions (pseudo R squared 0.0931) (Carter, 2003); and another by Saver et al., which predicts several acute and chronic hospitalizations for a year-long interval (AUC 0.87) (Saver et al., 2014). While these studies have longer prediction intervals and a smaller study population, they used logistic regression and a similar set of variables.

The remaining of this paper is organized into four sections (Fig. 1). In Sect. 2 we discuss the data used in this study. In Sect. 3 we provide details about our hierarchical clustering method. Section 4 presents a summary of results and comparison with a baseline model. Lastly, we conclude the paper with a summary of findings.

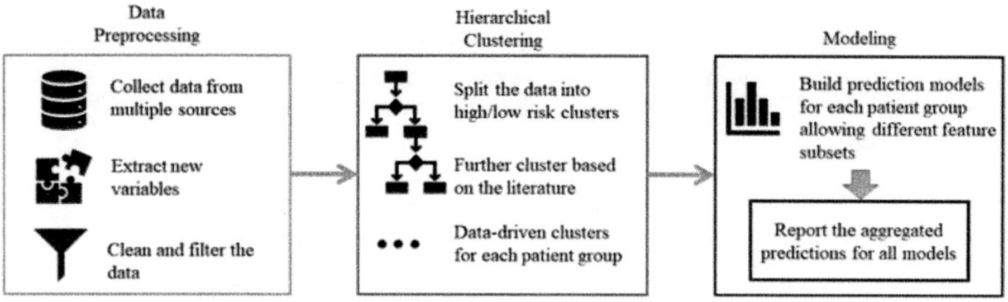

Fig. 1 Road map of the modeling procedure

2 Data Collection and Variable Quantification

This study uses 2008–2012 Medicare Limited Data Sets (LDS) which contain a 5% random sample of seven types of fee-for-service claims annually. Data from 2008–2011 are used to capture patient clinical history; data from April to November of 2011 are used for model training; and from 2012 are used for model testing and performance evaluation. In addition to medical claims, we collect multiple public data (CMS' SSA, 2011; Immunization, 2011; County Health Rankings, 2011; Population Census Elderly Living, 2011; Population Census, 2011; Hospital Compare Dataset, 2011; Nursing Home Compare Dataset, 2011; Public Use Files HRR Table for Beneficiaries, 2011; Diabetes Atlas, 2011; Weekly U.S. Influenza Surveillance Report, 2011) to create relevant predictors. From the prepossessing, we obtain a patient-month data set (where rows correspond to patient's data for each month of the year) with 784 features including patient demographics, clinical history, healthcare utilization and spending, provider quality metrics, and community safety metrics (Table 1) for beneficiaries who had at least one inpatient, outpatient, or Skilled Nursing Facility (SNF) claim during the year. These features were identified from many studies, more detail is provided below.

To define our target event, unplanned hospitalization for UTI, we analyze whether a patient's inpatient claim satisfies the Prevention Quality Indicators (PQI) criteria put forth by Agency for Healthcare Research and Quality (Indicators, 2001). We use the Clinical Classification System (CCS) developed by AHRQ to compute 285 CCS variables based on ICD-9 diagnosis codes. The data for acute CCS conditions are transformed into two binary indicators of length since the first diagnosis (less than 6 months ago or not). We aggregate specific CCS variables to alleviate data sparsity while making the presentation clinically meaningful (Table 1).

Table 1 Summary of the predictors considered in the model

Demographics	Age, gender, race, low income, managed care, supplemental insurance, socioeconomic status, smoking, obesity
Clinical History	Acute and chronic CCS conditions and their aggregates (neuro, heart, diabetes, cognitive, alcohol substance abuse, cancer), ESRD, immunocompromised, post transplant, number of CCS conditions
Healthcare Utilization	Number of inpatient, outpatient, SNF, carrier, Durable Medical Equip-ment (DME), homehealth, hospice claims in the last 1, 3, 6, and 12 months (Ma et al., 2015); past and current nursing homestay; elixhauser comorbidity index (Will et al., 2014); number of specialty visits in the last month (allergy, neurology, endocrinology, car-diology); number of emergency room, physician, hospitalization, ICU, CCU, and Oncology stays in the last 1 and 3 months; length of stay in hospital and SNF in the last 1 and 3 months (Moghadamyeghaneh et al., 2019; Chan et al., 2018)
Healthcare Spending	Medicare and non-medicare paid costs of inpatient, outpatient, SNF, carrier, DME, homehealth, hospice claims in the last 1, 3, 6, and 12months (Moghadamyeghaneh et al., 2019; Chan et al., 2018)
Most Recent Provider's Quality Metrics	Hospital overall rating, number of beds, count of outpatient procedures, emergency room volume (Carter, 2003); several disease-specific death rates, complications rates, postoperative complication rates, infection rates, and readmission rates (Hospital Compare Dataset, 2011; Nursing Home Compare Dataset, 2011)
Community Quality Metrics	Rural indicator, household income (Saver et al., 2014); state-level flu activity, vaccine effectiveness, air quality (Immunization, 2011; Weekly U.S. Influenza Surveillance Report, 2011); region safety scores; population statistics about race, education, income, access to care and food, etc. (County Health Rankings, 2011; Population Census Elderly Living, 2011; Public Use Files HRR Table for Beneficiaries, 2011)

3 Method

In this section, we discuss the hierarchical clustering modeling approach and the evaluation criteria employed.

3.1 Hierarchical Clustering Approach

As discussed in Sect. 1, the main challenge of developing predictive models using healthcare data is the heterogeneity of patterns coupled with scarcity of events. We propose a novel approach to address this challenge, referenced as hierarchical clustering. This approach partitions patient-month data points into more uniform groups, then builds targeted prediction models with coefficients and feature sets unique to each group. The key advantage of this approach is the ability to use known relationships identified from literature and domain knowledge to categorize archetypical patient groups meaningful for providers and based on their resemblance in risk of UTI hospitalization.

The model building process involves four steps:

Step 1. Overall partitioning into high versus low percentage of event occurrence Since data is highly imbalanced, the desired effect is to first separate the population into two groups such that the majority of event occurrences are concentrated in a small group. To identify the best partition rule, we use the R implementation of Classification and Regression Trees (CART) in the *caret* library with adjusted cost functions to emphasize more on correctly identifying events than non-events (Kuhn, 2008). We include high level variables that indicate healthcare utilization and UTI history so that the results are applicable to a larger population. The variables we include in CART are the number of inpatient, outpatient, carrier, SNF, hospice, homehealth, and DME claims in the previous 3, 6, and 12 months; the Medicare and non-Medicare paid costs of these seven types of claims in the previous 3, 6, and 12 months; and previous UTI.

Step 2. Categorizing archetypical patient groups meaningful for providers For the subset with the highest prevalence of events from Step 1, we categorize archetypical patient groups intended to be meaningful for providers. The goal is to define types of patient populations that have fundamentally different conditions that may drive differences in regression models. To promote understandability, the choice and order of these branches are based on domain knowledge and results from the literature.

The first differentiating group we define is those who are on Medicare because they have End-Stage Renal Disease (ESRD). These individuals may be younger than 65, and they have been identified to have a significant disease that may relate to UTI risk (Walsh et al., 2012; Saver et al., 2014; Naqvi & Collins, 2006). In the next level of the hierarchy, we consider people who have been in a nursing home (Wald et al., 2008; Grabowski et al., 2007) identified through the algorithm developed by Koroukian et al. (Koroukian et al., 2008). Nursing home residents are likely those who need help with activities of daily living (ADLs) and/or have difficulties with walking, hearing or seeing (Nursing Homes, 2020). These conditions cause them to be at greater risk for admissions or adverse events (Unroe et al., 2018). This setting is different from Skilled Nursing Facilities (SNF) because the residents may be self-financing, and nursing homes tend to be for longer occupation. Literature has also shown that UTIs can be associated with urinary-related cancer (Walsh et al., 2012; Saver et al., 2014; Moghadamyeghaneh et al., 2019) and with mental conditions such as dementia, delirium, and Parkinson's disease (Saver et al., 2014; Grabowski et al., 2007; Willis et al., 2012; Dharmarajan et al., 2013; Sampson et al., 2009); so these patient groups are identified for lower levels of the hierarchy.

Step 3. Data-driven clustering to improve the predictive power of the models We apply CART on each patient group identified in Step 2 to obtain candidate data-driven clusters. The variables we include in CART consists of the set of features discussed in Step 1, which capture general healthcare utilization. To decide whether to employ a cluster and when to stop branching, we build brute-force regression models for the two children nodes and the parent node before branching and choose the option that achieves higher AUC on the hold-out data set.

Step 4. Regularized regression model to provide monthly risk predictions We build prediction models for each resulting cluster. We choose the Lasso-Logistic Regression model (LLR) because it has been shown effective in largely imbalanced data (Wang et al., 2015). The penalty parameter is tuned separately for each cluster using three-fold cross-validation such that the chosen parameter minimizes the deviance of the predicted values from the logistic regression model (P. D. Allison and Others, 2014). The 784 features from Table 1 are provided to all levels of the hierarchy, and L1 norm regularization selects the most relevant features in each cluster. Youden's index is used to select the probability threshold for each cluster (Mokyr Horner & Cullen, 2015).

3.2 *Baseline Approach*

As a baseline model, we run LLR using the same settings described in Step 4 on the training data before clustering. In other words, the baseline approach builds one prediction model for the entire population. Parameter tuning and threshold selection are also performed only once. This is the most relevant benchmark, considering modeling procedure and the data, that we found in the literature as described in Sect. 1.

3.3 *Evaluation Metrics*

We use a combination of discrimination and calibration measures to assess model performance. The Area Under the Curve (AUC) evaluates the likelihood that the predicted probability of an event instance is higher than that of a non-event instance. However, AUC fails to measure the goodness of fit when data is imbalanced (Homer et al., 2013). Therefore, we include the TPR and FPR, which are especially useful for imbalanced class problems (Raschka, 2014). The former indicates the percent of events *correctly* predicted by the model out of the total event instances, and the latter measures the percent of non-events that the model *incorrectly* predicts as positive. We use TPR and FPR to understand the unplanned admissions that are captured by the model (TPR), as well as the potential cost associated if interventions are used unnecessarily (FPR). In addition, we report the Sensitivity at Low Alert Rates (SLA) at 1%, which measures the TPR for instances that are given the highest risks. Lastly, we include accuracy, which is another common metric to measure the percentage of correctly predicted event and non-event over all data points.

4 Results

Our predictions focus on patients who had at least one inpatient or SNF claim during the year. To ensure complete health profiles, we exclude beneficiaries who are Medicare part A and B enrollees for only part of the year, enrolled in managed care, with supplemental insurance, or with disability (Mokyr Horner & Cullen, 2015). Table 2 shows summary statistics of the preprocessed data.

We obtain a hierarchical structure with 12 knowledge-based and data-driven clusters from training data, as visualized in Fig. 2. The most important partition based on event prevalence identified by CART (Sect. 3.1 step 1) is whether the patient had a historical diagnosis of UTI or any medical claims in the past year. For the lower event occurrence group, data-driven clustering suggests assessing whether the beneficiary had been admitted to SNF in the last month or to ICU in the last 3 months. For people who had no urinary-related cancer but had cognitive conditions, this step suggests grouping patients based on inpatient costs in the last 12 months. Similarly, for the beneficiaries who did not have an inpatient or SNF visit in the last 12 months, this step suggests using more than 10 carrier claims (which include physician visits) in the last year as the best split criteria.

For each cluster, the model intercept proxies the base risk level, and the set of selected features shows which factors are most likely to be associated with UTI hospitalization for that particular patient group. The features' estimated coefficients indicate their quantified impact towards the base risk of their cluster. A positive coefficient indicates an increase in risk, and a negative coefficient is associated with a decrease in risk. Note that the base risk should be interpreted in combination with the variability of the coefficients; a cluster with few features, that have a small effect on the predicted probability, may have a lower risk than a cluster with highly variable coefficients. Therefore, we calculate the summation of the intercept and coefficient variance as the underlying risk of each cluster, which is shown with color-coding in Fig. 2; the darker the cluster, the higher the risk.

The most selected features across clusters include a previous month visit to oncology, ICU, or CCU, and whether the patient had a recent UTI. Other variables

Table 2 Descriptive statistics of the 2011 and 2012 study population

	2011		2012	
	Count	Percentage	Count	Percentage
Total beneficiaries	1,257,485	100%	1,274,142	100%
Age above 65	1,133,412	90%	1,149,054	90%
Disability	229,377	18%	245,031	19%
Male	518,698	41%	528,557	41%
Previous unplanned admission due to UTI	10,203	1%	10,099	1%
Had at least an inpatient claim	348,866	28%	327,198	26%
Had at least an SNF claim	96,163	8%	90,572	7%
After exclusions	237,675		230,042	

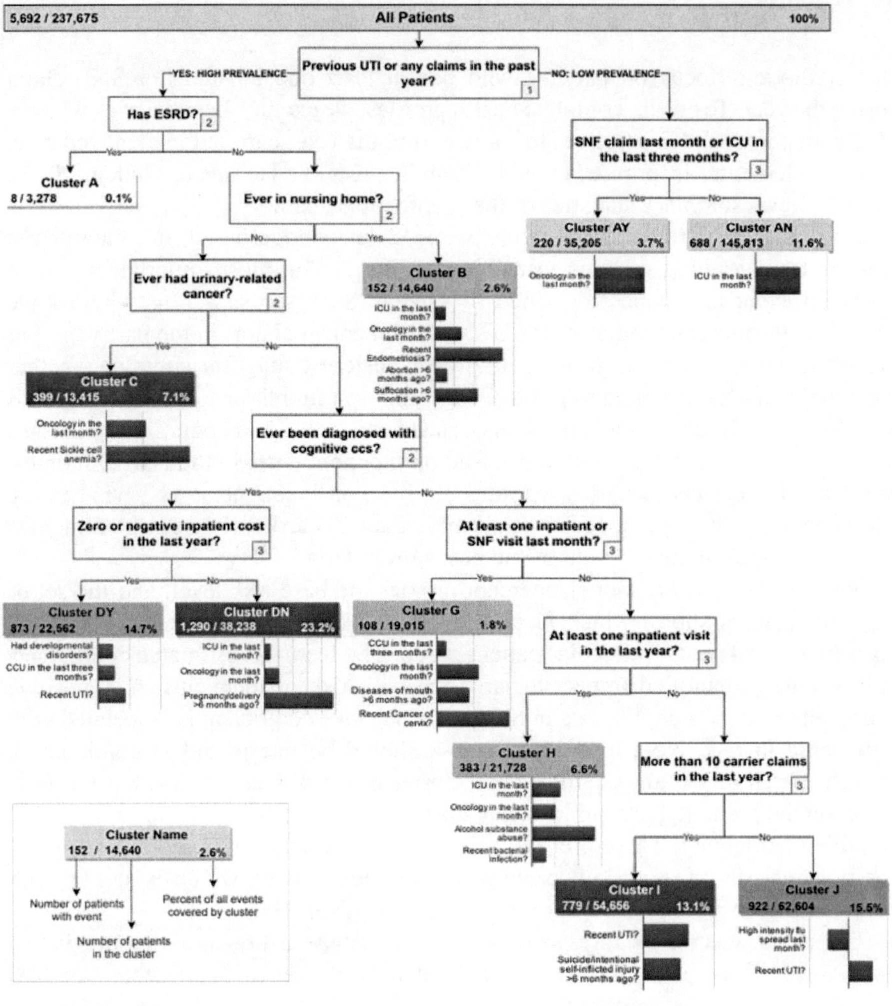

Fig. 2 Hierarchical structure is visualized in the form of a tree. For each cluster, the ratio of the number of patients with the event to the number of patients in the cluster is shown on the lower-left corner of the box; the percentage of all event rows contained in the cluster is shown on the lower right corner. The small square in the lower right corner of each decision node indicates which step (defined in Sect. 3.1) it belongs to. The plots display the relative coefficient values for the selected features (note that only features with a coefficient greater than 0.5 are displayed for visualization). Clusters are color-coded based on the average risk obtained from the model intercept and coefficient variance; darker colors correspond to higher risk. Cluster A received zero predictions due to its low event occurrence rate and low event coverage percentage

specific to each cluster include endometriosis, developmental disorders, pregnancy or delivery, abortion, cancer of the cervix, bacterial infection, alcohol substance abuse, sickle cell anemia, self-inflicted injury, suffocation, and flu intensity last

Table 3 Comparing results between baseline and hierarchical clustering-based LLR. The latter achieves higher AUC, SLA, accuracy, and a lower FPR than the former. Slightly higher TPR in baseline LLR is due to classifying a lot of data points as positive, indicated by high FPR

	AUC	SLA 1%	TPR	FPR	Accuracy
Baseline LLR	0.63	0.03	0.84	0.67	0.33
Hierarchical clustering-based LLR	**0.72**	**0.04**	0.77	**0.43**	**0.57**

month. Additional considerations should be taken into account while interpreting the selected features. Some of these variables may not be directly related to UTI but are proxies for their underlying health and/or environmental conditions that are associated with a risk to UTI. For example, the flu intensity variable can be interpreted as a proxy for weather seasonality, which is shown to be correlated with UTI hospitalizations in previous studies (Anderson, 1983; Hsu et al., 2019). Another example is suffocation, which may be an indicator of specific patient characteristics or behaviors such as intentional injuries (Sasso et al., 2018). The recency in the wording of these variables relies on health history recorded by Medicare claims. If a patient had their diagnosis more than 6 months ago but is added to the system not long ago, our data will still indicate that their first diagnosis was recent. Therefore, conditions that have very low probabilities of development after the age of 65, such as endometriosis or pregnancy, may be aliases for a recent addition of the patient's data or diagnosis code to the system. These variables can still be proxies for recent visits with providers who recorded previous health history.

In the baseline model, the variables with the highest coefficients selected are previous month visits to oncology and UTI history, which coincide with the top features selected from the hierarchical clustering approach. One advantage of the latter is providing more personalized feature importance summaries for each group of patients. The prediction performance of these two modeling approaches is summarized in Table 3. The clustering-based approach achieves a higher AUC (0.72) than the baseline (0.63), which means the model is more likely to predict a higher risk for instances that UTI admission actually occurred than a non-event instance. The higher accuracy score also indicates that the clustering-based approach predicts both events and non-events more accurately than the baseline approach. Although the baseline model achieves a slightly higher TPR (0.84), the model overpredicts many patients resulting in a high percentage of false positives (0.67). The clustering-based approach significantly reduces false positives (FPR 0.43) while maintaining a reasonable TPR (0.77). Therefore, we conclude that the hierarchical clustering approach achieves more accurate and precise predictions than the approach without clustering.

5 Conclusion

One of the main challenges of predictive healthcare analytics is the large heterogeneity of patient patterns coupled with high data imbalance. The hierarchical clustering approach proposed in this paper tackles this challenge by leveraging existing knowledge about UTI as well as data-driven algorithms to identify representative patient groups, then building personalized prediction models for each group. This approach starts by separating patients into two major clusters differentiated by high and low event prevalence. Then knowledge from literature and domain are used to define archetypical patient groups intended to be meaningful to providers. These rules include whether a patient has ESRD, nursing home residence, urinary-related cancer, and cognitive diseases. These are either disease-based characteristics that are often positively correlated with risk to UTI hospitalizations or frailty indicators that suggest the patient's vulnerability. The lower levels of the hierarchy are data-driven clusters that are associated with general healthcare utilization, such as inpatient, SNF, and carrier visits.

The prediction performance shows that the hierarchical clustering-based models achieve more accurate and precise predictions than the approach without clustering. Another advantage of this approach is to provide more personalized insight on which factors are most relevant to each patient group, instead of a single feature importance summary for the entire population. The variables most associated with UTI hospitalizations amongst all patient groups are whether the patient had a recent UTI diagnosis, as identified by previous studies (Walsh et al., 2012); or at least one oncology, ICU, or CCU visit in the previous month. This result agrees with studies that showed that about 15% of the patients admitted to acute hospitals receive a urinary catheter during their stay, after which infection frequently occurs, as ICU and CCU visits proxy the use of catheters (Saint, Meddings, et al., 2009). Additional feature insights for each of the 12 patient groups are discussed in Sect. 4.

The structure we have chosen for the tree is subject to our literature review and domain knowledge. For instance, we locate the nursing home variable at a higher level than urinary-related cancer because we believe that the frailty condition associated with nursing home residency dominates the specific health characteristics of urinary cancer. Other researchers could make different choices based on the knowledge they gather. In future studies, we suggest this framework to be used with more rigorous causal tools. The key contribution of the hierarchical clustering approach is providing a framework that can leverage existing knowledge to identify target groups meaningful to practitioners and that can be integrated with data-driven algorithms to build personalized prediction models for each representative group.

Although the LLR model was used to compare the performance of the hierarchical clustering approach with the non-clustering approach, other machine learning models may be used with the hierarchical clustering framework by modifying Step 4 in Sect. 3. The hierarchical clustering approach can also be applied to non-healthcare problems where data is highly heterogeneous and imbalanced, and domain knowledge is available to guide focused modeling.

6 Limitations and Future Research

In this study, data were limited to Medicare fee-for-service beneficiaries so health insights may not apply to all populations. We also rely on the accuracy and completeness of diagnosis from the claims to compute our predictors. In addition, studies have shown that the usage of urinary catheter is closely associated with UTI (Saint, Meddings, et al., 2009; Wald et al., 2008). Usage is not indicated in all claims like inpatient, however, adding an indicator for catheter would further improve the model predictions. Future studies may take these into account to build more accurate models for UTI.

Acknowledgements The authors gratefully acknowledge the Centers for Medicare and Medicaid Services (CMS) for providing medical claims data used in this study. Partial support was provided by Dr. Joseph Agor from Oregon State University, and graduate students James McKenna from Oregon State University, and Mina Mohammadi, Akash Pateria, and Prasanth Yadla from North Carolina State University for data preprocessing.

References

Anderson, J. E. (1983). Seasonality of symptomatic bacterial urinary infections in women. *Journal of Epidemiology and Community Health, 37*(4), 286–290.

Bertsimas, D., et al. (2008). Algorithmic prediction of health-care costs. *Operations Research, 56*(6), 1382–1392.

Beyan, C., & Fisher, R. (2015). Classifying imbalanced data sets using similarity based hierarchical decomposition. *Pattern Recognition, 48*(5), 1653–1672.

Billings, J., Zeitel, L., Lukomnik, J., Carey, T. S., Blank, A. E., & Newman, L. (1993). Impact of socioeconomic status on hospital use in new York City. *Health Affairs, 12*(1), 162–173.

Carter, M. W. (2003). Factors associated with ambulatory care—Sensitive hospitalizations among nursing home residents. *Journal of Aging and Health, 15*(2), 295–331.

Chan, J. K., Gardner, A. B., Mann, A. K., & Kapp, D. S. (2018). Hospital-acquired conditions after surgery for gynecologic cancer—An analysis of 82,304 patients. *Gynecologic Oncology, 150*(3), 515–520.

Choi, E., Bahadori, M. T., Kulas, J. A., Schuetz, A., Stewart, W. F., & Sun, J. (2016). RETAIN: An interpretable predictive model for healthcare using reverse time attention mechanism. *Advances in Neural Information Processing Systems, Nips*, 3512–3520.

"CMS' SSA to FIPS State and County Crosswalk," *The National Bureau of Economic Research*. 2011, [Online]. Available: https://data.nber.org/data/ssa-fips-state-county-crosswalk.html

"County Health Rankings," *County Health Rankings and Roadmaps*. 2011, [Online]. Available: https://www.countyhealthrankings.org/

Dharmarajan, K., et al. (2013). Diagnoses and timing of 30-day readmissions after hospitalization for heart failure, acute myocardial infarction, or pneumonia. *JAMA, 309*(4), 355–363.

"Diabetes Atlas," *United States Diabetes Surveillance System (USDSS)*. 2011, [Online]. Available: https://gis.cdc.gov/grasp/diabetes/DiabetesAtlas.html

Donzé, J., Aujesky, D., Williams, D., & Schnipper, J. L. (2013). Potentially avoidable 30-day hospital readmissions in medical patients: Derivation and validation of a prediction model. *JAMA Internal Medicine, 173*(8), 632–638.

Elbattah, M., & Molloy, O. (2017). *Clustering-aided approach for predicting patient outcomes with application to elderly healthcare in Ireland*.

Grabowski, D. C., O'Malley, A. J., & Barhydt, N. R. (2007). The costs and potential savings associated with nursing home hospitalizations. *Health Affairs, 26*(6), 1753–1761.

Hasan, O., et al. (2010). Hospital readmission in general medicine patients: a prediction model. *Journal of General Internal Medicine, 25*(3), 211–219.

Homer, D. W., Lemeshow, S., & Sturdivant, R. X. (2013). *Applied logistic regression* (3rd ed.). Wiley.

"Hospital Compare Dataset," *Centers for Medicare & Medicaid Services*. 2011, [Online]. Available: https://data.medicare.gov/data/hospital-compare

Hsu, P.-C., Lo, Y.-C., Wu, P.-Y., Chiu, J.-W., & Jeng, M.-J. (2019). The relationship of seasonality and the increase in urinary tract infections among hospitalized patients with spinal cord injury. *Journal of the Chinese Medical Association, 82*(5), 401–406.

"Immunization," *Centers for Disease Control and Prevention, Behavioral Risk Factor Surveillance System*. 2011, [Online]. Available: https://www.cdc.gov/brfss/index.html

Indicators, A. Q. (2001). Prevention Quality Indicators Technical Specifications. In *Department of Health and Human Services. Agency for Healthcare Research and Quality*.

Koroukian, S. M., Xu, F., & Murray, P. (2008). Ability of Medicare claims data to identify nursing home patients: a validation study. *Medical Care, 46*(11), 1184.

Kuhn, M. (2008). Building predictive models in R using the caret package. *Journal of Statistical Software, Articles, 28*(5), 1–26.

Ma, C., McHugh, M. D., & Aiken, L. H. (2015). Organization of hospital nursing and 30-day readmissions in Medicare patients undergoing surgery. *Medical Care, 53*(1), 65.

Moghadamyeghaneh, Z., Stamos, M. J., & Stewart, L. (2019). Patient Co-morbidity and functional status influence the occurrence of hospital acquired conditions more strongly than hospital factors. *Journal of Gastrointestinal Surgery, 23*(1), 163–172.

Mokyr Horner, E., & Cullen, M. R. (2015). Linking individual medicare health claims data with work-life claims and other administrative data. *BMC Public Health, 15*, 995.

Naqvi, S. B., & Collins, A. J. (2006). Infectious complications in chronic kidney disease. *Advances in Chronic Kidney Disease, 13*(3), 199–204.

Nithya, R., Manikandan, P., & Ramyachitra, D. (2015). Analysis of clustering technique for the diabetes dataset using the training set parameter. *International Journal of Advanced Research in Computer and Communication Engineering, 4*(9), 166–169.

"Nursing Home Compare Dataset," *Centers for Medicare & Medicaid Services*. 2011, [Online]. Available: https://data.medicare.gov/data/nursing-home-compare

"Nursing Homes." https://www.healthinaging.org/age-friendly-healthcare-you/care-settings/nursing-homes. Accessed 15 Oct 2020.

Ogbuabor, G., & Ugwoke, F. N. (2018). Clustering algorithm for a healthcare dataset using silhouette score value. *International Journal of Computer Science & Information Technology, 10*(2), 27–37.

P. D. Allison and Others. (2014). Measures of fit for logistic regression. In *Proceedings of the SAS global forum 2014 conference* (pp. 1–13).

"Population Census," *United States Census Bureau*. 2011, [Online]. Available: https://data.census.gov/cedsci/

"Population Census Elderly Living alone," *United States Census Bureau*. 2011, [Online]. Available: https://data.census.gov/cedsci/

"Public Use Files HRR Table for Beneficiaries 65 and older," *Centers for Medicare & Medicaid Services*. 2011, [Online]. Available: https://www.cms.gov/Research-Statistics-Data-and-Systems/Statistics-Trends-and-Reports/Medicare-Geographic-Variation/GV_PUF

Raschka, S. (2014). An overview of general performance metrics of binary classifier systems. *arXiv [cs.LG]*, Oct. 17.

Saint, S., Meddings, J. A., Calfee, D., Kowalski, C. P., & Krein, S. L. (2009). *Catheter-associated urinary tract infection and the Medicare rule changes*. American College of Physicians.

Saint, S., et al. (2009). Translating health care–associated urinary tract infection prevention research into practice via the bladder bundle. *Joint Commission Journal on Quality and Patient Safety, 35*(9), 449–455.

Sampson, E. L., Blanchard, M. R., Jones, L., Tookman, A., & King, M. (2009). Dementia in the acute hospital: Prospective cohort study of prevalence and mortality. *The British Journal of Psychiatry, 195*(1), 61–66.

Sasso, R., Bachir, R., & El Sayed, M. (2018). Suffocation injuries in the United States: Patient characteristics and factors associated with mortality. *The Western Journal of Emergency Medicine, 19*(4), 707–714.

Saver, B. G., Wang, C.-Y., Dobie, S. A., Green, P. K., & Baldwin, L.-M. (2014). The central role of comorbidity in predicting ambulatory care sensitive hospitalizations. *European Journal of Public Health, 24*(1), 66–72.

Unroe, K. T., Carnahan, J. L., Hickman, S. E., Sachs, G. A., Hass, Z., & Arling, G. (2018). The complexity of determining whether a nursing home transfer is avoidable at time of transfer. *Journal of the American Geriatrics Society, 66*(5), 895–901.

Wald, H. L., Ma, A., Bratzler, D. W., & Kramer, A. M. (2008). Indwelling urinary catheter use in the postoperative period: Analysis of the national surgical infection prevention project data. *Archives of Surgery, 143*(6), 551–557.

Walsh, E. G., Wiener, J. M., Haber, S., Bragg, A., Freiman, M., & Ouslander, J. G. (2012). Potentially avoidable hospitalizations of dually eligible Medicare and Medicaid beneficiaries from nursing facility and home-and community-based services waiver programs. *Journal of the American Geriatrics Society, 60*(5), 821–829.

Wang, H., Xu, Q., & Zhou, L. (2015). Large unbalanced credit scoring using lasso-logistic regression ensemble. *PLoS One, 10*(2), e0117844.

"Weekly U.S. Influenza Surveillance Report," *Centers for Disease Control and Prevention*. 2011, [Online]. Available: https://www.cdc.gov/flu/weekly/index.htm

Will, J. C., Nwaise, I. A., Schieb, L., & Zhong, Y. (2014). Geographic and racial patterns of preventable hospitalizations for hypertension: Medicare beneficiaries, 2004–2009. *Public Health Reports, 129*(1), 8–18.

Willis, A. W., et al. (2012). Neurologist-associated reduction in PD-related hospitalizations and health care expenditures. *Neurology, 79*(17), 1774–1780.

Risks Brought by Competition: Investment and Merger of Internet Enterprises

Ye Nan and Xu Runjie

1 Introduction

In the current environment, a large number of Internet companies compete with each other for user group traffic. This type of competition includes monopolizing the original business market and investing in new business markets. This has caused some businesses to be unprofitable or to put themselves at risk of future unprofitability. As a result, the financial industry has a greater risk exposure, and the risk value of Internet companies is much higher than that of traditional industries. This competitive business model is particularly evident in the Internet finance industry, and the role of Internet finance in the entire economy is becoming increasingly important (Meeker, 2018; Segal, 2016).

The entire Internet industry adopts this unique method to reinforce its strength in continuous financing and acquisitions. This business model believes that value is created not only by producers, but also by customers and other members of their value creation ecosystem. From this point of view, a company in the Internet industry only needs to make strategic investments and acquisitions in a range of fields to obtain business and users in this field, and to use these businesses to deploy infrastructure to serve original users, thus strengthening its position in the digital economy (Warnick, 2018; Havrylchyk et al., 2017; Whaley, 1993).

However, these investments and acquisitions do not take profit as the primary purpose, but rather depend on whether the field is related to the current main business or whether it provides infrastructure, technology, services or products for its business development. This expands the business, facilities and workforce but, at the same time, it also brings huge risk exposure, and involves more

Y. Nan · X. Runjie (✉)
College of Economics and Management, Nanjing University of Aeronautics and Astronautics, Nanjing, China

H. Yang et al. (eds.), *AI and Analytics for Public Health*, Springer Proceedings in Business and Economics, https://doi.org/10.1007/978-3-030-75166-1_35

other unpredictable cost risks, such as potential labour disputes, compliance costs, unprofitability, etc. On the other hand, existing unprofitable businesses may also be part of the Internet ecosystem, or they may be laying out strategies for future ecological structures. This business model concept may challenge the assumptions of traditional value creation and value acquisition theories (Barney, 1991; Peteraf, 1993). According to the Internet industry, value creation is provided by both the supply side and the demand side. Value is created not only by producers but also by customers and other members of their value creation ecosystem. From this perspective, competitive positions can be obtained from multiple perspectives, such as based on resource supply and user activity. This concept has led to frequent acquisitions and investment activities among the Internet industry. At present, there are few papers on the model, risks, or characteristics of Internet in major economic systems. The purpose of this article is to study risk fluctuations based on the consideration of corporate value disturbances (Bruton et al., 2015; Franks et al., 2016).

First, we explain the tendency of Internet companies' business income (Internet finance also belongs to the Internet industry). This tendency comes from a mode of competition in the Internet industry: in simple terms, Internet companies do not consider whether their business is profitable and continue to increase new investment acquisitions. Its purpose is to achieve Internet user group competition, infrastructure construction and data collection (Whaley, 1993).

We take into account that some Internet companies have excellent risk control capabilities, which makes them able to withstand the fluctuation of risks in a higher position while building their position in the digital economy. At this time, if the angle of risk mass is used to measure risk, distortion will often occur.

To address this challenge and mitigate the impact of diversified factors on risk, our goal is to propose an empirical method to measure the risk level of Internet companies – by generating a risk index to observe the risk level of individual companies and the risk level of the entire macro market. Conceptually, this calculation is similar to the stress test routinely applied to financial companies, but here it uses only publicly available information and is fast and cheap.

Our results show that companies with large risk values do not necessarily have the largest risk fluctuations. The value of the risk can only reflect the volume of the risk, but the risk fluctuation can show changes in investor confidence within a certain time frame, so it can better target Internet companies. Through this channel, Internet technology companies can learn about their own risk position in the industry, and regulatory agencies can observe the industry's overall risk dynamics in a timely manner in order to prevent and deal with the corresponding problems. We believe that the risk exposure of Internet technology companies is caused by a variety of factors. This factor not only includes multiple influences such as business models, market environment, and macroeconomics, but also is affected by corporate strategy.

We have further proved that our results are effective for the risk check of Internet technology companies through comparative experiments. The results of this research can be used for risk analysis in the same type of industry, which can have a good overview of the cyclical overlap of risks between industries, and thus have a clearer understanding of the overall risk of the entire industry.

2 Risk of Strategic Acquisition

In the Internet industry, the exploration of business models directly determines competitiveness and becomes a strategic focus for managers in different industries. In recent years, the Internet industry believes that a model different from traditional industries can stimulate user interest and may become a source of excess returns. Rumors of exceptional profitability from innovative business models are not uncommon. Take Google as an example: the company went to prosperity with a paid listing advertising business model. Xerox, meanwhile, chose to lease its copy machine instead of selling it, making the company one of the most profitable companies at the time (Chesbrough, 2007; Rometty, 2006; Ireland et al., 2001; Johnson et al., 2008).

As Internet technology blurs the differences between industries, lowers the barriers to entry and leads to more intense competition, Internet companies are forced to choose to directly obtain the right to use a technology or a business through investment and acquisition. This trend has created an ecological environment for investment and mergers in the existing Internet industry.

We use the financial statements of Internet companies to explore the relationship between investment M & A models and risks. As shown in Fig. 1 below, we have selected Alibaba as the reference object (the other 13 Internet companies are Ebay, Facebook, Paypal, Google, Apple, Twitter, Amazon, BIDU, JD, Tencent, YRD, PPDF, and DNJR,).

In Alibaba's financial report analysis, we can clearly find that research, development, sales and other expenses have steadily increased, while interest and investment income and income from operations have declined. This seems to indicate that the company's investment cannot be translated into actual returns. However, the company's revenue has risen at a faster rate, which indicates that the overall revenue and expenditure situation is more optimistic.

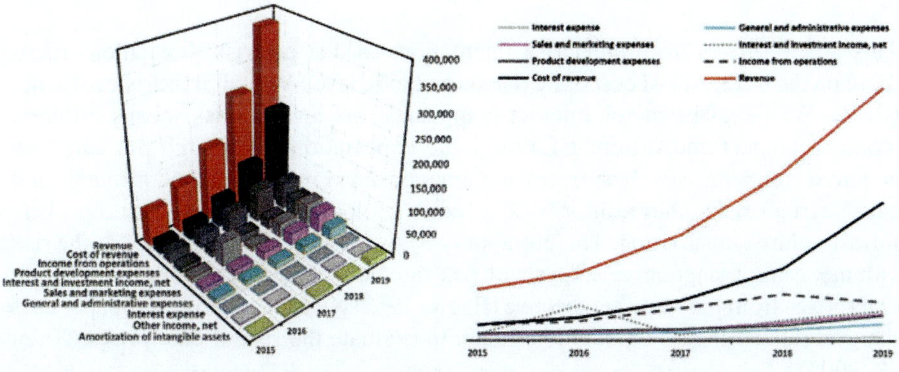

Fig. 1 Data from Alibaba's operating statements for 2015–2019 financial years

Alibaba, as China's largest e-commerce company, is also a representative Internet company in China. This type of company attaches great importance to the long-term benefits of the digital economy, as well as the acquisition of network traffic, so it continues to increase expenditures on business, investment acquisitions and strategy. Judging from the financial reports in recent years, many of the company's newly invested businesses have low or negative profit margins, and the invested businesses are often in the early stages of exploration. Many of these business models are more efficient in attracting and converting paying merchants low. These investments and acquisitions did not increase Alibaba's revenue. Instead, from FY 2018 to FY 2019, Alibaba's adjusted EBITDA profit margin fell from 42% to 32%. The types of these investments and initiatives include:

(a) Commercial products capable of expanding and strengthening core competitiveness, including supporting Alibaba's logistics network, local service business, new retail plans, direct sales, and cross-border e-commerce
(b) Expanding construction of various facilities and increasing the number of employees.
(c) Researching and developing new technologies to improve technological infrastructure and cloud computing capabilities;
(d) Innovative measures for digital media and entertainment business.

We see Alibaba's strategic investment and acquisitions in a range of areas as strengthening its digital economy leadership in China. These investments and acquisitions do not take profit as the primary purpose, but rather depend on whether the field is related to the current main business or whether it provides infrastructure, technology, services or products for its business development. These products can promote user activities and continue to create value for the entire ecosystem.

3 Risk Volatility Model (RFR)

This article differs from other risk fluctuations in that our risk fluctuations focus more on the trajectory of corporate risk on a certain level. We call it the risk volatility (RFR). We believe that for Internet companies, the level of risk means different strategic layouts and different future value expectations. Although the currently acquired companies and businesses and other business models are not profitable and cause certain risks, they can provide a future ecological environment and provide infrastructure construction. Therefore, only by excluding the interference of the risk volume, can a comparative analysis of risk fluctuations be carried out for different enterprises from the same perspective (Engle, 1982; Xu et al., 2005). This paper uses the generalized Pareto distribution model to evaluate the risk of enterprises (Xu et al., 2005).

First, according to the classical generalized Pareto model, set the function as

$$F(x;\mu,\sigma,k)=\begin{cases}1-\left(1-k\frac{x}{\sigma}\right)^{\frac{1}{k}} & k\neq 0\\ 1-e^{-\frac{x}{\sigma}} & k=0\end{cases} \tag{1}$$

where σ is scale parameter of distribution, and k is the shape parameter of distribution. $\sigma > 0$ and when $k \le 0$, $x \ge 0$; when $k > 0$, $0 < x < \frac{\sigma}{k}$; and when $k = 0$, the distribution is exponential.

Among them, the role of threshold is very important. In the POT model, by setting a threshold in advance, all observed data exceeding this threshold are constituted into a data group, and the data group is taken as the object of modeling and applied to the generalized Pareto distribution to calculate the risk value (Roth et al., 2016).

Because of the quantity of data in this paper and experiments, the threshold u is finally selected at the confidence level of 80% in this paper.

We use the excess rate of return to process the time series data:

$$AR_{i,t}=R_{i,t}-R_t=\frac{a_{i,t}-a_{i,t-1}-a_{i,t-1}R_t}{a_{i,t-1}} \tag{2}$$

Where $AR_{i,t}$ is the excess return rate of the company i at time t, $R_{i,t}$ is the stock return rate of the company i at time t; R_t is the risk-free rate of the market at time t; $a_{i,t}$ is the stock closing price of the company i at time t .Secondly, through the maximum likelihood estimation of the density function of generalized Pareto distribution, the excess return rate can estimate the scale parameters σ and shape parameters k of generalized Pareto distribution in different time periods. Since the scale parameters σ and shape parameters k in generalized Pareto distribution are determined by the maximum likelihood method, the maximum likelihood estimate $\hat{\sigma}$ and $\hat{k}$ is obtained. Setting the time period r, the VaR calculation formula is:

$$\mathrm{VaR}_{i,m}=\mu+\hat{b}\left[\frac{n}{N_u}(1-p)^{\frac{1}{n}\sum_{r=m'}^{m''}\ln\left(1-\hat{b}\frac{a_{i,r}-a_{i,r-1}-a_{i,r-1}R_t}{a_{i,r-1}}\right)}-1\right] \tag{3}$$

where μ is the threshold, $\hat{b}$ is the estimate, n is the total number of samples, N_u is the number of samples exceeding the thresholdμ, P is the confidence level selected, and then $VaR_{i,m}$ is the risk value of the company i in the month m. $R_{i,r}$ is the stock return of the company i at time r, R_t is the risk-free rate in the market at time t, $a_{i,r}$ is the stock closing price of the company i at time r. r and r − 1 time points are included in the month m. m' is the start of the month m. m'' is the end of the month m.

On this basis, the rise and fall of the risk value in a specified period of time can be converted into the way of slope. The positive or negative slope indicates whether the risk increases or decreases in the corresponding time series. That is:

$$RFR=\frac{\hat{b}\frac{n}{N_u}\left[(1-p)^{\frac{1}{n}\sum_{r=(m+1)'}^{(m+1)''}\ln\left(1-\hat{b}\frac{a_{i,r}-a_{i,r-1}-a_{i,r-1}R_t}{a_{i,r-1}}\right)}-(1-p)^{\frac{1}{n}\sum_{r=m'}^{m''}\ln\left(1-\hat{b}\frac{a_{i,r}-a_{i,r-1}-a_{i,r-1}R_t}{a_{i,r-1}}\right)}\right]}{t} \tag{4}$$

4 Assessment Strategies

According to different main businesses and the degree of attention given by the market, we selected 14 Chinese and American technology companies: Alibaba, JD, Facebook, PayPal, Ebay, Google, Apple, Twitter, Amazon, BIDU, Tencent, YRD, PPDF, and DNJR. According to the securities market, where the company is listed and the important securities markets of the main business countries, we have added S&P 500 (Standard & Poor's 500), DJIA (Dow Jones Industrial Average), National Association of Securities Dealers Automated Quotations (Nasdaq), Shanghai Securities Composite Index (SSE), Hang Seng Index (HIS), and Shenzhen Securities Component Index (SZI). Internet technology companies are similar in business model, operation model, profit model, user characteristics, etc. Therefore, we suspect that their risk fluctuation model is also correlated to some extent.

With the mutual influence and infiltration of financial and economic activities, as well as the massive transmission and exchange of market information, the interaction, behaviour and mutual correlation among financial markets also show a significantly rising trend. The interactive behaviour of the financial market promotes the optimal allocation of financial and economic resources, but also leads to the frequent outbreak of risk in recent years. This is due to the fact that economic and financial development in all countries of the world are closely connected, whether by each country's financial markets of the global financial system, or a country. Even in the financial system, there are indivisible and complex relationships among many financial individuals, and they eventually constitute the complex financial systems of various sizes.

Internet technology companies are similar in business model, operation model, profit model, user characteristics, etc. Therefore, we suspect that their risk fluctuation model is also correlated to some extent.

The Fig. 2 shows the risk fluctuation amplitude of 14 companies under the RFR method. After observation, it can be found that the companies that generated huge amplitude after time series 2018.2 are Internet finance companies, namely YRD, PPDF and DNJR.

The following picture considers the risk accumulation area after Internet finance companies. After observation, it can be found that the dark part after time series 2018.2 represents the fact that Internet finance companies deviate from the overall industry trend and generate huge risk fluctuations.

The Fig. 3 shows the risk volatility of 14 companies under the RFR method. The red dotted line is the Internet financial enterprise DNJR, which has repeatedly occurred risk events after listing. The other two dotted lines represent Internet finance companies and risk volatility is far greater than other Internet technology companies in the industry. The picture below shows the risk accumulation area after the Internet finance company is not considered. The Internet finance companies represented by the dotted line not only deviate from the cycle in risk volatility, but also far higher than other technology companies in risk level.

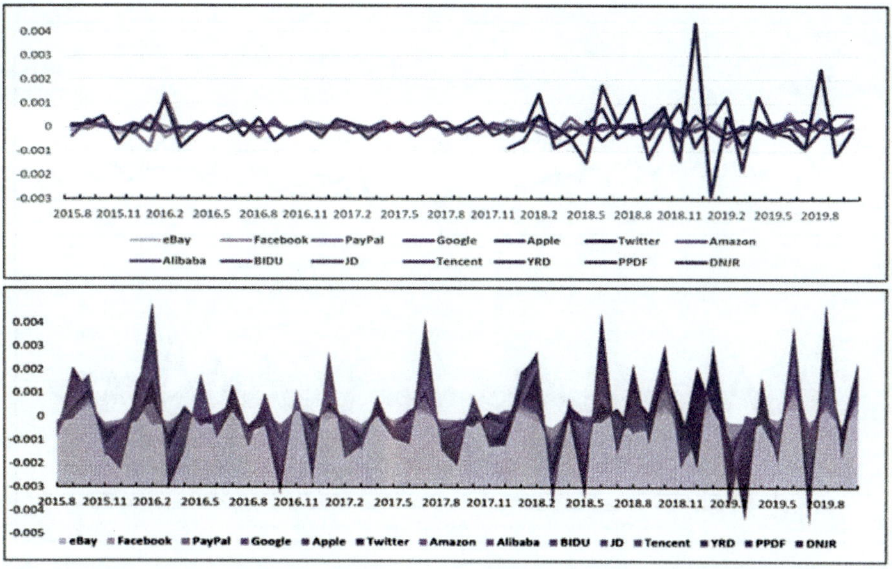

Fig. 2 Risk fluctuation amplitude analysis of 14 companies under the RFR method

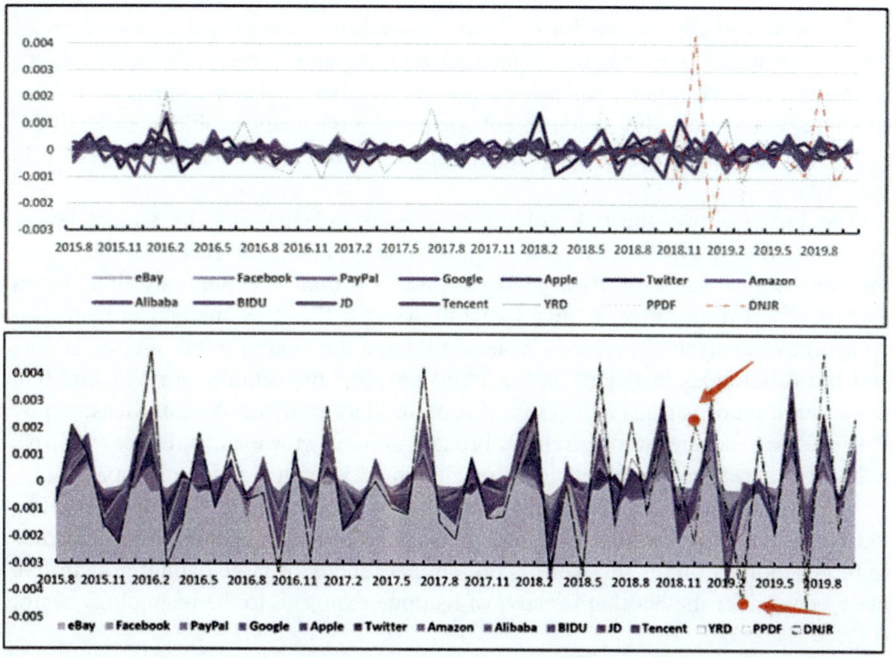

Fig. 3 Influence of three Internet Financial Enterprises on Risk Assessment

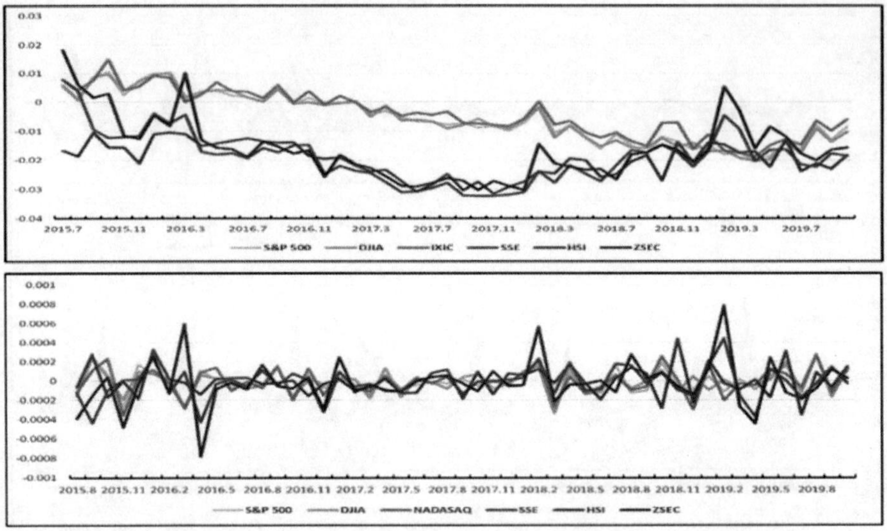

Fig. 4 Risk assessment of composite index

We explained the reasons for choosing these three Internet financial enterprises before. Because a large number of technology companies invest or directly engage in Internet finance-related businesses, but the number of listed companies is very small, so we can only choose the parent company as the analysis object. YRD, PPDF and DNJR are pure Internet financial enterprises, which are more representative in the analysis.

The Fig. 4 shows the risk volume of various indexes, and the picture below shows the risk fluctuation range of various indexes under the RFR method. Since the time series value of RFR is determined by both the previous time series and the fluctuation range, it therefore follows that the RFR method is better for comparative analysis, which can be used to judge the overall trend of a set of data and the differences in details in the trend process. In addition, we can find that the correlation of various indexes on risk periodicity is also different. Measured by the value at risk, the overall trend is broadly similar. However, in the case of risk volatility, there is no strong cyclical correlation between technology companies.

Due to the large number of Internet technology companies listed on Nasdaq, the Nasdaq is an emerging high-tech index that covers companies from telecoms to biotechnology and is the world's pre-eminent index of large capital growth. We have to consider the Nasdaq because of its huge exposure to the technology sector (Fig. 5).

Obviously, the risk volatility for tech companies is much larger, but the risk trends are almost similar across data values. This proves our conjecture that the securities market where enterprises are listed will lead to the convergence effect of risk changes.

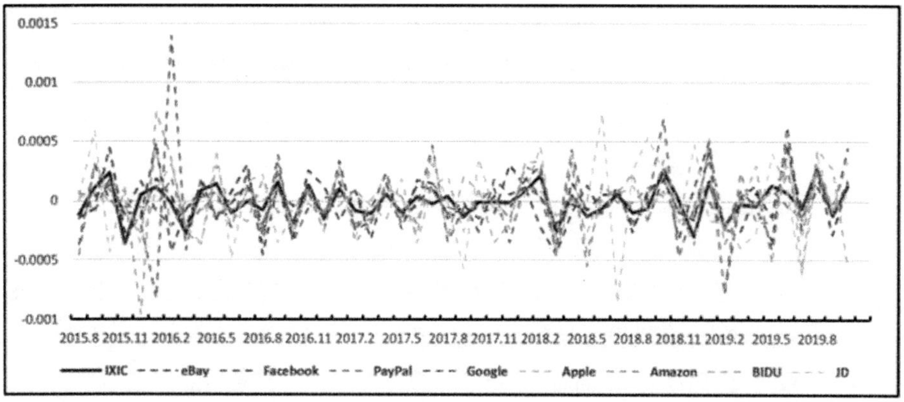

Fig. 5 Nasdaq versus the Risk of Technology Stocks in its Market

In addition, this also confirms that technology companies do not rule out the possibility of having a higher level of risk control. For example, we can find in the figure that after the risk fluctuates to a very high value, the next time series is a plummeting risk value. There are very few instances of continuous risk escalation.

For this purpose, we counted the number of times when RFR was above and below 0 value, which was used to measure whether the interval of risk trajectory was high risk or low risk in a certain time range (Fig. 6).

The riskiest company in the rankings is DNJR, an Internet finance company that has been delisted because of late payments. DNJR is a Chinese Internet finance company, while all other Chinese tech companies tend to be at the bottom of the list. This has to do with the intensity of regulation in Chinese and American markets. The acquisition behaviours of the 14 companies from 2015 to 2019, it seems that the more acquisitive the company, the better the risk control level remained.

In this regard, we have a potential concern. For technology industries with frequent acquisitions and investments, the important source of risk is the "external effects" of microeconomic risk-taking activity, which is a single company. The losses imposed on the society by the major risks of the institution are far greater than the losses suffered by investors themselves. In practical terms, powerful companies often pay great attention to their own risk prevention and control, but ordinary platforms that have direct business dealings with these large platforms do not pay enough attention to their own risks. If a powerful company acquires Internet financial companies only to improve its position in a new field or for its own ecological construction, then Internet financial companies are very likely to risk (Xu et al., 2020).

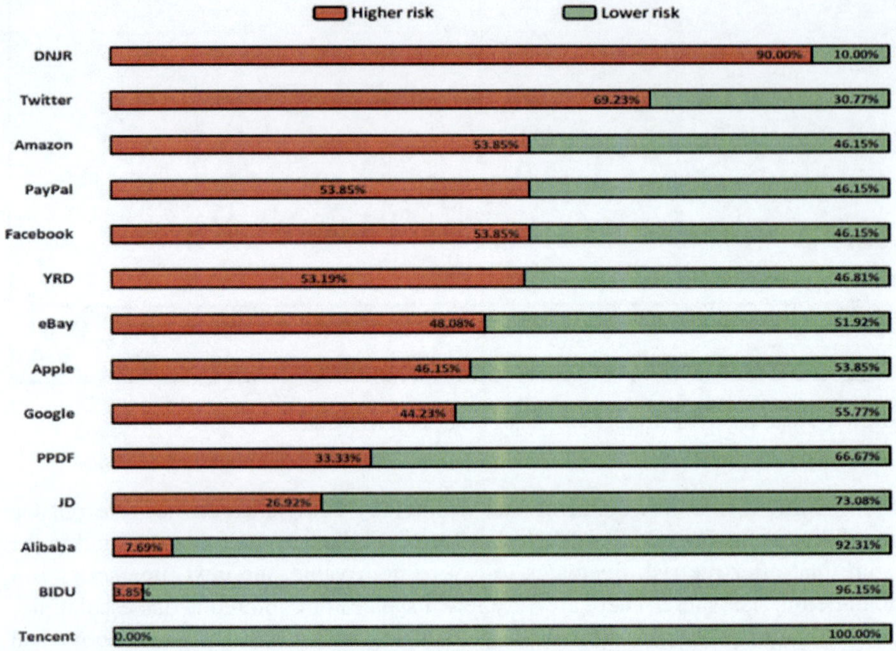

Fig. 6 The percentage of Internet technology companies above and below the RFR value

5 Conclusion

This paper uses the stock price data of Internet finance and Internet technology companies whose business includes Internet finance to construct a risk fluctuation model (RFR). This model can flexibly switch the required observation scales, such as day, week, month, etc., according to demand. We show that this new risk volatility model performs well in comparative risk analysis.

The experimental methods we use are the risk mass method and the risk fluctuation method (RFR). Through a large number of comparative experiments, we found that according to the main business countries of different companies, there will be differences in the movement model of risk mass. Countries with the same main business have a high correlation in the fluctuation pattern of risk volume. We also found that under the RFR method, the risk fluctuation amplitude of the entire Internet industry has a cycle-like characteristic, which is affected by the Nasdaq index. The Nasdaq index has an impact on the trend of the Internet industry in terms of risk volume and risk fluctuation range RFR.

We also discuss the business types of Internet finance and study the frequent acquisitions and investment activities in the entire Internet industry. We explain the purpose of this behavior and analyze its impact on the ecological value and risk exposure of the entire Internet industry.

Finally, we raise some concerns about the Internet industry. For the acquisition and investment of the Internet industry, the important source of risk is the "externality" of microeconomic risk-taking activity. If a large Internet company only considers the future ecological value and does not care about the profitability of the acquired enterprise in the process of acquiring the Internet enterprise, then the acquired Internet enterprise is prone to risk exposure of crisis events and various uncertainties.

Acknowledgements This study is supported by National Social Science Fund Project (17BGL055) Innovation Research Support Program of Nanjing University of Aeronautics and Astronautics(2019EC01)(2020CX00904), Top-quality Project of Social Science Application Research in Jiangsu Province (20SYC-132).

References

Barney, J. B. (1991). Firm resources and sustained competitive advantage. *Journal of Management, 17*(1), 99–120.

Bruton, G., Khavul, S., Siegel, D., & Wright, M. (2015). New financial alternatives in seeding entrepreneurship: Microfinance, crowdfunding, and peer-to-peer innovations. *Entrepreneurship Theory and Practice, 39*(1), 9–26.

Chesbrough, H. W. (2007). Business model innovation: It's not just about technology anymore. *Strategy & Leadership, 35*(6), 12–17.

Engle, R. F. (1982). Autoregressive conditional heteroscedasticity with estimates of the variance of United Kingdom inflation. *Econometrica: Journal of the Econometric Society*, 987–1007.

Franks, J. R., Serrano-Velarde, N. A. B., & Sussman, O. (2016). Marketplace lending, information aggregation, and liquidity. *Information Aggregation, and Liquidity.*

Havrylchyk, O., Mariotto, C., Rahim, T., & Verdier, M. (2017). *What drives the expansion of the peer-to-peer lending*. URL https://doi.org/10.2139/ssrn, p. 2841316.

Ireland, R. D., Hitt, M. A., Camp, M., & Sexton, D. L. (2001). Integrating entrepreneurship and strategic management actions to create firm wealth. *Academy of Management Executive, 15*(1), 49–63.

Johnson, M. W., Christensen, C. M., & Kagermann, H. (2008). Reinventing your business model. *Harvard Business Review, 86*(12), 50–59.

Meeker, M. (2018). *Internet trends 2018.*

Peteraf, M. A. (1993). The cornerstones of competitive advantage: A resource-based view. *Strategic Management Journal, 14*(3), 179–191.

Rometty, V. G. (2006). *Expanding the innovation horizon: The global CEO study 2006*. IBM Business Service.

Roth, M., Jongbloed, G., & Buishand, T. A. (2016). Threshold selection for regional peaks-over-threshold data. *Journal of Applied Statistics, 43*(7), 1291–1309.

Segal, M. (2016). *What is alternative finance?* US Small Business Administration, Office of Advocacy. Available at: https://www.sba.gov/sites/default/files/advocacy/What-Is-Alt-Fi.pdf

Warnick, B. (2018). Rhetorical criticism of public discourse on the internet: Eoretical implications. In *Fifty years of rhetoric society quarterly* (pp. 98–109).

Whaley, R. E. (1993). Derivatives on market volatility: Hedging tools long overdue. *The journal of Derivatives, 1*(1), 71–84.

Xu, L., Ivanov, P. C., Hu, K., Chen, Z., Carbone, A., & Stanley, H. E. (2005). Quantifying signals with power-law correlations: A comparative study of detrended fluctuation analysis and detrended moving average techniques. *Physical Review E, 71*(5), 051101.

Xu, R., Mi, C., Mierzwiak, R., & Meng, R. (2020). Complex network construction of internet finance risk. *Physica A: Statistical Mechanics and Its Applications*, 122930.